Infinite-dimensional Lie algebras

Infinite-dimensional Lie algebras

Ralph K. Amayo

Lecturer in mathematics

Ian Stewart

Lecturer in mathematics
University of Warwick, England

Noordhoff International Publishing
Leyden, The Netherlands

ISBN: 90 286 0144 9
Library of Congress Catalog Card Number: 73-9427 0

Printed in The Netherlands by Koninklijke Drukkerij Van de Garde B.V.

Contents

Preface

It is only in recent times that infinite-dimensional Lie algebras have been the subject of other than sporadic study, with perhaps two exceptions: Cartan's simple algebras of infinite type, and free algebras. However, the last decade has seen a considerable increase of interest in the subject, along two fronts: the topological and the algebraic. The former, which deals largely with algebras of operators on linear spaces, or on manifolds modelled on linear spaces, has been dealt with elsewhere*). The latter, which is the subject of the present volume, exploits the surprising depth of analogy which exists between infinite-dimensional Lie algebras and infinite groups.

This is not to say that the theory consists of groups dressed in Lie-algebraic clothing. One of the tantalising aspects of the analogy, and one which renders it difficult to formalise, is that it extends to theorems better than to proofs. There are several cases where a true theorem about groups translates into a true theorem about Lie algebras, but where the group-theoretic proof uses methods not available for Lie algebras and the Lie-theoretic proof uses methods not available for groups. The two theories tend to differ in fine detail, and extra variations occur in the Lie algebra case according to the underlying field. Occasionally the analogy breaks down altogether. And of course there are parts of the Lie theory with no group-theoretic counterpart.

Broadly speaking, the subject-matter divides into six sections, spread across eighteen chapters. The first is the study of subideals, which are analogous to subnormal subgroups, and the related coalescence phenomena. The second deals with locally nilpotent algebras and the Mal'cev correspondence between groups and Lie algebras. The third discusses finiteness conditions, especially chain conditions. The fourth is a fairly detailed development of properties of finitely generated soluble Lie algebras, along the lines of a famous series of

---

*) An extensive bibliography may be found in de la Harpe, Banach-Lie algebras and Banach-Lie groups of operators, Springer, Berlin 1972.

papers by P. Hall on soluble groups. The fifth gives an infinite-dimensional analogue of the classical structure theory of finite-dimensional Lie algebras. The sixth covers varieties, the finite basis problem, Engel conditions; the theorem of Kostrikin on the restricted Burnside problem; and the recent example of Razmyslov which yields non-nilpotent groups of prime exponent.

We have of necessity assumed a certain familiarity with infinite group theory, finite-dimensional and free Lie algebras, and parts of ring theory; but any results quoted without proof are given adequate references. The references themselves divide into two parts: the first a fairly comprehensive (though doubtless idosyncratic) list of some 250 items of more or less direct relevance to infinite-dimensional Lie algebras; the second containing all other material referred to in the text.

A short list of unsolved problems is included, in an effort to encourage further research. They are probably of extremely variable difficulty.

It is a pleasure to thank those of our colleagues in the mathematical community who, by the provision of information, discussion, and criticism, have contributed so much to the present work. Especially we thank B. Hartley, who started us out along the Lie-theoretic track; J. E. Roseblade, whose help with the 'Derived Join Theorem' provided the key to much of the work on coalescent classes; and J. Wiegold who explained Kostrikin's theorem to us, lent us a rough translation of Razmyslov's paper, and graciously allowed us to quote from his notes on the subject. Thanks are also due to R. W. Carter, J. Eells, J. A. Green, C. R. Hajarnavis, P. de la Harpe, H. Heineken, I. Kaplansky, R. Kibler, F. P. Lockett, P. J. McInerney, B. H. Neumann, P. M. Neumann, B. I. Plotkin, S. E. Stonehewer, D. A. Towers, W. Unsin, M. R. Vaughan-Lee, and E. Wallace; and to many others who we trust will forgive a collective, rather than an individual, mention.

R.K.A.
I.N.S.
November 1973.

Organisational notes

Each chapter is divided into numbered sections; and theorems, propositions, lemmas, corollaries, and other items of that ilk are numbered in the form $x.y$, where x is the section number and y runs consecutively throughout the section. Cross-references to theorems within a chapter are also given in the form $x.y$. Cross-references between different chapters are given in the form $x.y.z$, which represents theorem $y.z$ of chapter x. If theorem $y.z$ has several parts (a), (b), (c) the reference may be to $x.y.z$(b).

Displayed items are numbered (n) consecutively through each chapter, the numbering starting again at (1) in the next chapter; occasionally they may be marked in more cryptic fashion, such as (*), (**), (***), (!), which latter symbols have purely local validity.

The end of a proof (and hence the absence of a proof when the one given is empty) is signalled by a box □. In most cases the proof follows immediately after the statement of the theorem, although it is sometimes deferred; on rare occasions it may be found immediately *before* the statement of the theorem.

An index of notation may be found on page 415.

References of the form 'Cosgrove and Plojhar [m]' are to the paper of Cosgrove and Plojhar to be found as item m in the bibliography, which is numbered consecutively starting at page 398.

Chapter 1

Basic concepts

1. Preliminaries

The main purpose of this chapter is to introduce certain basic concepts of the theory of Lie algebras in a form suited to later developments. We also take the opportunity to standardise notation and terminology.

Let $\mathfrak{k}$ be a (commutative) field. A *Lie algebra* over $\mathfrak{k}$ is a vector space L over $\mathfrak{k}$ equipped with a bilinear multiplication

$$[,]:L\times L\to L$$

satisfying

$$[x, x]=0 \qquad (x\in L) \tag{1}$$

$$[[x, y], z]+[[y, z], x]+[[z, x], y]=0 \qquad (x, y, z\in L). \tag{2}$$

Equation (2) is the *Jacobi identity*. Equation (1) implies that

$$[x, y]=-[y, x] \qquad (x, y\in L) \tag{3}$$

and (3) implies (1) except when $\mathfrak{k}$ has characteristic 2.

We emphasise that *a Lie algebra may be of finite or infinite dimension*, and that *the underlying field* $\mathfrak{k}$ *may have arbitrary characteristic*. The word 'algebra' will mean 'Lie algebra' unless the contrary is indicated. We write char($\mathfrak{k}$) for the characteristic of $\mathfrak{k}$.

If A is an associative algebra over $\mathfrak{k}$ and we define

$$[a, b]=ab-ba \qquad (a, b\in A)$$

then A acquires the structure of a Lie algebra, which we denote by $[A]$.

We shall use standard set-theoretic notation. In particular we write $X\subseteq Y$ if X is a subset of Y. Further,

$\mathbb{N}$ denotes the set of natural numbers 0, 1, 2, ...
$\mathbb{Z}$ denotes the set of integers
$\mathbb{Q}$ denotes the set of rational numbers
$\mathbb{R}$ denotes the set of real numbers
$\mathbb{C}$ denotes the set of complex numbers.

Let L be a Lie algebra, with $X \subseteq L$, $Y \subseteq L$. Then

$$X+Y = \{x+y:\ x \in X,\ y \in Y\}$$

while

$$[X, Y]$$

is the subspace spanned by all $[x, y]$ $(x \in X,\ y \in Y)$. If X_α $(\alpha \in A)$ are subsets of L, then

$$\sum_{\alpha \in A} X_\alpha$$

is the set of finite sums $x_1 + \ldots + x_n$ where each $x_i \in \bigcup_{\alpha \in A} X_\alpha$.

A subspace $H \subseteq L$ is a *subalgebra* if $[H, H] \subseteq H$. In this case H is a Lie algebra under the operations induced from L. We write $H \leqq L$ if H is a subalgebra of L, and $H < L$ if $H \leqq L$ and $H \neq L$.

A subspace $I \subseteq L$ is an *ideal* of L if $[I, L] \subseteq I$; or equivalently by (3) if $[L, I] \subseteq I$. In this case we write $I \lhd L$. Clearly every ideal is a subalgebra.

If $I \lhd L$ we may form the *quotient algebra* (or *factor* algebra) L/I whose elements are the cosets of I in L with operations induced from L. The notions of *homomorphism* and *isomorphism* are defined in the obvious way, and we write $H \cong K$ if H is isomorphic to K. Since a Lie algebra may be considered as an abelian group with operators, the Noether isomorphism theorems and their consequences (in particular Zassenhaus's lemma (see Zassenhaus [352] p. 58) and the Jordan-Hölder theorem) hold good.

A Lie algebra L is *simple* if its only ideals are $\{0\}$ and L.

Left-normed products

$$[x_1, \ldots, x_n] \quad (x_1, \ldots, x_n \in L)$$

are defined recursively by

$$[x_1, \ldots, x_{n+1}] = [[x_1, \ldots, x_n], x_{n+1}].$$

If $x_1 = x$, $x_2 = x_3 = \ldots = x_{n+1} = y$, we write

$$[x, {}_n y] = [x_1, \ldots, x_{n+1}].$$

Similarly we define, for subsets $X_i \subseteq L$,

$$[X_1, \ldots, X_{n+1}] = [[X_1, \ldots, X_n], X_{n+1}]$$

and if $X_1 = X, X_2 = X_3 = \ldots = X_{n+1} = Y$, we write

$$[X, {}_nY] = [X_1, \ldots, X_{n+1}].$$

We extend this notation in an obvious way to define

$$[x, {}_{n_1}y_1, \ldots, {}_{n_t}y_t]$$

if $n_1, \ldots, n_t \in \mathbb{N}$.

By abuse of notation we shall often write x instead of $\{x\}$: for example 0 will refer to the subalgebra $\{0\}$ of L, and $[A, x] = [A, \{x\}]$.

If $X \subseteq L$ then $\langle X \rangle$ denotes the subalgebra *generated by* X, that is, the smallest subalgebra of L containing X. It consists of all elements obtainable from X by a finite sequence of vector space operations and Lie multiplications. However, we can prove a refinement of this:

LEMMA 1.1 *If $X \subseteq L$ then $\langle X \rangle$ is the subspace spanned by all left-normed products*

$$[x_1, \ldots, x_n] \quad (x_i \in X,\ n \in \mathbb{N}).$$

Proof: It is sufficient to prove that if

$$a = [x_1, \ldots, x_n]$$

$$b = [y_1, \ldots, y_m]$$

where $x_i, y_j \in X$ and $m, n \in \mathbb{N}$, then $[a, b]$ is a linear combination of left-normed products of elements of X. We do this by induction on m. If $m = 0$ or 1 the result is clear. Now if $y = y_{m+1}$ the Jacobi identity implies that

$$\begin{aligned} [a, [b, y]] &= -[b, [y, a]] - [y, [a, b]] \\ &= -[[a, y], b] + [[a, b], y] \end{aligned}$$

which is of the required form by the inductive hypothesis. □

A *set of generators* for L is a subset $X \subseteq L$ such that $L = \langle X \rangle$. If L has a finite set of generators we say it is *finitely generated.*

We extend the triangular bracket notation as follows:

$$\langle X_\alpha : \alpha \in A \rangle = \langle \bigcup_{\alpha \in A} X_\alpha \rangle$$

$$\langle X, Y, Z, \ldots \rangle = \langle X \cup Y \cup Z \cup \ldots \rangle$$

$$\langle x, y, z, \ldots \rangle = \langle \{x, y, z, \ldots\} \rangle$$

$$\langle X, x \rangle = \langle X \cup \{x\} \rangle$$

where X, Y, Z, and the X_α are subsets of L, while $x, y, z, \ldots$ are elements of L.

If X and Y are subsets of L we say that X is *Y-invariant* if whenever $x \in X$ and $y \in Y$ then $[x, y] \in X$. Alternatively we say that Y *idealises* X. Then

$$X^Y = \sum_{n \in \mathbb{N}} [X, {}_nY]$$

is the smallest Y-invariant subspace of L containing X. It follows from the Jacobi identity that $\langle X^Y \rangle$ is also Y-invariant, and is thus the smallest Y-invariant subalgebra containing X.

The following lemma is immediate from the definitions:

LEMMA 1.2 (a) $H \lhd L$ *and* $K \leqq L \Rightarrow H + K \leqq L$

(b) $H, K \lhd L \Rightarrow H + K \lhd L$

(c) $H, K \lhd L \Rightarrow [H, K] \lhd L$

(d) $X_\alpha \lhd L \ (\alpha \in A) \Rightarrow \sum_{\alpha \in A} X_\alpha \lhd L$. □

A vector space V is the *direct sum* of two subspaces U and W if $V = U + W$ and $U \cap W = 0$. We write

$$V = U \dot{+} W.$$

A Lie algebra L is the *direct sum* of two subalgebras H and K if as vector spaces $L = H \dot{+} K$ and if further $[H, K] = 0$. This latter condition is equivalent to assuming that $H, K \lhd L$. We write

$$L = H \oplus K.$$

The terminology (and notation) are extended to direct sums of sets of subspaces (subalgebras) in the usual way.

Given Lie algebras L_α ($\alpha \in A$) we can form the *Cartesian sum*

$$\mathrm{Cr}_{\alpha \in A} L_\alpha.$$

This consists of all functions $f: A \to \bigcup_{\alpha \in A} L_\alpha$ such that for all $\alpha \in A$, $f(\alpha) \in L_\alpha$, with Lie operations defined pointwise. The (*external*) *direct sum*

$$\mathrm{Dr}_{\alpha \in A} L_\alpha$$

is the subalgebra of $\mathrm{Cr}_{\alpha \in A} L_\alpha$ consisting of those f such that $f(\alpha) = 0$ for all but a finite set of α. It is a direct sum of subalgebras isomorphic to the L_α.

2. Nilpotency and solubility

We define the *lower central series* $(L^n)_{n\in\mathbb{N}}$ of L recursively by:

$$L^1 = L$$

$$L^{n+1} = [L^n, L].$$

The *derived series* $(L^{(n)})_{n\in\mathbb{N}}$ is defined by:

$$L^{(0)} = L$$

$$L^{(n+1)} = [L^{(n)}, L^{(n)}].$$

We call $L^2 = L^{(1)}$ the *derived algebra* of L.

It follows from lemma 1.2(c) that L^n and $L^{(n)}$ are ideals of L. If $L^{n+1} = 0$ we say that L is *nilpotent* (of *class* $\leqq n$) and if $L^{(n)} = 0$ then L is *soluble* (of *derived length* $\leqq n$). If $L^2 = 0$ then L is *abelian*. Any vector space V over $\mathfrak{k}$ becomes an abelian Lie algebra if we define $[u, v] = 0$ $(u, v \in V)$, and every abelian algebra arises in this way from its underlying vector space.

There is a connection between solubility and nilpotency:

LEMMA 2.1 *Let L be a Lie algebra. Then*

(a) $[L^m, L^n] \subseteq L^{m+n}$ *for all* $m, n \in \mathbb{N}\backslash\{0\}$.

(b) $L^{2^n} \supseteq L^{(n)}$ *for all* n.

(c) *If L is nilpotent of class c then it is soluble of derived length $\leqq n$, where n is the smallest integer $\geqq \log_2(c+1)$.*

Proof: (a) is proved in the same way as lemma 1.1.

(b) follows from (a) by induction.

(c) follows from (b). □

For example the Lie algebra of *zero-triangular* $n\times n$ matrices

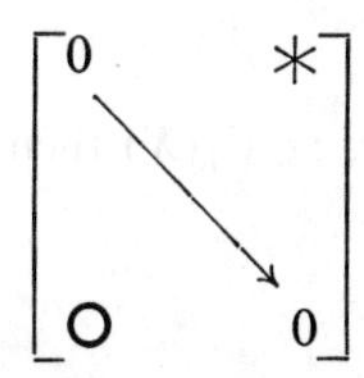

over $\mathfrak{k}$ is nilpotent (of class $n-1$), since the cth lower central term comprises those matrices of the form

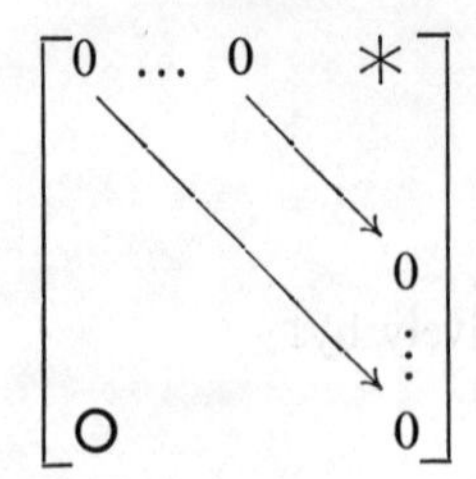

in which there are c diagonals all zero. The Lie algebra of all *triangular* $n \times n$ matrices

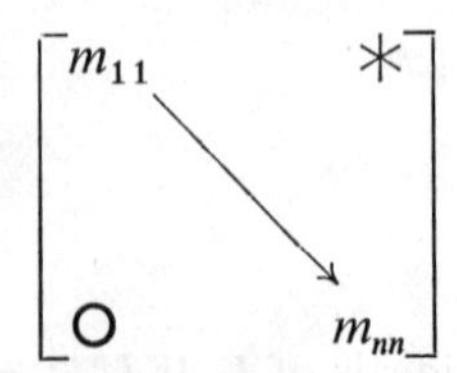

is soluble but not nilpotent. The former example shows on computation that the bound on derived length in terms of nilpotency class of lemma 2.1(c) is best possible in general.

If $X \subseteq L$ then the *centraliser* of X in L is

$$C_L(X) = \{y \in L: [X, y] = 0\}$$

and the *idealiser* of X in L is

$$I_L(X) = \{y \in L: [X, y] \subseteq X\}$$

so consists precisely of those elements y which idealise X (if X is a subspace of L).

LEMMA 2.2 *Let L be a Lie algebra. Then*

(a) *If* $X \subseteq L$ *then* $C_L(X) \leqq L$

(b) *If* $X \lhd L$ *then* $C_L(X) \lhd L$

(c) *If X is a subspace of L then* $I_L(X) \leqq L$

(d) *If X is a subspace of L then* $C_L(X) \lhd I_L(X)$.

Proof: (a) Clearly $C_L(X)$ is a subspace. Now if $x \in X$, $y, z \in C_L(X)$ then

$$\begin{aligned}[x, [y, z]] &= -[y, [z, x]] - [z, [x, y]] \\ &= 0\end{aligned}$$

so $[y, z] \in C_L(X)$.

(b) If $x \in X$, $y \in C_L(X)$, $z \in L$ the same equation yields $[x, [y, z]] = 0$, since $[z, x] \in X$.

(c) If $y, z \in I_L(X)$ then

$$\begin{aligned}[X, [y, z]] &\subseteq [y, [z, X]] + [z, [X, y]] \\ &\subseteq [y, X] + [z, X] \\ &\subseteq X + X \\ &= X.\end{aligned}$$

(d) Put $I_L(X)$ for L in (b). □

In view of (c), if X is a subspace, and we wish to show that some subalgebra Y idealises X, it is sufficient to show that some set of generators of Y idealises X.

The *centre* of L is defined to be

$$\zeta_1(L) = C_L(L)$$

and is therefore an ideal of L. The *upper central series* $(\zeta_n(L))_{n \in \mathbb{N}}$ is defined recursively by

$$\zeta_0(L) = 0$$

$$\zeta_{n+1}(L)/\zeta_n(L) = \zeta_1(L/\zeta_n(L)).$$

In general a (finite) *central series* for L is a series of subalgebras

$$0 = L_0 \leqq L_1 \leqq \ldots \leqq L_n = L \qquad (*)$$

such that $n \in \mathbb{N}$ and

$$[L_{i+1}, L] \subseteq L_i \quad (i = 0, \ldots, n-1).$$

The terms 'upper and lower central series' are explained by:

LEMMA 2.3 *Suppose L has a central series* (*). *Then*

(a) $L^i \leqq L_{n-i+1} \quad i = 1, \ldots, n+1$

(b) $L_i \leqq \zeta_i(L) \quad i = 0, \ldots, n$

(c) *L is nilpotent of class* $\leqq n$.

Conversely if L is nilpotent then the upper and lower central series of L are central series in the above sense.

Proof: A simple induction argument suffices. □

Next we establish two important properties of nilpotent Lie algebras.

THEOREM 2.4 *A finitely generated nilpotent Lie algebra is finite-dimensional.*

Proof: Let $L = \langle x_1, \ldots, x_r \rangle$ $(r \in \mathbb{N})$ and suppose that $L^n = 0$. By lemma 1.1 L is spanned by all products

$$[x_{i_1}, \ldots, x_{i_t}]$$

where $t < n$, and the x_{i_j} run through the x_i. There are at most

$$r+r^2+\ldots+r^{n-1}$$

such products. □

Note that the dimension is bounded in terms of the nilpotency class and the number of generators.

THEOREM 2.5 (*Fitting's theorem*) *Let H and K be ideals of the Lie algebra L, and suppose that H and K are nilpotent of class h and k respectively. Then $H+K$ is nilpotent of class $\leqq h+k$.*

Proof: We have $H^{h+1} = 0 = K^{k+1}$. Consider $(H+K)^{h+k+1}$. This is a sum of subspaces of the form

$$[X_1, \ldots, X_{h+k+1}]$$

where each X_i is either H or K. Either there are $\geqq h+1$ terms $X_i = H$, or $\geqq k+1$ terms $X_i = K$. In the first case, since each H^c is an ideal of L (by lemma 1.2(c) and induction) the subspace is 0. Similarly in the second case. □

There are transfinite analogues of the upper central, lower central, and derived series, defined for all ordinals α by

$$\zeta_{\alpha+1}(L)/\zeta_\alpha(L) = \zeta_1(L/\zeta_\alpha(L))$$

$$\zeta_\lambda(L) = \bigcup_{\alpha<\lambda} \zeta_\alpha(L) \qquad \text{for limit ordinals } \lambda$$

$$L^{\alpha+1} = [L^\alpha, L]$$

$$L^\lambda = \bigcap_{\alpha<\lambda} L^\alpha \qquad \text{for limit ordinals } \lambda$$

$$L^{(\alpha+1)} = [L^{(\alpha)}, L^{(\alpha)}]$$

$$L^{(\lambda)} = \bigcap_{\alpha<\lambda} L^{(\alpha)} \qquad \text{for limit ordinals } \lambda.$$

By set-theoretic considerations each of these series *terminates* for some ordinal, in the sense that from that ordinal onwards all terms are equal. For the transfinite upper central series this terminal $\zeta_\alpha(L)$ is called the *hypercentre* of L and denoted by $\zeta_*(L)$. If $L = \zeta_*(L) = \zeta_\alpha(L)$ we say that L is *hypercentral* of *central height* $\leqq \alpha$. If $H \leqq \zeta_*(L)$ we say that H is *hypercentral in L*. These algebras are analogues of hypercentral or ZA groups (Robinson [331] p. 14, Kuroš [303] p. 218) and share many of their elementary properties.

Finally we list some useful properties of soluble algebras.

LEMMA 2.6 *Let L be a Lie algebra,* $H \lhd L$.

(a) *If H and L/H are soluble, then L is soluble.*

(b) *L is soluble if and only if there exist finitely many subalgebras* $H_0, \ldots, H_n$ *of L such that*

$$0 = H_0 \lhd H_1 \lhd \ldots \lhd H_n = L$$

and such that each H_{i+1}/H_i *is abelian* $(0 \leqq i \leqq n-1)$.

(c) *If H and K are soluble ideals of L, of derived lengths c and d respectively, then* $H+K$ *is soluble of derived length* $\leqq c+d$.

Proof: (a) We have $L^{(m)} \leqq H$ for some m, so $L^{(m+n)} \leqq H^{(n)}$ which is 0 for suitable n.

(b) In one direction let the H_i be the derived series of L. In the other, use part (a) and induction.

(c) Apply part A to $(H+K)/K \cong H/H \cap K$, and to $H \cap K$. □

3. Subideals

Many concepts of the theory of infinite-dimensional Lie algebras are motivated by analogies with group theory. Thus the Lie product is analogous to the group *commutator* $(g, h) = g^{-1}h^{-1}gh$, subalgebras are analogous to subgroups, ideals to normal subgroups. It is remarkable (and this will be a major theme in what follows) just how far the analogies can be stretched.

Wielandt [348] introduced the concept of a subnormal subgroup, which has proved particularly fruitful as a source of group-theoretic problems. The analogous object for Lie algebras is a *subideal*, where H is a subideal of the Lie algebra L if there is a finite series

$$H = H_0 \lhd H_1 \lhd \ldots \lhd H_n = L.$$

We write

$$H \text{ si } L.$$

To emphasize the rôle of the integer n we say that H is an *n-step* subideal of L, and write

$$H \lhd^n L.$$

We sometimes refer to n as the *subideal index* of H. Subideals were introduced by Schenkman [185] under the title of 'subinvariant subalgebras'.

There is also a transfinite generalisation: for any ordinal σ we say that H is a (σ-*step*) *ascendant subalgebra* of L if there is a series $(H_\alpha)_{\alpha \leqq \sigma}$ of subalgebras of L, such that

(i) $H_0 = H$, $H_\sigma = L$
(ii) $H_\alpha \lhd H_{\alpha+1}$ if $\alpha < \sigma$
(iii) $H_\lambda = \bigcup_{\alpha<\lambda} H_\alpha$ if $\lambda \leqq \sigma$ is a limit ordinal. We write

$$H \operatorname{asc} L, \quad H \lhd^\sigma L.$$

Ascendant subalgebras were introduced by Hartley [76].

We note some elementary properties of subideals and ascendant subalgebras:

Lemma 3.1 *Let L be a Lie algebra. Then*

(a) *If $H \lhd^\sigma L$ and $K \leqq L$ then $H \cap K \lhd^\sigma K$.*

(b) *If $H \lhd^\sigma L$ and $K \lhd L$ then $H+K \lhd^\sigma L$.*

(c) *Let $(H_\alpha)_{\alpha \leqq \sigma}$ be a tower of subalgebras of L, indexed by ordinals $\alpha \leqq \sigma$, such that if $\alpha < \sigma$ then $H_\alpha \operatorname{asc} H_{\alpha+1}$, $H_\lambda = \bigcup_{\alpha<\lambda} H_\alpha$ for limits $\lambda \leqq \sigma$, and $H_0 = H$, $H_\sigma = L$. Then $H \operatorname{asc} L$.*

(d) *If $H \lhd^m K \lhd^n L$ then $H \lhd^{m+n} L$.*

(e) *The relation* 'si' *is transitive.*

Proof: (a) Let $(H_\alpha)_{\alpha \leqq \sigma}$ be an ascendant series for H in L. Then $(H_\alpha \cap K)_{\alpha \leqq \sigma}$ is an ascendant series for $H \cap K$ in K.

(b) $(H_\alpha + K)_{\alpha \leqq \sigma}$ is an ascendant series for $H+K$ in L.

(c) follows from properties of the ordinals.

(d) This is a special case of (c), and is obvious from the definitions in any case.

(e) follows from (d). □

An important but simple result, for which the corresponding group-theoretic statement is *false*, is due to Schenkman [185, 186].

Lemma 3.2 *If L is a Lie algebra and H* si *L, then $H^\omega \lhd L$ where ω is the first infinite ordinal.*

Proof: Induction and the Jacobi identity shows that for all $n \in \mathbb{N}$

$$[L, H^n] \leqq [L, {}_nH].$$

Suppose that $H \lhd^p L$. Then clearly $[L, {}_pH] \leqq H$, so for all n

$$[L, H^{n+p}] \leqq H^{n+1}$$

from which it follows that $H^\omega \lhd L$. □

There is a corresponding theorem for finite-dimensional ascendant subalgebras, due to Simonjan [192] (see also Tôgô and Kawamoto [231]). First we prove:

LEMMA 3.3 *Suppose K is a finite-dimensional ascendant subalgebra of L, and let M be any finite-dimensional subspace of L. Then there exists $t = t(M)$ such that*

$$[M, {}_tK] \leqq K.$$

Proof: Let $(K_\alpha)_{\alpha \leqq \sigma}$ be an ascending series for K in L. Let $e_1, \ldots, e_n$ be a basis for K, and $x_1, \ldots, x_m$ be a basis for M. For each integer $t \in \mathbb{N}$ let μ_t be the first ordinal such that

$$[x_i, e_{j_1}, \ldots, e_{j_t}] \in K_{\mu_t}$$

for all choices of indices such that $1 \leqq i \leqq m$, $1 \leqq j_k \leqq n$. For any given t the set of elements considered is finite, so μ_t is not a limit ordinal. Since $[K_{\alpha+1}, K] \leqq K_\alpha$ for any $\alpha < \sigma$, we have $\mu_{t+1} < \mu_t$ unless $\mu_t = 0$. The ordinals $\leqq \sigma$ are well-ordered, so we must have $\mu_t = 0$ for some $t = t(M)$. Then $[M, {}_tK] \leqq K$. □

LEMMA 3.4 *If K is a finite-dimensional ascendant subalgebra of L, then $K^\omega \lhd L$.*

Proof: Using lemma 3.3 and arguing as in lemma 3.2 it follows that $[M, H^\omega] \leqq H^\omega$ for any finite-dimensional subspace M of L. So $[L, H^\omega] \leqq H^\omega$ whence $H^\omega \lhd L$. □

We say that the Lie algebra L is *perfect* if $L^2 = L$.

PROPOSITION 3.5 *Every perfect ascendant subalgebra of a Lie algebra L is an ideal of L.*

Proof: Let $H = H^2$ and suppose that $(H_\alpha)_{\alpha \leqq \sigma}$ is an ascending series for H in L. Using lemma 3.2 and transfinite induction we prove that $H \lhd H_\alpha$ for any α. In consequence $H \lhd L$. □

For any subalgebra $H \leqq L$ we define the *ideal closure series* $(H_i)_{i \in \mathbb{N}}$ recursively by

$$H_0 = L$$
$$H_{i+1} = \langle H^{H_i} \rangle.$$

LEMMA 3.6 *H is an n-step subideal of L if and only if* $H_n = H$.

Proof: If $H_n = H$ then $H = H_n \lhd H_{n-1} \lhd \ldots \lhd H_0 = L$. Conversely if

$$H = K_0 \lhd K \lhd \ldots \lhd K_n = L$$

then inductively we have $H_i \leqq K_{n-i}$, so that $H \leqq H_n \leqq H$. □

A characterisation of the ideal closure series in a form more suited to computations is given in the next chapter.

The next lemma is of considerable use in the study of nilpotent subideals:

LEMMA 3.7 *If L is a nilpotent Lie algebra of class c, and if* $H \leqq L$, *then* $H \lhd^c L$.

Proof: We have

$$H = H + L^{c+1} \lhd H + L^c \lhd H + L^{c-1} \lhd \ldots \lhd H + L^1 = L.$$ □

4. Derivations

If $x \in L$ we define the *adjoint map* $x^*: L \to L$ by

$$yx^* = [y, x] \quad (y \in L).$$

(Many authors write ad(x) for x^*, or occasionally $\dot{x}$.) From the Jacobi identity it follows that for any $y, z \in L$

$$[y, z]x^* = [yx^*, z] + [y, zx^*].$$

Any linear map $\delta : L \to L$ such that for all $y, z \in L$

$$[y, z]\delta = [y\delta, z] + [y, z\delta]$$

is called a *derivation* of L. Thus for any $x \in L$ the adjoint map x^* is a derivation, the *inner derivation* induced by x. We write Der(L) for the set of derivations of L, and Inn(L) for the set of inner derivations. These are subsets of the associative algebra End(L) of all linear transformations of L. Further:

LEMMA 4.1 (a) $\mathrm{Inn}(L) \lhd \mathrm{Der}(L) \leq [\mathrm{End}(L)]$
(b) $\mathrm{Inn}(L) \cong L/\zeta_1(L)$.

Proof: First we show that $\mathrm{Der}(L)$ is a subalgebra of $[\mathrm{End}(L)]$. Let $\delta, \varepsilon \in \mathrm{Der}(L)$. Then for any $x, y \in L$

$$\begin{aligned}
[x, y](\delta\varepsilon - \varepsilon\delta) &= [x\delta, y]\varepsilon + [x, y\delta]\varepsilon - [x\varepsilon, y]\delta - [x, y\varepsilon]\delta \\
&= [x\delta\varepsilon, y] + [x\delta, y\varepsilon] + [x\varepsilon, y\delta] + [x, y\delta\varepsilon] - \\
&\quad - [x\varepsilon\delta, y] - [x\varepsilon, y\delta] - [x\delta, y\varepsilon] - [x, y\varepsilon\delta] \\
&= [x(\delta\varepsilon - \varepsilon\delta), y] + [x, y(\delta\varepsilon - \varepsilon\delta)]
\end{aligned}$$

so that $\delta\varepsilon - \varepsilon\delta \in \mathrm{Der}(L)$ as claimed.

Now let $x, y \in L$, $\delta \in \mathrm{Der}(L)$. Then

$$\begin{aligned}
x(y^*\delta - \delta y^*) &= [x, y]\delta - [x\delta, y] \\
&= [x, y\delta]
\end{aligned}$$

so that $y^*\delta - \delta y^* = (y\delta)^*$ and $\mathrm{Inn}(L) \lhd \mathrm{Der}(L)$.

Finally, note that the map $x \mapsto x^*$ defines a Lie homomorphism $L \to \mathrm{Der}(L)$, with image $\mathrm{Inn}(L)$ and kernel $\zeta_1(L)$, so that $\mathrm{Inn}(L) \cong L/\zeta_1(L)$. □

There is a useful result of 'orbit-stabiliser' type:

LEMMA 4.2 *Let X be a subalgebra of L. Then $I_L(X)/C_L(X)$ is isomorphic to a subalgebra of* $\mathrm{Der}(X)$.

Proof: Let $i \in I_L(X)$. Then $i^*|_X$ is a derivation of X. The map $i \mapsto i^*|_X$ is a homomorphism $I_L(X) \to \mathrm{Der}(X)$ with kernel $C_L(X)$. □

This is most useful in two situations:

COROLLARY 4.3 *If $H \lhd L$ and H is finite-dimensional* (*resp.* 1-*dimensional*) *then $L/C_L(H)$ is finite-dimensional* (*resp.* 0- or 1-*dimensional*).

Proof: If L is finite-dimensional (1-dimensional) then so is $\mathrm{Der}(L)$. □

A subspace K of L invariant under all derivations of L is said to be a *characteristic ideal.* It is clearly an ideal since it is invariant under inner derivations. We write

K ch L.

LEMMA 4.4 *If K ch $H \lhd L$ then $K \lhd L$.*

Proof: If $x \in L$ then $x^*|_H \in \mathrm{Der}(H)$ so leaves K invariant. Therefore $Kx^* \subseteq K$, whence $K \lhd L$. □

LEMMA 4.5 *Suppose that C, D, and E_α ($\alpha \in A$) are characteristic ideals of L. Then*

(a) $[C, D]$ ch L.
(b) $C_L(D)$ ch L.
(c) *If* $C \leq P$ *and* P/C ch L/C *then* P ch L.
(d) $\bigcap_{\alpha \in A} E_\alpha$ ch L.
(e) $\sum_{\alpha \in A} E_\alpha$ ch L.

Proof: Let $c \in C$, $d \in D$, $e_\alpha \in E_\alpha$, $\delta \in \mathrm{Der}(L)$.

(a) $[c, d]\delta = [c\delta, d] + [c, d\delta] \in [C, D]$.
(b) Let $x \in C_L(D)$. Then $[d, x\delta] = [d, x]\delta - [d\delta, x] = 0$.
(c) Use the fact that δ induces a derivation on L/C.
(d) This is clear.
(e) We have $(\sum e_\alpha)\delta = \sum(e_\alpha\delta) \in \sum E_\alpha$. □

COROLLARY 4.6 *For all ordinals α and Lie algebras L, each of*

$$L^\alpha, \quad L^{(\alpha)}, \quad \zeta_\alpha(L)$$

is a characteristic ideal of L. □

We have the *Leibniz formula*

$$[x, y]\delta^n = \sum_{i=0}^{n} \binom{n}{i}[x\delta^i, y\delta^{n-i}]$$

for $x, y \in L$, $\delta \in \mathrm{Der}(L)$, $n \in \mathbb{N}$. This is easily proved by induction.

A considerable obstacle in translating from group theory to Lie theory arises from the fact that conjugation in a group defines an automorphism, which preserves all manner of structures associated with the group; whereas the corresponding object in the Lie theory is a derivation, which does not preserve the corresponding structures. We can get round this, on occasion, by modifying derivations to obtain automorphisms.

Let L be a Lie algebra over $\mathfrak{k}$, let S be a subset of $\mathrm{Der}(L)$, and let $n \in \mathbb{N}$. We say that S is *n-nilpotent* if

$$LS^n = 0$$

and is *nil* if for each finite-dimensional subspace M of L there exists $n = n(M) \in \mathbb{N}$ such that

$$MS^n = 0,$$

where the powers S^n are defined in the associative algebra End(L). A single derivation δ is n-nilpotent (or nil) if $\{\delta\}$ is n-nilpotent (or nil).

Given a derivation δ of L we can define the *exponential*

$$\exp(\delta) = 1+\delta+\frac{1}{2!}\delta^2+\frac{1}{3!}\delta^3+\ldots$$

provided either

(i) char($\mathfrak{k}$) $= 0$ and δ is nil,

(ii) char($\mathfrak{k}$) $= p > 0$ and δ is p-nilpotent.

This is because, given any $x \in L$, the series $\sum \frac{1}{n!} xd^n$ terminates after finitely many steps, and since in case (ii) $\frac{1}{n!}$ is defined provided $n < p$. We then have:

THEOREM 4.7 *Suppose that* $\delta \in \mathrm{Der}(L)$, *and that either*

(a) char($\mathfrak{k}$) $= 0$ *and* δ *is nil, or*

(b) char($\mathfrak{k}$) $= p$ *and* δ *is* $\left[\frac{p}{2}\right]$*-nilpotent.*

Then $\exp(\delta)$ *is an automorphism of* L.

Proof: Certainly $\exp(\delta)$ is linear. We must show it preserves products. First we deal with case (b). Let $x, y \in L$ and put $N = \left[\frac{p}{2}\right]$. Then

$$e = \exp(\delta) = 1+\delta+\ldots+\frac{1}{(N-1)!}\delta^{N-1}$$

so that

$$[xe, ye] = \left[\sum_{i=0}^{N-1} \frac{1}{i!} x\delta^i, \sum_{j=0}^{N-1} \frac{1}{j!} y\delta^j\right]$$

$$= \sum_{n=0}^{2N-2}\left(\sum_{i=0}^{n}\left[\frac{1}{i!} x\delta^i, \frac{1}{(n-i)!} y\delta^{n-i}\right]\right).$$

We may use the Leibniz formula, and the fact that $2N-2 < p$, to rewrite this in the form

$$\sum_{n=0}^{2N-2} \frac{1}{n!} [x, y]\delta^n$$

which equals

$$[x, y]e$$

as required.

For case (a) we take $x, y \in L$, and choose N such that δ^N annihilates x, y, and $[x, y]$; the subsequent argument is exactly as above. $\square$

Note that in case (b) it is *not* sufficient to assume merely p-nilpotence of δ, even though this ensures that $\exp(\delta)$ is defined. For example, let $\text{char}(\mathfrak{k}) = 2$, and consider the Lie algebra L with basis $\{u, v, x, y, z\}$ such that $[u, v] = [v, x] = [u, y] = [x, z] = [y, z] = 0$; $[u, x] = v$; $[v, y] = u$; $[u, z] = u$; $[v, z] = v$; $[x, y] = z$; and where all other products are uniquely defined by antisymmetry and bilinearity. It is easy to check that the Jacobi identity holds, and that $x^{*2} = 0$. But

$$[u, y](1+x^*) = 0(1+x^*) = 0,$$

whereas

$$\begin{aligned}[u(1+x^*), y(1+x^*)] &= [u+v, y-z] \\ &= -v.\end{aligned}$$

We adopt the convention (Hartley [76], p. 261) that the image of an element x of a Lie algebra L under an automorphism α is written

$$x^\alpha$$

to distinguish automorphisms from other linear maps.

Our next task is to connect up δ-invariance and $\exp(\delta)$-invariance.

LEMMA 4.8 *Let L be a Lie algebra over $\mathfrak{k}$, M a subspace of L, and $\delta \in \text{Der}(L)$. If either*

(a) $\text{char}(\mathfrak{k}) = 0$ *and δ is nil, or*

(b) $\text{char}(\mathfrak{k}) = p$ *and δ is p-nilpotent*

then

$$M\delta \subseteq \textstyle\sum_{n=1}^{\infty} M^{\exp(n\delta)}.$$

Proof: Let $x \in M$. Then $x\delta^k = 0$ where in case (b) $k \leqq p$. We have

$$x^{\exp(n\delta)} = x + nx\delta + \ldots + \frac{n^{k-1}}{(k-1)!}\, x\delta^{k-1}$$

for any integer n. Now the matrix (a_{ij}) with

$$a_{ij} = i^j/j!$$

$(1 \leqq i, j \leqq k)$ is nonsingular, so we can find $\alpha_1, \ldots, \alpha_k \in \mathfrak{k}$ such that

$$\textstyle\sum_{n=1}^{k} \alpha_n n^i/i! = \delta_{1,i}.$$

Therefore

$$\sum_{n=1}^{k} \alpha_n x^{\exp(n\delta)} = x\delta$$

and the lemma is proved. □

If T is a set of automorphisms of a Lie algebra L and M is a subspace of L we let

$$M^T$$

be the smallest subspace of L containing M and t-invariant for all $t \in T$. Clearly if T is a subgroup of $\mathrm{Aut}(L)$ then

$$M^T = \sum_{t \in T} M^t.$$

If S is a set of derivations of L such that $\exp(s)$ is defined for all $s \in S$, we let $\exp(S) = \{\exp(s) : s \in S\}$.

LEMMA 4.9 *Let L be a Lie algebra over $\mathfrak{k}$, M a subspace of L, and $S \subseteq \mathrm{Der}(L)$. If either*

(a) $\mathrm{char}(\mathfrak{k}) = 0$ *and S is nil, or*

(b) $\mathrm{char}(\mathfrak{k}) = p$ *and each element of S is p-nilpotent*

then

$$M^S = M^{\exp(S)}.$$

Proof: Since M^S is s-invariant for any $s \in S$ it is certainly $\exp(s)$-invariant. Conversely if we let $M_1 = M^{\exp(S)}$ then for any $s \in S$

$$M_1 s \subseteq \sum_{n=1}^{\infty} M_1^{\exp(ns)} \subseteq M_1$$

since $\exp(ns) = \exp(s)^n$. □

LEMMA 4.10 *Let L be a Lie algebra over a field $\mathfrak{k}$ of characteristic* 0. *Let M be a finite-dimensional subspace of L and S a finite-dimensional nil subspace of* $\mathrm{Der}(L)$. *Then M^S is finite-dimensional and there exist finitely many elements $\alpha_1, \ldots, \alpha_r$ of $\langle \exp(S) \rangle$, the group of automorphisms generated by* $\exp(S)$, *such that*

$$M^S = \sum_{i=1}^{r} M^{\alpha_i}.$$

Proof: Let $s_1, \ldots, s_n$ be a basis for S. There exists m such that $MS^m = 0$, so that

$$M^S = \sum M s_{i_1} \ldots s_{i_k}$$

where $i_1, \ldots, i_k$ are all between 1 and n, and $1 \leqq k \leqq m-1$. Obviously M^S is finite-dimensional. The existence of the α_i is then clear. □

Theorems 4.7 onwards, for case (a), can be found in Hartley [76]. We shall not need case (b) until chapter 17, but we include it for completeness.

5. Classes and closure operations

The 'calculus of classes and closure operations' was introduced into group theory by Hall [288] (and is expounded in Robinson [331]). It provides a very compact symbolism in which to express certain kinds of result. Its use for Lie algebras stems from [205, 208].

We choose a fixed but arbitrary field $\mathfrak{k}$. A *class of Lie algebras over* $\mathfrak{k}$ is a class $\mathfrak{X}$ in the usual sense of set theory, whose members are Lie algebras over $\mathfrak{k}$, satisfying two mild conditions:

(i) $\mathfrak{X}$ contains some 0-dimensional subalgebra,
(ii) If $H \cong K \in \mathfrak{X}$ then $H \in \mathfrak{X}$.

We shall usually use Gothic letters for classes. We reserve $\mathfrak{X}$ and $\mathfrak{Y}$ for general classes, or for temporary notation, and attempt to adopt a fixed symbolism for the most common classes. In particular we make the following definitions:

$\mathfrak{D}$: all Lie algebras
$\mathfrak{F}$: finite-dimensional
$\mathfrak{F}_d$: of dimension $\leqq d$ ($d \in \mathbb{N}$)
$\mathfrak{G}$: finitely generated
$\mathfrak{G}_r$: generated by $\leqq r$ elements ($r \in \mathbb{N}$)
$\mathfrak{A}$: abelian
$\mathfrak{N}$: nilpotent
$\mathfrak{N}_c$: nilpotent of class $\leqq c$ ($c \in \mathbb{N}$)
$\mathfrak{Z}$: hypercentral
$\mathfrak{Z}_\sigma$: hypercentral of central height $\leqq \sigma$ (σ an ordinal).

Notation for other classes (including soluble algebras) will follow.

A (non-commutative non-associative) operation on classes is defined as follows: $\mathfrak{X}\mathfrak{Y}$ is the class of algebras L having an ideal $I \in \mathfrak{X}$ such that $L/I \in \mathfrak{Y}$. We define left-normed products recursively by

$$\mathfrak{X}_1 \ldots \mathfrak{X}_{n+1} = (\mathfrak{X}_1 \ldots \mathfrak{X}_n)\mathfrak{X}_{n+1}.$$

If $\mathfrak{X}_1 = \ldots = \mathfrak{X}_n = \mathfrak{X}$ we define

$$\mathfrak{X}^n = \mathfrak{X}_1 \ldots \mathfrak{X}_n.$$

It is easy to obtain the following characterisation:

$\mathfrak{A}^d$: soluble algebras of derived length $\leqq d$ $(d \in \mathbb{N})$.

Algebras belonging to $\mathfrak{X}$ will be called *$\mathfrak{X}$-algebras*; subalgebras (ideals, subideals, etc.) of L which lie in $\mathfrak{X}$ will be called *$\mathfrak{X}$-subalgebras* (*$\mathfrak{X}$-ideals*, *$\mathfrak{X}$-subideals*, etc.) of L. Algebras in the class $\mathfrak{X}\mathfrak{Y}$ will be called *$\mathfrak{X}$-by-$\mathfrak{Y}$-algebras*, so that (for example) abelian-by-nilpotent algebras are those in the class $\mathfrak{A}\mathfrak{N}$.

If L is a Lie algebra then we write (L) for the minimal class containing L, which consists of all 0-dimensional algebras and all algebras isomorphic to L. With this notation we have:

(0): 0-dimensional algebras.

We use the symbol '$\leqq$' for inclusion of classes, and otherwise use the standard set-theoretic symbolism.

A *closure operation* A assigns to each class $\mathfrak{X}$ a class $\mathrm{A}\mathfrak{X}$ in such a way that for all classes $\mathfrak{X}$ and $\mathfrak{Y}$ the following conditions hold:

(i) $\mathrm{A}(0) = (0)$
(ii) $\mathfrak{X} \leqq \mathrm{A}\mathfrak{X}$
(iii) $\mathrm{A}(\mathrm{A}\mathfrak{X}) = \mathrm{A}\mathfrak{X}$
(iv) $\mathfrak{X} \leqq \mathfrak{Y} \Rightarrow \mathrm{A}\mathfrak{X} \leqq \mathrm{A}\mathfrak{Y}$.

We say that $\mathfrak{X}$ is A-*closed* if $\mathfrak{X} = \mathrm{A}\mathfrak{X}$.

It is often easier to define a closure operation by specifying the closed classes. Suppose that $\mathscr{S}$ is a collection of classes (to make this respectable we should perhaps use Grothendieck set theory, in which everything happens inside a suitably large universe which is itself a set) such that $(0) \in \mathscr{S}$, $\mathfrak{D} \in \mathscr{S}$ and the intersection of any subset of $\mathscr{S}$ belongs to $\mathscr{S}$. For any class $\mathfrak{X}$ we define

$$\mathrm{A}\mathfrak{X} = \bigcap \{\mathfrak{Y} \in \mathscr{S} : \mathfrak{X} \leqq \mathfrak{Y}\}.$$

Then A is a closure operation, and the A-closed classes are precisely those in $\mathscr{S}$. Conversely if A is any closure operation then the set $\mathscr{S}$ of A-closed classes contains (0) and is closed under intersections. Each of A or $\mathscr{S}$ determines the other uniquely.

We list some standard closure operations:

S: $\mathrm{S}\mathfrak{X}$ consists of all subalgebras of $\mathfrak{X}$-algebras.
I: $\mathrm{I}\mathfrak{X}$ consists of all subideals of $\mathfrak{X}$-algebras.
Q: $\mathrm{Q}\mathfrak{X}$ consists of all quotients of $\mathfrak{X}$-algebras.

E: $\mathrm{E}\mathfrak{X}$ consists of all algebras L having a finite series

$$0 = L_0 \lhd \ldots \lhd L_n = L$$

whose factors $L_{i+1}/L_i \in \mathfrak{X}$ for $0 \leqq i \leqq n-1$.

L: $\mathrm{L}\mathfrak{X}$ consists of those algebras L such that every finite subset of L is contained in an $\mathfrak{X}$-subalgebra of L.

N_0: A class $\mathfrak{X}$ is N_0-closed if whenever $H, K \lhd L$ and $H, K \in \mathfrak{X}$ then $H+K \in \mathfrak{X}$.

R: $\mathrm{R}\mathfrak{X}$ consists of those algebras L having a family $(I_\alpha)_{\alpha \in A}$ of ideals such that $L/I \in \mathfrak{X}$ for all $\alpha \in A$ and $\bigcap_{\alpha \in A} I_\alpha = 0$.

D_0: $\mathfrak{X}$ is D_0-closed if $H, K \in \mathfrak{X}$ implies $H \oplus K \in \mathfrak{X}$.

D: $\mathfrak{X}$ is D-closed if $H_\alpha \in \mathfrak{X}$ $(\alpha \in A)$ implies $\mathrm{Dr}_{\alpha \in A} H_\alpha \in \mathfrak{X}$.

C: $\mathfrak{X}$ is C-closed if $H_\alpha \in \mathfrak{X}$ $(\alpha \in A)$ implies $\mathrm{Cr}_{\alpha \in A} H_\alpha \in \mathfrak{X}$.

The operations E, L, R are read as 'poly', 'locally', 'residually' respectively. In particular we note the classes

$\mathrm{E}\mathfrak{A}$: soluble
$\mathrm{L}\mathfrak{N}$: locally nilpotent
$\mathrm{L}\mathfrak{F}$: locally finite (-dimensional)
$\mathrm{R}\mathfrak{N}$: residually nilpotent
$\mathrm{R}\mathfrak{F}$: residually finite (-dimensional).

If $\mathrm{A}_1, \ldots, \mathrm{A}_n$ are closure operations we define

$$\{\mathrm{A}_1, \ldots, \mathrm{A}_n\}\mathfrak{X}$$

to be the smallest class containing $\mathfrak{X}$ which is A_i-closed for $i = 1, \ldots, n$. This makes $\{\mathrm{A}_1, \ldots, \mathrm{A}_n\}$ a closure operation. If A and B are operations we define AB by

$$\mathrm{AB}\mathfrak{X} = \mathrm{A}(\mathrm{B}\mathfrak{X}).$$

In general AB need not be a closure operation, in that (iii) may fail. For example $\mathrm{EL}\mathfrak{A} = \mathrm{E}\mathfrak{A}$, whereas $\mathrm{ELEL}\mathfrak{A} = \mathrm{ELE}\mathfrak{A}$ contains non-soluble algebras – for example the direct sum of a set of soluble algebras of unbounded derived length. However, let us define an ordering on operations by

$$\mathrm{A} \leqq \mathrm{B} \Leftrightarrow \mathrm{A}\mathfrak{X} \leqq \mathrm{B}\mathfrak{X} \text{ for all classes } \mathfrak{X}.$$

If $\mathrm{BA} \leqq \mathrm{AB}$ then it is easy to see that AB is a closure operation, and is in fact equal to $\{\mathrm{A}, \mathrm{B}\}$. In particular the operations

ES	QS	RS	LS
EI	QI	RI	LI
	EQ	LQ	

are closure operations.

If $L \in \text{R}\mathfrak{X}$, having a set of ideals $(I_\alpha)_{\alpha \in A}$ as above, there is a natural injection

$$L \to \text{Cr}_{\alpha \in A}\, L/I_\alpha$$

so that $L \in \text{SC}\mathfrak{X}$ and $\text{R} \leq \text{SC}$. From this we easily see that $L \in \text{R}\mathfrak{N}$ if and only if $L^\omega = 0$, and $L \in \text{RE}\mathfrak{A}$ if and only if $L^{(\omega)} = 0$.

A class $\mathfrak{X}$ which is $\{\text{Q}, \text{R}\}$-closed is called a *variety*. It is well known (see Birkhoff [263], Cohn [271] p. 169) that varieties are precisely the classes of algebras L satisfying some given set of identities

$$f_\alpha(x_1, \ldots, x_n) = 0$$

for all $x_1, \ldots, x_n \in L$. It follows at once that varieties are also $\{\text{L}, \text{S}, \text{C}\}$-closed. In particular $\mathfrak{A}$, $\mathfrak{A}^d$, and $\mathfrak{N}_c$ are varieties.

6. Representations and modules

Let L be a Lie algebra over $\mathfrak{k}$, and V a vector space over $\mathfrak{k}$ (possibly of infinite dimension). A *representation* of L on V is a homomorphism

$$\rho: L \to [\text{End}_\mathfrak{k}(V)].$$

Thus for all $x, y \in L$ and $\alpha, \beta \in \mathfrak{k}$ we have

$$\rho(\alpha x + \beta y) = \alpha\rho(x) + \beta\rho(y)$$

$$\rho([x, y]) = \rho(x)\rho(y) - \rho(y)\rho(x).$$

If we define, for $x \in L$, $v \in V$

$$vx = v\rho(x) \tag{*}$$

we have

$$\left.\begin{aligned} (\alpha v + \beta w)x &= \alpha vx + \beta wx \\ v(\alpha x + \beta y) &= \alpha vx + \beta vy \\ v[x, y] &= (vx)y - (vy)x, \end{aligned}\right\} \tag{**}$$

for all $v, w \in V$, $x, y \in L$, $\alpha, \beta \in \mathfrak{k}$.

A map $V \times L \to V$, $(v, x) \mapsto vx$ satisfying (**) is called an *L-action* on V. A vector space V together with an L-action on V is said to be an *L-module*. Using (*) we can pass at will from a representation to a module and back again.

Submodules W of V and *quotient modules* V/W, along with the accompanying paraphernalia of homomorphisms, isomorphisms, etc. are defined for modules in the obvious way. We write $W \cong_L V$ if W and V are isomorphic L-modules. An L-module can be considered as an abelian group with operators, so the Noether isomorphism theorems and the Jordan-Hölder theorem are true for modules. A module V is said to be *irreducible* if it has no submodules other than 0 and V itself.

Modules arise naturally as follows: suppose that L is a Lie algebra and $I \lhd L$. We define an L-action on I by

$$ix = [i, x]$$

$(i \in I,\ x \in L)$. Then I becomes an L-module, because of the Jacobi identity. In the same way, if also $J \lhd L$ and $J \leqq I$ then I/J is an L-module.

Further, if $H \leqq L$, then I or I/J become H-modules by restricting the L-action to H. And we can relax the requirement that I and J be ideals: all we then need is that H idealises each of them.

If V is an L-module we let

$$C_L(V) = \{x \in L \colon vx = 0 \text{ for all } v \in V\}.$$

Thus $C_L(I/J)$ is defined if $I \geqq J$ and both are ideals of L. The module V (or the associated representation ρ) is *faithful* if $C_L(V) = 0$. In general $C_L(V)$ is the kernel of ρ.

Split extensions

We now consider the situation of a Lie algebra L having an ideal I and a subalgebra K such that $I + K = L$, $I \cap K = 0$. Such an L is said to be a *split extension* of I by K.

As vector spaces we have $L = I \dot{+} K$. Each $k \in K$ induces a derivation $\delta(k) = k^*|_I$ of I, and we obtain a homomorphism

$$\delta \colon K \to \mathrm{Der}(I).$$

Further, if $i, j \in I$, $k, l \in K$, we have

$$\begin{aligned}[i+k, j+l] &= [i, j] + [i, l] + [k, j] + [k, l] \\ &= ([i, j] + i\delta(l) - j\delta(k)) + [k, l]. \qquad (*)\end{aligned}$$

Conversely, given any Lie algebras I and K and any homomorphism δ: $K \to \mathrm{Der}(I)$ we can use (*) to define a Lie product on $I \dotplus K$ making it into a Lie algebra which is a split extension of I by K. There is a bijection between split extensions and homomorphisms δ.

We can use this idea to construct split extensions, if we specify I, K, and δ. We speak of the split extension $I \dotplus K$ and omit δ from the notation. If this omission would cause confusion we write

$$I \dotplus_\delta K.$$

Usually we have $K \leqq \mathrm{Der}(I)$ and δ is the inclusion map. In particular if I is abelian we can take for K any subalgebra of $[\mathrm{End}(I)]$. Further, if we have an L-module M, we can think of M as an abelian Lie algebra and use the L-action to define δ.

We can use this construction to give a converse to lemma 4.4.

LEMMA 6.1 *Let $K \leqq H$, and suppose that whenever $H \lhd L$ it follows that $K \lhd L$. Then K is characteristic in H.*

Proof: Let $\delta \in \mathrm{Der}(H)$ and form the split extension $H \dotplus \langle\delta\rangle$. Since K is an ideal of this, we have $[K, \delta] \leqq K$. But $[K, \delta] = K\delta$. Therefore $K \operatorname{ch} H$. □

We use the split extension to construct a Lie algebra L whose lower central series first becomes stationary with $L^\omega \neq L^{\omega+1} = 0$. We take an abelian Lie algebra X with basis $\{x_{ij}\ (i, j \in \mathbb{N}), z\}$ and define a derivation σ by

$$\sigma\colon x_{ij} \mapsto x_{i+1,j-1} \qquad j \neq 0$$

$$x_{i0} \mapsto z$$

$$z \mapsto 0.$$

Then $L^\omega = \langle z\rangle$, $L^{\omega+1} = 0$. Note that L is *metabelian*, that is, soluble of derived length $\leqq 2$.

By modifying this procedure we can obtain examples of metabelian algebras whose lower central series first becomes zero at any given ordinal. This shows that there is little connection between the lower central and derived series. It also shows that algebras L with $L^\alpha = 0$ for $\alpha > \omega$ need not be residually nilpotent.

7. Chain conditions

Often it is not possible to make progress on a problem about *all* Lie algebras: it becomes necessary to impose extra conditions, obtained by relaxing to some extent the requirement that the dimension be finite. For example, we might ask that the algebra be finitely generated, or locally finite. A common kind of condition is the chain condition, which asserts that certain sets of subalgebras have maximal or minimal elements.

Let V be a vector space, and $\mathscr{S}$ a collection of subsets of V. We say that V *has* (or *satisfies*) Max-$\mathscr{S}$ if $\mathscr{S}$ satisfies the *maximal condition:* every ascending chain

$$S_0 \subseteq S_1 \subseteq \dots$$

of elements $S_i \in \mathscr{S}$ terminates finitely; so that $S_r = S_{r+1} = \dots$ for some $r \in \mathbb{N}$. Similarly V has Min-$\mathscr{S}$ if $\mathscr{S}$ satisfies the *minimal condition:* every descending chain

$$S_0 \supseteq S_1 \supseteq \dots$$

terminates.

If V is a Lie algebra L and $\mathscr{S}$ is respectively the set of subalgebras, ideals, subideals, n-step subideals, ascendant subalgebras, or α-step ascendant subalgebras of L we write in place of Max-$\mathscr{S}$

Max, Max-$\lhd$, Max-si, Max-$\lhd^n$, Max-asc, Max-$\lhd^\alpha$

and for Min-$\mathscr{S}$ we write

Min, Min-$\lhd$, Min-si, Min-$\lhd^n$, Min-asc, Min-$\lhd^\alpha$.

We use the same notation for the classes of Lie algebras satisfying the corresponding conditions.

Similarly if V is an L-module M and $\mathscr{S}$ is the set of submodules we write

Max-L Min-L

in place of Max-$\mathscr{S}$, Min-$\mathscr{S}$.

The following result is well known and easy to prove:

LEMMA 7.1 *Let L be a Lie algebra, M an L-module. Then*

(a) $L \in$ Max *if and only if every subalgebra of L is finitely generated,*

(b) $L \in \text{Max-}\triangleleft$ *if and only if every ideal of L is finitely generated as an ideal: i.e. is of the form* $\langle x_1^L \rangle + \ldots + \langle x_m^L \rangle$ *for finitely many* $x_i \in L$,

(c) $M \in \text{Max-}L$ *if and only if every submodule of M is finitely generated as an L-module.* □

We also recall the *Modular Law:*

LEMMA 7.2 *Let V be a vector space, having subspaces A, B, C such that* $B \subseteq A$. *Then*

$$A \cap (B+C) = B+(A \cap C).$$

Proof: Let $x \in A \cap (B+C)$. Then $x = b+c$ ($b \in B$, $c \in C$) and $x \in A$. Then $c = x-b \in A$, so $c \in A \cap C$, and $x \in B+(A \cap C)$. The converse implication is similar. □

The next theorem is a general formulation of a large number of useful results which all have essentially the same proof.

THEOREM 7.3 *Let* $\mathscr{S}$ *be a set of subspaces of a vector space V. Let K be a subspace of V, and define*

$$\mathscr{T} = \{S \cap K : S \in \mathscr{S}\}$$

$$\mathscr{U} = \{S+K/K : S \in \mathscr{S}\}.$$

Suppose that $K \in \text{Max-}\mathscr{T}$ *and* $V/K \in \text{Max-}\mathscr{U}$. *Then* $V \in \text{Max-}\mathscr{S}$. *If instead* $K \in \text{Min-}\mathscr{T}$ *and* $V/K \in \text{Min-}\mathscr{U}$ *then* $V \in \text{Min-}\mathscr{S}$.

Proof: Let

$$S_1 \subseteq S_2 \subseteq \ldots$$

be an ascending chain of elements of $\mathscr{S}$. Then

$$S_1 \cap K \subseteq S_2 \cap K \subseteq \ldots$$

and

$$S_1+K/K \subseteq S_2+K/K \subseteq \ldots$$

are ascending chains in $\mathscr{T}$, $\mathscr{U}$ respectively. Therefore there exists $R \in \mathbb{N}$ such that

$$S_r \cap K = S_R \cap K \qquad (r \geqq R)$$

$$S_r+K = S_R+K \qquad (r \geqq R)$$

and then for all $r \geqq R$

$$\begin{aligned} S_R &= S_R + (S_R \cap K) \\ &= S_R + (S_r \cap K) \\ &= S_r \cap (S_R + K) \\ &= S_r \cap (S_r + K) \\ &= S_r. \end{aligned}$$

The proof for Min-$\mathscr{S}$ is similar. □

Making the obvious choices for $\mathscr{S}$ we obtain:

THEOREM 7.4 *For any ordinal α the classes*

Max	Max-$\lhd^\alpha$	Max-si	Max-asc
Min	Min-$\lhd^\alpha$	Min-si	Min-asc

are E*-closed.*

If M is an L-module, K a submodule of M, and if both K and M/K satisfy Max-L (*resp.* Min-L) *then M satisfies* Max-L (*resp.* Min-L). □

A *composition series* for a Lie algebra L is a series

$$0 = L_0 \lhd L_1 \lhd \ldots \lhd L_n = L$$

with $n \in \mathbb{N}$ and each L_{i+1}/L_i simple.

PROPOSITION 7.5 *The Lie algebra L has a composition series if and only if* $L \in$ (Max-si) $\cap$ (Min-si).

Proof: Let $\mathfrak{X}$ be the class of simple algebras. Trivially $\mathfrak{X} \leqq$ (Max-si) $\cap$ (Min-si). By theorem 7.4 the inclusion remains valid if we pass to E$\mathfrak{X}$.

Conversely suppose $L \in$ (Max-si) $\cap$ (Min-si). We put $M_0 = L$, and recursively let M_{i+1} be a maximal ideal of M_i, which exists by Max-si unless $M_i = 0$. By Min-si the series terminates finitely with $M_n = 0$. Then

$$0 = M_n \lhd M_{n-1} \lhd \ldots \lhd M_0 = L$$

is a composition series. □

8. Series

The following concept is analogous to that introduced for groups by Hall [288], which is closely related to the concept of a 'normal system' suggested by Kuroš [303]. If H is a subalgebra of the Lie algebra L and Σ is a totally ordered set then a *series* from H to L of *type* Σ is a set

$$\mathscr{S} = \{\Lambda_\sigma, V_\sigma : \sigma \in \Sigma\}$$

of subalgebras of L, such that

(i) For all σ, $H \leqq \Lambda_\sigma$ and $H \leqq V_\sigma$,
(ii) $L \backslash H = \bigcup_{\sigma \in \Sigma} (\Lambda_\sigma \backslash V_\sigma)$,
(iii) $\Lambda_\tau \leqq V_\sigma$ if $\tau < \sigma$,
(iv) $V_\sigma \lhd \Lambda_\sigma$.

The quotient algebras $\Lambda_\sigma / V_\sigma$ are the *factors* of the series. If they all lie in some class $\mathfrak{X}$ we say that $\mathscr{S}$ is an $\mathfrak{X}$-*series* from H to L. If $H = 0$ we talk of series or $\mathfrak{X}$-series of L.

A series $\mathscr{S}'$ is a *refinement* of $\mathscr{S}$ if it can be obtained by extending Σ to a totally ordered set Σ' and inserting new Λ_σ's and V_σ's for $\sigma \in \Sigma' \backslash \Sigma$.

It is a consequence of the definition that

$$\Lambda_\sigma = \bigcap_{\tau > \sigma} V_\tau \text{ and } V_\sigma = \bigcup_{\tau < \sigma} \Lambda_\tau. \qquad (*)$$

We shall not need the full generality of this concept very often. We shall, however, need the more specialised ideas of an *ascending* or *descending* series, in which Σ is well-ordered, or its reverse is well-ordered. We may assume Σ to be an ordinal (or its reverse). Now (*) shows that in an ascending series the terms $\Lambda_\sigma = V_{\sigma+1}$. So an equivalent, and more intuitively accessible, definition of an ascending series would be a set $(V_\sigma)_{\sigma \leqq \rho}$ (ρ an ordinal) of subalgebras of L, such that

(i) $V_0 = H$, $V_\rho = L$,
(ii) $V_\lambda = \bigcup_{\sigma < \lambda} V_\sigma$ if λ is a limit ordinal,
(iii) $V_\sigma \lhd V_{\sigma+1}$ if $\sigma < \rho$.

There is a corresponding definition of a descending series: now we have

(i) $V_\rho = H$, $V_0 = L$,
(ii) $V_\lambda = \bigcap_{\sigma < \lambda} V_\sigma$ if λ is a limit ordinal,
(iii) $V_{\sigma+1} \lhd V_\sigma$ if $\sigma < \rho$.

Obviously H is ascendant in L if and only if it belongs to some ascending series. Dually, we say that H is a *descendant subalgebra* of L if it belongs to some descending series.

We can now introduce three new closure operations, generalising E.

$\hat{\mathrm{E}}$: $L \in \hat{\mathrm{E}}\mathfrak{X}$ if L has an $\mathfrak{X}$-series
$\acute{\mathrm{E}}$: $L \in \acute{\mathrm{E}}\mathfrak{X}$ if L has an ascending $\mathfrak{X}$-series
$\grave{\mathrm{E}}$: $L \in \grave{\mathrm{E}}\mathfrak{X}$ if L has a descending $\mathfrak{X}$-series.

If each Λ_σ or $V_\sigma \in \mathscr{S}$ is an ideal (subideal, etc.) of L we say that $\mathscr{S}$ is a *series of ideals* (subideals, etc.).

A detailed discussion of group-theoretic series may be found in Robinson [331] p. 9.

Chapter 2

Soluble subideals

It was for many years an open question whether finitely many soluble subnormal subgroups of a group necessarily generate a soluble subgroup. An affirmative answer was first given by Stonehewer [342]. Subsequently Roseblade [336] proved a theorem about the derived series of a join of subnormal subgroups which yields Stonehewer's result as a corollary.

It is possible to use Stonehewer's methods to prove the corresponding result for Lie algebras (cf. [1]). Here we shall obtain it as a corollary to analogues of Roseblade's theorems, which we refer to as the *Derived Join Theorems*. In addition the Derived Join Theorems provide an important tool for the study of subideals as later chapters will demonstrate. The proofs, taken from [3], are closely related to Roseblade's; but in order to make them work we have to introduce a new idea, the 'circle product' of two subsets of a Lie algebra.

The first section of this chapter is devoted to a study of this circle product, and to various permutability properties of subideals. The second gives the proofs of the Derived Join Theorems.

1. The circle product

We recall that if X and Y are subsets of a Lie algebra L then X^Y is the smallest subspace of L containing X and left invariant under Lie multiplication by the elements of Y. Thus

$$X^Y = \sum_{n \in \mathbb{N}} [X, {}_n Y].$$

From this it is fairly easy to deduce that:

$$\text{if } X \subseteq Y \subseteq L, \text{ then } X^Y = \langle X^Y \rangle \lhd \langle Y \rangle, \tag{1}$$

$$\text{if } X \subseteq H = \langle Y \rangle \leqq L, \text{ then } X^Y = X^H = \langle X^H \rangle \lhd H. \tag{2}$$

Let A and B be subsets of a Lie algebra L. We define the *circle product* $A \circ B$ of A and B as

$$A \circ B = [A, B]^{A \cup B}.$$

We also let $A \circ_1 B = A \circ B$ and define recursively

$$A \circ_{m+1} B = (A \circ_m B) \circ B$$

for all positive integers m.

Then we have:

PROPOSITION 1.1 *Let L be a Lie algebra. Then*

(a) *If A and B are subsets of L and $C = \langle A, B \rangle$ then $A \circ B = B \circ A$ is the smallest ideal of C containing $[A, B]$.*

(b) *If A and B are subsets of L and $A_1 \subseteq A$ then*

$$A_1 \circ B \leqq A \circ B.$$

(c) *If A and B are subsets of L and S is a finite subset of $A \circ_m B$ (any $m > 0$), then there exist finite subsets A_1 of A and B_1 of B such that $S \subseteq A_1 \circ_m B_1$.*

(d) *Let A be the union of a totally ordered chain $\{A_i : i \in I\}$ of subsets of L and let B be the union of a similar chain $\{B_j : j \in J\}$. Then for every positive integer m,*

$$A \circ_m B = \bigcup_{i \in I, j \in J} A_i \circ_m B_j.$$

(e) *Let A and B be subsets of L and let $H = \langle A \rangle$ and $K = \langle B \rangle$. Then $A \circ B = H \circ K$.*

(f) *If $H \leqq L$, $K \leqq L$ and $J = \langle H, K \rangle$ then*

$$\langle H^J \rangle = H^J = H + J \circ H = H + K \circ H = \langle H^K \rangle.$$

(g) *Let $H \leqq L$, $K \leqq L$ and $J = \langle H, K \rangle$. If H_m denotes the m-th term of the ideal closure series of H in J then*

$$H_m = H + K \circ_m H$$

for every positive integer m. Hence H si *J if and only if $K \circ_m H \leqq H$ for some m; and $H \lhd^m J$ if and only if $m = 0$ or $K \circ_m H \leqq H$.*

(h) *If A and B are subsets of L then $I_L(A) \cap I_L(B) \subseteq I_L(A \circ B)$. Hence $I_L(A) \cap I_L(B) \subseteq I_L(A \circ_m B)$ for every positive integer m.*

(i) *If $H \leqq L$ and H_m denotes the m-th term of the ideal closure series of H in L, then $I_L(H) \leqq I_L(H_m)$ for all m.*

(j) *Let A be a subset of L and $B \subseteq I_L(A)$. If $\{B_j : j \in J\}$ is any collection of subsets whose union is B then*

$$A \circ B_j \lhd A \circ B \quad (\text{all } j \in J)$$

and

$$A \circ B = \sum_{j \in J} A \circ B_j.$$

(k) *Let $H \lhd^m L$ and $K \lhd^n L$. Then $H \cap K \lhd^r L$, where $r \leqq \max\{m, n\}$.*

(l) *Let $m \in \mathbb{N}$. The union of any totally ordered chain of m-step subideals of L is an m-step subideal of L. The intersection of an arbitrary collection of m-step subideals of L is an m-step subideal of L.*

Proof: (a) Since $[A, B] = [B, A]$ it follows by definition that $A \circ B = B \circ A$. Now $C = \langle A, B \rangle = \langle A \cup B \rangle$ and so by equation (2), $A \circ B$ is the smallest ideal of C containing $[A, B]$.

(b) follows from the second part of (a) and the fact that $[A_1, B] \subseteq \subseteq [A, B] \subseteq A \circ B$.

(c) By definition every element x of $A \circ B$ is a linear combination of elements of the form

$$x_i = [a_i, b_i, c_{i_1}, \ldots, c_{i_n}] \quad (n \in \mathbb{N})$$

where $a_i \in A$, $b_i \in B$ and $c_{i_1}, \ldots, c_{i_n} \in A \cup B$. Suppose that S is a finite subset of $A \circ B$. If $x \in S$ let $A_1(x)$ be the subset of elements of A which occur in the x_i of which x is a linear combination. Define the subset $B_1(x)$ of B similarly. Now let $A_1 = \bigcup_{x \in S} A_1(x)$ and let $B_1 = \bigcup_{x \in S} B_1(x)$. Clearly A_1 is a finite subset of A, B_1 is a finite subset of B and for each $x \in S$, $x \in A_1(x) \circ B_1(x)$. Hence $S \subseteq A_1 \circ B_1$, by (b).

The general case follows by induction on m and the use of (b).

(d) It follows from (b) and by induction on m that $A_i \circ_m B_j \leqq A \circ_m B$ for all $i \in I$ and $j \in J$. Conversely let $x \in A \circ_m B$. From (c) we can find a finite subset A_1 of A and a finite subset B_1 of B such that $x \in A_1 \circ_m B_1$. Now the A_i form a totally ordered chain and so there is an i for which $A_1 \subseteq A_i$. Similarly there is a j for which $B_1 \subseteq B_j$. Therefore by (b) and the obvious induction on m,

$$A_1 \circ_m B_1 \subseteq A_i \circ_m B_1 \subseteq A_i \circ_m B_j,$$

whence $x \in A_i \circ_m B_j$. Therefore

$$A \circ_m B \leqq \bigcup_{i \in I, j \in J} A_i \circ_m B_j$$

and (d) is proved.

(e) From (b) we have $A \circ B \subseteq H \circ K$. If $C = \langle A, B \rangle$ then $C = \langle H, K \rangle$ and so by (a) we have $H \circ K$ as the smallest ideal of C containing $[H, K]$. Now $A \circ B \lhd C$ and so it is enough to show that $A \circ B$ contains $[H, K]$. But modulo $A \circ B$ we have $[A, B] = 0$ so that B and hence $K = \langle B \rangle$ centralises A. Thus A and so $H = \langle A \rangle$ centralises K. In other words, $[H, K] \subseteq A \circ B$.

(f) Let $X = H + K \circ H$. Clearly H and K idealise X and so $X \lhd J$. Thus $\langle H^J \rangle \leqq X$. But $H \circ K \leqq \langle H^K \rangle$ and so $X \leqq \langle H^K \rangle \leqq \langle H^J \rangle$. We also have $X \leqq H + J \circ H \leqq H^J$ and the result follows.

(g) From (f) we have $H_1 = \langle H^J \rangle = H + K \circ H$. Suppose inductively that $H_m = H + K \circ_m H$ so that $H_m = \langle H, K \circ_m H \rangle$. Replace the K of (f) by $K \circ_m H$. Then we have from (f), $H_{m+1} = \langle H^{H_m} \rangle = H + (K \circ_m H) \circ H =$ $= H + K \circ_{m+1} H$ and our induction is complete. Clearly $H_{m+1} \lhd H_m$ and by definition $H_0 = J$. Therefore $H_m \lhd^m J$ for all $m \in \mathbb{N}$. The rest now follows easily.

(h) Let $y, x_1, \ldots, x_n \in L$. By the Jacobi identity and by induction on n it follows that

$$[[x_1, \ldots, x_n], y] = \textstyle\sum_{i=1}^n [x_1, \ldots, x_{i-1}, [x_i, y], x_{i+1}, \ldots, x_n].$$

Thus if $x_1 \in A$, $x_2 \in B$, $x_3, \ldots, x_n \in A \cup B$ and $y \in I_L(A) \cap I_L(B)$ then for each $i = 1, \ldots, n$, $[x_i, y] \in A$ or B according as $x_i \in A$ or B. Therefore $[[x_1, \ldots, x_n], y]$ is a sum of elements of the same form as $[x_1, \ldots, x_n]$. But $A \circ B$ is spanned by these elements and so y idealises $A \circ B$. Therefore $I_L(A) \cap I_L(B) \subseteq I_L(A \circ B)$. In general $I_L(A \circ_m B) \cap I_L(B) \subseteq I_L(A \circ_{m+1} B)$ and the general result now follows by induction on m.

(i) Let $K = J = L$ in (g) so that $H_m = H + L \circ_m H$. Then it follows (on putting $A = L$ in (g)) that $I_L(H) \leqq I_L(L \circ_m H)$ and so $I_L(H) \leqq I_L(H_m)$; if $m = 0$ then $H_m = L$ and there is nothing to prove.

(j) Let $H = \langle A \rangle$. Then $A \circ B \leqq H$ since B idealises A and so H. Let $j \in J$. Then $A \circ B_j \leqq H$ and $A \subseteq I_H(A \circ B_j) \leqq H$; hence $H = \langle A \rangle$ idealises $A \circ B_j$ and in particular $A \circ B$ idealises $A \circ B_j$. From (b) we have $A \circ B_j \leqq$ $\leqq A \circ B$ and so $A \circ B_j \lhd A \circ B$. Define $X = \sum_{j \in J} A \circ B_j$. Then we have $X \lhd H$. Let $b \in B$ so that $b \in B_j$ for some $j \in J$. Then $[A, b] \subseteq A \circ B_j \subseteq X$. Therefore $(B+X)/X \subseteq C_{H/X}((A+X)/X)$. This implies that $[A, B] \subseteq X \lhd H$. Furthermore for any $j, k \in J$ we have $[A \circ B_j, B_k] \subseteq [H, B_k] \subseteq H \circ B_k =$ $= A \circ B_k \leqq X$ (making use of (e)). So each B_k and therefore B idealises X. Hence

$$A \circ B = [A, B]^{A \cup B} \subseteq X^{A \cup B} = X.$$

Finally $X \subseteq A \circ B$ since by (b) we have $A \circ B_j \subseteq A \circ B$ for each $j \in J$. Therefore $A \circ B = X = \sum_{j \in J} A \circ B_j$.

(k) Let $Y = H \cap K$. If $m = 0$ or $n = 0$ then $Y = K$ or H and there is nothing to prove. Assume that $m > 0$ and $n > 0$. It follows from (b) and by induction on i that

$$L \circ_i Y \leqq (L \circ_i H) \cap (L \circ_i K).$$

Let $s = \max\{m, n\}$. Then $H \lhd^s L$ and $K \lhd^s L$ and therefore using (g) we have

$$L \circ_s Y \leqq (L \circ_s H) \cap (L \circ_s K) \leqq H \cap K = Y$$

and so again by (g) we have $Y \lhd^s L$.

(l) If $m = 0$ there is nothing to prove; so let $m > 0$. Let $\{H_j : j \in J\}$ be a totally ordered set of m-step subideals of L and let H be their union. Clearly $H \leqq L$. From (g) we have $L \circ_m H_j \leqq H_j$ for each $j \in J$. Therefore using (d) we have $L \circ_m H = \bigcup_{j \in J} L \circ_m H_j \subseteq \bigcup_{j \in J} H_j = H$; it follows from (g) that $H \lhd^m L$.

Now suppose that $\{K_j : j \in J\}$ is a collection of m-step subideals of L so that (by (g)) $L \circ_m K_j \leqq K_j$ for each $j \in J$. Let $K = \bigcap_{j \in J} K_j$. Using (b) and the obvious induction on i we have $L \circ_i K \leqq L \circ_i K_j$ for every $j \in J$. In particular we have $L \circ_m K \leqq \bigcap_{j \in J} L \circ_m K_j \leqq \bigcap_{j \in J} K_j = K$; therefore by (g) we have $K \lhd^m L$. □

Next we have a simple result due to Hartley [76].

Lemma 1.2 (*Hartley*) *Let L be a Lie algebra and suppose that $H \lhd^m L$, $K \lhd^n L$ and $J = \langle H, K \rangle$. If $H \lhd J$ then $J \lhd^{mn} L$.*

Proof: Let $H = H_m \lhd H_{m-1} \lhd \ldots \lhd H_0 = L$ be the ideal closure series of H in L. By proposition 1.1(i) we have K idealises each H_r since K idealises H. Thus $H_r \lhd H_{r-1} + K$ and $K \lhd^n H_{r-1} + K$ for $r = 1, \ldots, m$. It follows from lemma 1.3.1 that $H_r + K \lhd^n H_{r-1} + K$. So we have

$$J = H + K \lhd^n H_{m-1} + K \lhd^n \ldots \lhd^n H_0 + K = L,$$

whence we deduce that $J \lhd^{mn} L$. □

Two subspaces A and B of a Lie algebra L are said to be *permutable* if and only if

$$[A, B] \subseteq A + B.$$

In other words given $a \in A$ and $b \in B$ we can find $a' \in A$ and $b' \in B$ such that

$$[a, b] = a' + b'.$$

We also say that *A permutes with B*.

Evidently if $H \leqq L$ and A and B are subspaces of H such that $H = A+B$ then A permutes with B.

Conversely suppose that A permutes with B. From lemma 1.1.1 we know that $\langle A \rangle$ is the subspace spanned by the products $[a_1, \ldots, a_n]$ with $a_i \in A$, $n \in \mathbb{N}$. Let $a_1, \ldots, a_{n+1} \in A$ and $b \in B$. Since A permutes with B we can find $a' \in A$ and $b' \in B$ such that $[a_{n+1}, b] = a'+b'$. Using this and the Jacobi identity we have

$$[a_1, \ldots, a_{n+1}, b] = [a_1, \ldots, a_n, b, a_{n+1}]+[a_1, \ldots, a_n, a'+b'].$$

Noting also that $[\langle A \rangle+B, A] \subseteq \langle A \rangle+B$ it follows that by induction on n

$$[a_1, \ldots, a_n, b] = \text{element of } \langle A \rangle + \text{element of } B.$$

Therefore $\langle A \rangle$ permutes with B. By a similar argument we now have that $\langle A \rangle$ permutes with $\langle B \rangle$.

So we have proved:

LEMMA 1.3 *Let A and B be permutable subspaces of a Lie algebra L. Then $\langle A \rangle$ and $\langle B \rangle$ are permutable and*

$$\langle A, B \rangle = \langle A \rangle+\langle B \rangle.$$

Conversely if $H \leqq L$ and there exist subspaces C and D of H such that $H = C+D$, then C permutes with D. □

We note that if L is a Lie algebra and H is an ideal of L or a subalgebra of codimension 1 then H permutes with every subspace of L. It turns out that except over fields of characteristic 2 the only subalgebras of a Lie algebra L which permute with every subspace are ideals or subalgebras of codimension 1 or certain subalgebras H for which $L^{(2)} \subseteq H$; however we shall not pursue the matter here.

Let A and B be subspaces of a Lie algebra. We define the *permutiser* $P_B(A)$ *of A in B* as

$$P_B(A) = \sum\{C \subseteq B : [A, C] \subseteq A+C\}.$$

Thus $P_B(A)$ is the sum of all subspaces of B which permute with A. Evidently $P_B(A)$ permutes with A and so it is the largest subspace of B which permutes with A. From this and lemma 1.3 it follows quite easily that

$$P_B(A) \subseteq P_B(\langle A \rangle) \cap P_{\langle B \rangle}(A) \subseteq P_{\langle B \rangle}(\langle A \rangle).$$

In particular if $B = \langle B \rangle$ then

$$P_B(A) = \langle P_B(A) \rangle \leqq B.$$

LEMMA 1.4 *Let L be a Lie algebra and let $H \lhd^m L$, $K \lhd^n L$ and $J = \langle H, K \rangle$. Then there exists $g_1 = g_1(r)$ such that*

$$J \lhd^{g_1} L$$

whenever H and K are permutable and $m+n \leqq r$.

Proof: Define $g_1(r) = 1$ if $r \leqq 2$ and $g_1(r) = [r!/2^{r-2}]$ if $r > 2$. If $r \leqq 2$ or if $m = 0$ or $n = 0$ there is nothing to prove. Let $r > 1$ and assume that $m > 0$, $n > 0$ and the result is true for $r-1$ in place of r. Let $H_1 = H^L$ and $K_1 = K^L$. Now $H_1 \lhd^{m-1} H_1$ and $H_1 \cap K \lhd^n H_1$. Furthermore $J = H+K$ and so by the modular law (lemma 1.7.2) we have

$$H_1 \cap J = H + H_1 \cap K$$

so that (by lemma 1.3) H and $H_1 \cap K$ are permutable. Therefore by induction $H_1 \cap J \lhd^s H_1 \lhd L$ where $s = g_1(r-1)$. Now $H_1 \cap J \lhd J$ and clearly $J = H_1 \cap J + K$. So it follows from lemma 1.2 that $J \lhd^{n(s+1)} L$. Similarly working with $K_1 \cap J$ we have $J \lhd^{m(s+1)} L$. Therefore $J \lhd^{g_1(r)} L$ where

$$g_1(r) \leqq (m+n)/2.(g_1(r-1)+1).$$

Thus

$$g_1(r) \leqq r(g_1(r-1)+1)/2.$$

From this and the fact that $g_1(2) = 1$ we have

$$\begin{aligned} g_1(r) &\leqq (r!/2^{r-1})(1+2/2!+2^2/3!+\ldots+2^{r-1}/(r-1)!) \\ &< (r!/2^{r-1})(\mathrm{e}^2-2) \qquad (\mathrm{e} \sim 2.718) \\ &< r!/2^{r-2}. \end{aligned}$$

□

Remark 1 Under the same conditions as lemma 1.4 it can be shown that if $H \lhd^s J$ then

$$J \lhd^{mn(n+1)\ldots(n+s-1)} L.$$

On some occasions the subideal index of J in L so obtained may be better than $g_1(m+n)$. Some other bounds on the subideal index of J in L are

$$n+n^2+\ldots+n^m \text{ (if } n \neq 0) \text{ and } n+\ldots+n^{m-2}+2n^{m-1} \text{ (if } m > 1).$$

In future we shall employ the bound which best fits the occasion.

LEMMA 1.5 *Let L be a Lie algebra and let $H \lhd^m L$ and $K \lhd^n L$. Then there exists $g_2 = g_2(r)$ such that*

$$P_H(K) \lhd^{g_2(r)} L$$

whenever $m+n \leqq r$.

Proof: Define $g_2(r) = 1$ if $r \leqq 2$ and $g_2(r) = g_1(r+g_2(r-1)-1)+1$ if $r > 1$. We proceed by induction on r. If $m = 0$ there is nothing to prove. Assume that $m > 0$ and that the result is true for $r-1$ in place of r. Let $H_1 = H^L$, $P = P_H(K)$ and $K_1 = P^L \cap K$ and $P_1 = P_H(K_1)$. Now $H \lhd^{m-1} H_1 \lhd L$ and $K_1 \lhd^n L$, (by proposition 1.1(k): for $P^L \lhd L$ and we may assume that $n \neq 0$), so that $K_1 \lhd^n H_1$ since $H_1 \lhd L$ and $P \leqq H \leqq H_1$. Therefore by induction $P_1 \lhd^s H_1$ ($s = g_2(r-1)$). Define $Q = \langle P_1, K_1 \rangle$ so that $Q = P_1+K_1$; then since P_1 permutes with K_1 it follows from lemma 1.4 that

$$Q \lhd^{g_1(n+s)} H_1.$$

Now let $U = \langle P^K, K_1 \rangle$. Then U is idealised by K and $U \leqq \langle P, K \rangle = P+K$. Since also $U = \langle P^K \rangle + K_1$ we have

$$\langle P^K \rangle = P + \langle P^K \rangle \cap K \subseteq P+K_1$$

and so $U \subseteq P+K_1$. This implies that $U = P+K_1$, so P permutes with K_1 and therefore

$$P \subseteq U \cap P_1.$$

From this we have $U = P+K_1 \subseteq P_1+K_1 = Q$. Thus if $t = 1+g_1(n+s)$ and U_t is the t-th term of the ideal closure series of U in L then $U_t \leqq Q$ and K idealises U_t (by proposition 1.1). Therefore $U_t \subseteq P_1+K_1$ and so by the modular law we have $U_t = U_t \cap P_1+K_1$. But then $\langle U_t, K \rangle = U_t+K =$ $= U_t \cap P_1+K$. In other words $U_t \cap P_1$ permutes with K and so

$$U_t \cap P_1 \subseteq P \subseteq U \cap P_1 \subseteq U_t \cap P_1.$$

Therefore $U_t \cap P_1 = P$. Finally we note that $t > s+1$ and so by proposition 1.1(k) we have

$$P = U_t \cap P_1 \lhd^t L$$

and as $m > 0$ we have $n \leqq r-1$. This completes our proof. □

Lemmas 1.4 and 1.5 are counterparts of the corresponding results for groups (cf. Roseblade and Stonehewer [337]). However our next result is peculiar to Lie algebras in that the group-theoretic analogue is false in general.

Let L be a Lie algebra and let A and B be subsets of L. Then

$$I_L(A) \cap I_L(B) \subseteq I_L([A, B]).$$

This follows easily by the Jacobi identity. Furthermore if $H \leqq L$ then

$$[A, H^i] \subseteq [A, {}_iH]$$

for all $i \in \mathbb{N}\setminus\{0\}$. Now suppose that $H \lhd^m L$, $K \lhd^n L$ where $m > 1$ and $n > 1$. Then for any $i, j > 0$ we have

$$H^i \lhd^m L, \quad K^j \lhd^n L.$$

Suppose that $n \leqq m$. Then by proposition 1.1 we have

$$H \cap K \lhd^m L$$

and since also $H \cap K$ idealises K^n it follows by lemma 1.2 that

$$C = H \cap K + K^n \lhd^{mn} L.$$

Evidently H^m permutes with C and so it follows from lemma 1.4 that

$$H^m + C = H^m + H \cap K + K^n \lhd^r L$$

where $r = g_1(m+mn)$. We note that if $i > m-1$ and $j > n-1$ then $[H^i, K^j] \subseteq$ $\subseteq [H^m, K^n] \subseteq H \cap K$; therefore in much the same way as before we have

$$H^i + H \cap K + K^j \lhd^r L.$$

Thus if we define $F_{i,j} = \langle H^i, K^j \rangle$ then we have

$$F_{i,j} \lhd H^i + H \cap K + K^j \lhd^r L$$

and

$$F_{i,j} = H \cap F_{i,j} + K \cap F_{i,j}.$$

In case $m \leqq n$ we obtain the same results but this time with r replaced by $g_1(n+nm)$.

Now let H_{m-1} be the $(m-1)$-th ideal closure of H in L and let K_{n-1} be the $(n-1)$-th ideal closure of K in L. Then (with $n \leqq m$)

$$H_{m-1} \cap K_{n-1} \lhd^{m-1} L.$$

Clearly $[H^{m-1}, K^{n-1}] \subseteq H_{m-1} \cap K_{n-1}$ and we have

$$H_{m-1} \cap K_{n-1} + K^{n-1} \lhd^{(m-1)n} L.$$

Therefore

$$H^{m-1}+H_{m-1}\cap K_{n-1}+K^{n-1}\lhd^{s} L$$

and so

$$F_{m-1,n-1} = \langle H^{m-1}, K^{n-1}\rangle \lhd^{s+1} L$$

where $s = g_1(m+mn-n) < r$.

Finally we note that $H \lhd H_{m-1}$ and $[L, {}_iH] \subseteq H_{m-1} \lhd^{m-1} L$ and $[L, H^i] \subseteq H_{m-1}$ for all $i \geqq m-1$. Therefore

$$[L, {}_iH] \lhd H_{m-1} \lhd^{m-1} L$$

and

$$[L, H^i] \lhd H_{m-1} \lhd^{m-1} L$$

for all $i \geqq m-1$.

So we have proved:

PROPOSITION 1.6 *Let L be a Lie algebra and let $H \lhd^m L$ and $K \lhd^n L$, where $m, n > 1$. Then for any $i > m-1$ and $j > n-1$ we have*

$$F_{i,j} = \langle H^i, K^j\rangle \lhd^{g_1(2mn)} L$$

and $F_{i,j} = H_i \cap F_{i,j} + K_j \cap F_{i,j}$, where H_i is the i-th ideal closure of H in L and K_j is the j-th ideal closure of K in L. Furthermore

$$[L, {}_iH] \lhd^m L$$

and

$$[L, H^i] \lhd^m L. \qquad \square$$

The next two results give us a taste of things to come – namely, the effect subideals have on the derived series of their join. Both results are counterparts of group-theoretic ones (cf. Roseblade [336]).

PROPOSITION 1.7 *Let H and K be subalgebras of a Lie algebra L. Then for any positive integer m,*

$$(H \circ K)^{(m)} \leqq K \circ_2 H + \{(K \circ_2 H) \cap (H \circ_m K)\}^K.$$

Proof: Let $J = \langle H, K\rangle$ so that $J = H+K+H\circ K$. Put $X = H \circ K$, $Y = K \circ_2 H$ and $Z = Y^K$. Now $[Z, H] \subseteq [H \circ K, H] \subseteq Y \subseteq Z$. Therefore $Z \lhd J$ and $Z = Y + Y \circ K$. We also have $X \lhd J$ and hence $C = C_J(X/Z) \lhd J$.

Obviously $H \leqq C$ and so

$$X = H \circ K \leqq C \circ K \leqq C.$$

Therefore

$$X^{(1)} \leqq [X, C] \leqq Z = Y^K$$

and the required result holds for the case $m = 1$ (for $Y \leqq X$).

Let $m > 1$ and assume that the result is true for the case $m-1$. Let $M = Y \cap (H \circ_{m-1} K)$. Then we have by induction,

$$X^{(m-1)} \leqq Y + M^K \subseteq Y + M + M \circ K = Y + M \circ K,$$

since $M \leqq Y$. Since $Y \lhd X$ and $Y + M \circ K \leqq X$ we have

$$X^{(m)} \leqq (Y + M \circ K)^{(1)} \subseteq Y + (M \circ K)^{(1)}.$$

We apply the case $m = 1$ to the pair M, K to obtain

$$(M \circ K)^{(1)} \leqq (K \circ_2 M)^K.$$

Therefore

$$X^{(m)} \leqq Y + (K \circ_2 M)^K.$$

Now $M \leqq Y$ and $Y \lhd X \lhd J$ and so $K \circ_2 M \leqq J \circ_2 Y \leqq Y$. We also have $M \leqq H \circ_{m-1} K$ and so

$$K \circ_2 M \leqq ((H \circ_{m-1} K) \circ K) \circ M \leqq H \circ_m K,$$

since $H \circ_{m-1} K$ idealises $H \circ_m K$. From this we have

$$X^{(m)} \leqq Y + \{Y \cap (H \circ_m K)\}^K$$

and the induction is complete. □

Corollary 1.8 *Let L be a Lie algebra and H and K subalgebras of L. If* $H_1 = \langle H^K \rangle$ *and* $H_2 = H^{H_1}$, *then*

$$H_1^{(m)} \leqq H_2 + H \circ_m K$$

for any positive integer m.

Proof: Now $H_2 \lhd H_1$ and $(H_1/H_2)^{(m)} = (H_2 + (H \circ K)^{(m)})/H_2$. Thus as K idealises $H \circ_m K$, and $K \circ_2 H \leqq H_2$ and

$$(H_2 + H_1^{(m)})/H_2 = (H_1/H_2)^{(m)},$$

the required result follows from proposition 1.7. □

It is clear that if H is a proper subideal of a Lie algebra L, then $H^L \neq L$. With this observation we can now prove:

LEMMA 1.9 *Let H and K be subalgebras of a Lie algebra L such that $L = H+K^2$. Then*

$$L = H+K^{n+1}$$

for any $n \in \mathbb{N}$. If in addition $H \cap K$ si K then for any $n \in \mathbb{N}$,

$$L = H+K^{(n)}.$$

Proof: Both results are trivially true for $n \leqq 1$. Let $n > 1$ and suppose that $L = H+K^n$. By the modular law we have $K = K \cap L = K \cap H + K^n$. Thus $K^2 = (K \cap H + K^n)^2 \leqq (K \cap H)^2 + K^{n+1}$. Therefore $L = H+K^2 \leqq H+K^{n+1} \subseteq L$. So the first result holds for all n.

Now suppose that $K \cap H$ si K. By the modular law we have $K = K \cap L = K \cap H + K^2$. Let $X = K \cap H + K^{(n)}$. Then X si K since $K^{(n)} \lhd K$. Let $Y = X^K$. Then $Y \lhd K$ and K/Y is soluble since $K^{(n)} \leqq X \leqq Y$. However

$$K/Y = (K \cap H + K^2)/Y = (Y+K^2)/Y = (K/Y)^2.$$

Therefore K/Y is trivial and so $K = Y$. So by our remark above we must have $X = K$. Hence

$$L = H+K^2 = H+K = H+X = H+K \cap H + K^{(n)}$$
$$= H+K^{(n)}. \qquad \square$$

One of the nice properties of ideals is that their join with any ascendant subalgebra is always ascendant. We will show that 2-step subideals have this same property, but not 3-step subideals.

Let L be a Lie algebra and suppose that

$$H \lhd K \lhd L.$$

If $x_1, \ldots, x_n \in L$ we define the subspace $H(x_1, \ldots, x_n)$ of L by $H(x_1, \ldots, x_n) = H + \sum_{i=1}^{n} [H, x_i] + \sum_{1 \leqq i_1 < i_2 \leqq n} [H, x_{i_1}, x_{i_2}] + \ldots + [H, x_1, \ldots, x_n]$.

Thus $H(x_1) = H + [H, x_1]$ and it is clear that for $n > 1$, we have

$$H(x_1, \ldots, x_n) = [H, x_1, \ldots, x_n] + \sum H(x_{j_1}, \ldots, x_{j_r})$$

where the summation is taken over all r-tuples with $1 \leqq r \leqq n-1$ and $1 \leqq j_1 < j_2 < \ldots < j_r \leqq n$.

Clearly $H(x_1, \ldots, x_n) \subseteq K$. If $z \in H$ and $y \in K$ then we have by the Jacobi identity,

$$[z, x_i, y] = [z, y, x_i] + [z, [x_i, y]]$$

and

$$[z, x_1, \ldots, x_n, y] = [z, x_1, \ldots, x_{n-1}, y, x_n] + [z, x_1, \ldots, x_{n-1}, [x_n, y]].$$

From these relations and the fact that $K \lhd L$ it follows by induction on n that

$$H(x_1, \ldots, x_n) \lhd K.$$

So we deduce:

Proposition 1.10 *Suppose that L is a Lie algebra, A is a subset of L and $H \lhd K \lhd L$. Then*

$$\langle H^A \rangle = H^A = \sum [H, a_1, \ldots, a_r]$$

where the summation is taken over all $r \geqq 0$ and all choices of $a_i \in A$; and

$$H^A \lhd K.$$

In particular if σ is an ordinal and $A \lhd^\sigma L$ then

$$\langle H, A \rangle = H^A + A \lhd^\sigma K + A \lhd^\sigma L. \qquad \square$$

But for 3-step subideals we have:

Lemma 1.11 (*Hartley*) *Over any field $\mathfrak{k}$ there exists a Lie algebra L with the following properties:*

(a) *there exist infinite dimensional abelian subalgebras C, H and K of L such that*

$$C \lhd L, \; H \lhd^3 L, \; K \lhd^3 L.$$

(b) *If $J = \langle H, K \rangle$ then $J \in \mathfrak{N}_2$, and*

$$C \cap J = 0, \; L = C \dotplus J.$$

(c) *$I_L(J) = J$ and $J^L = L$ so that J is not an ascendant nor a descendant subalgebra of L.*

Proof: Let $\mathscr{S}$ be the set of all infinite sequences $P = (p_0, p_1, \ldots)$ of integers and let $\mathscr{T}$ be the subset of $\mathscr{S}$ consisting of the strictly increasing sequences. Let U and V be vector spaces over $\mathfrak{k}$ with bases $u(P)$ $(P \in \mathscr{S})$ and $v(P)$ $(P \in \mathscr{S})$.

Define a linear map $f\colon U \to U$ by

$f\colon u(P) \mapsto (-1)^r u(P')$ if $P' \in \mathcal{T}$ and P can be transformed into P' by a finite number r of transpositions; and

$f\colon u(P) \mapsto 0$ otherwise.

Thus if at least two members of P are equal then $u(P) \in \ker f$. Define another linear map $g\colon V \to V$ in exactly the same way.

Let A be the quotient space $U/\ker f$ and B the quotient space $V/\ker g$. Clearly if $a(P) = u(P) + \ker f$ and $b(P) = v(P) + \ker g$ then it follows from above that

$$a(P) = (-1)^r a(P') \text{ and } b(P) = (-1)^r b(P') \tag{3}$$

for any $P, P' \in \mathcal{S}$ such that P can be transformed into P' by a finite number, r, of transpositions. We also have

$$a(P) = b(P) = 0 \tag{4}$$

if P cannot be transformed into an element of $\mathcal{T}$ by a finite number of transpositions (all transpositions are to involve only the elements of the sequence P). It is not hard to see that A and B have bases $a(P)$ $(P \in \mathcal{T})$ and $b(P)$ $(P \in \mathcal{T})$. Let C be the vector space direct sum

$$C = A \oplus B.$$

Clearly C is infinite dimensional and we may regard C as an abelian Lie algebra over $\mathfrak{k}$.

If m is an integer and $P \in \mathcal{S}$ we denote by (m, P) the sequence consisting of m followed by the elements of P in order. For any integer m, we define linear transformations x_m and y_m of C by the rules

$$\begin{aligned} a(P)x_m &= b(m, P), \quad b(P)x_m = 0 \qquad (P \in \mathcal{S}) \\ a(P)y_m &= 0, \qquad b(P)y_m = a(m, P) \qquad (P \in \mathcal{S}). \end{aligned} \tag{5}$$

Clearly x_m and y_m are well defined linear transformations of C and $x_m, y_m \in \mathrm{Der}(C)$.

It is not hard to see from (3), (4) and (5) that

$$[x_m, x_n] = x_m x_n - x_n x_m = 0 = y_m y_n - y_n y_m = [y_m, y_n]$$

for all integers m and n. Therefore if

$$H = \textstyle\sum_m \mathfrak{k} x_m \text{ and } K = \sum_m \mathfrak{k} y_m$$

then H and K are abelian and infinite dimensional subalgebras of $\mathrm{Der}(C)$.

Define $z_{mn} = [x_m, y_n]$ for all integers m, n. Then from (3), (4) and (5) we have

$$a(P)z_{mn} = a(P)(x_m y_n - y_n x_m) = a(n, m, P)$$

and

$$b(P)z_{mn} = -b(m, n, P) = b(n, m, P)$$

for all $P \in \mathscr{S}$. Thus

$$a(P)[z_{mn}, x_k] = b(k, n, m, P) - b(n, m, k, P) = 0$$

since $b(n, m, k, P) = b(k, n, m, P)$ by (3). We also have

$$b(P)[z_{mn}, x_k] = 0 = a(P)[z_{mn}, y_k] = b(P)[z_{mn}, y_k].$$

Hence we obtain for all integers m, n, k,

$$[z_{mn}, x_k] = [z_{mn}, y_k] = 0.$$

Now let $Z = \sum_{m,n} \mathfrak{k} z_{mn}$ and define $J = H \oplus Z \oplus K$. Then $J \leqq \mathrm{Der}(C)$ and J is nilpotent of class 2; and clearly $J = \langle H, K \rangle$.

We define L to be the split extension of C by J,

$$L = C \dotplus J; \; C \lhd L \text{ and } C \cap J = 0.$$

Evidently $[A, H] \subseteq B$, $[B, H] = 0$ and $Z = [H, K]$. Hence

$$H \lhd B + H \lhd C + Z + H \lhd C + J = L,$$

since $[A, Z] \subseteq A$ and $[B, Z] \subseteq B$. Similarly we have $[A, K] = 0$ and

$$K \lhd A + K \lhd C + Z + K \lhd C + J = L.$$

So J is the join of two infinite dimensional and abelian 3-step subideals of L. We have now proved (a) and (b).

(c) Let $w \in C \cap I_L(J)$. Then $[J, w] \subseteq C \cap J = 0$. Suppose that

$$w = \sum_{i=1}^{n} (\lambda_i a(P_i) + \mu_i b(P_i)) \quad (\lambda_i, \mu_i \in \mathfrak{k} \text{ and } P_i \in \mathscr{T}).$$

Let m_i denote the first member of P_i and choose $m = r - 1$ where $r = \min\{m_i : i = 1, \ldots, n\}$. Clearly $(m, P_i) \in \mathscr{T}$ for each i. But we have

$$0 = [w, x_m] = \sum \lambda_i b(m, P_i)$$

and

$$0 = [w, y_m] = \sum \mu_i a(m, P_i)$$

and therefore $\lambda_i = \mu_i = 0$ for all i; so $w = 0$. Therefore $I_L(J) = J$ and this implies that J cannot be a subideal nor an ascendant subalgebra of L.

Finally we note that if $P = (p_0, p_1, p_2, \ldots) \in \mathscr{T}$, then

$$b(P) = a(p_1, p_2, \ldots)x_{p_0} \text{ and } a(P) = b(p_1, p_2, \ldots)y_{p_0}.$$

From this it is not hard to see that

$$H^L = C+Z+H \text{ and } K^L = C+Z+K.$$

Thus $L = H^L+K = J^L$. Hence if ρ is any ordinal there cannot be a descending series $(J_\sigma : \sigma \leqq \rho)$ with $J_0 = L$, $J_\rho = J$. That is, J is not a descendant subalgebra of L.

We note also that H^L is not nilpotent. □

2. The Derived Join Theorems

The results of chapter 1 tell us that in any Lie algebra the join of a pair of (and so of finitely many) soluble ideals is always soluble. One of the aims of this section is to show that the same conclusion holds for the join of finitely many soluble subideals. To do this we first investigate the derived series of a join of two subideals; and then that of a finite number of subideals. As stated earlier in the introduction our mode of attack is based on the one employed by Roseblade [336] in solving the corresponding problem for groups.

First we formalise our remark on soluble ideals with:

LEMMA 2.1 *Suppose that J is a Lie algebra and $J = \langle H_1, \ldots, H_n\rangle$ with $H_i \lhd J$ for $i = 1, \ldots, n$. If $r_1, \ldots, r_n \in \mathbb{N}$ and $r = r_1+\ldots+r_n$ then*

$$J^{(r)} \leqq H_1^{(r_1)}+\ldots+H_n^{(r_n)}.$$

Proof: We use induction on r. It is trivial for $r = 0$. Let $r > 0$ and assume that the result holds for $r-1$. Since $r > 0$ we can find an i such that $r_i > 0$. Let $K = \sum_{j \neq i} H_j^{(r_j)}$. By the induction we have $J^{(r-1)} \leqq H_i^{(r_i-1)}+K$. Clearly $K \lhd J$ and so $(H_i^{(r_i-1)}+K)^2 \leqq H_i^{(r_i)}+K$, and the result follows. □

If in the statement of lemma 2.1 we replace '$H_i \lhd J$' by 'H si J', then $H_1^{(r_1)}+\ldots+H_n^{(r_n)}$ is still a subspace but not necessarily a subalgebra of J. However we will prove that we can replace r by some r^* so that the inequality holds. Moreover if L is a Lie algebra and H_i si L for all i then r^* can be so chosen that in addition $J^{(r^*)}$ si L.

As a first step in this direction we have

LEMMA 2.2 *Let L be a Lie algebra and $L = H+K$ where $H \lhd L$ and $K \lhd^m L$. Then for each $n \in \mathbb{N}$,*

$$L^{(mn)} \leqq H^{(n)}+K.$$

Proof: Clearly if $X = H^{(n)}+K$ then $X \lhd^m L$. Evidently if X_i is the ith ideal closure of X in L then $X_i = X_i \cap H + K = X_i \cap H + X_{i+1}$. Thus X_i/X_{i+1} is soluble of derived length $\leqq n$ since $H^{(n)} \leqq X \leqq X_{i+1}$. It follows then that

$$L^{(mn)} \leqq X = H^{(n)}+K. \qquad \square$$

As is to be expected the next stage of generalisation will be to look at permutable subideals; and to help us along the way we have:

LEMMA 2.3 *Let A and B be permutable subspaces of a Lie algebra L and let X be a subset of L. Then*

$$X^{\langle A, B\rangle} = X^{A+B} = X^{A\cup B} = (X^A)^B = (X^B)^A.$$

Proof: The first two equalities follow trivially from the definition of X^Y when $X, Y \subseteq L$. By symmetry we need only show that $X^{A\cup B} = (X^A)^B$. Clearly $(X^A)^B \subseteq X^{A\cup B}$ and the reverse inclusion will follow (by the definition of $X^{A\cup B}$) if we can show that A idealises $(X^A)^B$ (for B idealises it as well). Now $(X^A)^B$ is spanned by products of the form

$$x_r = [x, b_1, \ldots, b_r] \quad (r > 0)$$

where $x \in X^A$ and $b_1, \ldots, b_r \in B$. Let $a \in A$ and $b \in B$. By permutability we can find $a' \in A$ and $b' \in B$ such that

$$[a, b] = a'+b'.$$

Hence by Jacobi,

$$[x_r, b, a] = [x_r, a, b]-[x_r, a']-[x_r, b'].$$

If $r = 0$ then $x_r = x$ and $[x_r, a] \in X^A \subseteq (X^A)^B$. Suppose that for some r we have $[x_r, a] \in (X^A)^B$ for all choices of $x \in X^A$, $b_1, \ldots, b_r \in B$ and $a \in A$; then our last equation shows that the same is true for $r+1$, since each x_{r+1} is of the form $[x_r, b]$ for some x_r, and some $b \in B$. Therefore the conclusion is true for all r and so A idealises $(X^A)^B$. $\square$

THEOREM 2.4 (*First Derived Join Theorem*) *Suppose that J is a Lie algebra,* $H_1 \lhd^{h_1} J$, $H_2 \lhd^{h_2} J$, *and* $J = \langle H_1, H_2\rangle$. *Then there exists* $\lambda_1 = \lambda_1(h)$ *such that*

$$J^{(\lambda_1)} \leqq H_1+H_2$$

whenever $h_1+h_2 \leqq h$; *and we may take* $\lambda_1(h) = 4^{h-2}((h-2)!)$ *if* $h > 1$.

Proof: Put $\lambda_1(h) = 0$ if $h \leqq 1$. The result is obvious for $h \leqq 2$, and for $h_1 \leqq 1$ or $h_2 \leqq 1$. Assume that $h > 2$ and $h_1 > 1$ and $h_2 > 1$, and proceed by induction on h.

Put $\{i, j\} = \{1, 2\}$. Clearly there exist subalgebras L_i and K_i of J such that

$$H_i \lhd K_i \lhd^{h_i - 2} L_i \lhd J.$$

Let $m = \lambda_1(h-1)$ and define $X = (H_1 \circ H_2)^{(m)}$. Now $J = \langle H_j, K_i \rangle$ and so the inductive hypothesis applied to the pair H_j, K_i (for $K_i \lhd^{h_i - 1} J$) yields

$$J^{(m)} \leqq K_i + H_j,$$

whence

$$X \leqq J^{(m)} \cap L_i = K_i + H_j \cap L_i.$$

Now $X \lhd J$ and so if $Y = \langle X, H_j \cap L_i \rangle$ then $Y = X + H_j \cap L_i$. Thus $Y \leqq K_i + H_j \cap L_i$ and it follows by the modular law that

$$Y = (K_i + H_j \cap L_i) \cap Y = K_i \cap Y + H_j \cap L_i.$$

By definition K_i idealises H_i and so by lemma 2.3 we have

$$\langle H_i^Y \rangle = \langle H_i^{H_j \cap L_i} \rangle \leqq \langle H_i, H_j \cap L_i \rangle.$$

In particular

$$X \circ H_i \leqq \langle H_i^X \rangle \leqq \langle H_i^Y \rangle \leqq \langle H_i, H_j \cap L_i \rangle.$$

Clearly $H_i \lhd^{h_i - 1} L_i$ and $H_j \cap L_i \lhd^{h_j} L_i$ and so applying the inductive hypothesis to the pair H_i, $H_j \cap L_i$ we have

$$(X \circ H_i)^{(m)} \leqq J_i^{(m)} \leqq H_i + H_j \cap L_i \subseteq H_1 + H_2,$$

where $J_i = \langle H_i, H_j \cap L_i \rangle \leqq L_i$. Thus

$$(X \circ H_1)^{(m)} + (X \circ H_2)^{(m)} \subseteq H_1 + H_2.$$

Since $X \lhd J$ and $X \circ J = X \circ (H_1 \cup H_2)$ (by lemma 1.1(e)) it follows by lemma 1.1(j) that

$$X \circ H_i \lhd X \circ J = X \circ H_1 + X \circ H_2.$$

By lemma 2.1 we have

$$(X \circ J)^{(2m)} \leqq (X \circ H_1)^{(m)} + (X \circ H_2)^{(m)} \leqq H_1 + H_2.$$

Let $U = H_1 \circ H_2$, so that $X = U^{(m)}$. Then

$$U^{(1+3m)} \leqq [U^{(m)}, J]^{(2m)} \leqq (X \circ J)^{(2m)} \leqq H_1 + H_2.$$

Clearly we can choose $L_i = H_i + U$. So by lemma 2.2 we have

$$L_i^{(\{1+3m\}\{h_i-1\})} \leqq U^{(1+3m)} + H_i \leqq H_1 + H_2.$$

Finally $L_i \lhd J$ and $J = L_1 + L_2$ and so by lemma 2.1 we have

$$J^{(\{1+3m\}\{h_1+h_2-2\})} \leqq L_1^{(\{1+3m\}\{h_1-1\})} + L_2^{(\{1+3m\}\{h_2-1\})}$$

$$\leqq H_1 + H_2.$$

Clearly $\{1+3m\}\{h_1+h_2-2\} \leqq 4m(h-2) = \lambda_1(h)$. Therefore

$$J^{(\lambda_1(h))} \leqq H_1 + H_2 \qquad \square$$

If A and B are subspaces of a Lie algebra L then by $A \leqq B$ we mean that $A \leqq L$ and $A \subseteq B$. The symbols $\lambda_i(m, n, \ldots)$ for $i = 1, 2, \ldots$ will denote non-negative integers depending only on the arguments shown.

COROLLARY 2.5 *Let J be a Lie algebra and suppose that $J = \langle H_1, H_2 \rangle$ with $H_1 \lhd^{h_1} J$ and $H_2 \lhd^{h_2} J$. Then*

$$H_1^{(\lambda_1)} \leqq P_{H_1}(H_2).$$

Proof: Let $K = \langle J^{(\lambda_1)}, H_2 \rangle$. Since $J^{(\lambda_1)} \lhd J$ we have $K = J^{(\lambda_1)} + H_2$, whence by theorem 2.4, $K \leqq H_1 + H_2$ and $K = (H_1 + H_2) \cap K = H_1 \cap K + H_2$. Therefore $H_1 \cap K$ permutes with H_2 and so $H_1 \cap K \leqq P_{H_1}(H_2)$. Hence

$$H_1^{(\lambda_1)} \leqq H_1 \cap J^{(\lambda_1)} \leqq H_1 \cap K \leqq P_{H_1}(H_2). \qquad \square$$

We observe that the subalgebra K defined in the proof of corollary 2.5 is the join of two permutable subideals. It is easy to see that J would be soluble if the result was true for the join of two soluble and permutable subideals. This is what our next result shows.

LEMMA 2.6 *Let $J = \langle H_1, H_2 \rangle$ be a Lie algebra with $H_1 \lhd^{h_1} J$ and $H_2 \lhd^{h_2} J$. If H_1 permutes with H_2 then there exists $\lambda_2 = \lambda_2(h, r)$ such that*

$$J^{(\lambda_2)} \leqq H_1^{(r_1)} + H_2^{(r_2)}$$

whenever $h_1 + h_2 \leqq h$ and $r_1 + r_2 \leqq r$; and we may take $\lambda_2(h, r) = 2^{h-2} r$ if $h > 1$.

Proof: Put $\lambda_2(h, r) = r$ if $h \leqq 2$, then $H_1 \lhd J$ and $H_2 \lhd J$ (or one of H_1, H_2 equals J) and the result follows by lemma 2.1. Let $h > 2$ and assume inductively that the result is true for $h-1$. Clearly we may also assume that $h_1 > 0$ and $h_2 > 0$.

Let $m = \lambda_2(h-1, r)$ and set $\{i, j\} = \{1, 2\}$. Clearly there exist L_i such that

$$H_i \lhd^{h_i-1} L_i \lhd J.$$

By hypothesis $J = H_1 + H_2 = H_i + H_j$ and so by the modular law,

$$L_i = H_i + H_j \cap L_i.$$

This implies that H_i and $H_j \cap L_i$ are permutable. We also have $H_j \cap L_i \lhd^{h_j} L_i$. Thus the inductive hypothesis applied to the pair H_i, $H_j \cap L_i$ gives

$$\begin{aligned} L_i^{(m)} &\leqq H_i^{(r_i)} + (H_j \cap L_i)^{(r_j)} \\ &\subseteq H_1^{(r_1)} + H_2^{(r_2)}. \end{aligned}$$

Finally $J = L_1 + L_2$, $L_1 \lhd J$ and $L_2 \lhd J$. Therefore by lemma 2.1 we have

$$\begin{aligned} J^{(2m)} &\leqq L_1^{(m)} + L_2^{(m)} \\ &\leqq H_1^{(r_1)} + H_2^{(r_2)}. \end{aligned}$$

Clearly $2m = \lambda_2(h, r)$. □

From this and the First Derived Join Theorem we have:

THEOREM 2.7 (*Second Derived Join Theorem*) *Let L be a Lie algebra and let $H_1 \lhd^{h_1} L$, $H_2 \lhd^{h_2} L$ and $J = \langle H_1, H_2 \rangle$. Then there exists $\lambda_3 = \lambda_3(h, r)$ such that*

$$J^{(\lambda_3)} \leqq H_1^{(r_1)} + H_2^{(r_2)}$$

and

$$J^{(\lambda_3)} \lhd^{\lambda_3} L$$

whenever $h_1 + h_2 \leqq h$ and $r_1 + r_2 \leqq r$; and we may take $\lambda_3(h, r) = \lambda_1(h) + \lambda_2(h, r)$.

Proof: Let $p = \lambda_3(h, r)$ and $m = \lambda_1(h)$. If $h \leqq 1$ then $p = r$ and $J = L$ and clearly $J^{(p)} \lhd^p L$. If $h = 2$, then $m = 1$ and $J \lhd L$ so that $J^{(p)} \lhd^p L$. If $h = 3$ then $m = 4$ and clearly $J \lhd^2 L$ so that $J^{(p)} \lhd^p L$. Hence by lemma 2.6 the theorem holds for $h \leqq 3$. Assume then that $h > 3$, so that

$$m = 4^{h-2}((h-2)!) > h!.$$

Let $M = J^{(m)} \leqq H_1 + H_2$ (by theorem 2.4) and let $K = \langle M, H_2 \rangle$. First we will show that $K \lhd^{h!} L$.

Now $M \lhd J$ and so by lemma 1.2 we have $K = M + H_2 \lhd^{h_2} J$. Therefore $K \cap H_1 \lhd^{h_2} H_1 \lhd^{h_1} L$; we also have $K \cap H_1 \lhd^{h_1} K$. Finally we have $K = M + H_2 \leqq H_1 + H_2$ and so by the modular law $K = K \cap H_1 + H_2$; thus

$K \cap H_1$ and H_2 are permutable. Therefore (by the remark after lemma 1.4) $K \lhd^n L$ where $n = (h_2+h_1)h_2(h_2+1)\ldots(h_2+h_1-1) \leqq h! < m$. Hence since $J^{(p)} \lhd J$ we have

$$J^{(p)} \lhd K \lhd^n L, \text{ so } J^{(p)} \lhd^p L.$$

Now $K = K \cap H_1 + H_2$, $H_2 \lhd^{h_2} K$, $K \cap H_1 \lhd^{h_1} K$ and $K \cap H_1$ permutes with H_2. Thus if $q = \lambda_2(h, r)$ then by lemma 2.6 we have

$$K^{(q)} \leqq (K \cap H_1)^{(r_1)} + H_2^{(r_2)}$$

$$\subseteq H_1^{(r_1)} + H_2^{(r_2)}.$$

Therefore $J^{(p)} = (J^{(m)})^{(q)} \leqq K^{(q)} \leqq H_1^{(r_1)} + H_2^{(r_2)}$. □

Corollary 2.8 *The join of two soluble subideals of a Lie algebra is soluble.* □

The bound $\lambda_1(h) + \lambda_2(h, r)$ on the derived length of this join given by theorem 2.7 is astronomical. A better bound can be obtained using the methods of Stonehewer [342], and details can be found in Amayo [1].

In general the inequality $J^{(\lambda_3(h,r))} \leqq H_1^{(r_1)} + H_2^{(r_2)}$ of theorem 2.7 is strict. However we have:

Corollary 2.9 *Let H_1 and H_2 be subideals of a Lie algebra and let $J = \langle H_1, H_2 \rangle$. If $H_1 \cap H_2 = 0$ then for any non-zero limit ordinal α,*

$$J^{(\alpha)} = H_1^{(\alpha)} + H_2^{(\alpha)}.$$

Proof: Clearly $H_1^{(\alpha)} + H_2^{(\alpha)} \subseteq J^{(\alpha)}$ for any ordinal α. Suppose that α is a limit of limit ordinals and the result is true for all non-zero limit ordinals $\beta < \alpha$. Then

$$J^{(\alpha)} = \bigcap\nolimits_{\gamma<\alpha} J^{(\gamma)} = \bigcap\nolimits_{\{\text{limit ordinals } \beta<\alpha\}} J^{(\beta)}$$

$$\subseteq \bigcap\nolimits_{\beta<\alpha} (H_1^{(\beta)} + H_2^{(\beta)})$$

$$\subseteq \bigcap\nolimits_{\beta<\alpha} H_1^{(\beta)} + \bigcap\nolimits_{\beta<\alpha} H_2^{(\beta)}$$

(since $H_1 \cap H_2 = 0$)

$$\subseteq H_1^{(\alpha)} + H_2^{(\alpha)}.$$

If α is not a limit of limit ordinals then $\alpha = \beta + \omega$ (ω being the first infinite ordinal) where β is a limit ordinal. Thus $J^{(\beta)} = \langle H_1^{(\beta)}, H_2^{(\beta)} \rangle$ (by definition if $\beta = 0$ and by induction otherwise). Now each $H_i^{(\beta)}$ is a subideal of J, so theorem 2.7 applied to the pair $H_1^{(\beta)}$, $H_2^{(\beta)}$ yields

$$J^{(\beta+\omega)} = \bigcap_{n<\omega} J^{(\beta+n)}$$

$$\subseteq \bigcap_{n<\omega} (H_1^{(\beta+n)} + H_2^{(\beta+n)})$$

$$\subseteq \bigcap_n H_1^{(\beta+n)} + \bigcap_n H_2^{(\beta+n)}$$

(since $H_1 \cap H_2 = 0$)

$$\subseteq H_1^{(\beta+\omega)} + H_2^{(\beta+\omega)}.$$

So the result is true in all cases. □

We remark that the hypothesis $H_1 \cap H_2 = 0$ above cannot be omitted; but we will not prove this here.

As we have seen (lemma 1.11) the join of a pair of subideals of a Lie algebra is not in general a subideal. Thus we cannot employ a straightforward induction to extend theorem 2.7 to a join of finitely many subideals. However this extension is still true.

First, it is easy to see from lemma 1.2 and by induction on r that if Y_i is an n_i-step subideal of a Lie algebra L for $i = 1, \ldots, r$ and Y_i is an ideal of $Y = \langle Y_1, \ldots, Y_r \rangle$ for each i, then

Y is an $(n_1 \ldots n_r)$-step subideal of L. (6)

THEOREM 2.10 (*Third Derived Join Theorem*) *Let L be a Lie algebra and suppose that $J = \langle H_1, \ldots, H_n \rangle$ with $H_i \lhd^{h_i} L$ for $i = 1, \ldots, n$. Then there exists $\lambda_4 = \lambda_4(h, r)$ such that*

$$J^{(\lambda_4)} \leqq H_1^{(r_1)} + \ldots + H_n^{(r_n)}$$

and

$$J^{(\lambda_4)} \lhd^{\lambda_4} L$$

whenever $h_1 + \ldots + h_n \leqq h$ and $r_1 + \ldots + r_n \leqq r$.

Proof: The case $n \leqq 2$ is theorem 2.7; so assume that $n > 2$. Define the vector subspace U of J by

$$U = H_1^{(r_1)} + \ldots + H_n^{(r_n)}.$$

If $h_i = 0$ for some i, then $H_i = L$ and so $L^{(r)} \leqq U$ and $L^{(r)} \lhd L$. Thus if $h < n$, we define $\lambda_4(h, r) = r$. Now assume that no h_i is zero (so that $h \geqq n$), and that $\lambda_4(h-1, r)$ has been defined for all r so as to satisfy all requirements.

Let

$$K_i = \langle H_j : j \neq i \rangle \qquad 1 \leqq i \leqq n, \tag{7}$$

and

$$l = \lambda_4(h-1, r). \tag{8}$$

Now $h_i > 0$, so $\sum_{j \neq i} h_j = h - h_i \leqq h-1$. So by induction on h,

$$K_i^{(l)} \leqq \sum_{j \neq i} H_j^{(r_j)} \tag{9}$$

and

$$K_i^{(l)} \lhd^l L \qquad 1 \leqq i \leqq n. \tag{10}$$

For each i, $h_i > 0$ so there exists L_i such that

$$H_i \lhd L_i \lhd^{h_i - 1} L.$$

Put

$$J_i = \langle L_i, K_i \rangle = \langle L_i, \{H_j : j \neq i\} \rangle$$

and

$$m = \lambda_4(h-1, (h-1)l). \tag{11}$$

Consider J_i as the join of L_i and the H_j for j different from i as shown above. Then by the induction on h we have

$$J_i^{(m)} \leqq L_i + \sum_{j \neq i} H_j^{(l)} \subseteq L_i + K_i^{(l)}. \tag{12}$$

Let

$$M_i = \langle H_i, K_i^{(l)} \rangle$$

and

$$V = \langle J_i^{(m)}, K_i^{(l)} \rangle.$$

Now $V \leqq J_i$ and $J_i^{(m)} \lhd J_i$ and so $V = J_i^{(m)} + K_i^{(l)}$. It follows from (12) that $V \leqq L_i + K_i^{(l)}$ and so by the modular law we have

$$V = V \cap L_i + K_i^{(l)}.$$

Now $H_i \lhd L_i$ and $J_i^{(m)} \leqq V$, and from above $V \cap L_i$ permutes with $K_i^{(l)}$. Therefore by lemma 2.3 we have

$$J_i^{(m)} \circ H_i \leqq \langle H_i^V \rangle = \langle H_i^{\{K_i^{(l)}\}} \rangle \leqq M_i. \tag{13}$$

Define

$$p = \lambda_3(h+l-1, r). \tag{14}$$

Now $h_i \leqq h-1$ and so using (10) and (14) we apply theorem 2.7 to the pair H_i, $K_i^{(l)}$ to get

$$M_i^{(p)} \leqq H_i^{(r)} + K_i^{(l)} \tag{15}$$

and

$$M_i^{(p)} \lhd^p L. \tag{16}$$

So we have from (9), (15) and the definition of U,

$$M_i^{(p)} \leqq U. \tag{17}$$

Let

$$X = J^{(m)} \circ J \text{ and } X_i = J^{(m)} \circ H_i, \qquad 1 \leqq i \leqq n.$$

Since $J^{(m)} \lhd J$ it follows from proposition 1.1(j) that

$$X = X_1 + \ldots + X_n \text{ and } X_i \lhd J^{(m)} \lhd J, \qquad 1 \leqq i \leqq n.$$

Clearly $J \leqq J_i$ for each i and so $X_i \leqq J_i^{(m)} \circ H_i \leqq M_i$. Let

$$Y_i = X_i^{(p)} \text{ and } Y = Y_1 + \ldots + Y_n.$$

Then $Y_i \lhd^2 J$ and $Y_i \leqq M_i^{(p)}$. So using (16) and (17) we get $Y_i \lhd^{2+p} L$ for each $i = 1, \ldots, n$. We also have $Y_i \operatorname{ch} X_i \lhd X$ and $Y \subseteq X$; so $Y_i \lhd Y \leqq X$ for every i. Apply (6) to get

$$Y \lhd^{(2+p)^h} L,$$

since $n \leqq h$. By lemma 2.1 (for each $X_i \lhd X$) it follows that

$$X^{(hp)} \leqq X_1^{(p)} + \ldots + X_n^{(p)} = Y.$$

Therefore $J^{(1+hp+m)} \leqq (J^{(m)} \circ J)^{(hp)} = X^{(hp)} \leqq Y$. Using (17) and the fact $Y_i \leqq M_i^{(p)}$ we now have

$$J^{(1+hp+m)} \leqq Y \leqq U = H_1^{(r_1)} + \ldots + H_n^{(r_n)}.$$

But $Y \leqq J$ and so

$$J^{(1+hp+m)} \lhd Y \lhd^{(2+p)^h} L.$$

Clearly $1+hp+m \leqq 1+m+(2+p)^h$; so we may define

$$\begin{aligned}\lambda_4(h, r) &= 1+m+(2+p)^h \\ &= 1 + \lambda_4(h-1, (h-1).\lambda_4(h-1, r)) \\ &\quad + \{2+\lambda_3(h-1+\lambda_4(h-1, r), r)\}^h. \qquad (18)\end{aligned}$$

□

COROLLARY 2.11 *In any Lie algebra the join of finitely many soluble subideals is soluble and its derived length is bounded in terms of the number of subideals, their derived lengths and subideal indices.* □

THEOREM 2.12 (*Fourth Derived Join Theorem*) *Let L be a Lie algebra and suppose that $J = \langle H_1, \ldots, H_n \rangle$ with $H_i \lhd^{h_i} L$ for $i = 1, \ldots, n$. Then there exist subideals $P_1, \ldots, P_n$ of $H_1, \ldots, H_n$ respectively and $\lambda_5 = \lambda_5(h)$ such that*

$$\langle P_1, \ldots, P_i \rangle = P_1+\ldots+P_i \text{ for } i = 1, \ldots, n$$

and

$$J^{(\lambda_5)} \leqq P_1+\ldots+P_n$$

and

$$J^{(\lambda_5)} \lhd^{\lambda_5} L$$

whenever $h_1+\ldots+h_n \leqq h$.

Proof: We define recursively subalgebras P_i, Q_i of J and integers p_i, q_i for $1 \leqq i \leqq n$ as follows.

$$P_1 = Q_1 = H_1 \qquad (19)$$

$$P_{i+1} = P_{H_{i+1}}(Q_i) \qquad 1 \leqq i \leqq n-1 \qquad (20)$$

$$Q_{i+1} = \langle Q_i, P_{i+1} \rangle = Q_i+P_{i+1}. \qquad (21)$$

By induction on i it follows that

$$Q_i = \langle P_1, \ldots, P_i \rangle = P_1+\ldots+P_i \qquad 1 \leqq i \leqq n. \qquad (22)$$

Let

$$p_1 = q_1 = h \qquad (23)$$

$$p_{i+1} = g_2(h+q_i) \qquad 1 \leqq i \leqq n-1 \qquad (24)$$

$$q_{i+1} = g_1(q_i+p_{i+1}). \qquad (25)$$

Suppose inductively that

$$P_i \lhd^{p_i} L \text{ and } Q_i \lhd^{q_i} L. \tag{*}$$

Then it follows from (20), lemma 1.5 and (24) that

$$P_{i+1} \lhd^{p_{i+1}} L.$$

By definition (20) P_{i+1} and Q_i permute. Therefore from (21), lemma 1.4 and (*) and (25) it follows that

$$Q_{i+1} \lhd^{q_{i+1}} L.$$

Thus (*) (it is true for $i = 1$) is true for all $i = 1, \ldots, n$. Define

$$r_1 = 0,\ r_{i+1} = \lambda_1(h+q_i) \qquad 1 \leqq i \leqq n-1. \tag{26}$$

If we apply corollary 2.5 to the pair H_{i+1}, Q_i we obtain

$$H_{i+1}^{(r_{i+1})} \leqq P_{H_{i+1}}(Q_i) = P_{i+1}. \tag{27}$$

Define

$$\lambda_5(h) = \lambda_4(h, r_1+\ldots+r_n). \tag{28}$$

Then it follows from (19), (28) and theorem 2.10 that

$$J^{(\lambda_5)} \lhd^{\lambda_5} L$$

and

$$\begin{aligned} J^{(\lambda_5)} &\leqq H_1^{(r_1)}+\ldots+H_n^{(r_n)} \\ &\subseteq P_1+\ldots+P_n. \end{aligned}$$

□

For the rest of this section we will be concerned with some applications of the Derived Join Theorems.

In general if $\mathfrak{X}$ is $\{\mathrm{N}_0, \mathrm{I}\}$-closed the join of two $\mathfrak{X}$-subideals of a Lie algebra may not be an $\mathfrak{X}$-algebra. However we have:

THEOREM 2.13 *Let $\mathfrak{X}$ be an $\{\mathrm{N}_0, \mathrm{I}\}$-closed class of Lie algebras and suppose that H and K are $\mathfrak{X}$-subideals of a Lie algebra L and $J = \langle H, K\rangle$. If H and K are permutable then J is an $\mathfrak{X}$-subideal of L.*

Proof: By hypothesis $J = H+K$ and by lemma 1.4 we have J si L. So it is enough to show that $J \in \mathfrak{X}$. Suppose then that $H \lhd^m J$ and $K \lhd^n J$ and use induction on $m+n$. If m or n is zero the result is trivial; so assume that $m > 0$

and $n > 0$. If $m+n \leqq 2$, then J is the sum of two $\mathfrak{X}$-ideals and so $J \in \mathfrak{X}$. Let $m+n > 2$ and suppose that the required result is true for $m+n-1$.

By the modular law $H_1 = H^J = H+H_1 \cap K$. This implies that H and $H_1 \cap K$ are permutable. We have $H \lhd^{m-1} H_1$ and $H_1 \cap K \lhd^n H_1$. Finally $H_1 \cap K \lhd K$ and so $H_1 \cap K \in \mathrm{I}\mathfrak{X} = \mathfrak{X}$ and so by induction $H_1 = H^J \in \mathfrak{X}$. Similarly $K^J \in \mathfrak{X}$ and so $J = H^J + K^J \in \mathrm{N}_0\mathfrak{X} = \mathfrak{X}$. □

We know from chapter 1 that the class $\mathrm{E}\mathfrak{A}$ of soluble Lie algebras is $\{\mathrm{N}_0, \mathrm{I}\}$-closed, and so is the class $\mathfrak{N}$ of nilpotent Lie algebras.

COROLLARY 2.14 *Let $\mathfrak{X}$ be an $\{\mathrm{N}_0, \mathrm{I}\}$-closed class and let L be any Lie algebra. Then $L \in \mathfrak{X} \cap \mathrm{E}\mathfrak{A}$ if and only if every term of the derived series of L is the join of finitely many $\mathfrak{X} \cap \mathrm{E}\mathfrak{A}$-subideals.*

Proof: The 'only if' part is clear since every derived term of L is an ideal and $\mathfrak{Y} = \mathfrak{X} \cap \mathrm{E}\mathfrak{A}$ is $\{\mathrm{N}_0, \mathrm{I}\}$-closed. Conversely suppose that every derived term of L is the join of finitely many $\mathfrak{Y}$-subideals. By corollary 2.11 L is soluble of derived length d for some d. We use induction on d to show that $L \in \mathfrak{Y}$; so we may inductively assume that $L^2 = L^{(1)} \in \mathfrak{Y}$. Suppose that $L = \langle H_1, \ldots, H_n \rangle$ where H_i si L and $H_i \in \mathfrak{Y}$, for $i = 1, \ldots, n$. By theorem 2.13 we have $H_i + L^2 \in \mathfrak{Y}$ and clearly $H_i + L^2 \lhd L$ for each i. Thus $L = (H_1 + L^2) + \ldots + (H_n + L^2)$ is the join of finitely many $\mathfrak{Y}$-ideals and so $L \in \mathfrak{Y}$. □

We have constructed (see lemma 1.11) a non-nilpotent Lie algebra which is generated by three abelian subideals. So we deduce from this and corollary 2.14 that the class of Lie algebras which can be generated by finitely many nilpotent subideals is not I-closed.

Theorem 2.13 implies the more general:

THEOREM 2.15 *Let $\mathfrak{X}$ be an $\{\mathrm{N}_0, \mathrm{I}\}$-closed class of Lie algebras and let L be a Lie algebra with $H \lhd^m L$, $K \lhd^n L$ and $J = \langle H, K \rangle$. If $H, K \in \mathfrak{X}$ then there exists $\lambda_6 = \lambda_6(m, n)$ such that to any pair of permutable subspaces A and B of H, K respectively, there corresponds an $\mathfrak{X}$-subalgebra X of J with*

$$X \lhd^{\lambda_6} L$$

and

$$A + B \subseteq X \leqq H + K.$$

Proof: Define $\lambda_6(0, n) = 0$. Let $m > 0$ and assume that $\lambda_6(m-1, n)$ has been defined to satisfy all requirements. Let $p = \lambda_6(m-1, n)$. By lemma 1.3 we know that $\langle A \rangle$ and $\langle B \rangle$ are permutable and $\langle A, B \rangle = \langle A \rangle + \langle B \rangle$. Put

$A_0 = \langle A \rangle$, $B_0 = \langle B \rangle$, $Q = \langle A^B \rangle$ and $H_1 = H^L$. Then $A_0 \leqq H$, $B_0 \leqq K$ and $Q \leqq H_1$. By the modular law $Q = Q \cap (A_0 + B_0) = A_0 + Q \cap B$, so A_0 permutes with $Q \cap B_0$.

Now $H \lhd^{m-1} H_1$, $H_1 \cap K \lhd^n H_1$ and $Q \cap B_0 \leqq H_1 \cap K$; and $H_1 \cap K \lhd K$ so that $H_1 \cap K \in \mathrm{I}\mathfrak{X} = \mathfrak{X}$. So by the inductive hypothesis there exists X_1 such that $X_1 \in \mathfrak{X}$, $X_1 \lhd^p H_1 \lhd L$ and

$$Q = \langle A^B \rangle = A_0 + Q \cap B_0 \subseteq X_1 \leqq H + H_1 \cap K.$$

Let Q_{p+1} be the $(p+1)$th ideal closure of Q in L. Then $Q_{p+1} \leqq X_1$ and $Q_{p+1} \lhd^{p+1} L$. But B idealises Q and so by lemma 1.1, $B \subseteq I_L(Q_{p+1})$. Thus B_0 idealises Q_{p+1} so

$$B_0 \leqq P = P_K(Q_{p+1}).$$

By lemma 1.5 we have $P \lhd^r L$ where $r = g_2(n+p+1)$. Thus $P \in \mathrm{I}\mathfrak{X} = \mathfrak{X}$ (for P si K) and $Q_{p+1} \in \mathrm{I}\mathfrak{X} = \mathfrak{X}$ (for Q_{p+1} si X_1) and P permutes with Q_{p+1}. By theorem 2.13 we have

$$X = Q_{p+1} + P \in \mathfrak{X}$$

and

$$A + B \subseteq Q + B_0 \subseteq X \subseteq X_1 + K \subseteq (H + H_1 \cap K) + K = H + K.$$

Finally from lemma 1.4 we have $X \lhd^r L$ where $r = g_1(1 + p + g_2(n+p+1))$. So we define

$$\lambda_6(m, n) = g_1(1 + \lambda_6(m-1, n) + g_2(1 + n + \lambda_6(m-1, n))). \tag{29}$$

□

Evidently the subideals X of theorem 2.15 can be chosen to lie in the $\{\mathrm{N}_0, \mathrm{I}\}$-closure of $(H) \cup (K)$.

Clearly theorem 2.15 can be extended in the obvious manner to any finite collection of subideals.

Let L be a Lie algebra and $H \leqq L$. Then we define

$$\lim H = \lim_L H = \bigcap_{n \in \mathbb{N}} H_n$$

where H_n is the nth ideal closure of H in L. Then $H \leqq \lim H$ and $H \leqq K \leqq L$ implies that $\lim H \leqq \lim K$. We do not know (though it seems unlikely) whether $\lim H$ si L in general, however

$$(\lim H)^\omega \lhd L$$

since for any n, $[(\lim H)^\omega, L] \subseteq [H_n^{n+1}, L] \subseteq H_n$.

We recall that Min-si is the class of Lie algebras satisfying the minimal condition on subideals. It follows from theorem 1.7.4 that Min-si is $\{\mathrm{N}_0, \mathrm{I}\}$-closed. Suppose that in theorem 2.15 A and B are permutable subalgebras of L and $L \in$ Min-si. Put $H = \lim A = A_m$ (for some m) and $K = \lim B = B_n$ (for some n) since $L \in$ Min-si. Let $\mathfrak{X}$ = Min-si. Then the subideal X contains $A+B$ and so

$$\lim A + \lim B \subseteq \lim(A+B) \subseteq X.$$

But $X \leqq H+K$, so $X \subseteq \lim A + \lim B$ (for $H, K \in$ Min-si). Thus

$$\lim A + \lim B = \lim(A+B).$$

So we have:

Corollary 2.16 *Suppose that $L \in$ Min-si and A and B are permutable subalgebras of L. Then $\lim A$ and $\lim B$ are permutable and*

$$\lim A + \lim B = \lim(A+B). \qquad \square$$

Let $\mathfrak{X}$ be $\{\mathrm{N}_0, \mathrm{I}\}$-closed. Suppose that in theorem 2.12 each $H_i \in \mathfrak{X}$. Since each P_i si H_i then $P_i \in \mathfrak{X}$. Therefore by theorem 2.13 it follows by induction on i that each $Q_i \in \mathfrak{X}$. In particular $Q_n = P_1 + \ldots + P_n \in \mathfrak{X}$. As $J^{(\lambda_s)} \lhd Q_n$ we have $J^{(\lambda_s)} \in \mathfrak{X}$. So we have proved:

Corollary 2.17 *Let L be a Lie algebra and $J = \langle H_1, \ldots, H_n \rangle$ with $H_i \lhd^{h_i} L$ for $i = 1, \ldots, n$. If each H_i lies in an $\{\mathrm{N}_0, \mathrm{I}\}$-closed class $\mathfrak{X}$ then*

$$J^{(\lambda_s)} \in \mathfrak{X} \text{ and so } J \in \mathfrak{X}\mathfrak{A}^{\lambda_s}. \qquad \square$$

Lemma 2.18 *Let $\mathfrak{X} = \{\mathrm{I}, \mathrm{Q}\}\mathfrak{X}$ be a class of Lie algebras and let H and K be permutable $\mathfrak{X}$-subalgebras of some Lie algebra. If $H \lhd^m J = \langle H, K \rangle$ then $J \in \mathfrak{X}^{1+m}$. If also $K \lhd^n J$ then $J \in \mathfrak{X}^{1+r}$ where $r = \min\{m, n\}$.*

Proof: By hypothesis $J = H+K$. So if $H = H_m \lhd \ldots \lhd H_0 = J$ is the ideal closure series of H in J then

$$H_i = H + H_i \cap K = H_{i+1} + H_i \cap K.$$

Thus since $H_i \cap K$ si K so that $H_i \cap K \in \mathrm{I}\mathfrak{X} = \mathfrak{X}$ we have

$$H_i/H_{i+1} = (H_{i+1} + H_i \cap K)/H_{i+1} \cong H_i \cap K / H_{i+1} \cap K \in \mathrm{Q}\mathfrak{X} = \mathfrak{X}.$$

From this we deduce that $J \in \mathfrak{X}^{1+m}$. If also $K \lhd^n J$ then $J \in \mathfrak{X}^{1+n}$ and so $J \in \mathfrak{X}^{1+r}$ with $r = \min\{m, n\}$. $\square$

We note that if $\mathfrak{X} = \{\mathrm{I}, \mathrm{Q}\}\mathfrak{X}$ then $\mathfrak{X}^s = \{\mathrm{I}, \mathrm{Q}\}\mathfrak{X}^s$ for all $s > 0$.

COROLLARY 2.19 *Let* $\mathfrak{X} = \{\mathrm{I}, \mathrm{Q}\}\mathfrak{X}$ *be a class of Lie algebras. Suppose that* L *is a Lie algebra and* $J = \langle H_1, \ldots, H_n\rangle$ *with* $H_i \lhd^{h_i} L$ *and* $H_i \in \mathfrak{X}$ *for* $i = 1, \ldots, n$. *Then there exists* $\lambda_7 = \lambda(h_7)$ *such that*

$$J^{(\lambda_5)} \in \mathfrak{X}^{\lambda_7} \text{ and so } J \in \mathfrak{X}^{\lambda_7}\mathfrak{A}^{\lambda_5}$$

whenever $h_1 + \ldots + h_n \leqq h$.

Proof: Using the notation of the proof of theorem 2.12 we have $P_1 = Q_1 = H_1 \in \mathfrak{X}$. Define recursively

$$m_1 = 1,\ m_{i+1} = (1+q_i)m_i \qquad 1 \leqq i \leqq n-1. \tag{29}$$

Assume inductively that $Q_i \in \mathfrak{X}^{m_i}$ and apply lemma 2.8 to the pair Q_i, P_{i+1} (for P_{i+1} si L and $P_{i+1} \leqq H_{i+1} \in \mathfrak{X} \leqq \mathfrak{X}^{m_i} = \{\mathrm{I}, \mathrm{Q}\}\mathfrak{X}^{m_i}$). From (24) and (29) we have $q_i \leqq p_{i+1}$ and

$$Q_{i+1} = \langle Q_i, P_{i+1}\rangle \in (\mathfrak{X}^{m_i})^{1+q_i} \leqq \mathfrak{X}^{m_{i+1}}.$$

Hence $Q_i \in \mathfrak{X}^{m_i}$ for all i and in particular $Q_n \in \mathfrak{X}^{m_n}$. But $J^{(\lambda_5)} \lhd Q_n$ and so $J^{(\lambda_5)} \in \mathrm{I}\mathfrak{X}^{m_n} = \mathfrak{X}^{m_n}$. Se wo define

$$\lambda_7(h) = m_n = (1+q_1)\ldots(1+q_{n-1}). \tag{30}$$

□

We define the closure operation J_s by

$$L \in \mathrm{J}_s\mathfrak{X}$$

if L can be generated by $\mathfrak{X}$-subideals.

It follows from corollaries 2.17 and 2.19 that

$$\text{If } \mathfrak{X} = \{\mathrm{I}, \mathrm{N}_0\}\mathfrak{X} \text{ then } \mathrm{J}_s\mathfrak{X} \leqq \mathrm{L}(\mathfrak{X}\,\mathrm{E}\mathfrak{A}). \tag{31}$$

$$\text{If } \mathfrak{X} = \{\mathrm{I}, \mathrm{Q}\}\mathfrak{X} \text{ then } \mathrm{J}_s\mathfrak{X} \leqq \mathrm{L}(\mathrm{E}\mathfrak{X}\,\mathrm{E}\mathfrak{A}). \tag{32}$$

Where necessary theorems 2.4, 2.7, 2.10 and 2.12 collectively will be referred to as 'the Derived Join Theorem'.

Chapter 3

Coalescent classes of Lie algebras

We have seen (lemma 2.1.11) that the join of two subideals of a Lie algebra need not be a subideal. This raises the question of finding conditions under which the join is a subideal. The same question arises in group theory. One way to answer it involves the phenomenon of 'coalescence': for suitable classes $\mathfrak{X}$ two subnormal $\mathfrak{X}$-groups always generate a subnormal $\mathfrak{X}$-group. Known coalescent classes of groups include finite and polycyclic groups (Wielandt [348]); maximal condition for subgroups (Baer [262]); finitely generated nilpotent (Baer [260]); maximal or minimal condition on subnormal subgroups (Robinson [326], Roseblade [333, 334]); minimax groups (Roseblade, unpublished); groups which are finitely generated modulo their derived group (Roseblade [336]); certain classes of finitely generated groups (Roseblade and Stonehewer [337]); groups of finite rank (Drukker, Robinson and Stewart [273]). These results suggest an attack on the Lie-theoretic analogues. First we show (using an example due to Hartley [76]) that over fields of characteristic $p > 0$ there is no hope of success. Then we restrict ourselves to fields of characteristic zero. Following Hartley [76] we prove coalescence of $\mathfrak{F}$ and $\mathfrak{F} \cap \mathfrak{N}$. By modifying his methods we prove coalescence for certain classes with minimal conditions.

Next we apply the Derived Join Theorem to a wide variety of classes, including Max-si (see Amayo [5]). In the fourth section we introduce the idea of local coalescence (cf. Roseblade and Stonehewer [337]) and use it to prove that $\mathfrak{G}$ is coalescent in characteristic zero. This latter result, taken from Amayo [6], solves a problem due to Hartley.

A coalescent class has two properties. It is 'subjunctive' (our use of this word differs from the usage in Roseblade and Stonehewer [337]) and 'persistent'. We study these two phenomena and give an example to show that in characteristic $p > 0$ it is possible for two finite-dimensional nilpotent subideals to generate an infinite dimensional subalgebra (cf. Amayo [4]).

1. An example

LEMMA 1.1 (*Hartley*) *Let $\mathfrak{k}$ be a field of characteristic $p > 0$. Then there exists a Lie algebra L defined over $\mathfrak{k}$ such that:*

(a) $L \in \mathfrak{G}_3 \cap \mathfrak{A}\mathfrak{N}_2 \cap \mathfrak{F}$;

(b) *there exist $x, y \in L$ such that $\langle x \rangle \lhd^{p+1} L$ and $\langle y \rangle \lhd^{p+1} L$;*

(c) *If $Q = \langle x, y \rangle$ then $Q \in \mathfrak{F}_3 \cap \mathfrak{N}_2$, $I_L(Q) = Q$ and $Q^L = L$ so that Q is neither an ascendant nor a descendant subalgebra of L.*

Proof: Let P be the associative and commutative algebra over $\mathfrak{k}$ with basis $1, t, \ldots, t^{p-1}$, where $t^p = 0$ (i.e. $P = \mathfrak{k}[T]/(T^p)$). Define linear transformations x and y of P by

$$x: f(t) \mapsto tf(t) \qquad (f(t) \in P)$$

$$y: f(t) \mapsto \mathrm{d}f(t)/\mathrm{d}t.$$

Consider P as an abelian Lie algebra over $\mathfrak{k}$. Then $x, y \in \mathrm{Der}(P)$ and

$$z = [x, y] = xy - yx : f(t) \mapsto f(t)$$

and so z is the identity transformation of P. We also have

$$[x, z] = [y, z] = 0.$$

Thus if $Q = \langle x, y \rangle \leqq \mathrm{Der}(P)$ then $Q = \langle x \rangle + \langle y \rangle + \langle z \rangle$ and $Q \in \mathfrak{F}_3 \cap \mathfrak{N}_2$.

Form the split extension

$$L = P \dotplus Q, \; P \lhd L, \; P^2 = 0 = P \cap Q.$$

Now $1 = [t, y] \in Q^L$ and $t^i = [1, {}_i x] \in Q^L$. Thus $P \leqq Q^L$ and so $L = P + Q = Q^L$. Let $u \in I_L(Q)$. Then $u = a + b$ for some $a \in P$ and $b \in Q$. Since $[Q, z] = 0$ we have $[u, z] = [a, z] = a \in P \cap Q = 0$. Thus $a = 0$ and $u = b \in Q$; so $I_L(Q) = Q$.

Let $X = \langle x \rangle$, $Y = \langle y \rangle$ and $Z = \langle z \rangle$. Then we have

$$X \lhd X + [P, {}_{p-1}x] \lhd \ldots \lhd X + [P, x] \lhd X + P + Z \lhd L$$

and

$$Y \lhd Y + [P, {}_{p-1}y] \lhd \ldots \lhd Y + [P, y] \lhd Y + P + Z \lhd L.$$

Thus $X \lhd^{p+1} L$ and $Y \lhd^{p+1} L$. □

We say that a class $\mathfrak{X}$ of Lie algebras is *subjunctive* if in any Lie algebra the join of two $\mathfrak{X}$-subideals is always a subideal.

So lemma 1.1 says that over any field of characteristic $p > 0$, the classes $\mathfrak{F}$ and $\mathfrak{F} \cap \mathfrak{N}$ are not subjunctive. We note also that the example of lemma 2.1.11 says that over any field the class $\mathfrak{A}$ of abelian Lie algebras is not subjunctive.

It is clear that if $\mathfrak{X} \leqq \mathfrak{Y}$ then $\mathfrak{Y}$ is subjunctive only if $\mathfrak{X}$ is subjunctive. Thus over any field the classes $\mathfrak{N}$, $\text{E}\mathfrak{A}$, $\text{L}\mathfrak{N}$ are not subjunctive.

These results suggest that we restrict ourselves to fields of characteristic zero and that some sort of finiteness conditions be placed on the classes we want to be subjunctive.

2. Coalescence of classes with minimal conditions

If L is a Lie algebra and M is an L-module we denote by

$$\text{ad}_L(M)$$

the set of linear transformations of M induced by the elements of L. Clearly

$$\text{ad}_L(M) \leqq [\text{End}(M)].$$

If also M is a Lie algebra and the elements of L induce derivations of M then $\text{ad}_L(M) \leqq \text{Der}(M)$. In particular if $M \lhd L$ then $\text{ad}_L(M) \cong L/C_L(M)$.

LEMMA 2.1 (*Hartley*) *Let K be a nilpotent subalgebra of a Lie algebra L. Then*

(a) *If K si L then* $\text{ad}_K(L)$ *is nilpotent.*

(b) *If K is finite-dimensional and ascendant in L then* $\text{ad}_K(L)$ *is a nil subset of* $\text{Der}(L)$.

Proof: (a) Suppose that $K \lhd^n L$ and $K \in \mathfrak{N}_c$. Then $[L, {}_nK] \leqq K$ and so $[L, {}_{n+c}K] \leqq K^{c+1} = 0$. Thus if $d_1, \ldots, d_{n+c} \in \text{ad}_K(L)$ then $d_1 \ldots d_{n+c} = 0$ and so $[d_1, \ldots, d_{n+c}] = 0$. Thus $\text{ad}_K(L) \in \mathfrak{N}_{n+c-1}$.

(b) Let M be a finite dimensional subspace of L. By lemma 1.3.3 there exists $t = t(M)$ such that $[M, {}_tK] \subseteq K$. Therefore if $K \in \mathfrak{N}_c$ then $[M, {}_{t+c}K] = 0$. In other words if $d_1, \ldots, d_{t+c} \in \text{ad}_K(L)$ then $Md_1 \ldots d_{t+c} = 0$. So $\text{ad}_K(L)$ is a nil subset of $\text{Der}(L)$. □

Clearly if $K \leqq L$ then $\text{Inn}(K) \leqq \text{Inn}(L)$. It follows in the same way as (b) above was proved that:

COROLLARY 2.2 *Let K be an ascendant subalgebra of a Lie algebra L. If S is a finite dimensional nil subspace of* $\mathrm{Inn}(K)$ *then S is a finite dimensional nil subspace of* $\mathrm{Der}(L)$. □

LEMMA 2.3 *Let L be a Lie algebra defined over a field of characteristic zero. If $H \operatorname{asc} L$, $K \operatorname{asc} L$, $[H, K] \subseteq H$ and $K \in \mathfrak{F} \cap \mathfrak{N}$ then $H+K \operatorname{asc} L$.*

Proof: By lemma 2.1 $\mathrm{ad}_K(L)$ is a nil subspace of $\mathrm{Der}(L)$. Let $A = \exp(\mathrm{ad}_K(L)) = \text{group}\,\langle \exp \delta : \delta \in \mathrm{ad}_K(L)\rangle$ and let $(H_\sigma : \sigma \leqq \rho)$ be an ascending series from H to L. For each ordinal σ put $N_\sigma = \bigcap_{\alpha \in A} H_\sigma^\alpha$; then $N_0 = H$ and $N_\rho = L$. Suppose that $\sigma < \rho$. For each $\alpha \in A$ we have $[N_{\sigma+1}, H_\sigma^\alpha] = [N_{\sigma+1}^{\alpha^{-1}}, H_\sigma]^\alpha \subseteq [H_{\sigma+1}, H_\sigma]^\alpha \subseteq H_\sigma^\alpha$. Therefore $N_\sigma \lhd N_{\sigma+1}$. Let μ be a limit ordinal, $\mu \leqq \rho$. Evidently $\bigcup_{\sigma<\mu} N_\sigma \subseteq N_\mu$. Let $y \in N_\mu$; then $\langle y^A \rangle \leqq N_\mu$ since N_μ is A-invariant. By lemma 1.4.10 $\langle y^A \rangle = \langle y^S \rangle$ (where $S = \mathrm{ad}_K(L)$) is finite-dimensional. Therefore since $N_\mu \leqq H_\mu = \bigcup_{\sigma<\mu} H_\sigma$ we can find $\sigma < \mu$ such that $\langle y^A \rangle \leqq H_\sigma$. Thus if $\alpha \in A$ then $\langle y^A \rangle = \langle y^A \rangle^\alpha \leqq H_\sigma^\alpha$; hence $\langle y^A \rangle \leqq N_\sigma$ and so $y \in N_\sigma$. Therefore $N_\mu \leqq \bigcup_{\sigma<\mu} N_\sigma \subseteq N_\mu$. Thus $(N_\sigma : \sigma \leqq \rho)$ is an ascending series from H to L and each term is idealised by K. So it follows from lemma 1.3.1 that

$$H+K \operatorname{asc} N_1+K \operatorname{asc} N_2+K \operatorname{asc} \dots N_\rho+K = L.$$

We also have at limit ordinals $\mu \leqq \rho$, $N_\mu+K = \bigcup_{\sigma<\mu}(N_\sigma+K)$ and so by lemma 1.3.1 we have $H+K \operatorname{asc} L$. □

Let $\mathfrak{X}$ be a class of Lie algebras. We say that $\mathfrak{X}$ is *coalescent* (resp. *ascendantly coalescent*) if in any Lie algebra L the join of two $\mathfrak{X}$-subideals (resp. ascendant $\mathfrak{X}$-subalgebras) of L is always an $\mathfrak{X}$-subideal (resp. ascendant $\mathfrak{X}$-subalgebra) of L.

Clearly a coalescent class is necessarily subjunctive. From lemma 1.1 we see that over any field of characteristic $p > 0$ the class $\mathfrak{F} \cap \mathfrak{N}$ is neither coalescent nor ascendantly coalescent. However we have:

THEOREM 2.4 (*Hartley*) *Over any field of characteristic zero the class $\mathfrak{F} \cap \mathfrak{N}$ is coalescent and ascendantly coalescent.*

Proof: We will prove that $\mathfrak{F} \cap \mathfrak{N}$ is ascendantly coalescent. The proof of coalescence is similar but easier.

Let $H, K \in \mathfrak{F} \cap \mathfrak{N}$ and suppose that $H \lhd^\rho L$, $K \operatorname{asc} L$ and $J = \langle H, K \rangle$. We use transfinite induction on ρ to show that $J \operatorname{asc} L$ and $J \in \mathfrak{F} \cap \mathfrak{N}$.

If $\rho \leqq 1$ the result follows from lemma 2.3.

Let $(H_\sigma : \sigma \leqq \rho)$ be an ascending series from H to L. Assume that the result holds for all ordinals $< \rho$.

If ρ is a limit ordinal then as $L = \bigcup_{\sigma<\rho} H_\sigma$ we can find $\sigma < \rho$ such that $K \leqq H_\sigma$ (for $K \in \mathfrak{F}$), and so by induction J asc H_σ (for $H \lhd^\sigma H_\sigma$), and finally J asc L (H_σ asc L).

If ρ is not a limit ordinal then $\rho = \lambda+1$ for some λ. Now $H \lhd^\lambda H_\lambda \lhd L$ and by lemma 1.4.10 we can find $\alpha_1, \ldots, \alpha_n \in A$ (where $A = \exp(\mathrm{ad}_K(L))$) such that

$$\langle H^K \rangle = \langle H^{\alpha_1}, \ldots, H^{\alpha_n} \rangle.$$

For each i, $1 \leqq i \leqq n$, it is clear that $(H_\sigma^{\alpha_i}: \sigma \leqq \lambda)$ is an ascending series from H^{α_i} to $H_\lambda^{\alpha_i} = H_\lambda$. By the inductive hypothesis on λ and a second finite induction on i it follows that

$$\langle H^{\alpha_1}, \ldots, H^{\alpha_i} \rangle \text{ asc } H_\lambda \text{ and } \langle H^{\alpha_1}, \ldots, H^{\alpha_i} \rangle \in \mathfrak{F} \cap \mathfrak{N}$$

for every i, $1 \leqq i \leqq n$. In particular

$$\langle H^K \rangle = \langle H^{\alpha_1}, \ldots, H^{\alpha_n} \rangle \text{ asc } H_\lambda \lhd L \text{ and } \langle H^K \rangle \in \mathfrak{F} \cap \mathfrak{N}.$$

Therefore by lemma 2.3 we have $J = \langle H^K \rangle + K$ asc L and clearly $J \in \mathfrak{F}$. Put $H_1 = \langle H^K \rangle$. By lemma 1.3.3 there is a $t = t(H_1)$ such that $[H_1, {}_t K] \subseteq K$; so if $K \in \mathfrak{N}_c$ then $[H_1, {}_r K] = 0$ where $r = t+c$. An easy induction on m gives:

$$J^{m+1} \leqq H_1^2 + [H_1, {}_m K] + K^{m+1}.$$

Thus $J^{r+1} \leqq H_1^2$. More generally we have for any d,

$$J^{dr+m+1} \leqq H_1^{d+1} + [H_1^d, {}_m K] + K^{m+1}.$$

Therefore since $H_1 \in \mathfrak{N}_d$ for some d then $J \in \mathfrak{N}$. □

Let M be a finite-dimensional subalgebra of a Lie algebra L and let H asc L. Looking at the proof of lemma 1.3.3 it is easy to see that for any finite-dimensional subspace N of H there exists $m = m(M, N)$ such that $[M, {}_m N] \subseteq H$. Hence by the Jacobi identity it follows that if $N_0 = \langle N \rangle$ then $[M, {}_m N_0] \subseteq H$. In particular if $N_0 = H$, then $[M, {}_m H] \subseteq H$. So we have:

$$\text{If } H \in \mathfrak{G} \text{ and } H \text{ asc } L \text{ then } H^\omega \lhd L \text{ and } H^{(\omega)} \lhd L. \qquad (1)$$

THEOREM 2.5 *Over any field of characteristic zero the class* Min-$\lhd$ *is coalescent and the classes* $\mathfrak{F}$, $\mathfrak{F} \cap \mathrm{E}\mathfrak{A}$, Min, Min-si, Min-asc, Min-$\lhd^\alpha$ $(\alpha > 1)$, $\mathfrak{G} \cap$ Min-$\lhd$ *and* Min-$\lhd$ $\cap$ Max-$\lhd$ *are coalescent and ascendantly coalescent.*

Proof: We observe that all the classes are $\{\mathrm{Q}, \mathrm{E}\}$-closed. Let $\mathfrak{X}$ ($\neq$ Min-$\lhd$) be any one of the classes above. Let L be a Lie algebra (over a field of characteristic zero) and let H, K be ascendant $\mathfrak{X}$-subalgebras of L and $J = \langle H, K \rangle$.

(a) $\mathfrak{X} = \mathfrak{F}$; then by (1) above $H^{\omega}, K^{\omega} \lhd L$. Clearly there exist c, d with $H^{\omega} = H^{c}$ and $K^{\omega} = K^{d}$. Thus if $M = H^{c}+K^{d}$ then $M \in \mathfrak{F}$ and $M \lhd L$. Clearly J/M is the join of two ascendant $\mathfrak{F} \cap \mathfrak{N}$-subalgebras of L/M and so by theorem 2.4 we have $J/M \text{ asc } L/M$ and $J/M \in \mathfrak{F} \cap \mathfrak{N}$. Thus $J \text{ asc } L$ and $J \in \mathfrak{F}$. The proof of coalescence is similar.

(b) $\mathfrak{X} = \mathfrak{F} \cap \text{E}\mathfrak{A}$; from (a) $J \in \mathfrak{F}$ and $J \text{ asc } L$. Since $J \in \mathfrak{F}$ we must have $H \text{ si } J$ and $K \text{ si } J$ and so by the Derived Join Theorem $J \in \text{E}\mathfrak{A}$. Coalescence is deduced in the same way.

(c) $\mathfrak{X} =$ Min, Min-si, Min-asc or Min-$\lhd^{\alpha}$ $(\alpha \geqq \omega)$; then $\mathfrak{X}$ is $\{\text{I}, \text{Q}, \text{E}\}$-closed. By the minimal condition $H^{(\omega)} = H^{(c)}$ and $K^{(\omega)} = K^{(d)}$ for some c and d. Thus $H^{(c)}$ and $K^{(d)}$ are perfect subalgebras and ascendant in L and so by proposition 1.3.5 we have $H^{(c)} \lhd L$ and $K^{(d)} \lhd L$. So $M = H^{(c)}+K^{(d)} \in \mathfrak{X}$ and $M \lhd L$. Now J/M is the join of two $\mathfrak{X} \cap \text{E}\mathfrak{A}$-subalgebras which are ascendant in L/M. Furthermore $\mathfrak{X} \cap \text{E}\mathfrak{A} = \mathfrak{F} \cap \text{E}\mathfrak{A}$ (which follows easily by induction on derived lengths) and so from (b) above $J/M \in \mathfrak{F} \cap \text{E}\mathfrak{A}$ and $J/M \text{ asc } L/M$. Thus $J \in \mathfrak{X}$ (by E-closure) and $J \text{ asc } L$. Coalescence follows in the same way.

(d) $\mathfrak{X} = \mathfrak{G} \cap \text{Min-}\lhd$; from (1) we have $H^{\omega} \lhd L$ and $K^{\omega} \lhd L$. By Min-$\lhd$ there exist m, n such that $H^{\omega} = H^{m}$ and $K^{\omega} = K^{n}$. Put $M = H^{m}+K^{n}$ so that $M \lhd L$. Then J/M is the join of two $\mathfrak{G} \cap \mathfrak{N} = \mathfrak{F} \cap \mathfrak{N}$-ascendant subalgebras of L/M and so by theorem 2.4 $J/M \in \mathfrak{F} \cap \mathfrak{N} = \mathfrak{X} \cap \mathfrak{N}$ and $J/M \text{ asc } L/M$; thus $J \text{ asc } L$. Now $H^{m} \lhd J$ and H^{m} has Min-H (the minimal condition for ideals of H inside H^{m}) and therefore H^{m} has Min-J. Similarly K^{n} has Min-J and since $H^{m} \cap K^{n} \lhd J$ then $K^{n}/H^{m} \cap K^{n}$ has Min-J. But $M/H^{m} \cong K^{n}/H^{m} \cap K^{n}$ (as a J-module) and so M/H^{m} has Min-J. Thus as H^{m} has Min-J then M has Min-J. Finally $J \in \text{Min-}\lhd$ (for $J/M \in \text{Min-}\lhd$). Clearly $J \in \mathfrak{G}$. The proof for coalescence is similar.

(e) $\mathfrak{X} = \text{Min-}\lhd \cap \text{Max-}\lhd$; by the minimal condition $H^{(\omega)} = H^{(c)}$ and $K^{(\omega)} = K^{(d)}$ for some c and d. Thus $H^{(c)}$ and $K^{(d)}$ are perfect and ascendant subalgebras of L; so by proposition 1.3.5 we have $H^{(c)}, K^{(d)} \lhd L$. Arguing in the same way as the last half of (d) above we can show that $N = H^{(c)}+K^{(d)}$ has Min-J; and in a similar way that N has Max-J (the maximal condition for ideals of J contained in N). Since also $N \lhd L$ we may put $N = 0$ (for if $J/N \in \mathfrak{X}$ and $J/N \text{ asc } L/N$ then $J \in \mathfrak{X}$ and $J \text{ asc } L$). Then $H, K \in \mathfrak{X} \cap \text{E}\mathfrak{A} = \mathfrak{G} \cap \mathfrak{X} \cap \text{E}\mathfrak{A}$ (it is easy to verify by induction on derived lengths and is proved in lemma 8.6.1 that $\text{Max-}\lhd \cap \text{E}\mathfrak{A} \leqq \mathfrak{G} \cap \text{E}\mathfrak{A}$). Therefore by equation (1) we have $H^{\omega} \lhd L$ and $K^{\omega} \lhd L$ and for some m and n we have $H^{\omega} = H^{m}$ and $K^{\omega} = K^{n}$. Let $M = H^{m}+K^{n}$ so that $M \lhd L$. Then arguing as in (d) we get $J/M \in \mathfrak{F} \cap \mathfrak{N} = \mathfrak{X} \cap \mathfrak{N}$ and $J/M \text{ asc } L/M$. We also have $M \in \text{Min-}J \cap \text{Max-}J$, since $H^{m}, K^{n} \in \text{Min-}J \cap \text{Max-}J$. Thus $J \in \mathfrak{X}$ and $J \text{ asc } L$. Coalescence follows in the same way.

(f) $\mathfrak{X} = \text{Min-}\lhd^{\alpha}$ $(2 \leqq \alpha < \omega)$; then $\mathfrak{X} \cap \text{E}\mathfrak{A} = \mathfrak{F} \cap \text{E}\mathfrak{A}$ (induction on derived lengths). As in (e) we have $H^{(c)} \lhd L$ and $K^{(d)} \lhd L$ for some c and d. Put $N = H^{(c)} + K^{(d)}$ so that $N \lhd L$. Then J/N is the join of two ascendant $\mathfrak{F} \cap \text{E}\mathfrak{A}$-subalgebras of L/N and so by (b) $J/N \in \mathfrak{F} \cap \text{E}\mathfrak{A}$ and $J/N \text{ asc } L/N$. Now $H^{(c)}$ has $\text{Min-}\lhd^{\alpha}(H)$ (the minimal condition for α-step subideals of H contained in $H^{(c)}$) and so $H^{(c)}$ has $\text{Min-}\lhd^{\alpha}(J)$. Similarly $K^{(d)}$ has $\text{Min-}\lhd^{\alpha}(J)$ and since $H^{(c)} \cap K^{(d)} \lhd J$ then $K^{(d)}/H^{(c)} \cap K^{(d)}$ has $\text{Min-}\lhd^{\alpha}(J/H^{(c)} \cap K^{(d)})$. But $M/H^{(c)} \cong K^{(d)}/H^{(c)} \cap K^{(d)}$ (as a J-module) and so M has $\text{Min-}\lhd(J)$. Therefore $J \in \mathfrak{X}$ and $J \text{ asc } L$. Coalescence follows similarly.

Finally suppose that $H \text{ si } L$, $K \text{ si } L$ and $H, K \in \text{Min-}\lhd$ and let $J = \langle H, K \rangle$. By lemma 1.3.2 we have $H \lhd L$ and $K \lhd L$. By $\text{Min-}\lhd$ we have $H^{\omega} = H^{m}$ and $K^{\omega} = K^{n}$ for some m, n. Put $M = H^{m} + K^{n}$. The usual argument shows that M has Min-J. Now $\text{Min-}\lhd \cap \mathfrak{N} = \mathfrak{F} \cap \mathfrak{N}$ (for $P \in \text{Min-}\lhd$ implies that $\zeta_1(P) \in \mathfrak{F}$ and the general result follows by induction on the nilpotency class). Therefore J/M is the join of two $\mathfrak{F} \cap \mathfrak{N}$-subideals of L/M and so by theorem 2.4 we have $J/M \in \mathfrak{F} \cap \mathfrak{N}$ and $J/M \text{ si } L/M$. Thus $J \in \text{Min-}\lhd$ and $J \text{ si } L$. □

Very little is known concerning the ascendant coalescence of classes with the maximal condition. However most of the usual classes are coalescent (of course over fields of characteristic zero) as we show in the next section.

3. Coalescence of classes with maximal conditions

LEMMA 3.1 *Let $\mathfrak{X} = \text{Q}\mathfrak{X}$ be a class of Lie algebras such that $\mathfrak{X} \cap \text{E}\mathfrak{A}$ is subjunctive. Suppose that L is a Lie algebra and H si L, K si L and $J = \langle H, K \rangle$. If $A \lhd J$ such that A si L, $(H+A)/A \in \mathfrak{X}$, and $(K+A)/A \in \mathfrak{X}$ then J si L.*

In particular a Q*-closed class $\mathfrak{X}$ of Lie algebras is subjunctive if and only if $\mathfrak{X} \cap \text{E}\mathfrak{A}$ is subjunctive.*

Proof: By the Derived Join Theorem there is a finite m such that $J^{(m)}$ si L. Let $B = A + J^{(m)}$ so that $B \lhd J$ and by lemma 2.1.2 we have $B \lhd^{n} L$ for some n. Let

$$B = B_n \lhd B_{n-1} \lhd \ldots \lhd B_0 = L$$

be the ideal closure series of B in L. By proposition 2.1.1 each B_i is idealised by J and so $\langle B_i, J \rangle = J + B_i$. Furthermore $B_{i+1} \lhd J + B_i$ and we have $H/H \cap A \cong (H+A)/A \in \mathfrak{X}$; thus since

$$(H + B_{i+1})/B_{i+1} \cong H/H \cap B_{i+1} \cong (H/H \cap A)/(H \cap B_{i+1}/H \cap A),$$

then $(H+B_{i+1})/B_{i+1} \in \mathrm{Q}\mathfrak{X} = \mathfrak{X}$. So $(H+B_{i+1})/B_{i+1} \in \mathfrak{X} \cap \mathrm{E}\mathfrak{A}$ and similarly $(K+B_{i+1})/B_{i+1} \in \mathfrak{X} \cap \mathrm{E}\mathfrak{A}$ (for $J^{(m)} \leqq B \leqq B_{i+1}$). Therefore $(J+B_{i+1})/B_{i+1}$ is the join of two $\mathfrak{X} \cap \mathrm{E}\mathfrak{A}$-subideals (namely $(H+B_{i+1})/B_{i+1}$ and $(K+B_{i+1})/B_{i+1}$) of $(J+B_i)/B_{i+1}$ and so $(J+B_{i+1})/B_{i+1}$ si $(J+B_i)/B_{i+1}$ (for $\mathfrak{X} \cap \mathrm{E}\mathfrak{A}$ is subjunctive). So $J+B_{i+1}$ si $J+B_i$ for all i, $0 \leqq i < n$, and we have

$$J = J+B_n \text{ si } J+B_{n-1} \text{ si } \ldots \text{ si } J+B_0 = L. \qquad \square$$

In lemma 3.1 (as in the rest of this section) there is no restriction on fields (unless the contrary is explicitly stated).

We remark that the group-theoretic analogue of lemma 3.1 is also true. So is that of the following result:

THEOREM 3.2 *Let $\mathfrak{X} = \{\mathrm{I}, \mathrm{Q}, \mathrm{E}\}\mathfrak{X}$ be a class of Lie algebras. Then $\mathfrak{X}$ is coalescent if and only if $\mathfrak{X} \cap \mathrm{E}\mathfrak{A}$ is coalescent.*

Proof: Since the join of two soluble subideals is always soluble, the implication one way is clear.

Suppose that $\mathfrak{X} \cap \mathrm{E}\mathfrak{A}$ is coalescent. Then if L is a Lie algebra and H and K are $\mathfrak{X}$-subideals of L it follows by lemma 3.1 that $J = \langle H, K \rangle$ si L. Now $\mathfrak{X} = \mathrm{N}_0\mathfrak{X}$ and so by corollary 2.2.17 we have $J^{(m)} \in \mathfrak{X}$ for some m. Then as $\mathfrak{X} = \mathrm{Q}\mathfrak{X}$ we have $J/J^{(m)}$ as a join of two $\mathfrak{X} \cap \mathrm{E}\mathfrak{A}$-subideals and so $J/J^{(m)} \in \mathfrak{X} \cap \mathrm{E}\mathfrak{A}$. Thus $J \in \mathfrak{X}^2 = \mathfrak{X}$. □

From this result we deduce the coalescence of a wide variety of classes. We define

$$\mathfrak{G}^{\mathrm{I}}$$

to be the class of Lie algebras in which every subideal is finitely generated.

Now let $\mathfrak{X}$ be any one of the classes

Min, Min-si, Min-asc, Max, Max-si, Max-asc, $\mathfrak{G}^{\mathrm{I}}$.

Then

$$\mathfrak{X} = \{\mathrm{I}, \mathrm{Q}, \mathrm{E}\}\mathfrak{X} \text{ and } \mathfrak{X} \cap \mathrm{E}\mathfrak{A} = \mathfrak{F} \cap \mathrm{E}\mathfrak{A}.$$

From this and theorems 2.5 and 3.2 we have:

THEOREM 3.3 *Over any field of characteristic zero the classes*

Min, Min-si, Min-asc, Max, Max-si, Max-asc, $\mathfrak{G}^{\mathrm{I}}$

are all coalescent. □

Let $(\mathfrak{X}_i : i \in I)$ be any collection of classes of Lie algebras. We define

$$\sum_{i \in I} \mathfrak{X}_i$$

to be the class of Lie algebras L which have a finite series

$$0 = L_0 \lhd L_1 \lhd \ldots \lhd L_m = L$$

such that $L_{r+1}/L_r \in \bigcup_{i \in I} \mathfrak{X}_i$ for each $r = 0, 1, \ldots, m$. In case we have a finite number of classes say $\mathfrak{X}_1, \ldots, \mathfrak{X}_n$ we also write

$$\mathfrak{X}_1 + \ldots + \mathfrak{X}_n \text{ for } \sum_{i \in \{1,\ldots,n\}} \mathfrak{X}_i.$$

Evidently the operation '+' on classes is commutative and associative. The following facts are easy to verify.

(a) $\sum_{i \in I} \mathfrak{X}_i = \mathrm{E}(\bigcup_{i \in I} \mathfrak{X}_i)$ is the smallest E-closed class containing all the $\mathfrak{X}_i$; so $\sum_{i \in I} \mathfrak{X}_i = \sum_{i \in I} \mathrm{E}\mathfrak{X}_i$.

(b) $\mathrm{A}(\sum_{i \in I} \mathfrak{X}_i) \leqq \sum_{i \in I} \mathrm{A}\mathfrak{X}_i$ for $\mathrm{A} = \mathrm{I}, \mathrm{Q}$, or S.

(c) $\mathrm{E}\mathfrak{A} \cap \sum_{i \in I} \mathfrak{X}_i = \sum_{i \in I} (\mathrm{E}\mathfrak{A} \cap \mathfrak{X}_i)$.

It is also clear that $\mathfrak{X} + \mathfrak{X} = \mathrm{E}\mathfrak{X}$ for any class $\mathfrak{X}$. We also have the analogue of theorem 3.2:

THEOREM 3.4 *Let* $(\mathfrak{X}_i : i \in I)$ *be any collection of* $\{\mathrm{I}, \mathrm{Q}\}$*-closed classes of Lie algebras. Then* $\sum_{i \in I} \mathfrak{X}_i$ *is coalescent if and only if* $\sum_{i \in I} (\mathrm{E}\mathfrak{A} \cap \mathfrak{X}_i)$ *is coalescent.* □

From this we deduce:

COROLLARY 3.5 *Let* $\mathfrak{X}_1, \ldots, \mathfrak{X}_n$ *be any of the classes*

Min, Min-si, Min-asc, Max, Max-si, Max-asc, $\mathfrak{G}^{\mathrm{I}}$.

Then over any field of characteristic zero,

$\mathfrak{X}_1 + \ldots + \mathfrak{X}_n$ *is coalescent.* □

We observe that Min-si + Max-si is the Lie-theoretic analogue of the class of minimax groups, which is itself coalescent as noted in the introduction.

4. The local coalescence of $\mathfrak{D}$

We say that a class $\mathfrak{X}$ is *locally coalescent* if and only if whenever H and K are $\mathfrak{X}$-subideals of a Lie algebra L, then to every finitely generated subalgebra C of $J = \langle H, K \rangle$ there corresponds an $\mathfrak{X}$-subideal X of L such that $C \leqq X \leqq J$. (For each C there may be several X's). See figure 1.

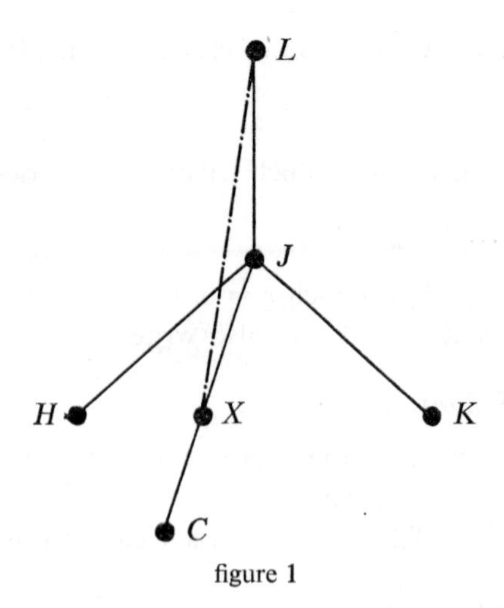

figure 1

Now every subalgebra of a nilpotent algebra is a subideal, and it follows from theorem 2.4 that the class $\mathfrak{N}$ is locally coalescent over fields of characteristic zero.

Similarly we say that $\mathfrak{X}$ is *locally ascendantly coalescent*. Then $\mathfrak{N}$ is locally ascendantly coalescent in characteristic zero.

Clearly if $\mathfrak{X}$ is a locally coalescent class then $\mathfrak{G} \cap \mathfrak{X}$ is coalescent.

A result of Roseblade and Stonehewer [337] states that if a class $\mathfrak{X}$ of groups is closed under the taking of subnormal subgroups and products of pairs of normal $\mathfrak{X}$-subgroups, then $\mathfrak{X}$ is a locally coalescent class. We have not been able to obtain such a sweeping result for Lie algebras. However we will prove:

THEOREM 4.1 *Over any field of characteristic zero the class $\mathfrak{D}$ of all Lie algebras is locally coalescent.*

By our remarks above and the fact the join of a pair of soluble subideals is always soluble we now have:

COROLLARY 4.2 *Over any field of characteristic zero the class* $\mathrm{E}\mathfrak{A}$ *is locally coalescent and the classes* $\mathfrak{G}$ *and* $\mathfrak{G} \cap \mathrm{E}\mathfrak{A}$ *are coalescent.* □

We denote by

$\mathfrak{C}$

the class of Lie algebras L such that $L/L^2 \in \mathfrak{G}$ (or equivalently $\mathfrak{F}$).

COROLLARY 4.3 *Suppose that* $\mathfrak{X}$ *is an* $\{\mathrm{I}, \mathrm{N}_0\}$*-closed and locally coalescent class over a field of characteristic zero. Let H and K be* $\mathfrak{X}$*-subideals of a Lie algebra L and let* $J = \langle H, K\rangle$.

If $J \in \mathfrak{C}$ *then* J si L *and* $J \in \mathfrak{C} \cap \mathfrak{X}$.
If $J^2 \in \mathfrak{C}$ *then* J si L, $J \in \mathfrak{X}$ *and* $J^2 \in \mathfrak{C} \cap \mathfrak{X}$.

In particular the classes $\mathfrak{C}$ *and* $\mathfrak{C} \cap \mathfrak{X}$ *are coalescent.*

Proof: Let $M \in \mathfrak{C}$ where M denotes J or J^2. Now there exists a finitely generated subalgebra C of M such that $M = C+M^2$. By the local coalescence of $\mathfrak{X}$ there exists an $\mathfrak{X}$-subideal X of L with $C \leqq X \leqq J$. Thus if $N = X \cap M$ then $N \lhd X$ si L and so N si L, $N \in \mathrm{I}\mathfrak{X} = \mathfrak{X}$ and $M = N+M^2$. From lemma 2.1.9 we have $M = N+M^{(r)}$ for all r. By corollary 2.2.17 we have $J^{(r)} \in \mathfrak{X}$ for some r and so $M^{(r)} \in \mathrm{I}\mathfrak{X} = \mathfrak{X}$. Finally by theorem 2.2.13 we have $M = N+M^{(r)} \in \mathfrak{X}$ and M si L (for $J^{(r)}$ and so $M^{(r)}$ si L for some r by the Derived Join Theorem). We also have by theorem 2.2.13 that $H+M$, $K+M \in \mathfrak{X}$ and $H+M$, $K+M$ si L, and so by the same result $J = H+M + + K+M \in \mathfrak{X}$ and J si L.

To get the coalescence of $\mathfrak{C}$ we put $\mathfrak{X} = \mathfrak{D}$ and note that if $H, K \in \mathfrak{C}$ then $J \in \mathfrak{C}$. □

We remark that the result above will hold if J is the join of $\mathfrak{X}$-subideals and $J \in \mathfrak{C}$; in particular it will hold if J is the join of finitely many $\mathfrak{X}$-subideals. The group-theoretic analogue is also true and part of it is proved in Roseblade [336].

We say that a class $\mathfrak{X}$ is *persistent* if in any Lie algebra the join of a pair of $\mathfrak{X}$-subideals is always an $\mathfrak{X}$-algebra.

So by the Derived Join Theorem the class $\mathrm{E}\mathfrak{A}$ is persistent. We deduce from theorem 4.1 that

COROLLARY 4.4 *Over any field of characteristic zero every* I*-closed and persistent class is locally coalescent.* □

Evidently a class $\mathfrak{X}$ is coalescent if and only if it is subjunctive and persistent.

LEMMA 4.5 *Let A be a Lie algebra with a finite chain*

$$A_n \leqq A_{n-1} \leqq \ldots \leqq A_1 \leqq A_0 = A$$

of subalgebras. If there exist subalgebras B_i *such that*

$$B_i \leqq A_i \text{ and } B_i \text{ si } A_{i-1} \qquad 1 \leqq i \leqq n,$$

Then $B = \bigcap_{i=1}^{n} B_i$ si A.

Proof: Use induction on n. □

Proof of theorem 4.1:

We want to establish the truth (for any Lie algebra L defined over a field of characteristic zero) of the statement:

If $H \lhd^m L$, $K \lhd^n L$, $J = \langle H, K \rangle$ and $C \in \mathfrak{C}$ and $C \leqq J$ then there exists X si L with $C \leqq X \leqq J$. (*)

We use induction on $m+n$. If $m+n \leqq 2$ or if $m = 0$ or $n = 0$ then clearly we may take J for X (by lemma 2.1.4). So assume that $m > 0$, $n > 0$, $m+n > 2$ and that the result is true for $m+n-1$ in place of $m+n$.

Now we consider some special cases.

Case 1: $L = A+J$ with $A^2 = 0 = A \cap J$ and $A \lhd L$.
Then $[A, H^m] \leqq [A, {}_mH] \leqq A \cap H = 0$. So $H^m \leqq C_L(A) \lhd L$ and hence $(H^m)^J \leqq C_L(A)$. Similarly $(K^n)^J \leqq C_L(A)$. Thus if $N = \langle H^m, K^n \rangle^J$ then $N \lhd L$. Now $\mathfrak{N}$ is locally coalescent and J/N is the join of two $\mathfrak{N}$-subideals of L/N. So there exists X/N si L/N with $(C+N)/N \leqq X/N \leqq J/N$. So X si L and $C \leqq X \leqq J$ and (*) is true in case 1.

Case 2: $L = B+J$, with $B^2 = 0$ and $B \lhd L$.
Then $M = B \cap J \lhd L$ and L/M satisfies the hypothesis of case 1. So we can find X/M si L/M with $(C+M)/M \leqq X/M \leqq J/M$. Then X si L and $C \leqq X \leqq J$ so (*) is true for case 2.

Case 3: $L^{(d)} = 0$ for some d.
Use induction on d. If $d = 0$ or 1 take J for X. Let $d > 1$ and assume that (*) is true for $d-1$ in place of d. Put $B = L^{(d-1)}$. By induction there exists Y/B si L/B with $(C+B)/B \leqq Y/B \leqq (J+B)/B$. So Y si L and $C \leqq Y \leqq J+B$. Now $B^2 = 0$ and so by case 2 there exists X_1 si $J+B$ with $C \leqq X_1 \leqq J$. Let $X = Y \cap X_1$. Then $C \leqq X$ and X si Y si L. So (*) is true in case 3 for all d.

Case 4: $J \in \mathrm{E}\mathfrak{A}$.
We have $m, n > 0$, so there exist M and N such that

$$H \lhd M \lhd^{m-1} L \text{ and } K \lhd N \lhd^{n-1} L.$$

Let $D = \langle H, N \rangle$ and $E = \langle M, K \rangle$. Using the Derived Join Theorem and the fact that $H, K \in \mathrm{E}\mathfrak{A}$, we can find r such that

$$D^{(r)} \leqq N \text{ and } E^{(r)} \leqq M.$$

Let $F = D \cap E$ so that $J \leqq F$ and $F^{(r)} \leqq M \cap N \leqq I_L(J)$. By the inductive hypothesis on $m+n-1$ we can find X_1 si L and X_2 si L such that

$$C \leqq X_1 \leqq D \text{ and } C \leqq X_2 \leqq E.$$

So $X_1 \cap X_2$ si L and $C \leqq X_1 \cap X_2 \leqq D \cap E = F$. Apply case 3 to $F/F^{(r)}$ to obtain a subideal $Y/F^{(r)}$ of $F/F^{(r)}$ such that

$$(C+F^{(r)})/F^{(r)} \leqq Y/F^{(r)} \leqq (J+F^{(r)})/F^{(r)}.$$

Now $F^{(r)}$ idealises J so $J \cap Y \lhd Y$ and so $J \cap Y$ si F. Define $X = X_1 \cap X_2 \cap J \cap Y$. Then $C \leqq X \leqq J$ and

$$X \text{ si } X_1 \cap X_2 \text{ si } L.$$

So (*) is true in case 4.

Case 5: the general case.
By the Derived Join Theorem there exists r such that $J^{(r)} \lhd^r L$. Let $D = J^{(r)}$ and let $D = D_r \lhd D_{r-1} \lhd \ldots \lhd D_0 = L$ be the ideal closure series of D in L. By Proposition 2.1.1 J idealises each D_i and so $D_i \lhd J+D_{i-1}$. Apply case 4 to the pair $(J+D_i)/D_i$, $(J+D_{i-1})D_i$. Then there exists X_i/D_i such that

$$X_i/D_i \text{ si } (J+D_{i-1})/D_i$$

and

$$(C+D_i)/D_i \leqq X_i/D_i \leqq (J+D_i)/D_i.$$

So X_i si $J+D_{i-1}$ and $C \leqq X_i \leqq J+D_i$ $(i = 1, \ldots, r)$. By lemma 4.5 we have $X = \bigcap_{i=1}^r X_i$ si $J+D_0 = L$ and clearly $C \leqq X \leqq J+D_r = J$.

This completes the induction and the proof of theorem 4.1. □

The proof of case 5 above enables us to deduce a generalisation of lemma 3.1.

PROPOSITION 4.6 *Let* $\mathfrak{X} = \{\mathrm{I}, \mathrm{Q}, \mathrm{E}\}\mathfrak{X}$ *be a class of Lie algebras. Then* $\mathfrak{X}$ *is subjunctive, persistent, coalescent or locally coalescent if and only if* $\mathfrak{X} \cap \mathrm{E}\mathfrak{A}$ *is subjunctive, persistent, coalescent or locally coalescent respectively.*

Proof: Since $\mathrm{E}\mathfrak{A}$ is persistent, then $\mathfrak{X} \cap \mathrm{E}\mathfrak{A}$ has the properties above whenever $\mathfrak{X}$ does. If $\mathfrak{X} \cap \mathrm{E}\mathfrak{A}$ is subjunctive, then so is $\mathfrak{X}$ (by lemma 3.1 since $\mathfrak{X}$ is Q-closed).

Suppose that $\mathfrak{X} \cap \mathrm{E}\mathfrak{A}$ is persistent or locally coalescent. Let L be any Lie algebra with H si L, K si L, $H, K \in \mathfrak{X}$ and $J = \langle H, K \rangle$. By the Derived Join Theorem there exists r such that $J^{(r)} \in \mathfrak{X}$ (for $\mathfrak{X} = \{\mathrm{I}, \mathrm{N}_0\}\mathfrak{X}$) and $J^{(r)} \lhd^r L$. Put $D = J^{(r)}$. Then J/D is the join of two $\mathfrak{X} \cap \mathrm{E}\mathfrak{A}$ subideals and so $J/D \in \mathfrak{X} \cap \mathrm{E}\mathfrak{A}$, in case $\mathfrak{X} \cap \mathrm{E}\mathfrak{A}$ is persistent. In the other case we consider the ideal closure series of D in L and let C be a finitely generated subalgebra of J. Since $(J+D_i)/D_i$ is the join of two $\mathfrak{X} \cap \mathrm{E}\mathfrak{A}$-subideals of $(J+D_{i-1})/D_i$, we can choose

$X_i/D_i \in \mathfrak{X} \cap \mathrm{E}\mathfrak{A}$ and such that $(C+D_i)/D_i \leqq X_i/D_i \leqq (J+D_i)/D_i$ and X_i/D_i si $(J+D_{i-1})/D_i$. So we have for $X = \bigcap_{i=1}^r X_i$, that X si L and $C \leqq X \leqq J$. But $D_r \in \mathfrak{X}$ and $X_r/D_r \in \mathfrak{X} \cap \mathrm{E}\mathfrak{A}$, so by E-closure we have $X_r \in \mathfrak{X}$. Finally $X \leqq X_r$ and X si X_r so $X \in \mathrm{I}\mathfrak{X} = \mathfrak{X}$.

If $\mathfrak{X} \cap \mathrm{E}\mathfrak{A}$ is coalescent then it is subjunctive and persistent and so from above $\mathfrak{X}$ is subjunctive and persistent and hence coalescent. □

The following result is now clear:

COROLLARY 4.7 *Let $(\mathfrak{X}_i : i \in I)$ be a collection of $\{\mathrm{I}, \mathrm{Q}\}$-closed classes of Lie algebras. Then $\sum_{i\in I} \mathfrak{X}_i$ is subjunctive, persistent, coalescent or locally coalescent if and only if $\sum_{i\in I}(\mathfrak{X}_i \cap \mathrm{E}\mathfrak{A})$ is subjunctive, persistent, coalescent or locally coalescent respectively.* □

In the next section we give an example to show that $\mathfrak{F}$ is not persistent over any field of non-zero characteristic. Hence (proposition 4.6) neither is $\mathfrak{F} \cap \mathrm{E}\mathfrak{A}$ over such fields and so (again by proposition 4.6) neither are any of the classes $\mathfrak{X}_1 + \ldots + \mathfrak{X}_n$ where the $\mathfrak{X}_i$'s come from the collection Min, Min-si, Min-asc, Max, Max-si, Max-asc, $\mathfrak{G}^1$.

We observed also that since $\mathrm{E}\mathfrak{A}$ is persistent then so is any $\{\mathrm{I}, \mathrm{Q}, \mathrm{E}\}$-closed class containing $\mathrm{E}\mathfrak{A}$; and over fields of characteristic zero such a class is locally coalescent.

5. A counterexample

Let p be a prime number. We saw in lemma 1.1 that over any field of characteristic p, the class $\mathfrak{F}$ is not subjunctive. Here we will give a counterexample (see Amayo [4]) to show that over any field of characteristic p, $\mathfrak{F}$ is not persistent. The example that we give was arrived at by looking at the general situation:

$$H, K \in \mathfrak{F},\ H \lhd^m L,\ K \lhd^n L \text{ and } J = \langle H, K\rangle.$$

Since $H^\omega \lhd L$ and $K^\omega \lhd L$ we may assume that $H, K \in \mathfrak{F} \cap \mathfrak{N}$. Thus $J \in \mathfrak{A}^d$ for some d. It turns out that if $d \leqq 2$, or $m \leqq 2$ or $n \leqq 2$, then $J \in \mathfrak{F}$ (see Amayo [2] for details).

Construction

Let p be a prime number and let $\mathfrak{k}$ be a field of characteristic p. Define W to be the vector space over $\mathfrak{k}$ with basis

$$\{a(m, n, r) : m, n, r \in \mathbb{Z}\}.$$

Define linear transformations x and y of W by

$$x: a(m, n, r) \leftrightarrow a(m+1, n, r) - na(m, n-1, r+1). \tag{2}$$

$$y: a(m, n, r) \leftrightarrow a(m, n+1, r). \tag{3}$$

Consider W as an abelian Lie algebra over $\mathfrak{k}$ so that $x, y \in \mathrm{Der}(W)$. It follows easily from (2) and (3) that

$$z = [x, y] = xy - yx: a(m, n, r) \leftrightarrow a(m, n, r+1) \tag{4}$$

and

$$[x, z] = xz - zx = 0 = yz - zy = [y, z]. \tag{5}$$

It is clear that if $m, n, r \geqq 0$ then $x^m, y^n, z^r \in \mathrm{Der}(W)$. We have

$$y^n x = xy^n - ny^{n-1}z \tag{6}$$

and

$$y^n x^m = \sum_{r=0}^{\infty}(-1)^r(r!)\binom{m}{r}\binom{n}{r}x^{m-r}y^{n-r}z^r, \tag{7}$$

where $\binom{i}{j}$ is the binomial coefficient and we take x^{m-r} to be zero if $r > m$, and y^{n-r} to be zero if $r > n$.

Equation (6) will follow by induction on n and noting from (4) that $z = xy - yx$ and from (5) that every power of z commutes with every power of x or y.

Now (7) is obviously true for $m = 0$. If it is true for some m then using (6) we have

$$\begin{aligned} y^n x^{m+1} &= (y^n x^m)x \\ &= \sum_{r=0}^{\infty}(-1)^r(r!)\binom{m}{r}\binom{n}{r}x^{m-r}(xy^{n-r} - (n-r)y^{n-r-1}z)z^r. \end{aligned}$$

We note that

$$x^{m-r}y^{n-r-1}z^{r+1} = x^{m+1-(r+1)}y^{n-(r+1)}z^{r+1}$$

and that the coefficient of $x^{m+1-r}y^{n-r}z^r$ in the summation above is

$$(-1)^r(r!)\binom{m}{r}\binom{n}{r} - (-1)^{r-1}((r-1)!)\binom{m}{r-1}\binom{n}{r-1}(n-r+1) =$$

$$= (-1)^r\binom{m+1}{r}\binom{n}{r}(r!).$$

Therefore

$$y^n x^{m+1} = \sum_{r=0}^{\infty}(-1)^r(r!)\binom{m+1}{r}\binom{n}{r}x^{m+1-r}y^{n-r}z^r,$$

which is of the same form as (7). So (7) is true for all $m, n \geqq 0$. In particular as $\mathfrak{k}$ has characteristic p we have

$$y^n x^p = x^p y^n \text{ and } y^p x^m = x^m y^p$$

for all $m, n \geqq 0$. Thus for all $m, n, r \geqq 0$ we have

$$x^m y^n z^r x^p = x^{m+p} y^n z^r; \; y^p x^m y^n z^r = x^m y^{n+p} z^r. \tag{8}$$

Now let U be the subspace of W spanned by the basis elements

$$\{a(m, n, r) : m, n, r \geqq 0\}$$

and let V be the subspace of U spanned by the basis elements

$$\{a(m, n, r) : r \geqq 0, \text{ and } m \geqq p \text{ or } n \geqq p\}.$$

It follows from (2) and (3) that $Ux, Uy \subseteq U$ and so $Uz \subseteq U$. We note that if $a(m, n, r)$ is a basis element for U then

$$a(m, n, r) = a(0, 0, 0)x^m y^n z^r. \tag{9}$$

From this and equations (2), (3) and (8) we get $Vx, Vy \subseteq V$ and so $Vz \subseteq V$.
Let $J = \langle x, y\rangle \leqq \mathrm{Der}(W)$.
It follows from (5) that $J = \langle x\rangle + \langle y\rangle + \langle z\rangle$ and $J \in \mathfrak{F}_3 \cap \mathfrak{N}_2$. We have U and V as J-submodules of W and $Ux^p, Uy^p \subseteq V$. Define A to be the quotient module U/V and let

$$a = a(0, 0, 0) + V.$$

Then

$$Ax^p = Ay^p = 0.$$

Let B be an isomorphic copy of A and make it into a J-module via this isomorphism. Then $Bx^p = By^p = 0$. Let b be the image of a under the isomorphism $A \cong B$. Define C to be the vector space direct sum

$$C = A \oplus B$$

and make C into a J-module via the action on A and B. Then

$$Cx^p = Cy^p = 0 \tag{10}$$

and every element of C is a linear combination of the basis elements $ax^m y^n z^r$, $bx^{m'} y^{n'} z^{r'}$ where $r, r' > 0$ and $0 \leqq m, n, m', n' \leqq p-1$.

We consider C as an abelian Lie algebra and form the split extension

$$L = C \dotplus J, \; C \lhd L, \; C \cap J = 0.$$

Let $X = \langle x\rangle$, $Y = \langle y\rangle$ and $Z = \langle z\rangle$. Put $H = \langle a, x\rangle$, $K = \langle a, y\rangle$ and $M = \langle H, K\rangle$. Evidently

$$L = \langle a, b, x, y\rangle \text{ and } L \in \mathfrak{G}_4 \cap (\mathfrak{A}(\mathfrak{F}_3 \cap \mathfrak{N}_2)).$$

We have

$$C+H = C+X \lhd C+X+Z \lhd C+X+Y+Z = L$$

and

$$C+K = C+Y \lhd C+Y+Z \lhd C+Y+X+Z = L.$$

Now $C^2 = 0$ so we have $(C+X)^{p+1} = Cx^p = 0$ (by (10)).

Thus $C+X \in \mathfrak{N}_p$ and so $H \lhd^p C+X \lhd^2 L$. Similarly we have $K \lhd^{p+2} L$. Evidently $H = \langle a, ax, \ldots, ax^{p-1}\rangle + X \in \mathfrak{F}_{p+1} \cap \mathfrak{N}_p$ and similarly $K \in \mathfrak{F}_{p+1} \cap \mathfrak{N}_p$. We have $M = \langle H, K\rangle = \langle a, x, y\rangle = A+J$ and $M \notin \mathfrak{F}$ since the elements $a, az, az^2, \ldots$ are linearly independent and contained in $A \subseteq M$. Furthermore $A \lhd M$ and $Az^r \lhd M$ for all $r > 0$. So M has a strict descending chain $A > Az > Az^2 > \ldots$ of ideals. Thus $M \notin \text{Min-}\lhd$. We also have $M \in \mathfrak{G}_3 \cap (\mathfrak{A}\mathfrak{N}_2)$.

Finally we will show that

$$I_L(M) = M.$$

Now $L = C+J = B+M$ and clearly $B \cap M = 0$ (for $C = A \oplus B$ and $M = A+J$ and $C \cap J = 0$); and $B \lhd L$.

Let $u+w \in I_L(M)$, where $u \in B$ and $w \in M$. Then $u = (u+w)-w \in I_L(M) \cap B$. So $[u, z] = uz \in B \cap M = 0$. Hence $u = 0$, since the action of z on A and so B, is such that no non-zero element is mapped to zero (by equation (4) and the definition of A). Thus $I_L(M) \leqq M$.

Gathering the results above together we can now state:

THEOREM 5.1 *Let $\mathfrak{k}$ be a field of characteristic $p > 0$. Then there is a Lie algebra L defined over $\mathfrak{k}$ such that*

(a) $L \in \mathfrak{G}_4 \cap (\mathfrak{A}(\mathfrak{F}_3 \cap \mathfrak{N}_2))$;

(b) *there exist subalgebras H and K of L with $H, K \in \mathfrak{F}_{p+1} \cap \mathfrak{N}_p$ and $H \lhd^{p+2} L$, $K \lhd^{p+2} L$;*

(c) *if $M = \langle H, K\rangle$ then $M \notin \mathfrak{F}$, $M \notin \text{Min-}\lhd$ and $I_L(M) = M$ so that M is not a subideal of L.* □

We have now shown that over any field of characteristic $p > 0$, the classes Min-$\lhd$ and $\mathfrak{F}$ are neither subjunctive nor persistent; hence neither are the classes Min, Min-si, …, occuring in theorem 3.3.

It is not very hard to show that if $L \in \mathfrak{G}$ and $L^2 > L^3$ then (in characteristic $p > 0$) L can be embedded in a Lie algebra L_1 in such a way that there is some ideal A of L_1 contained in L, modulo which L is the join of two $\mathfrak{F}$-subideals of L_1 and yet $I_{L_1}(L) = L$ (see Amayo [2] for details).

In the notation of our construction above we note that A, B and C are faithful but not irreducible J-modules. Let $U(J)$ be the universal enveloping algebra of J (see [98]) so that U, A, B are all cyclic and unital $U(J)$-modules. Then

$$U \cong U(J) \quad \text{as } U(J)\text{-modules}$$

under the map

$$a(m, n, r) \leftrightarrow x^m y^n z^r.$$

Let $d = a(0, 0, 0) - a(0, 0, 1)$ and let $D = dU(J)$. Then D is a $U(J)$-submodule of U and (by equation (7))

$$U/D \cong A_1,$$

where A_1 is the *Weyl algebra* (see McConnell [160]); the ring of polynomials $\mathfrak{k}[x][y]$ over $\mathfrak{k}[x]$ in an indeterminate y subject to the relation

$$fy - yf = \mathrm{d}f/\mathrm{d}x$$

for all $f \in \mathfrak{k}[x]$.

Now Max-$\lhd$ $\leq \mathfrak{C}$ and so over fields of characteristic zero Max-$\lhd$ is subjunctive (by corollary 4.3). Our example above also shows that over fields of characteristic $p > 0$, Max-$\lhd$ is not subjunctive. However we have not been able to decide whether or not Max-$\lhd$ is a persistent class (either in characteristic zero or $p > 0$). □

Chapter 4

Locally coalescent classes of Lie algebras

The concept of 'local coalescence' was first introduced by Roseblade and Stonehewer in [337], where they proved that any class of groups closed under the taking of subnormal subgroups and finite products of normal subgroups is locally coalescent. In section 3.4 we remarked that over fields of characteristic zero the class $\mathfrak{N}$ is locally coalescent and proved that the class $\mathfrak{D}$ of all groups is locally coalescent. Here we show that over fields of characteristic zero several other useful classes are locally coalescent: in particular $\mathfrak{Z}$, $\mathrm{L}\mathfrak{N}$, $\mathrm{L}\mathfrak{F}$. Our methods give us new proofs of the coalescence of $\mathfrak{F} \cap \mathfrak{N}$, $\mathfrak{F}$ and $\mathfrak{G}$ in characteristic zero.

The proof depends on the technique of 'formal power series algebras' due to Hartley [76], and the first section is devoted to setting up this technique.

1. The algebra of formal power series

Let $\mathfrak{k}$ be a field of characteristic zero and let $\mathfrak{k}_0 = \mathfrak{k}\langle t\rangle$ be the set of formal power series in the indeterminate t. A typical element α of $\mathfrak{k}_0$ has the form

$$\alpha = \sum_{r=n}^{\infty} \alpha_r t^r, \qquad n = n(\alpha) \in \mathbb{Z},\ \alpha_r \in \mathfrak{k}. \tag{1}$$

Let $\beta = \sum \beta_r t^r \in \mathfrak{k}_0$. We make $\mathfrak{k}_0$ into a field by defining: $\alpha+\beta = \sum(\alpha_r+\beta_r)t^r$; $\alpha\beta = \sum \gamma_r t^r$, where $\gamma_r = \sum_{i+j=r} \alpha_i\beta_j$; if $\alpha \neq 0$ and $\alpha = \alpha_n t^n + \ldots$ where $\alpha_n \neq 0$, then $\alpha^{-1} = \sum_{r=-n}^{\infty} \beta_r t^r$ where $\beta_{-n} = \alpha_n^{-1}$ and inductively β_{-n+j} is obtained by solving

$$\alpha_n\beta_{-n+j} + \alpha_{n+1}\beta_{-n+j-1} + \ldots + \alpha_{n+j}\beta_{-n} = 0, \text{ for } j > 0.$$

We have $\mathfrak{k}$ naturally embedded in $\mathfrak{k}_0$ by defining for $\alpha \in \mathfrak{k}$, $\alpha \mapsto \alpha + 0t + 0t^2 + \ldots$; and we identify $\mathfrak{k}$ with its image in $\mathfrak{k}_0$.

Now let A be a vector space over $\mathfrak{k}$, and let $A^{\dagger}$ be the set of formal power series

$$a = \sum_{r=n}^{\infty} a_r t^r, \qquad n = n(a) \in \mathbb{Z},\ a_r \in A. \tag{2}$$

We make $A^\dagger$ into a vector space over $\mathfrak{k}_0$ as follows. Let $b = \sum_{r=m}^{\infty} b_r t^r \in A^\dagger$ and let $\alpha \in \mathfrak{k}_0$. We define

$$a+b = \sum (a_r+b_r)t^r \tag{3}$$

and

$$\alpha a = \sum c_r t^r, \text{ where } c_r = \sum_{i+j=r} \alpha_i a_j. \tag{4}$$

It is easy to verify that (3) and (4) make $A^\dagger$ a vector space over $\mathfrak{k}_0$. If $a \in A$ we write $a = a+0t+\ldots$, so that A is embedded as a subset of $A^\dagger$.

If B is a subset of A we write $B^\dagger$ for the set of all $b = \sum b_r t^r$ such that each $b_r \in B$. In case B is a subspace of A then clearly $B^\dagger$ is a ($\mathfrak{k}_0$-) subspace of $A^\dagger$.

Suppose that B is a subspace of A and consider the quotient space A/B and the formal space $(A/B)^\dagger$. Then the map $A^\dagger \to (A/B)^\dagger$:

$$a = \sum a_r t^r \mapsto \sum (a_r+B)t^r$$

is a $\mathfrak{k}_0$-linear map with kernel $B^\dagger$. Thus

$$A^\dagger/B^\dagger \cong (A/B)^\dagger. \tag{5}$$

Suppose that A is finite-dimensional over $\mathfrak{k}$ with basis $u_1, \ldots, u_q$. Let $a = \sum a_r t^r \in A^\dagger$: for each r we have $a_r = \lambda_{r1}u_1+\ldots+\lambda_{rq}u_q$, where the $\lambda_{ri} \in \mathfrak{k}$. Hence

$$a = \alpha_1 u_1+\ldots+\alpha_q u_q, \text{ where } \alpha_i = \sum_r \lambda_{ri} t^r \in \mathfrak{k}_0. \tag{*}$$

Thus $A^\dagger$ is finite-dimensional over $\mathfrak{k}_0$ and has $u_1, \ldots, u_q$ as a basis. On the other hand if A is infinite dimensional and $u_0, u_1, \ldots$ form part of a basis for A then $u_0+u_1t+\ldots$, has no finite expression as in (*).

Now let L be a Lie algebra over $\mathfrak{k}$. Define $L^\dagger$ as above and make it a $\mathfrak{k}_0$-vector space using (3) and (4). Extend the Lie product [,] on L to $L^\dagger$ as follows: if $x, y \in L^\dagger$, with $x = \sum x_r t^r$ and $y = \sum y_r t^r$, where $x_r, y_r \in L$ for all r, then

$$[x, y] = \sum z_r t^r, \text{ where } z_r = \sum_{i+j=r} [x_i, y_j]. \tag{6}$$

It is fairly easy to verify that equations (3), (4) and (6) make $L^\dagger$ a Lie algebra over $\mathfrak{k}_0$; and L is embedded in $L^\dagger$ as a subset.

Suppose that L is a Lie algebra over $\mathfrak{k}$, A is an L-module and B is a submodule of A. Then we make $A^\dagger$ into an $L^\dagger$-module by defining, for $a =$

$= \sum a_r t^r \in A^\dagger$ and $x = \sum x_r t^r \in L^\dagger$,

$$ax = \sum c_r t^r, \text{ where } c_r = \sum_{i+j=r} a_i x_j. \tag{7}$$

Equations (6) and (7) ensure that $A^\dagger$ is indeed an $L^\dagger$-module; we also have $B^\dagger$ as a submodule of $A^\dagger$ and in this case the isomorphism of equation (5) is an L-module isomorphism. If we define $C_L(A/B) = \{v \in L : Av \subseteq B\}$ then we have

$$(C_L(A/B))^\dagger = C_{L^\dagger}((A/B)^\dagger) = C_{L^\dagger}(A^\dagger/B^\dagger), \tag{8}$$

using (5) and (7).

From equations (1)–(8) we deduce:

LEMMA 1.1 *Let $\mathfrak{k}$ be a field of characteristic zero. Then*

(a) *If A is a vector space over $\mathfrak{k}$ and B is a subspace of finite codimension m, then $B^\dagger$ is a $\mathfrak{k}_0$-subspace of $A^\dagger$ of codimension m.*

(b) *If L is a Lie algebra over $\mathfrak{k}$ and $K \lhd H \leqq L$ then $K^\dagger \lhd H^\dagger \leqq L^\dagger$. In particular if $H \lhd^m L$ then $H^\dagger \lhd^m L^\dagger$.*

(c) *If L is a Lie algebra over $\mathfrak{k}$ and A and B are subsets of L then $(A \cup B)^\dagger \supseteq A^\dagger \cup B^\dagger$ and $(A+B)^\dagger = A^\dagger + B^\dagger$.*

(d) *If L is a Lie algebra over $\mathfrak{k}$ and $\{A_i : i \in I\}$ is a family of subsets of L then $(\bigcap_{i\in I} A_i)^\dagger = \bigcap_{i\in I} A_i^\dagger$ and $\bigcup_{i\in I} A_i^\dagger \subseteq (\bigcup_{i\in I} A_i)^\dagger$.*

(e) *Let L be a Lie algebra over $\mathfrak{k}$. If A, B are subsets of L then $[A^\dagger, B^\dagger] \subseteq$ $\subseteq [A, B]^\dagger$. In particular if $H \leqq L$ then for every non-negative integer n, $H^{\dagger n+1} \subseteq H^{n+1\dagger}$ and $H^{\dagger(n)} \subseteq H^{(n)\dagger}$.*

(f) *$L \in \mathfrak{N}_c$ if and only if $L^\dagger \in \mathfrak{N}_c$; and $L \in \mathfrak{A}^d$ if and only if $L^\dagger \in \mathfrak{A}^d$ for any positive integers c, d.*

(g) *If L is a Lie algebra over $\mathfrak{k}$ then for any non-negative integer m, $\zeta_m(L^\dagger) = \zeta_m(L)^\dagger$; for any ordinal α, $\zeta_\alpha(L^\dagger) \leqq \zeta_\alpha(L)^\dagger$.* □

Let L be a Lie algebra over $\mathfrak{k}$ and let $a \neq 0$, $a \in L^\dagger$. Then a can be written uniquely in the form

$$a = t^m \sum_{r=0}^{\infty} a_r t^r, \text{ where } m = m(a) \in \mathbb{Z} \text{ and } a_0 \neq 0$$

is the first non-zero coefficient of a. We will call a_0 the *first coefficient* of a. Clearly every non-zero element α of the field $\mathfrak{k}_0$ has a similar expression: $\alpha = t^k \sum_{r=0}^{\infty} \alpha_r t^r$, $\alpha_0 \neq 0$, and we call α_0 the *first coefficient* of α.

Let $b \neq 0$, $b \in L^\dagger$ so that $b = t^n \sum_{r=0}^{\infty} b_r t^r$ where $b_0 \neq 0$. Clearly

$$[a, b] = t^{m+n}[a_0, b_0] + t^{m+n} \sum_{r=1}^{\infty} (\sum_{i+j=r} [a_i, b_j]) t^r, \tag{9}$$

and for any $\gamma, \delta \in \mathfrak{k}$,

$$\gamma a + t^{m-n}\delta b = t^m \sum_{r=0}^{\infty} (\gamma a_r + \delta b_r) t^r. \tag{10}$$

Suppose that $u_1, \ldots, u_{p+1}$ are non-zero elements of $L^\uparrow$ and $u_i = x_i + \sum_{r=1}^{\infty} u_{ir}t^r$, where $0 \neq x_i \in L$ $(i = 1, \ldots, p+1)$. Let $q \leqq p+1$ and let $\lambda_1, \ldots, \lambda_q$ be non-zero elements of $\mathfrak{k}_0$ so that $\lambda_i = t^{k(i)} \sum_{r=0}^{\infty} \lambda_{ir}t^r$, $0 \neq \lambda_{i0} \in \mathfrak{k}$ $(i = 1, \ldots, q)$; if $q < p+1$, put $\lambda_i = 0$ for all $i > q$. Now let $k = \min\{k(1), \ldots, k(q)\}$. Then clearly we have

$$\lambda_1 u_1 + \ldots + \lambda_{p+1} u_{p+1} = t^k \sum_{k(i)=k} \lambda_{i0} x_i + t^{k+1} w, \tag{**}$$

with w of the form $w = \sum_{r=0}^{\infty} w_r$ and the w_r certain ($\mathfrak{k}$-linear) combinations of the x_i and u_{ij}. Thus if $u_1, \ldots, u_{p+1}$ are linearly dependent over $\mathfrak{k}_0$ then by renaming the u_i if necessary and using (**) we find that $x_1, \ldots, x_{p+1}$ are linearly dependent over $\mathfrak{k}$.

Now let M be a subset of $L^\uparrow$. We define a subset $M^\downarrow$ of L by:

$$M^\downarrow = \{x \in L : x = 0 \text{ or } x \text{ is the first coefficient of some element of } M\}.$$

Equation (10) shows that if M is a ($\mathfrak{k}_0$-) subspace of $L^\uparrow$ then M is a ($\mathfrak{k}$-) subspace of L.

Suppose that M is a subspace of $L^\uparrow$ and N is a subspace of M of codimension $p > 0$. Then $N^\downarrow$ is a subspace of $M^\downarrow$ of codimension not exceeding p. For this it is enough to show that if $x_1, \ldots, x_{p+1}$ are any non-zero elements of $M^\downarrow$ then some non-trivial linear combination of the x_i belongs to $N^\downarrow$. Let x_i be the first coefficient of $a_i = t^{m(i)} \sum_{r=0}^{\infty} a_{ir}t^r \in M$; put $u_i = t^{-m(i)}a_i \in M$, so that x_i is the first coefficient of $u_i = x_i + \sum_{r=1}^{\infty} u_{ir}t^r$ $(u_{ir} = a_{ir})$. Since N has codimension p in M we can find $\lambda_1, \ldots, \lambda_{p+1} \in \mathfrak{k}_0$ and not all zero such that $\lambda_1 u_1 + \ldots + \lambda_{p+1} u_{p+1} \in N$. By renaming the u_i if necessary we may assume that for some q, $1 \leqq q \leqq p+1$, the first q of the λ_i are non-zero. Using (**) we see that some non-trivial $\mathfrak{k}$-linear combination of the x_i is either zero or the first coefficient of an element of N; i.e. this combination is in $N^\downarrow$.

Using (9), (10) and (**) we can deduce:

LEMMA 1.2 *Let $\mathfrak{k}$ be a field of characteristic zero and let L be a Lie algebra defined over $\mathfrak{k}$. Then*

(a) *If M is a subspace (resp. subalgebra) of $L^\uparrow$ then $M^\downarrow$ is a subspace (resp. subalgebra) of L. If N is a subspace of M then* $\dim_{\mathfrak{k}}(M^\downarrow/N^\downarrow) \leqq \dim_{\mathfrak{k}_0}(M/N)$.

(b) *If $N \lhd M \leqq L$, then $N^\downarrow \lhd M^\downarrow \leqq L$. In particular if $N \lhd^n L$ then $N^\downarrow \lhd^n L$.*

(c) *Let M and N be subsets of L. Then* $[M^\downarrow, N^\downarrow] \subseteq [M, N]^\downarrow$; $(M \cap N)^\downarrow \subseteq M^\downarrow \cap N^\downarrow$; $M^\downarrow + N^\downarrow \subseteq (M+N)^\downarrow$.

(d) *If $M \leqq L^\uparrow$ then for each non-negative integer n, $M^{\downarrow n+1} \subseteq M^{n+1\downarrow}$ and $M^{\downarrow(n)} \subseteq M^{(n)\downarrow}$.*

(e) *Let $\mathfrak{X}$ be any one of the classes $\mathfrak{N}_c$, $\mathfrak{A}^d$, $\text{L}\mathfrak{F}$, $\text{L}\mathfrak{N}$, $\text{LE}\mathfrak{A}$. If $M \leqq L^\uparrow$ and $M \in \mathfrak{X}$ (as a $\mathfrak{k}_0$-algebra) then $M^\downarrow \in \mathfrak{X}$ (as a $\mathfrak{k}$-algebra).*

(f) *If A is a subspace of L then $A^{\uparrow\downarrow} = A$; if M is any subspace of L such that $A^\uparrow \subseteq M$ and $A = M^\downarrow$, then $A^\uparrow = M$.*

(g) *If $\{M_i : i \in I\}$ is any family of subsets of L then*

$$\bigcup_{i\in I} M_i^\downarrow = (\bigcup_{i\in I} M_i)^\downarrow.$$

(h) *If $M \leqq L^\uparrow$, then for any ordinal α, $(\zeta_\alpha(M))^\downarrow \leqq \zeta_\alpha(M^\downarrow)$. In particular if $M \in \mathfrak{Z}$ then $M^\downarrow \in \mathfrak{Z}$.* □

There are several other properties that can be derived but the ones above are enough for our purposes. We remark that if M is a subspace of $L^\uparrow$ and M^* is the set of elements of M of the form $a = \sum_{r=0}^{\infty} a_r t^r$, then $M^\downarrow = M^{*\downarrow}$.

Let A be a vector space over $\mathfrak{k}$ with basis $a_1, a_2, \ldots$. Consider A as an abelian Lie algebra over $\mathfrak{k}$ and let d be the derivation of A defined by: $d: a_1 \mapsto 0$, $a_{i+1} \mapsto a_i$. Form the split extension $L = A \dotplus \langle d\rangle$. Then L is a $\mathfrak{Z}$-algebra of upper central height $\omega+1$. Yet $L^\uparrow \notin \text{L}\mathfrak{N}$ (consider $x = a_1 t + a_2 t^2 + \ldots$, and d; then $[x, {}_m d] \neq 0$ for any m).

Exponential mapping

Let L be a Lie algebra over the field $\mathfrak{k}$ of characteristic zero. Let d be a derivation of L. Define a mapping $\exp(td)$ of $L^\uparrow$ as follows: let $e = \exp(td)$ and let $x = \sum x_r t^r \in L^\uparrow$. Then

$$x^{\exp(td)} = \sum u_r t^r, \text{ where } u_r = \sum_{i+j=r} x_i d^j/j!. \tag{11}$$

Now let $\lambda = \sum \lambda_r t^r \in \mathfrak{k}_0$ and $y = \sum y_r t^r \in L^\uparrow$. Clearly

$$(x+y)^e = x^e + y^e;$$

and

$$\begin{aligned}(\lambda x)^e &= \textstyle\sum_r (\sum_{i+j=r} (\lambda x)_i d^j/j!) t^r = \sum_r (\sum_{l+k+j=r} \lambda_l x_k d^j/j!) t^r \\ &= \textstyle\sum_r (\sum_{l+p=r} \lambda_l (x^e)_p) t^r = \lambda(x^e).\end{aligned}$$

We also have

$$\begin{aligned}[x^e, y^e] &= \textstyle\sum_r (\sum_{i+j+k+l=r} [x_i d^j/j!, y_k d^l/l!]) t^r \\ &= \textstyle\sum_r (\sum_{i+k+p=r} [x_i, y_k] d^p/p!) t^r \qquad \text{(by Leibniz)} \\ &= \textstyle\sum_r (\sum_{n+p=r} ([x, y])_n d^p/p!) t^r \\ &= [x, y]^e.\end{aligned}$$

Hence $\exp(td)$ is a Lie homomorphism of $L^\dagger$. We note that if d_1, d_2 are derivations of L then

$$[d_1, d_2] = 0 \text{ implies } \exp(t(d_1+d_2)) = \exp(td_1)\exp(td_2).$$

In particular since $\exp(t.0) = \text{identity}$, then $\exp(t(-d))$ is the inverse of $\exp(td)$. Hence $\exp(td)$ is a Lie automorphism of $L^\dagger$. We note that if $E = [\text{End}_{\mathfrak{k}}(L)]$, the vector space of all linear transformations of L considered as a Lie algebra with the usual product, then L is an E-module. Thus $L^\dagger$ is an $E^\dagger$-module and $\exp(td)$ can be considered as an element of $E^\dagger$:

$$\begin{aligned}\exp(td) &= 1+dt+d^2t^2/2!+\ldots \\ &= \textstyle\sum_{r=0}^{\infty} d^r t^r/r!.\end{aligned}$$

Furthermore equation (11) is exactly the same as (7), which defines $L^\dagger$ as an $E^\dagger$-module.

If $A \subseteq L$ we denote by $\exp(tA)$ the group of automorphisms of $L^\dagger$ generated by the elements $\exp(tx^*)$, where $x \in A$ and x^* is the adjoint map of L induced by x. So

$$\exp(tA) = \text{group}\langle \exp(tx^*) : x \in A\rangle \subseteq E^\dagger, \tag{12}$$

for E is an E-module, so $E^\dagger$ is an $E^\dagger$-module in the natural way.

Extension of algebras

We recall (see Jacobson [98] p.26–28) that if L is a Lie algebra and C is a commutative and associative algebra over a field $\mathfrak{k}$, then the tensor product $L \otimes_{\mathfrak{k}} C$ can be made into a Lie algebra over $\mathfrak{k}$ as follows: if $x, y \in L$ and $a, b \in C$ then define

$$[xa, yb] = [x, y]ab$$

(where $xa = x \otimes a$, etc.). Furthermore the elements of $L \otimes_{\mathfrak{k}} C$ are finite sums of the form

$$x_1a_1+\ldots+x_ma_m, \quad x_i \in L,\ a_i \in C.$$

Now let A be a vector space over the field $\mathfrak{k}$ of characteristic zero and define

$$A^\wedge = A \otimes_{\mathfrak{k}} \mathfrak{k}_0.$$

For the sake of convenience identify $A^\wedge$ with $\mathfrak{k}_0 \otimes_{\mathfrak{k}} A$, so that a typical element x of $A^\wedge$ has the form

$$x = \alpha_1a_1+\ldots+\alpha_ma_m, \text{ where } \alpha_i \in \mathfrak{k}_0,\ a_i \in A. \tag{13}$$

Let $\alpha = \sum \alpha_r t^r \in \mathfrak{k}_0$ and let $a \in A$. We can consider $A^\wedge$ as embedded in $A^\uparrow$ under the natural map $\alpha a \leftrightarrow \sum (\alpha_r a) t^r$.

Hence $A^\wedge$ *consists precisely of those elements* $x = \sum x_r t^r$ *of* $A^\uparrow$, *all of whose coefficients* x_r *lie in some finite dimensional subspace of* A. In other words, if $\{A_i : i \in I\}$ is the family of all finite-dimensional subspaces of A then

$$A^\wedge = \bigcup_{i \in I} A_i^\uparrow = \sum_{i \in I} A_i^\uparrow.$$

So $A^\wedge$ is a ($\mathfrak{k}_0$-) vector subspace of $A^\uparrow$.

If L is a Lie algebra over $\mathfrak{k}$ we define $L^\wedge$ similarly and it is clear that the Lie structures obtained from above and by equation (6) are the same. Thus $L^\wedge$ is a subalgebra of $L^\uparrow$. If $A \subseteq L$ we put $A^\wedge = \{x = \sum x_r t^r \in L^\wedge : \text{each } x_r \in A\}$. Then $A^\wedge \subseteq A^\uparrow$; and $A^\wedge = A^\uparrow$ whenever A is a finite subset or a finite dimensional subspace of L.

If $M \subseteq L^\wedge$, we put $M^\vee = \{x \in L : x = 0 \text{ or } x \text{ is the first coefficient of some element of } M\}$; then $M^\vee = M^\downarrow$.

The analogues of lemmas 1.1 and 1.2 hold if we replace '$\uparrow, \downarrow$' by '$\wedge, \vee$'. Clearly if $\{A_i : i \in I\}$ is any family of subspaces of L then

$$\left(\bigcup_{i \in I} A_i\right)^\wedge = \bigcup_{i \in I} A_i^\wedge.$$

In particular for any ordinal λ,

$$\zeta_\lambda(L^\wedge) = \zeta_\lambda(L)^\wedge, \tag{14}$$

and so $L \in \mathfrak{Z}$ implies $L^\wedge \in \mathfrak{Z}$ (as a $\mathfrak{k}_0$-algebra).

Derivations

Let $d, d_1, \ldots, d_s$ be derivations of the Lie algebra L. Put $e = \exp(td)$ and $e_i = \exp(td_i)$, $i = 1, \ldots, s$. Let $z = t^n \sum_{r=0}^\infty z_r t^r \in L^\uparrow$, where $z_0 \neq 0$ is the first coefficient of z. Then we have

$$z^e - z = t^n\left(z_0 dt + \sum_{r=2}^\infty \left(\left(\sum_{i+j=r} z_i d^j / j!\right) - z_r\right) t^r\right). \tag{15}$$

Thus

$$z_0 d \in \{z^e - z\}^\downarrow, \tag{16}$$

i.e. $z_0 d$ is either zero or the first coefficient of $z^e - z$. Define for each m, $1 \leqq m \leqq s$,

$$w_m = \sum_{1 \leqq i_1 < \ldots < i_k \leqq m;\, 0 \leqq k \leqq m} (-1)^{m-k} z^{e_{i_1} \ldots e_{i_k}}. \tag{17}$$

It is clear that if $m > 0$ then

$$w_{m-1}^{e_m} - w_{m-1} = w_m. \tag{***}$$

LEMMA 1.3 *Let L be a Lie algebra over a field $\mathfrak{k}$ of characteristic zero and let $d_1, \ldots, d_s$ be derivations of L. Then*

(a) *If $z \in L^{\uparrow}$ and w_s is the element defined in equation* (17) *then $z_0 d_1 \ldots d_s \in \{w_s\}^{\downarrow}$ whenever $z_0 \in \{z\}^{\downarrow}$.*

(b) *Let $p = 2^s$ and let $\{e_1^*, \ldots, e_p^*\} = \{e_{i_1} \ldots e_{i_k} : 0 \leqq k \leqq s$ and $1 \leqq i_1 < \ldots < i_k \leqq s$, where $e_i = \exp(td_i)\}$. If $M \subseteq L^{\uparrow}$ then*

$$M^{\downarrow} d_1 \ldots d_s \subseteq (M^{e_1^*} + \ldots + M^{e_p^*})^{\downarrow}.$$

Proof: (a) If $z = 0$ there is nothing to prove. Suppose that $z \neq 0$. For $s = 1$, the result follows from equation (16). If the result is true for $s-1$, then it follows from (16) and (***) that the result holds for s. This proves (a).

(b) is an immediate consequence of (17), (a) and the definition of $\{e_1^*, \ldots, e_p^*\}$. □

Remark 1. If $M \subseteq L^{\uparrow}$ then $M^{\downarrow} = \bigcup_{z \in M} \{z\}^{\downarrow}$.

COROLLARY 1.4 *Let L be a Lie algebra over a field $\mathfrak{k}$ of characteristic zero. If $A \subseteq L$ and $B \subseteq L$, then*

$$A^B = (\textstyle\sum_{\theta \in \exp(tB)} A^{\theta})^{\downarrow}.$$

Proof: Let $N = \sum_{\theta \in \exp(tB)} A^{\theta}$. Then clearly N is a subset of $(A^B)^{\uparrow}$. So $N^{\downarrow} \subseteq A^B$. By lemma 1.3(b), since $A^{\downarrow} = \{0\} \cup A$, we have $A^B \subseteq N^{\downarrow}$. □

COROLLARY 1.5 *Let L be a Lie algebra over a field $\mathfrak{k}$ of characteristic zero and let H si L. Let $\mathfrak{X}$ be any of the classes $\mathfrak{F}$, $\mathfrak{N}$, $\text{E}\mathfrak{A}$. If $H \in \mathfrak{X}$ then $H^L \in \text{L}\mathfrak{X}$.*

Proof: Let $N = \langle H^{\theta} : \theta \in \exp(tL) \rangle \leqq L^{\uparrow}$. By lemma 1.1 we have that N is the join of $\mathfrak{X}$-subideals of $L^{\uparrow}$. Since $\mathfrak{F}$ is coalescent, $\mathfrak{N}$ and $\text{E}\mathfrak{A}$ are locally coalescent, then $N \in \text{L}\mathfrak{X}$ and so by lemma 1.2, $N^{\downarrow} \in \text{L}\mathfrak{X}$. But by corollary 1.4 we have $H^L \subseteq N^{\downarrow} \subseteq (H^L)^{\uparrow\downarrow} = H^L$. □

2. Complete and locally coalescent classes

When dealing with a class $\mathfrak{X}$ of Lie algebras we usually work over a fixed ground field $\mathfrak{k}$, tacitly included in the specification of $\mathfrak{X}$. However, when we use field-changing techniques, it becomes necessary to incorporate $\mathfrak{k}$ into the symbol for the class: for instance we might write $\mathfrak{X}_{\mathfrak{k}}$. For some classes, such as

$\mathfrak{N}$, $\mathrm{L}\mathfrak{N}$, $\mathrm{E}\mathfrak{A}$, which behave well under field extensions, no confusion arises in any case; but for classes such as $\mathfrak{F}$ the more precise notation may be needed. For instance, if $\mathfrak{k}$ has characteristic 0 and $L \in \mathfrak{F}_{\mathfrak{k}}$ then $L^{\wedge} \in \mathfrak{F}_{\mathfrak{k}_0}$ but $L^{\wedge} \notin \mathfrak{F}_{\mathfrak{k}}$. In future when we refer to a class $\mathfrak{X}$ the underlying field will be understood to be the obvious one: if any ambiguity is present we shall use the more precise notation.

Let $\mathfrak{X}$ be a class of Lie algebras defined over a field $\mathfrak{k}$ of characteristic zero. We say that the class $\mathfrak{X}$ is *complete* if and only if

(a) whenever $H \in \mathfrak{X}$ then $H^{\wedge} \in \mathfrak{X}$ (i.e. $H^{\wedge} \in \mathfrak{X}_{\mathfrak{k}_0}$) and

(b) for any Lie algebra L over $\mathfrak{k}$, if $M \in \mathfrak{X}_{\mathfrak{k}_0}$ and $M \leqq L^{\wedge}$ then $M^{\vee} \in \mathfrak{X}$.

Lemmas 1.1 and 1.2 show that (b) is satisfied for the classes $\mathfrak{F}$, $\mathfrak{N}$, $\mathrm{E}\mathfrak{A}$, $\mathfrak{Z}$, $\mathrm{L}\mathfrak{N}$, $\mathrm{L}\mathfrak{F}$. By lemma 1.1(a) is satisfied for $\mathfrak{F}$, $\mathfrak{N}$, $\mathrm{E}\mathfrak{A}$. By equation (14) we see that (a) is true for $\mathfrak{Z}$. Let $\mathfrak{X}$ be $\mathfrak{N}$ or $\mathfrak{F}$ and let $H \in \mathrm{L}\mathfrak{X}$. Let M be a finite subset of $H^{\wedge}$. By the definition of $H^{\wedge}$ we can find a finite-dimensional subspace A of H such that $M \subseteq B^{\wedge}$. Put $A = \langle B \rangle$. Then $A \in \mathfrak{X}$ and so by lemma 1.1. we have $A^{\wedge} \in \mathfrak{X}_{\mathfrak{k}_0}$; we also have $M \subseteq B^{\wedge} \subseteq A^{\wedge}$. Thus $H^{\wedge} \in (\mathrm{L}\mathfrak{X})_{\mathfrak{k}_0}$.

So the classes $\mathfrak{F}$, $\mathfrak{N}$, $\mathrm{E}\mathfrak{A}$, $\mathfrak{Z}$, $\mathrm{L}\mathfrak{N}$, $\mathrm{L}\mathfrak{F}$ are all complete. Evidently they are all I-closed. Now $\mathfrak{F}$, $\mathfrak{N}$, $\mathrm{E}\mathfrak{A}$ are N_0-closed and it is easy to prove that $\mathfrak{Z}$ is N_0-closed. By corollary 6.1.2, the classes $\mathrm{L}\mathfrak{F}$ and $\mathrm{L}\mathfrak{N}$ are N_0-closed. So we may state:

LEMMA 2.1 *Over any field of characteristic zero, the classes*

$$\mathfrak{F},\ \mathfrak{N},\ \mathrm{E}\mathfrak{A},\ \mathfrak{Z},\ \mathrm{L}\mathfrak{N},\ \mathrm{L}\mathfrak{F}$$

are all complete and $\{\mathrm{I}, \mathrm{N}_0\}$*-closed.* □

Let L be a Lie algebra over a field $\mathfrak{k}$ of characteristic zero. Let d be a derivation of L and let $e = \exp(td)$. In general $L^{\wedge}$ is not invariant under e.

However, make the following definition: a derivation d of a Lie algebra L (over any field) is a *locally finite derivation* if every finite subset of L is contained in a finite dimensional d-invariant subspace. Then we have:

LEMMA 2.2 *Let L be a Lie algebra over a field of characteristic zero and let d be a locally finite derivation of L. If A is a subspace of L which is d-invariant then*

$$A^{\wedge \exp(td)} = A^{\wedge}.$$

Proof: Let $e = \exp(td)$. It is enough to show that $A^{\wedge e} \subseteq A^{\wedge}$. Let $a \in A^{\wedge}$. Then there is a finite-dimensional subspace A_0 of A such that $a \in A_0^{\wedge}$. Since d is locally finite, we can find a finite-dimensional d-invariant subspace A_1 of A such that $A_0 \subseteq A_1$. Let $b = \sum b_r t^r \in A_1^{\wedge}$. Then $b^e = \sum_r (\sum_{i+j=r} b_i d^j/j!) t^r$ and so $b^e \in A_1^{\wedge}$ since $b_i d^j/j! \in A_1$ for all i, j. In particular $a^e \in A_1^{\wedge} \subseteq A^{\wedge}$. □

From the proof of lemma 1.3.3 we obtain the following assertion. Over any field, if $H \operatorname{asc} L$ and B is a finite-dimensional subspace of H and A a finite-dimensional subspace of L then there exists $m = m(A, B)$ such that $[A, {}_mB] \subseteq H$. In particular if $H \in \mathrm{L}\mathfrak{F}$, then $\langle [A, {}_mB], B \rangle \in \mathfrak{F}$ (for $[A, {}_mB]$ is a finite-dimensional subspace of H) and so $[A, {}_mB]^B$ is finite-dimensional. Thus

$$A^B = \sum_{i=0}^{m-1} [A, {}_iB] + [A, {}_mB]^B$$

is a finite-dimensional subspace in this case.

We note also that if the elements of H induce locally finite derivations of H, then they induce locally finite derivations of L. This is in particular true if $H \in \mathrm{L}\mathfrak{F}$.

LEMMA 2.3 *Let L be a Lie algebra defined over a field $\mathfrak{k}$ of characteristic zero and let H an ascendant $\mathrm{L}\mathfrak{F}$-subalgebra of L. Suppose that A is a finite-dimensional subspace of L and B is a finite-dimensional subspace of H. Then A^B is finite-dimensional and there exist $\theta_1, \ldots, \theta_n \in \exp(tB)$ such that*

$$A^B = (A^{\wedge\theta_1} + \ldots + A^{\wedge\theta_n})^\vee.$$

Proof: That A^B is finite-dimensional follows from the discussion above. We observe that if M is a subspace of $L^\wedge$ and C is a finite-dimensional subspace of $M^\vee$, then we can find a finite-dimensional subspace P of M such that $C \subseteq P^\vee$. Let $N = \sum_{\theta \in \exp(tB)} A^{\wedge\theta}$. Then as the elements of H induce locally finite derivations of L we have $N \subseteq L^\wedge$ (by lemma 2.2). By corollary 1.4 we have $A^B = N^\vee$, and so there is a finite-dimensional subspace P of N such that $A^B \subseteq P^\vee \subseteq N^\vee = A^B$. The result now follows. □

THEOREM 2.4 *Let $\mathfrak{X}$ be a complete and $\{\mathrm{I}, \mathrm{N}_0\}$-closed class of Lie algebras over a field of characteristic zero. If $\mathfrak{X} \leq \mathrm{L}\mathfrak{F}$ then $\mathfrak{X}$ is locally coalescent. In particular $\mathfrak{G} \cap \mathfrak{X}$ is a coalescent class.*

Proof: Let H and K be $\mathfrak{X}$-subideals of a Lie algebra L and let $J = \langle H, K \rangle$. Every $\mathfrak{G}$-subalgebra of J is contained in one of the form $C = \langle A, B \rangle$, where A and B are finite-dimensional subspaces of H and K respectively. So it is enough to show that for any such C we can find an $\mathfrak{X}$-subideal X of L such that $C \leq X \subseteq J$.

Let $H \lhd^m L$ and use induction on m. If $m \leq 1$, then $H \lhd L$ and so by theorem 2.2.13 we have that J is an $\mathfrak{X}$-subideal of L; so we may take J for X.

Suppose that $m > 1$ and the result is true for $m-1$. By lemma 2.3 $A^B = M^\vee$, where $M = A^{\wedge\theta_1} + \ldots + A^{\wedge\theta_n}$, for some $\theta_i \in \exp(tB)$ (for $H, K \in \mathrm{L}\mathfrak{F}$). Let

$H_1 = H^L$. By lemma 2.2, $H_1^{\wedge}$ is invariant under $\exp(tK)$. Define

$$M_i = \langle H^{\wedge\theta_1}, \ldots, H^{\wedge\theta_i} \rangle \qquad i = 1, \ldots, n.$$

Now $H \lhd^{m-1} H_1$ and so $H^{\wedge\theta} \lhd^{m-1} H_1^{\wedge\theta} = H_1^{\wedge}$, for $\theta \in \exp(tK)$. Furthermore $\mathfrak{X}$ is complete and so $H^{\wedge} \in \mathfrak{X}_{\mathfrak{t}_0}$; hence $H^{\wedge\theta} \in \mathfrak{X}_{\mathfrak{t}_0}$ for all $\theta \in \exp(tK)$.

Thus by the induction on $m-1$ and a second induction on i it follows that for any $\mathfrak{G}$-subalgebra C_i of M_i there exists an $\mathfrak{X}_{\mathfrak{t}_0}$-subideal X_i of $H_1^{\wedge}$ such that $C_i \leqq X_i \leqq M_i$. In particular we can find an $\mathfrak{X}_{\mathfrak{t}_0}$-subideal X_0 of $H_1^{\wedge}$ with

$$M \subseteq X_0 \subseteq M_n.$$

Since $\mathfrak{X}$ is complete we have $X_0^{\vee} \in \mathfrak{X}$ and (since $M_n \leqq \langle H^B \rangle^{\wedge}$)

$$A^B = M^{\vee} \subseteq M_n^{\vee} \leqq \langle H^B \rangle.$$

We also have $X_0^{\vee}$ si $H_1 \lhd L$. Finally $A^B + B = M^{\vee} + B$, so $M^{\vee}$ permutes with B. Therefore by theorem 2.2.15 we can find an $\mathfrak{X}$-subideal X of L with

$$M^{\vee} + B \subseteq X \subseteq X_0^{\vee} + K.$$

Now $C = \langle A, B \rangle = \langle A^B, B \rangle \leqq X$ and $X_0^{\vee} + K \subseteq \langle H^B \rangle + K \subseteq J$. This completes our induction on m and the proof of the theorem. □

From theorem 2.4 and lemma 2.1 there follows:

COROLLARY 2.5 *Over any field of characteristic zero the classes* $\mathfrak{N}$, $\mathfrak{Z}$, $\mathrm{L}\mathfrak{N}$ and $\mathrm{L}\mathfrak{F}$ *are locally coalescent. The classes* $\mathfrak{G} \cap \mathfrak{N} = \mathfrak{F} \cap \mathfrak{N}$ *and* $\mathfrak{G} \cap \mathrm{L}\mathfrak{F} = \mathfrak{F}$ *are coalescent.* □

We have defined the closure operation J_s by $L \in \mathrm{J}_s\mathfrak{X}$ if L can be generated by its $\mathfrak{X}$-subideals (see chapter 2, section 2). We now define a new closure operation

$$\mathrm{L}_s$$

by $L \in \mathrm{L}_s\mathfrak{X}$ if every $\mathfrak{G}$-subalgebra of L is contained in some $\mathfrak{X}$-subideal of L. It is clear that for any class $\mathfrak{X}$,

$$\mathrm{L}_s\mathfrak{X} \leqq \mathrm{J}_s\mathfrak{X}$$

and that if $\mathfrak{X}$ is a locally coalescent class then

$$\mathrm{L}_s\mathfrak{X} = \mathrm{J}_s\mathfrak{X} \leqq \mathrm{L}\mathfrak{X}. \tag{18}$$

So we have by theorem 2.4 and corollary 2.5:

COROLLARY 2.6 *If $\mathfrak{X}$ is a complete and $\{\mathrm{I}, \mathrm{N}_0\}$-closed class of Lie algebras over a field of characteristic zero then*

$$\mathrm{L}_s\mathfrak{X} = \mathrm{J}_s\mathfrak{X}$$

whenever $\mathfrak{X} \leqq \mathrm{L}\mathfrak{F}$.

In particular $\mathrm{L}\mathfrak{N} = \mathrm{L}_s\mathrm{L}\mathfrak{N} = \mathrm{J}_s\mathrm{L}\mathfrak{N}$ and $\mathrm{L}\mathfrak{F} = \mathrm{L}_s\mathrm{L}\mathfrak{F} = \mathrm{J}_s\mathrm{L}\mathfrak{F}$, over any field of characteristic zero. □

It is not very hard to show that

$$\mathfrak{N} < \mathrm{L}_s\mathfrak{N} = \mathrm{J}_s\mathfrak{N} \text{ and } \mathfrak{F} < \mathrm{L}_s\mathfrak{F} = \mathrm{J}_s\mathfrak{F}. \qquad (19)$$

Engel conditions

An element x of a Lie algebra L is called a *left Engel element* if the adjoint map x^* of L induced by x is a nil derivation. Equivalently for any $y \in L$ we can find $m = m(y)$ such that $[y, {}_m x] = 0$. We denote by

$$\mathfrak{E}_*$$

the class of Lie algebras which can be generated by left Engel elements. Clearly

$$\mathfrak{E}_* = \mathrm{J}_s\mathfrak{E}_*.$$

We also have $\mathrm{L}\mathfrak{N} \leqq \mathfrak{E}_*$. Properties of the class $\mathfrak{E}_*$ enable us to deduce the local coalescence of most $\{\mathrm{I}, \mathrm{N}_0\}$-closed classes (relative to it). Specifically we will prove:

THEOREM 2.7 *Let $\mathfrak{X}$ be an $\{\mathrm{I}, \mathrm{N}_0\}$-closed class of Lie algebras over a field of characteristic zero. Suppose that H and K are $\mathfrak{X} \cap \mathfrak{E}_*$-subideals of a Lie algebra L and $J = \langle H, K\rangle$. Then to any finitely generated subalgebra C of J there corresponds an $\mathfrak{X}$-subideal X of L with $C \leqq X \leqq J$.*

In particular every $\{\mathrm{I}, \mathrm{N}_0\}$-closed subclass of $\mathfrak{E}_*$ is locally coalescent.

Before proving theorem 2.7 we need a few elementary observations.

(a) if $L \in \mathrm{E}\mathfrak{A}$ and x is a left Engel element of L then $\langle x\rangle$ asc L. Hence over fields of characteristic zero,

$$\mathfrak{E}_* \cap \mathrm{E}\mathfrak{A} \leqq \mathrm{L}\mathfrak{N}.$$

The first part of (a) follows by induction on the derived length and noting that if A is an abelian ideal of L and $A_n = \{a \in A : [a, {}_n x] = 0\}$, then $A = \bigcup_n A_n$ and $A_n + \langle x\rangle \lhd A_{n+1} + \langle x\rangle$.

The second part follows from the fact that $\mathfrak{F} \cap \mathfrak{N}$ is ascendantly coalescent (theorem 3.2.4).

(b) Let $J = \langle H, K\rangle$ be the join of two $\mathfrak{E}_*$-subideals of L. By the Derived Join Theorem we can find r such that $J^{(r)}$ si L. Now $\mathfrak{E}_*$ is Q-closed so $J/J^{(r)} \in \mathfrak{E}_* \cap \mathrm{E}\mathfrak{A} \leqq \mathrm{L}\mathfrak{N}$, in characteristic zero. If also $\mathfrak{X} = \{\mathrm{I}, \mathrm{N}_0\}\mathfrak{X}$ and $H, K \in \mathfrak{X}$, then r can be so chosen that $J^{(r)} \in \mathfrak{X}$.

(c) If K si L and $x \in K$ induces a nil derivation of K then x induces a nil derivation of L. Thus in characteristic zero we may define the automorphism $\exp(x^*)$ as before. In particular suppose that A is a finite-dimensional subspace of L and B is a finite-dimensional subspace of K with a basis of elements which induce nil derivations of K. Let $B_0 = \{b_1, \ldots, b_k\}$ be this basis and let $\exp(\mathrm{ad}_{B_0}(L)) = \mathrm{group}\langle \exp(b_i^*) : i = 1, \ldots, k\rangle$. Then for any finite integer q we can find (using the methods of lemma 1.4.10) $\theta_1, \ldots, \theta_p \in$ $\in \exp(\mathrm{ad}_{B_0}(L))$ such that

$$\textstyle\sum_{i=0}^{q} [A, {}_iB] \subseteq \sum_{j=1}^{p} A^{\theta_j} \subseteq A^B. \qquad (!)$$

Proof of theorem 2.7

Every finitely generated subalgebra of J is contained in one of the form $C = \langle A, B\rangle$ where A and B are finite-dimensional subspaces spanned by left Engel elements of H and K respectively. Thus it is enough to find for such a C, an $\mathfrak{X}$-subideal X of L with $C \leqq X \leqq L$.

Let $H \lhd^m L$ and use induction on m. For $m \leqq 1$, the result follows by theorem 2.2.13 (take J for X). Let $m > 1$, and assume the result is true for $m-1$. Now by (b) there exists r such that $J^{(r)} \in \mathfrak{X}$, $J^{(r)}$ si L and $J/J^{(r)} \in \mathfrak{N}$. Therefore $C/C \cap J^{(r)} \in \mathfrak{F} \cap \mathfrak{N}$. So we can find q with

$$A^B = \textstyle\sum_{i=0}^{q} [A, {}_iB] + A^B \cap J^{(r)}.$$

Hence employing (!) we have for some $\theta_1, \ldots, \theta_p \in \exp(\mathrm{ad}_{B_0}(L))$,

$$A^B = \textstyle\sum_{j=1}^{p} A^{\theta_j} + A^B \cap J^{(r)}.$$

We note that if $H_1 = H^L$, then $H^\theta \lhd^{m-1} H_1^\theta = H_1$ for $\theta \in \{\theta_1, \ldots, \theta_p\}$. Define

$$U = \textstyle\sum_{j=1}^{p} A^{\theta_j}.$$

Arguing as in the proof of theorem 2.4 we get an $\mathfrak{X}$-subideal X_0 of H_1 such that

$$U \subseteq X_0 \leqq \langle H^B\rangle.$$

Now X_0 si $H_1 \lhd L$ and $J^{(r)} \lhd J$, $J^{(r)}$ si L and $J^{(r)} \in \mathfrak{X}$. So by theorem 2.2.13 we get $X_0 + J^{(r)} \in \mathfrak{X}$ and $X_0 + J^{(r)}$ si L. We also have

$$A^B = U + A^B \cap J^{(r)} \subseteq X_0 + J^{(r)} \subseteq J.$$

Since B idealises A^B we can apply theorem 2.2.15 (to the pair $X_0 + J^{(r)}$, K) to obtain an $\mathfrak{X}$-subideal X of L with

$$A^B + B \subseteq X \leqq X_0 + J^{(r)} + K \subseteq J.$$

Finally $C = \langle A, B \rangle \leqq X$. This completes our induction on m and proves theorem 2.7. □

3. Acceptable subalgebras

In this section we give another application of the algebra of formal power series: we prove a result which yields a new proof of the fact that the classes $\mathfrak{G}$ and $\mathfrak{C}$ are coalescent over fields of characteristic zero. For this we need the idea of 'acceptable subalgebras' first introduced by Hartley.

Let $\mathfrak{X}$ be a class of Lie algebras and let H be a subalgebra of a Lie algebra L. We say that H is an *$\mathfrak{X}$-acceptable subalgebra* of L if there is an ideal H_0 of H such that

$$H/H_0 \in \mathfrak{X} \text{ and } [H_0, L] \subseteq H.$$

Suppose that $H \leqq L$ and $A \subseteq H$ with $[A, L] \subseteq H$. For each $n \in \mathbb{N}$

$$[A, {}_{n+1}H, L] \subseteq [A, {}_nH, L, H] + [A, {}_nH, [H, L]]$$

and so it follows by induction that $[A, {}_nH, L] \subseteq H$ for all n. Hence

$$[A^H, L] = [\textstyle\sum_{n=0}^{\infty} [A, {}_nH], L] \subseteq H \text{ and } A^H \lhd H. \tag{20}$$

LEMMA 3.1 (a) *If H is an $\mathfrak{X}$-acceptable subideal of a Lie algebra and $\mathfrak{X}$ is* Q-*closed then H is an $\mathfrak{X} \cap \mathfrak{N}$-acceptable subideal of that algebra.*

(b) *If L is a Lie algebra over a field of characteristic zero and H is a $\mathfrak{G}$-acceptable subideal of L then $H^\dagger$ is a $\mathfrak{G}$-acceptable subideal of $L^\dagger$.*

Proof: (a) Let $H \lhd^n L$, and let $H_0 \lhd H$ with $H/H_0 \in \mathfrak{X}$ and $[H_0, L] \subseteq H$. We have $[H^{n+1}, L] \subseteq [L, {}_{n+1}H] \subseteq H^2$ and so $[H^{n+1} + H_0, L] \subseteq H$. By Q-closure $H/(H^{n+1} + H_0) \in \mathfrak{X} \cap \mathfrak{N}_n$.

(b) Since $\mathfrak{G}$ is Q-closed and $\mathfrak{G} \cap \mathfrak{N} = \mathfrak{F} \cap \mathfrak{N}$, we may assume, by (a), that H is an $\mathfrak{F} \cap \mathfrak{N}$-acceptable subideal of L. So let $H_0 \lhd H$ with $H/H_0 \in \mathfrak{F} \cap \mathfrak{N}$ and $[H_0, L] \subseteq H$. By equation (5) $H^\dagger/H_0^\dagger \cong (H/H_0)^\dagger$ and so by lemma 1.1 we have $H^\dagger/H_0^\dagger \in \mathfrak{F} \cap \mathfrak{N}$. We also have by lemma 1.1, $[H_0^\dagger, L^\dagger] \subseteq [H_0, L]^\dagger \subseteq H^\dagger$. So $H^\dagger$ is an $\mathfrak{F} \cap \mathfrak{N}$-acceptable subideal of $L^\dagger$ (for by lemma 1.1, $H^\dagger$ si $L^\dagger$). □

Clearly if $H \in \mathfrak{X}$ then H is an $\mathfrak{X}$-acceptable subalgebra (take $H_0 = 0$) of every Lie algebra in which it is embedded. (Though we do not prove the converse of this statement here it also is true). From equation (20) it is not hard to deduce that:

$$\text{if } H \leqq L \text{ and } H_0 = \{x \in H : [x, L] \subseteq H\} \text{ then } H_0 \lhd H. \tag{21}$$

Now $\mathfrak{G}$ is Q-closed and the join of finitely many $\mathfrak{G}$-subalgebras is also a $\mathfrak{G}$-subalgebra. So (21) yields:

LEMMA 3.2 *The join of finitely many $\mathfrak{G}$-acceptable subalgebras of a Lie algebra is always a $\mathfrak{G}$-acceptable subalgebra of that algebra.* □

Evidently if $H \leqq K \leqq L$ and H is an $\mathfrak{X}$-acceptable subalgebra of L then H is an $\mathfrak{X}$-acceptable subalgebra of K.

Let L be a Lie algebra over a field of characteristic zero and let K be a $\mathfrak{G}$-acceptable subideal of L. By lemma 3.1 we can find a finite-dimensional subspace B of K and $K_0 \lhd K$ such that $K = B+K_0$ and $[K_0, L] \subseteq K$. Suppose that $K \lhd^n L$ and H is a subspace of L. Then it is not hard to see that

$$H^K = X+H^K \cap K, \text{ where } X = \textstyle\sum_{i=0}^n [H, {}_iB].$$

Evidently X permutes with K and so $\langle X, K\rangle = \langle X\rangle+K$. Hence

$$\langle H, K\rangle = \langle X\rangle+K.$$

Suppose that $b_1, \ldots, b_k$ is a basis for B. Then

$$X = \textstyle\sum_{i=0}^n [H, {}_iB] = \sum_{s=0}^n [H, b_{i_1}, \ldots, b_{i_s}],$$

where the last sum is taken over all s-tuples with the $i_j \in \{1, \ldots, k\}$. So X is the sum of a finite number (namely $1+k+\ldots+k^n$) of subspaces of the form $[H, b_{i_1}, \ldots, b_{i_s}]$. We apply lemma 1.3(b) to each of these. Then we can find $\theta_1, \ldots, \theta_r \in \exp(tB) \subseteq \exp(tK)$ such that if

$$M = \langle H^{\uparrow\theta_1}, \ldots, H^{\uparrow\theta_r}\rangle \leqq L^{\uparrow}, \tag{22}$$

then

$$X \subseteq M^{\downarrow}.$$

We note that $M \leqq \langle H^B\rangle^{\uparrow}$, so that $M^{\downarrow} \leqq (\langle H^B\rangle^{\uparrow})^{\downarrow} = \langle H^B\rangle$. We also have $\langle X\rangle \leqq \langle M^{\downarrow}\rangle = M^{\downarrow}$ and so it follows from the above that

$$\langle H, K\rangle = M^{\downarrow}+K. \tag{23}$$

THEOREM 3.3 *Let L be a Lie algebra over a field of characteristic zero and let H and K be $\mathfrak{G}$-acceptable subideals of L. If $J = \langle H, K\rangle$ then J is a $\mathfrak{G}$-acceptable subideal of L.*

Proof: By lemma 3.1 we may assume that H and K are $\mathfrak{F} \cap \mathfrak{N}$-acceptable subideals of L. Let $H \lhd^m L$ and use induction on m. If $m \leqq 1$ then J si L (lemma 2.1.2) and by lemma 3.2 J is a $\mathfrak{G}$-acceptable subalgebra of L. Let $m > 1$, and assume the result holds for $m-1$. Let $H_1 = \langle H^L \rangle$.

Define the subalgebra M of $L^\uparrow$ as in equation (22) so that $J = M^\downarrow + K$ (by equation (23)).

If $\theta = \{\theta_1, \dots, \theta_r\}$ then $H^{\uparrow\theta} \lhd^{m-1} H_1^{\uparrow\theta} = H_1^\uparrow$ and $H^{\uparrow\theta}$ is a $\mathfrak{G}$-acceptable subalgebra of $L^\uparrow$ and so of $H_1^\uparrow$, by lemma 3.1. For each $i \in \{1, \dots, r\}$ let

$$M_i = \langle H^{\uparrow\theta_1}, \dots, H^{\uparrow\theta_i} \rangle.$$

Then the inductive hypothesis on $m-1$ and a second induction on i shows that each M_i is a $\mathfrak{G}$-acceptable subideal of $H_1^\uparrow$. In particular $M = M_r$ si $H_1^\uparrow \lhd L^\uparrow$ and so $M^\downarrow$ si L. Now $J = M^\downarrow + K$, a join of two permutable subideals of L, and so J si L (by lemma 2.1.4). By lemma 3.2 we have that J is a $\mathfrak{G}$-acceptable subalgebra of L. □

Evidently it follows from lemma 3.1 that if H is a $\mathfrak{C}$-subideal of a Lie algebra L then H is a $\mathfrak{G}$-acceptable subideal of L. So we have from theorem 3.3,

COROLLARY 3.4 *Over fields of characteristic zero the classes $\mathfrak{G}$ and $\mathfrak{C}$ are coalescent.* □

We also have the following analogue of corollary 3.4.3.

PROPOSITION 3.5 *Suppose that $\mathfrak{X}$ is a locally coalescent class of Lie algebras and H and K are $\mathfrak{X}$-subideals of a Lie algebra L and $J = \langle H, K \rangle$.*

(a) *If C is a $\mathfrak{G}$-acceptable subalgebra of L with $C \leqq J$ then there exists a* Q$\mathfrak{X}$*-acceptable subideal X of L with $C \leqq X \leqq J$.*

(b) *If C is a $\mathfrak{C}$-subalgebra of J then there exists an* {I, N$_0$}$\mathfrak{X}$*-subideal X of L with $C \leqq X \leqq J$.*

(c) *If C is a $\mathfrak{C}$-acceptable subalgebra of L with $C \leqq J$ then there exists a* Q{I, N$_0$}$\mathfrak{X}$*-acceptable subideal X of L with $C \leqq X \leqq J$.*

First we need a result about subideals. If $C_0 \lhd C \leqq L$ and $D \leqq L$ we say that D *covers* C/C_0 if $C = C \cap D + C_0$.

LEMMA 3.6 *Let L be a Lie algebra and let $C_0 \lhd C \leqq L$ such that $[L, C_0] \subseteq C$. If $D \leqq L$ and D covers C/C_0 then C permutes with D; if also $D \lhd^m L$ then $C + D \lhd^m L$.*

Proof: We have $C = C \cap D + C_0$ and so $[C, D] = [C \cap D, D] + [C_0, D] \subseteq$ $\subseteq D^2 + C$, whence C permutes with D (for $[C_0, L] \subseteq C$). Suppose that we have $D = D_m \lhd D_{m-1} \lhd \ldots \lhd D_0 = L$. For each i we have

$$[C, D_i] = [C \cap D, D_i] + [C_0, D] \subseteq D_{i+1} + C,$$

since $C \cap D \leqq D_{i+1} \lhd D_i$ and $[C_0, L] \subseteq C$. Therefore

$$C + D_{i+1} \lhd C + D_i \text{ for } i = 1, \ldots, m-1.$$

So $C + D \lhd C + D_{m-1} \lhd \ldots \lhd C + D_0 = L$. □

In the same way as above we could prove that if $C \leqq L$, D si L and $[C, L] \subseteq C + D$ then $C + D$ si L. (24)

Proof of proposition 3.5:

(a) We have $C = A + C_0$ for some $A \in \mathfrak{G}$ and $C_0 \lhd C$ with $[C_0, L] \subseteq C$. By local coalescence of $\mathfrak{X}$ we can find M si L with $M \in \mathfrak{X}$ such that $A \leqq M \leqq J$. So M covers C/C_0 and by lemma 3.6 we have $C + M$ si L. Put $X = C + M = C_0 + M$. By equation (20), $[C_0^X, L] \subseteq X$ and clearly $X/C_0^X \in \mathrm{Q}\mathfrak{X}$ and $C \leqq X \leqq J$.

(b) We have $C = A + C^2$ for some $A \in \mathfrak{G}$ (for $C/C^2 \in \mathfrak{G}$). By local coalescence of $\mathfrak{X}$ there exists $M \in \mathfrak{X}$, M si L such that $A \leqq M \leqq J$. Let $N = M \cap C$ so that N si C and $C = N + C^2$. Then from lemma 2.1.9 we have $C = N + C^{(r)}$ for all $r \in \mathbb{N}$. Put $\mathfrak{Y} = \{\mathrm{I}, \mathrm{N}_0\}\mathfrak{X}$. By corollary 2.2.17 we have for some r, $J^{(r)} \in \mathfrak{Y}$ and $J^{(r)}$ si L, whence by theorem 2.2.13, $M + J^{(r)} \in \mathfrak{Y}$ and $M + J^{(r)}$ si L. Put $X = M + J^{(r)}$ so that $C = M \cap C + C^{(r)} \leqq X$ and $X \leqq J$.

(c) Put $\mathfrak{Y} = \{\mathrm{I}, \mathrm{N}_0\}\mathfrak{X}$. For some $C_0 \lhd C$ we have $[C_0, L] \subseteq C$ and $C/C_0 \in \mathfrak{C}$. So we can find $A \in \mathfrak{G}$ such that $C = A + C_0 + C^2$ (for $C/(C_0 + C^2) \cong (C/C_0)/(C/C_0)^2 \in \mathfrak{G}$). By the local coalescence of $\mathfrak{X}$ we can find $M \in \mathfrak{X}$ with M si L and $A \leqq M \leqq J$. Let $N = M \cap C + C_0$; then $M \cap C$ si C and so N si C. We also have $C = N + C^2$ and hence (lemma 2.1.9) $C = N + C^{(r)}$ for all r. But by corollary 2.2.17 we have $J^{(r)} \in \mathfrak{Y}$ and $J^{(r)}$ si L for some r. Now by theorem 2.2.13, $M + J^{(r)} \in \mathfrak{Y}$ and $M + J^{(r)}$ si L. Clearly $C/C_0 = (M \cap C + C^{(r)} + C_0)/C_0$ is covered by $M + J^{(r)}$. So if $X = C + M + J^{(r)}$ then (lemma 3.6) X si L, $C \leqq X \leqq J$, $X = C_0 + M + J^{(r)}$ and so X is a $\mathrm{Q}\mathfrak{Y}$-acceptable subalgebra of L. For $[C_0, L] \subseteq C \leqq X$ and so by equation (20) $[C_0^X, L] \subseteq X$; finally $X/C_0^X \in \mathrm{Q}\mathfrak{Y}$. □

We remark that part (a) still holds if J is the join of an arbitrary collection of $\mathfrak{X}$-subideals of L; whilst parts (b) and (c) hold if J is the join of a finite number of $\mathfrak{X}$-subideals of L.

It is not hard to see that proposition 3.5 includes corollary 3.4.3. One of the many conclusions we can draw from proposition 3.5 (using the fact that the class $\mathfrak{L}$ of all Lie algebras is locally coalescent) is the rather striking

COROLLARY 3.7 *Let L be a Lie algebra over a field of characteristic zero and let J be a subalgebra of L generated by finitely many subideals of L. Then to each $\mathfrak{C}$-subalgebra C of J there corresponds a subideal X of L with $C \leqq X \leqq J$.* □

The group-theoretic analogue of this result is also true.

Chapter 5

The Mal'cev correspondence

A. I. Mal'cev [152] discovered a remarkable connection between locally nilpotent Lie algebras over $\mathbb{Q}$ and a certain class of locally nilpotent groups, whereby the ideal structure of the Lie algebra is related to the normality structure of the group. This *Mal'cev correspondence* allows us, under conditions which often hold, to pass at will from Lie-theoretic problems to group-theoretic ones, and vice versa. Essentially it is the usual correspondence between Lie groups and Lie algebras in an infinite-dimensional setting and without analytic considerations.

The correspondence can be approached in several ways. Mal'cev's original version [152] is topological, involving properties of nilmanifolds; Lazard [137] uses certain sequences of elements of a free group endowed with a topology induced by the lower central series; while Quillen [324] has pointed out the rôle played by Hopf algebras. The results can also be extended in various ways which we shall not go into. Since the main properties of the correspondence can be simply stated in algebraic terms it is of interest to give simple algebraic proofs, using as little 'machinery' as possible. The treatment we give here is taken from Stewart [208, 222] and is based on a few elementary properties of the representations of nilpotent groups and Lie algebras, together with the Campbell-Hausdorff formula. A topological version with a similar proof can be found in Gluškov [67].

1. The Campbell-Hausdorff formula

Let A be a locally nilpotent associative algebra over a field $\mathfrak{k}$ of characteristic zero. The direct sum $\mathfrak{k} \dotplus A$ has a natural associative algebra structure in which the elements of $\mathfrak{k}$ act on A by scalar multiplication. Further, the element 1 of $\mathfrak{k}$ is an identity for $\mathfrak{k} \dotplus A$.

The elements $1+a$ $(a \in A)$ form a group under multiplication, since

$$(1+a)(1+b) = 1 + (a+b+ab)$$
$$(1+a)^{-1} = 1 + (-a+a^2-a^3+\ldots)$$

and the series terminates by local nilpotency. This group is the *adjoint group* A^0 of A (also called the *associated* or *circle group*) and can be defined for more general rings (cf. Mal'cev [154], Kuroš [303] p. 39).

Let $L = [A]$ be the Lie algebra on A. We can define functions

$$\exp: L \to A^0$$
$$\log: A^0 \to L$$

by

$$\exp(a) = 1+a+\frac{1}{2!}a^2+\frac{1}{3!}a^3+\ldots \qquad (a \in L)$$
$$\log(1+a) = a-\tfrac{1}{2}a^2+\tfrac{1}{3}a^3-\ldots \qquad (a \in A)$$

and again the series terminate.

There is a connection between the operations in L and A^0 which is most easily stated if we introduce the formal power series algebra $\mathfrak{k}\langle X, Y\rangle$ in two noncommuting variables X and Y (cf. Jacobson [98] p. 171). If $\alpha(X, Y) \in \mathfrak{k}\langle X, Y\rangle$ then for any $x, y \in A$ the expression $\alpha(x, y)$ obtained by 'substituting' x and y for X and Y is well-defined, since A is locally nilpotent. We now have:

LEMMA 1.1 (*Campbell-Hausdorff Formula*). *With the foregoing notation there exists an element* $\mu \in \mathfrak{k}\langle X, Y\rangle$ *such that for all* $x, y \in A$

$$\log(\exp(x)\exp(y)) = \mu(x, y).$$

Further,

$$\mu(X, Y) = X+Y+\tfrac{1}{2}[X, Y]+\ldots$$

where each term on the right is a rational multiple of a Lie product $[Z_1, \ldots, Z_m]$ *where each* Z_i *is* X *or* Y, *and only finitely many terms with given length* m *occur.*

Proof: See Jacobson [98] p. 173. The theorem proved there is more general, in that log and exp also take values in power series algebras. The result stated follows by substituting x and y. □

(An explicit expression for μ is given in Jacobson [98] p. 173, but is unnecessary for our purposes.)

We note some simple properties of log and exp:

LEMMA 1.2 *With the foregoing notation,*

(a) *The functions* log *and* exp *are mutual inverses,*

(b) *If* $x, y \in L$ *and* $[x, y] = 0$ *then*

$$\log(\exp(x)\exp(y)) = x+y$$

(c) *If* $x \in A^0$, $n \in \mathbb{N}$ *then* $\log(x^n) = n \log(x)$.

Proof: All three parts follow by simple calculations. Part (b) is also a consequence of lemma 1.1. □

In the sequel we shall be most interested in the case where A is the algebra $U_n(\mathbb{Q})$ of $n \times n$ zero-triangular matrices over $\mathbb{Q}$ (as defined in chapter 1). In this case A^0 is isomorphic to the group $T_n(\mathbb{Q})$ of $n \times n$ *unitriangular* matrices

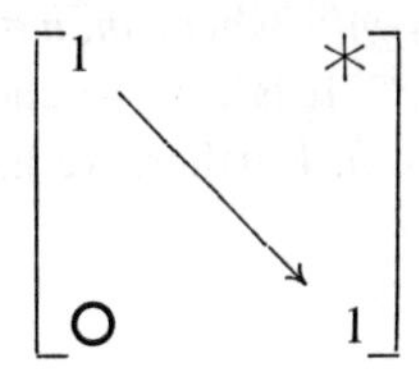

2. Complete groups

We introduce some group-theoretic notation and terminology.

Let G be a group. If $x, y \in G$ their *commutator* is

$$(x, y) = x^{-1}y^{-1}xy.$$

Iterated commutators

$$(x_1, \ldots, x_n)$$

are left-normed in the same way as for Lie algebras. If H and K are subgroups of G then the *commutator* (H, K) is the group generated by all (h, k) for $h \in H$, $k \in K$. The *lower central series* of G is defined recursively by

$$\gamma_1(G) = G$$

$$\gamma_{n+1}(G) = (\gamma_n(G), G).$$

The *upper central series* is defined by

$$\zeta_1(G) = \text{the centre of } G$$

$$\zeta_{n+1}(G)/\zeta_n(G) = \zeta_1(G/\zeta_n(G)).$$

Both of these series may be continued transfinitely as for Lie algebras, to define $\gamma_\alpha(G)$ and $\zeta_\alpha(G)$ for any ordinal α. We say that G is *nilpotent* (of *class* $\leqq n$) if $\gamma_{n+1}(G) = 1$ (or equivalently $\zeta_n(G) = G$); and G is *hypercentral* of *central height* $\leqq \alpha$ if $G = \zeta_\alpha(G)$. We say that G is *locally nilpotent* if every finitely generated subgroup of G is nilpotent. If G has no elements $\neq 1$ of finite order then G is *torsion-free*.

The group G is *complete* (in the sense of Kuroš [303] p. 233) if for every $n \in \mathbb{Z}\backslash\{0\}$, $g \in G$ there exists $h \in G$ with $h^n = g$. (Other authors use the terms *radicable*, *divisible*.) It is an *R-group* (Kuroš [303] p. 242) if for all $g, h \in G$ and $n \in \mathbb{Z}$ the equation $g^n = h^n$ implies that $g = h$. Thus in a complete group nth roots exist for all n: in an R-group they are unique if they exist. If G is a complete R-group nth roots exist *and* are unique. For any $q \in \mathbb{Q}$, $g \in G$ we can then define 'rational powers' g^q as follows. Let $q = m/n$ where $m, n \in \mathbb{Z}$, $n \neq 0$. Define g^q to be the unique $h \in G$ such that $h^n = g^m$. It is easy to check that common factors in m and n do not alter the value of h. Further, we have

$$g^q g^r = g^{q+r}$$

$$(g^q)^r = g^{qr}$$

for any $g \in G$, $q, r \in \mathbb{Q}$.

LEMMA 2.1 *The unitriangular group $T_n(\mathbb{Q})$ is a complete R-group for any $n \in \mathbb{N}$.*

Proof: Let $t \in T = T_n(\mathbb{Q})$. Let $m \in \mathbb{Z}\backslash\{0\}$. Define

$$s = \exp\left(\frac{1}{m}\log(t)\right).$$

By lemma 1.2(c) we have $s^m = t$, whence T is complete.

Now suppose that $s, t \in T$ and $s^m = t^m$. Then

$$m\log(s) = m\log(t)$$

so that $s = t$. Therefore T is an R-group. □

We shall call any complete locally nilpotent torsion-free group a *Mal'cev group*. From Kuroš [303] p. 248 we see that every Mal'cev group is an R-group. It is Mal'cev groups which arise in the Mal'cev correspondence.

If H is a subgroup of the complete group G we define the *completion* $\bar{H}$ *of H in G* to be the smallest complete subgroup of G which contains H.

Now let H be any locally nilpotent torsion-free group, not necessarily complete. We will sketch the construction of a *completion* H^* of H with the following properties:

(i) H^* is complete, locally nilpotent, and torsion-free (so is a Mal'cev group).

(ii) H^* has a certain universal property: given any homomorphism $\varphi : H \to G$, where G is a Mal'cev group, there is a unique extension to a homomorphism $\varphi^* : H^* \to G$ so that the diagram

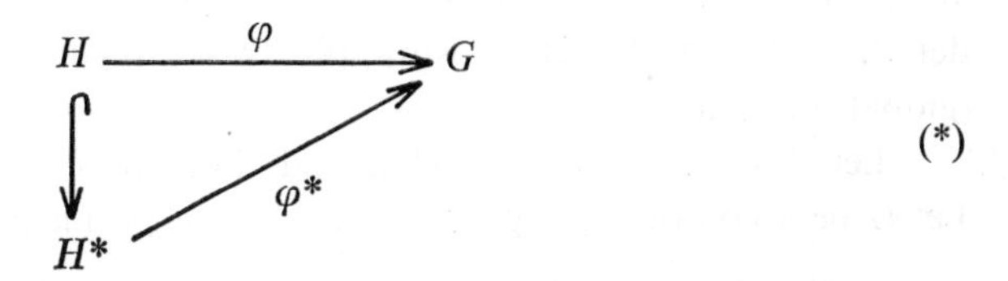

(*)

commutes.

From (ii) it is clear that if H^* exists it is unique up to isomorphism, and that the completion of H in H^* is the whole of H^*. It is also quite easy to see that we need consider only the case where H is finitely generated (and so nilpotent) by working locally and using (*) to fit the pieces together. Now a theorem of Hall [285] (proved also in Swan [346], Robinson [332] p. 159) shows that such a group embeds in $T_n(\mathbb{Z})$ for some $n \in \mathbb{N}$: this in turn embeds in the complete group $T_n(\mathbb{Q})$. We let H^* be the completion of H in $T_n(\mathbb{Q})$. Then H^* is complete and nilpotent. To show universality we must construct φ^*. The only possible choice is to put

$$\varphi^*(h^q) = (\varphi(h))^q$$

for $h \in H$, $q \in \mathbb{Q}$. So φ^* is unique if it exists. That this procedure defines a homomorphism can be proved as outlined by Hall [285] p. 45.

The existence of completions was first discovered by Mal'cev [152]. Another proof, of a more general result, is due to Kargapolov [297, 298].

A subgroup K of a group G is *isolated* in G if whenever $g^n \in K$ for some $g \in G$, $n \in \mathbb{Z}$ it follows that $g \in K$. The *isolator* of a subgroup F of G is the smallest isolated subgroup of G containing F. With these concepts defined we can state some useful properties of completions. The proofs are to be found in Kuroš [303] pp. 248–258.

LEMMA 2.2 *Let G be a Mal'cev group, with subgroups $H \leqq K \subseteq G$.*

(a) *If $H \lhd K$ then $\overline{H} \lhd \overline{K}$.*

(b) *The isolated subgroups of G are precisely the complete subgroups.*

(c) *$\overline{K}$ is equal to the isolator of K in G, which is equal to the set of all $g \in G$ such that $g^n \in K$ for some $n \in \mathbb{Z}$.*

(d) *If K is nilpotent of class c then so is $\overline{K}$.*

(e) *If M is any subgroup of K^* then $\overline{M} = \overline{\overline{M} \cap K}$.* □

3. The matrix version

We first develop a version of the Mal'cev Correspondence for the matrix algebras $U_n(\mathbb{Q})$ and groups $T_n(\mathbb{Q})$. This gives a 'local' form of the correspondence. In the next section we apply two embedding theorems to deduce a 'global' version.

Let $T = T_n(\mathbb{Q})$, $U = U_n(\mathbb{Q})$, and think of U as a Lie algebra over $\mathbb{Q}$. Let G be a complete subgroup of T, and define maps $\flat$, $\sharp$ as follows:

$$\flat : G \to U \qquad g^\flat = \log(g) \qquad (g \in G). \tag{1}$$

Let $L = G^\flat = \{g^\flat : g \in G\}$:

$$\sharp : L \to T \qquad l^\sharp = \exp(l) \qquad (l \in L). \tag{2}$$

The main result of this section will be:

THEOREM 3.1 *With the above notation,*

(a) *The maps $\flat$, $\sharp$ are mutual inverses.*

(b) *If H is a complete subgroup of G then $H^\flat$ is a Lie subalgebra of L.*

(c) *If M is a subalgebra of L then $M^\sharp$ is a complete subgroup of G.*

(d) *If H is a complete normal subgroup of a complete subgroup K of G, then $H^\flat$ is an ideal of $K^\flat$.*

(e) *If M is an ideal of a subalgebra N of L, then $M^\sharp$ is a complete normal subgroup of $N^\sharp$.*

Before embarking on the proof we make several remarks.

Remark 3.2 L is contained in a nilpotent Lie algebra, since U is nilpotent.

Remark 3.3 Let $g \in G$, $\lambda \in \mathbb{Q}$. Then $(g^\lambda)^\flat = \lambda g^\flat$. For if $\lambda = m/n$ $(m, n \in \mathbb{Z})$ then $(g^\lambda)^n = g^m$. Taking logarithms gives $n(g^\lambda)^\flat = mg^\flat$, whence $(g^\lambda)^\flat = \lambda g^\flat$.

Remark 3.4 By an induction argument (written out in full in Jennings [99]) it follows from the Campbell-Hausdorff formula that for all $g_1, \ldots, g_m \in G$

$$(g_1, \ldots, g_m)^\flat = [g_1^\flat, \ldots, g_m^\flat] + \textstyle\sum_w P_w$$

where each P_w is a rational linear combination of products $[g_{i_1}^\flat, \ldots, g_{i_w}^\flat]$ with $w > m$ and $i_\lambda \in \{1, \ldots, m\}$ for $1 \leqq \lambda \leqq w$, such that each of $1, \ldots, m$ occurs at least once among the i_λ $(1 \leqq \lambda \leqq w)$.

Remark 3.5 We now describe the *shuttle procedure*, a special way of manipulating expressions with terms of the form $h^\flat$, where h lies in some sub*set* H of G.

Suppose we have an expression

$$h^\flat + \sum \lambda_j C_j \qquad (\lambda_j \in \mathbb{Q}) \tag{3}$$

where each C_j is a Lie product of length $\geqq r$ of elements of $H^\flat$. We can split this up as

$$h^\flat + \sum \mu_j D_j + \sum \nu_i E_i \qquad (\mu_j, \nu_i \in \mathbb{Q})$$

where the D_j are of length r, the E_i of length $\geqq r+1$. Take one of the terms D_j, say

$$D = [h_1^\flat, \ldots, h_r^\flat].$$

By remark 3.4 we may replace D by the expression

$$(h_1, \ldots, h_r)^\flat + \sum \alpha_k F_k \qquad (\alpha_k \in \mathbb{Q})$$

where each F_k is a product of length $\geqq r+1$ of elements of $H^\flat$. Let

$$g = (h_1, \ldots, h_r) \in G.$$

By the Campbell-Hausdorff formula and remark 3.3

$$(hg^\lambda)^\flat = h^\flat + \lambda g^\flat + \sum \beta_l G_l \qquad (\lambda, \beta_l \in \mathbb{Q})$$

where the G_l are products of length $\geqq 2$ of elements $h^\flat$ or $g^\flat$. But $g^\flat = D - \sum \alpha_k F_k$, each term of which is a product of $\geqq r$ elements of $H^\flat$. Thus we may remove the terms D_j one by one to obtain a new expression for (3) of the form

$$(hg_1^{\lambda_1} \ldots g_s^{\lambda_s})^\flat + \sum \gamma_i H_i \qquad (\lambda_j, \gamma_i \in \mathbb{Q})$$

where the g_j are group commutators of length r in elements of H, and the H_i are products of length $\geqq r+1$ in elements of $H^\flat$.

Proof of theorem 3.1

(a) This is immediate from the definitions.

(b) Any element of the Lie algebra generated by $H^\flat$ is of the form (3) with $r = 1$, $h = 0$. By repeated use of the shuttle procedure we can express this element as

$$(h')^\flat + \sum \delta_i J_i \qquad (\delta_i \in \mathbb{Q})$$

where, since H is a complete subgroup of G, we have $h' \in H$; and the J_i are products of length $> c$, the nilpotency class of U. But this makes all $J_i = 0$, so the given element lies in $H^\flat$. Therefore $H^\flat$ is a Lie algebra. In particular $L = G^\flat$ is a Lie algebra.

(c) Let $m, n \in M$, $\lambda \in \mathbb{Q}$. We must show that $(m^\#)^\lambda$ and $m^\# n^\#$ are elements of $M^\#$. Now $(m^\#)^\lambda = (\lambda m)^\# \in M^\#$. Further, the Campbell-Hausdorff formula implies that $(m^\# n^\#)^\flat = \mu(m, n) \in M$. Hence by part (a) $m^\# n^\# \in M^\#$.

(d) Let $h \in H$, $k \in K$. We must show that $[h^\flat, k^\flat] \in H^\flat$. We prove by descending induction on r that any product of the form $[a_1^\flat, \ldots, a_r^\flat]$ with $a_j \in K$ for all j and at least one $a_i \in H$ is a member of $H^\flat$. This is trivially true for $r > c$, the class of nilpotency of U. The transition from $r+1$ to r follows from remark 3.4, noting that if a group commutator $(k_1, \ldots, k_m)$ with all $k_j \in K$ has some $k_i \in H$ then the whole commutator must lie in H, since H is a normal subgroup. The case $r = 2$ gives the desired result.

(e) Let $m \in M$, $n \in N$. Then $(m^\#, n^\#)^\flat = [m, n] +$ products of length $\geqq 3$ of elements of M and N, each term containing at least one element from M (remark 3.4). Since M is an ideal of N each such term lies in M, so that $(m^\#, n^\#)^\flat \in M$. Therefore $(m^\#, n^\#) \in M^\#$, so that $M^\#$ is normal in $N^\#$. □

4. Inversion of the Campbell-Hausdorff formula

In order to extend the results to locally nilpotent groups and algebras we must 'globalise' the matrix correspondence. To do this we effect what Lazard [138] calls 'inversion of the Campbell-Hausdorff formula'. To express the result concisely we must briefly discuss infinite products in locally nilpotent groups.

Suppose we have a *finite* set of variables $\{x_1, \ldots, x_f\}$. A formal infinite product

$$\omega(x_1, \ldots, x_f) = \prod_{i=0}^{\infty} K_i^{\lambda_i}$$

is an *extended word* in these variables if

(i) $\lambda_i \in \mathbb{Q}$ for all i,

(ii) Each K_i is a commutator word $K_i(x_1, \ldots, x_f) = (x_{j_1}, \ldots, x_{j_r})$ (r depending on i) in the variables $x_1, \ldots, x_f$,

(iii) Only finitely many terms K_i have a given length r.

Let G be a Mal'cev group, with $g_1, \ldots, g_f \in G$. Then G is a complete R-group so that

$$(K_i(g_1, \ldots, g_f))^{\lambda_i} = (g_{j_1}, \ldots, g_{j_r})^{\lambda_i}$$

is uniquely defined in G. Now the group H generated by $g_1, \ldots, g_f$ is nilpotent, of class c (say), so if K_i has length $> c$ it follows that $K_i(g_1, \ldots, g_f) = 1$. Hence we may define

$$\omega(g_1, \ldots, g_f)$$

to be the product of the finitely many terms $\neq 1$. Thus if ω is any extended word and G is a Mal′cev group, we can think of ω as a function

$$\omega : G \times \ldots \times G \to G.$$

Similarly we may define an *extended Lie word* to be a formal sum

$$\zeta(w_1, \ldots, w_e) = \sum_{i=0}^{\infty} \mu_j J_j$$

where

(i) $\mu_j \in \mathbb{Q}$ for all j,

(ii) Each J_j is a Lie product $J_j(w_1, \ldots, w_e) = [w_{i_1}, \ldots, w_{i_s}]$ (s depending on j) in the variables $w_1, \ldots, w_e$,

(iii) Only finitely many terms J_j have given length s. Then if L is any locally nilpotent Lie algebra over $\mathbb{Q}$ we may consider ζ as a function

$$\zeta : L \times \ldots \times L \to L.$$

In the context of the matrix correspondence of the previous section, we can 'lift' the Lie operations from L to G by defining

$$\lambda g = (\lambda g^{\flat})^{\#}$$
$$g + h = (g^{\flat} + h^{\flat})^{\#}$$
$$[g, h] = [g^{\flat}, h^{\flat}]^{\#}$$

($g, h \in G$; $\lambda \in \mathbb{Q}$). Then G itself becomes a Lie algebra which we denote by $\mathscr{L}(G)$. In the same way we can 'drop' the group operations from G to L by defining

$$lm = (l^{\#} m^{\#})^{\flat}$$
$$l^{\lambda} = (l^{\#\lambda})^{\flat}$$

($l, m \in L$; $\lambda \in \mathbb{Q}$). Now L becomes a complete group, which we denote by $\mathscr{G}(L)$. Clearly $\mathscr{L}(G)$ is isomorphic to L and $\mathscr{G}(L)$ is isomorphic to G.

The crucial point we must observe is that these new operations can be expressed by means of extended words or extended Lie words. This is Lazard's 'inversion'.

LEMMA 4.1 *Let G be a complete subgroup of T, and let $L = G^{\flat}$. Then there exist extended words $\varepsilon_\lambda(x)$ ($\lambda \in \mathbb{Q}$), $\sigma(x, y)$, $\pi(x, y)$ such that for $g, h \in G$ and $\lambda \in \mathbb{Q}$*

$$\lambda g = \varepsilon_\lambda(g)$$

$$g+h = \sigma(g, h)$$

$$[g, h] = \pi(g, h)$$

(where the operations on the left are lifted from L).

Further, there exist extended Lie words $\delta_\lambda(x)$ ($\lambda \in \mathbb{Q}$), $\mu(x, y)$, $\gamma(x, y)$ such that for all $l, m \in L$ and $\lambda \in \mathbb{Q}$

$$l^\lambda = \delta_\lambda(l)$$

$$lm = \mu(l, m)$$

$$(l, m) = \gamma(l, m)$$

(where the operations on the left are dropped from G).

The extended words may be chosen independently of G and L.

Proof:

(1) ε_λ: We have $(\lambda g^{\flat})^{\#} = \exp(\lambda . \log(g)) = g^\lambda$, so we may set

$$\varepsilon_\lambda(x) = x^\lambda.$$

(2) σ: Here we must do more work. We show that there exist words $\sigma_i(x, y)$ satisfying

$$\sigma_{i+1}(x, y) = \sigma_i(x, y)\gamma_{i+1}(x, y)$$

where γ_{i+1} is a word of the form

$$K_1^{\lambda_1} \dots K_u^{\lambda_u} \qquad (\lambda_j \in \mathbb{Q},\ 1 \leqq j \leqq u)$$

with each K_j a commutator $(z_{j_1}, \dots, z_{j_{i+1}})$ of length $i+1$ with all $z_{j_k} = x$ or y. These σ_i must also have the property that for any $c \geqq 1$, if G is a complete subgroup of $T_c(\mathbb{Q})$, then

$$g+h = \sigma_c(g, h) \qquad (g, h \in G).$$

The existence of the σ_i is a consequence of the shuttle procedure. This enables us to take an expression of the form

$$h^{\flat} + \sum \lambda_j C_j \qquad (\lambda_j \in \mathbb{Q})$$

where h lies in some subset H of G, and the C_j are products of length $\geqq r$ in elements of $H^\flat$; and replace it by an expression

$$(hg_1^{\mu_1}\dots g_m^{\mu_m})^\flat + \sum \gamma_i H_i \qquad (\mu_j, \gamma_i \in \mathbb{Q})$$

where the g_j are commutators in elements of H of length r, and the H_i are Lie products of elements of $H^\flat$ with length $\geqq r+1$.

To obtain the σ_i we apply the shuttle procedure to the expression $g^\flat + h^\flat$. We choose a total ordering $\preccurlyeq$ of the left-normed Lie products in x, y in such a way that the length is compatible with the ordering. With g playing the role of h, $\lambda_1 = 1$, $C_1 = h^\flat$, we apply the shuttle procedure and at each stage in the process we

(1) Express all Lie products in $g^\flat$, $h^\flat$ as sums of left-normed commutators in a canonical way,

(2) Collect together all multiples of the same left-normed commutator,

(3) Operate on that term D (in the notation of remark 3.5) which is smallest in the ordering $\preccurlyeq$.

At the ith stage we will have expressed $g^\flat + h^\flat$ in the form

$$(\sigma_i(g, h))^\flat + \sum \theta_k I_k \qquad (\theta_k \in \mathbb{Q})$$

where σ_i is a word in g, h and the terms I_k are Lie products in $g^\flat$, $h^\flat$ of length $> i$. At the $(i+1)$-th stage this will have been modified to

$$(\sigma_i(g, h) . g_1^{\lambda_1}\dots g_m^{\lambda_m})^\flat + \sum \varphi_l J_l \qquad (\varphi_l \in \mathbb{Q})$$

where the g_i are group commutators in g, h of length $i+1$, the $\lambda_i \in \mathbb{Q}$, and the J_l are Lie products in $g^\flat$, $h^\flat$ of length $> i+1$.

We put

$$\gamma_{i+1}(g, h) = g_1^{\lambda_1}\dots g_m^{\lambda_m}$$

$$\sigma_{i+1}(g, h) = \sigma_i(g, h)\gamma_{i+1}(g, h)$$

$$\sigma_0(g, h) = 1.$$

It is clear that the form of the words σ_i, γ_i can be made independent of G, and so correspond to words $\sigma_i(x, y)$ and $\gamma_i(x, y)$.

Now if G consists of $c \times c$ matrices then at the cth stage we have

$$g^\flat + h^\flat = (\sigma_c(g, h))^\flat + \sum \psi_p K_p \qquad (\psi_p \in \mathbb{Q})$$

where the terms K_p are of length $> c+1$ so are zero. Thus

$$g + h = (g^\flat + h^\flat)^\# = \sigma_c(g, h).$$

We now let

$$\sigma(x, y) = \prod_{i=1}^{\infty} \gamma_i(x, y)$$

and this will have the desired properties.
(3) π: Work similarly on the expression

$$1^\flat + [g^\flat, h^\flat]$$

(which equals $[g^\flat, h^\flat]$) with 1 playing the role of h in the shuttle procedure, $\lambda_1 = 1$, $C_1 = [g^\flat, h^\flat]$.
(4) δ_λ: Define

$$\delta_\lambda(x) = \lambda x.$$

(5) μ: Let

$$\mu(x, y) = x+y+\tfrac{1}{2}[x, y]+\ldots$$

as in the Campbell-Hausdorff formula.
(6) γ: The existence of γ follows at once from that of δ_λ and μ. □

In illustration we compute σ up to length 3. To this length the Campbell-Hausdorff formula reads

$$(gh)^\flat = g^\flat+h^\flat+\tfrac{1}{2}[g^\flat, h^\flat]+\tfrac{1}{12}([g^\flat, h^\flat, h^\flat]+[h^\flat, g^\flat, g^\flat]),$$

from which we get

$$(g, h)^\flat = [g^\flat, h^\flat]+\tfrac{1}{2}([g^\flat, h^\flat, g^\flat]+[g^\flat, h^\flat, h^\flat]).$$

We choose left-normed commutators as follows:

$$a^\flat \ll b^\flat \ll [a^\flat, b^\flat] \ll [a^\flat, b^\flat, a^\flat] \ll [a^\flat, b^\flat, b^\flat].$$

Now we have

$$\begin{aligned}(a+b)^\flat &= a^\flat+b^\flat \\ &= (ab)^\flat-\tfrac{1}{2}[a^\flat, b^\flat]-\tfrac{1}{12}([a^\flat, b^\flat, a^\flat]-[a^\flat, b^\flat, b^\flat]) \\ &= (ab)^\flat-\tfrac{1}{2}\{(a, b)^\flat-\tfrac{1}{2}([a^\flat, b^\flat, a^\flat]+[a^\flat, b^\flat, b^\flat])\} \\ &\quad +\tfrac{1}{12}([a^\flat, b^\flat, a^\flat]-[a^\flat, b^\flat, b^\flat]) \\ &= (ab(a, b)^{-1/2})^\flat-\tfrac{1}{2}([(ab)^\flat, (a, b)^{-1/2\flat}]) \\ &\quad +\tfrac{1}{4}([a^\flat, b^\flat, a^\flat]+[a^\flat, b^\flat, b^\flat]) \\ &\quad +\tfrac{1}{12}([a^\flat, b^\flat, a^\flat]-[a^\flat, b^\flat, b^\flat])\end{aligned}$$

$$= (ab(a,b)^{-1/2})^{\flat} - \tfrac{1}{2}([a^{\flat}+b^{\flat}, -\tfrac{1}{2}[a^{\flat}, b^{\flat}]])$$
$$+\tfrac{1}{4}([a^{\flat}, b^{\flat}, a^{\flat}]+[a^{\flat}, b^{\flat}, b^{\flat}])$$
$$+\tfrac{1}{12}([a^{\flat}, b^{\flat}, a^{\flat}]-[a^{\flat}, b^{\flat}, b^{\flat}])$$
$$= (ab(a,b)^{-1/2})^{\flat} + \tfrac{1}{12}[a^{\flat}, b^{\flat}, a^{\flat}] - \tfrac{1}{12}[a^{\flat}, b^{\flat}, b^{\flat}]$$
$$= (ab(a,b)^{-1/2}(a,b,a)^{1/12}(a,b,b)^{-1/12})^{\flat}.$$

Thus up to length 3,

$$\sigma(x,y) = xy(x,y)^{-1/2}(x,y,x)^{1/12}(x,y,y)^{-1/12}.$$

Similarly we find

$$\pi(x,y) = (x,y)(x,y,x)^{-1/2}(x,y,y)^{-1/2}.$$

5. The general version

Let G be any Mal'cev group. Then we can define Lie operations on G by setting

$$\lambda g = \varepsilon_{\lambda}(g)$$
$$g+h = \sigma(g,h)$$
$$[g,h] = \pi(g,h)$$

for $g, h \in G$ and $\lambda \in \mathbb{Q}$. Denote the result by $\mathscr{L}(G)$. In the same way for any locally nilpotent Lie algebra L over $\mathbb{Q}$ define group operations on L by

$$l^{\lambda} = \delta_{\lambda}(l)$$
$$lm = \mu(l,m)$$

for $l, m \in L$ and $\lambda \in \mathbb{Q}$. Denote the result by $\mathscr{G}(L)$.

LEMMA 5.1 *With G and L as above,*
(a) *$\mathscr{L}(G)$ is a locally nilpotent Lie algebra over $\mathbb{Q}$,*
(b) *$\mathscr{G}(L)$ is a Mal'cev group.*

Proof: The axioms for a Lie algebra can be expressed as certain relations between the functions ε_{λ}, σ, π involving at most 3 variables. Thus if these relations hold in the completion of any 3-generator subgroup of G they hold throughout G. Now an aforementioned theorem of Hall [285] (see also Swan [346] or Robinson [332] p. 159) shows that every finitely generated nilpotent torsion-free group can be embedded in the group of $c \times c$ unitriangular matrices over $\mathbb{Z}$ for some c; so its completion embeds in $T_c(\mathbb{Q})$. But the required relations certainly hold in $T_c(\mathbb{Q})$ by the results of section 3.

The local nilpotence of $\mathscr{L}(G)$ follows in the same manner, since $U_c(\mathbb{Q})$ is nilpotent.

The proof that $\mathscr{G}(L)$ is a Mal'cev group is similar, except that the relevant embedding theorem is that of Birkhoff [27]: every finite-dimensional nilpotent Lie algebra over $\mathbb{Q}$ can be embedded in $U_c(\mathbb{Q})$ for some c. □

Now we establish the fundamental relations between G and $\mathscr{L}(G)$, L and $\mathscr{G}(L)$.

THEOREM 5.2 *Let G and H be Mal'cev groups, L a locally nilpotent Lie algebra over $\mathbb{Q}$. Then*

(a) $\mathscr{G}(\mathscr{L}(G)) = G$, $\mathscr{L}(\mathscr{G}(L)) = L$.

(b) *H is a subgroup of G if and only if $\mathscr{L}(H) \leqq \mathscr{L}(G)$.*

(c) *H is a normal subgroup of G if and only if $\mathscr{L}(H) \lhd \mathscr{L}(G)$.*

(d) *A map $\varphi : G \to H$ is a group homomorphism if and only if $\varphi : \mathscr{L}(G) \to \mathscr{L}(H)$ is a Lie homomorphism. The kernel of φ is the same in both cases.*

(e) *If H is a normal subgroup of G then $\mathscr{L}(G/H) = \mathscr{L}(G)/\mathscr{L}(H)$.*

(Note that using part (a) we can recast (b)–(e) in a '$\mathscr{G}$' form rather than an '$\mathscr{L}$' form.)

Proof: (a) Let $g, h \in G$. We must show that for $\lambda \in \mathbb{Q}$

$$g^\lambda = \delta_\lambda(g)$$

$$gh = \mu(g, h)$$

where δ_λ, μ are defined in terms of the operations of $\mathscr{L}(G)$. Now $\delta_\lambda(g) = \lambda g = \varepsilon_\lambda(g) = g^\lambda$. To show that $gh = \mu(g, h)$ we may confine our attention to the completion of the group generated by g and h. Thus without loss of generality G is a subgroup of $T_c(\mathbb{Q})$ for some c. Now the operations in $\mathscr{L}(G)$ are defined by

$$g + h = (g^\flat + h^\flat)^\sharp$$

$$[g, h] = [g^\flat, h^\flat]^\sharp$$

so that

$$\mu(g, h)^\flat = g^\flat + h^\flat + \tfrac{1}{2}[g^\flat, h^\flat] + \ldots$$
$$= (gh)^\flat$$

by Campbell-Hausdorff: therefore $\mu(g, h) = gh$.

(b) and (c) follow at once from the form of the functions ε_λ, π, σ, δ_λ, μ, γ.

(d) Follows from the observation that group homomorphisms (resp. Lie homomorphisms) preserve extended words (resp. extended Lie words). The equality of kernels is because the identity of G is the zero of $\mathscr{L}(G)$.

(e) First we show that the H-cosets in G are setwise the same as the $\mathscr{L}(H)$-cosets in $\mathscr{L}(G)$. Let $x \in G$, $z \in Hx$. Then $z = hx$ for some $h \in H$, and

$$hx = h + x + \tfrac{1}{2}[h, x] + \ldots$$

belongs to $\mathscr{L}(H)+x$ since $\mathscr{L}(H)$ is an ideal of $\mathscr{L}(G)$. Therefore $Hx \subseteq \mathscr{L}(H)+x$.

Now let $y \in \mathscr{L}(H)+x$. Then $y = h+x$ for some $h \in H$, and

$$h + x = h.x.(h, x)^{-1/2}\ldots$$

which lies in Hx since H is normal in G. Therefore $\mathscr{L}(H)+x \subseteq Hx$.

The operations on cosets are obviously defined by the same extended words, and the result follows. □

If we let $\mathscr{C}_\mathscr{G}$ denote the category of Mal'cev groups and group homomorphisms, and $\mathscr{C}_\mathscr{L}$ denote the category of locally nilpotent Lie algebras over $\mathbb{Q}$ and Lie homomorphisms, then we have defined two mutually inverse exact covariant functors

$$\mathscr{L} : \mathscr{C}_\mathscr{G} \to \mathscr{C}_\mathscr{L}$$
$$\mathscr{G} : \mathscr{C}_\mathscr{L} \to \mathscr{C}_\mathscr{G}.$$

Thus we may refer to $\mathscr{L}$ and $\mathscr{G}$ as the *Mal'cev functors*.

(Note, however that in our definition the underlying *sets* of G and $\mathscr{L}(G)$, or of L and $\mathscr{G}(L)$, are equal.)

Next we establish some centraliser/idealiser properties of the Mal'cev functors. We write

$$C_L(X/H) = \{c \in L : [c, X] \subseteq H\}$$

whenever $H \subseteq X \subseteq L$, $H \leqq L$, and $[H, X] \leqq H$.

LEMMA 5.3 *Let G and H be Mal'cev groups with $H \subseteq X \subseteq G$, $H \leqq G$, $(H, X) \subseteq H$. Then*

$$\mathscr{L}(C_G(X/H)) = C_{\mathscr{L}(G)}(\mathscr{L}(X)/\mathscr{L}(H)).$$

Proof: Let $c \in C = C_G(X/H)$. Then for any $x \in X$, we have $[c, x] = (c, x)(c, x, c)^{-1/2} \ldots$ which belongs to H. This gives inclusion one way: the reverse inclusion is proved similarly. □

COROLLARY 5.4
(a) $\mathscr{L}(C_G(X)) = C_{\mathscr{L}(G)}(\mathscr{L}(X))$,
(b) $\mathscr{L}(N_G(H)) = I_{\mathscr{L}(G)}(\mathscr{L}(H))$,
(c) $\mathscr{L}(\zeta_\alpha(G)) = \zeta_\alpha(\mathscr{L}(G))$ *for any ordinal* α.

Proof: (a) put $H = 0$ in lemma 5.3. (b) put $X = H$. For (c) use transfinite induction. □

From (c) it follows that if either G or $\mathscr{L}(G)$ is nilpotent then so is the other, and their nilpotency classes are equal.

LEMMA 5.5 *Let G be a Mal'cev group, and H a complete subgroup of G. Then $H \lhd^\alpha G$ if and only if $\mathscr{L}(H) \lhd^\alpha \mathscr{L}(G)$.*

Proof: There is a normal series

$$H = H_0 \lhd H_1 \lhd \ldots \lhd H_\beta \lhd H_{\beta+1} \lhd \ldots H_\alpha = G$$

with appropriate behaviour at limit ordinals. Let $\overline{H}_\beta$ be the completion of H_β in G. Set $L_\beta = \mathscr{L}(\overline{H}_\beta)$. Then the L_β form an ascending series from $\mathscr{L}(H)$ to $\mathscr{L}(G)$. □

It can be shown that G is soluble of derived length d if and only if $\mathscr{L}(G)$ is soluble of derived length d (see [208]): in fact a more general result is true. Related matters are dealt with by Andreev [16, 17].

6. Explicit descriptions

Because $\mathscr{L}$ and $\mathscr{G}$ are defined locally, and because of the complexity of the functions σ, π, μ, γ it is seldom easy to obtain an explicit description of $\mathscr{G}(L)$ for a given L or $\mathscr{L}(G)$ for a given G. However, there is one situation with a pre-existing global structure where such a description can be given. Moreover, the situation arises sufficiently often for such a description to be useful.

The proof, taken from [222], is quite simple given our present line of development.

6. Explicit descriptions

THEOREM 6.1 *Let A be a locally nilpotent associative algebra over $\mathbb{Q}$. Then A^0 is a Mal'cev group, $[A]$ is a locally nilpotent Lie algebra over $\mathbb{Q}$, and there are natural isomorphisms*

$$\mathscr{L}(A^0) \cong [A]$$

$$\mathscr{G}([A]) \cong A^0.$$

Proof: Define a map

$$\varphi : \mathscr{G}([A]) \to A^0$$

by

$$\varphi(x) = \exp(x)$$

as in section 1. This is well-defined and its image lies inside $1+A = A^0$. Further, φ is a group homomorphism by Campbell-Hausdorff. Now φ is injective, for if $\varphi(x) = 1$ then $x = \log(1) = 0$. Also φ is surjective since if $a \in A$ then $1+a = \exp(\log(1+a))$. □

As an illustration of the use of this result we consider a famous group-theoretic construction of McLain [308, 309, 310] and its Lie analogue [213].

Let $\mathfrak{k}$ be a field, Σ a partially ordered set. Let $\mathscr{A}_{\mathfrak{k}}(\Sigma)$ be the associative algebra with basis

$$\{a_{\sigma\tau} : \sigma < \tau;\ \sigma, \tau \in \Sigma\}$$

and multiplication

$$a_{\sigma\tau} a_{\lambda\mu} = \delta_{\tau\lambda} a_{\sigma\mu}.$$

(Here δ_{**} is the Kronecker delta.) We define the *McLain group*

$$\mathscr{M}_{\mathfrak{k}}(\Sigma) = \mathscr{A}_{\mathfrak{k}}(\Sigma)^0$$

and the *McLain Lie algebra*

$$\mathscr{L}_{\mathfrak{k}}(\Sigma) = [\mathscr{A}_{\mathfrak{k}}(\Sigma)].$$

It is clear that these definitions accord with those in [213, 308, 309, 310]. By theorem 6.1, $\mathscr{M}_{\mathfrak{k}}(\Sigma)$ and $\mathscr{L}_{\mathfrak{k}}(\Sigma)$ correspond (up to isomorphism) under the Mal'cev functors, if $\mathfrak{k} = \mathbb{Q}$. This helps to explain the close analogy between them. For example, if Σ is the integers in natural order then $\mathscr{M}_{\mathfrak{k}}(\Sigma)$ is locally nilpotent with trivial centre: the same is therefore true of $\mathscr{L}_{\mathfrak{k}}(\Sigma)$ (cf. [213] p. 96).

McLain [310] has constructed torsion-free hypercentral groups of the form $\mathscr{M}_{\mathbb{Q}}(\Sigma)$ of arbitrary central height. The corresponding $\mathscr{L}_{\mathbb{Q}}(\Sigma)$ afford examples of hypercentral Lie algebras of arbitrary central height.

Chapter 6

Locally nilpotent radicals

The *Hirsch-Plotkin radical* of a group is its unique maximal locally nilpotent normal subgroup. Its existence was demonstrated by Hirsch [294]: Plotkin [322] proved that it contained every *ascendant* locally nilpotent subgroup. Subsequently other authors introduced a number of related radicals, of which we note in particular the *Fitting*, *Baer*, and *Gruenberg radicals*; generated respectively by the nilpotent normal, subnormal, and ascendant subgroups (cf. Baer [260], Gruenberg [279], Robinson [331]). All of these radicals are locally nilpotent characteristic subgroups.

Analogous definitions may be made for Lie algebras. However, the properties of and relations between the resulting objects depend on the ground field, and are not always the same as in group theory. The object of this chapter is to elucidate these relationships. Most of the basic results are due to Simonjan [192] and Hartley [76]: the main counterexamples come from Hartley [76], Levič and Tokarenko [145], and Stewart [207, 216, 222].

For finite-dimensional Lie algebras over fields of characteristic 0, the various radicals all coincide with the classical nil radical, of which they may be considered generalisations.

1. The Hirsch-Plotkin radical

We begin with a general result of Hartley [76]:

THEOREM 1.1 *Let* $\mathfrak{X}$ *be a class of Lie algebras such that*

(a) $\mathfrak{X} = \{\mathrm{N}_0, \mathrm{S}\}\mathfrak{X}$

(b) $\mathfrak{X} \leq \mathrm{Max}$.

Then $\mathrm{L}\mathfrak{X}$ *is* N_0*-closed.*

Proof: Let $L = H+K$ where $H, K \lhd L$ and $H, K \in \mathrm{L}\mathfrak{X}$. It is sufficient to prove that if $\{a_1, \ldots, a_m\}$ is a finite subset of H and if $\{b_1, \ldots, b_n\}$ is a finite subset of K, then

$$C = \langle a_1, \ldots, a_m, b_1, \ldots, b_n \rangle \in \mathfrak{X}.$$

Let A be the subalgebra of C generated by the a_i and by all products

$$[a_i, b_{j_1}, \ldots, b_{j_k}] \qquad (k \geqq 1). \tag{1}$$

Each b_j idealises A, so $A \lhd C$. The elements (1) all lie in the subalgebra of K generated by the b_k and the $[a_i, b_j]$. This is a finitely generated subalgebra of K, hence lies in $\mathfrak{X}$ and so in Max. Hence $A \in \mathfrak{G}$, and therefore $A \in \mathfrak{X}$. Similarly $\langle b_1, \ldots, b_n \rangle$ is contained in an $\mathfrak{X}$-ideal B of C. Then

$$C = A+B \in \mathrm{N}_0\mathfrak{X} = \mathfrak{X}$$

as required. □

COROLLARY 1.2 *The classes* $\mathrm{L}\mathfrak{F}$, $(\mathrm{LE}\mathfrak{A} \cap \mathrm{L}\mathfrak{F})$, *and* $\mathrm{L}\mathfrak{N}$ *are all* N_0*-closed.*

Proof: The classes $\mathfrak{F}$, $\mathrm{E}\mathfrak{A} \cap \mathfrak{F}$, and $\mathfrak{N} \cap \mathfrak{F}$ all satisfy the hypotheses on $\mathfrak{X}$ in theorem 1.1. Since $\mathfrak{G} \cap \mathfrak{N} \leqq \mathfrak{F}$ we have $\mathrm{L}\mathfrak{N} = \mathrm{L}(\mathfrak{N} \cap \mathfrak{F})$. Clearly $\mathrm{L}(\mathrm{E}\mathfrak{A} \cap \mathfrak{F}) = (\mathrm{LE}\mathfrak{A} \cap \mathrm{L}\mathfrak{F})$. □

THEOREM 1.3 *With the hypotheses of theorem 1.1, every Lie algebra L has a unique maximal* $\mathrm{L}\mathfrak{X}$*-ideal.*

Proof: By Zorn's lemma there exist maximal $\mathrm{L}\mathfrak{X}$-ideals: by theorem 1.1 these are unique. □

This unique maximal $\mathrm{L}\mathfrak{X}$-ideal is called the $\mathrm{L}\mathfrak{X}$*-radical* of L and is written $\rho_{\mathrm{L}\mathfrak{X}}(L)$. When $\mathfrak{X} = \mathfrak{N}$ we have the *Hirsch-Plotkin radical*, which for brevity we write $\rho(L)$.

2. Baer, Fitting, and Gruenberg radicals

For any Lie algebra L we define the *Fitting radical* $\nu(L)$ to be the sum of the nilpotent ideals of L. If the underlying field has characteristic zero we define the *Baer radical* $\beta(L)$ to be the subalgebra generated by the nilpotent subideals of L, and the *Gruenberg radical* $\gamma(L)$ to be the subalgebra generated by the nilpotent ascendant subalgebras of L.

Clearly $\nu(L) \lhd L$ and is locally nilpotent. The reason for defining $\beta(L)$ and $\gamma(L)$ only for characteristic zero is that we may use the coalescence theorems of chapter 3 to prove them locally nilpotent.

THEOREM 2.1 *If L is a Lie algebra over a field of characteristic zero then $\beta(L)$ and $\gamma(L)$ are locally nilpotent. Further, we can characterise $\beta(L)$ as*

(1) *The join of the $\mathfrak{N} \cap \mathfrak{F}$-subideals of L,*
(2) *The join of the 1-dimensional subideals of L,*
(3) *The set of elements $x \in L$ such that $\langle x \rangle$ si L,*

and $\gamma(L)$ as

(4) *The join of the ascendant $\mathfrak{N} \cap \mathfrak{F}$-subalgebras of L,*
(5) *The join of the ascendant 1-dimensional subalgebras of L,*
(6) *The set of elements $x \in L$ such that $\langle x \rangle$ asc L.*

Proof: We deal with $\gamma(L)$: the proof are similar but easier for $\beta(L)$. Local nilpotence follows immediately from theorem 3.2.4.

Let $\gamma_4(L)$, $\gamma_5(L)$, $\gamma_6(L)$ be the subsets of L defined by (4), (5), (6). Clearly

$$\gamma_6(L) \subseteq \gamma_5(L) \subseteq \gamma_4(L) \subseteq \gamma(L)$$

so it suffices to prove that $\gamma(L) \subseteq \gamma_6(L)$. Let $x \in \gamma(L)$. Then $x \in \langle N_1, \ldots, N_k \rangle$ where each N_i is nilpotent and ascendant in L. Hence there exist finitely generated subalgebras $M_i \leqq N_i$ such that $x \in \langle M_1, \ldots, M_k \rangle$. Now M_i si N_i asc L so that M_i asc L; further each $M_i \in \mathfrak{N} \cap \mathfrak{F}$. By theorem 3.2.4 again $\langle M_1, \ldots, M_k \rangle$ is nilpotent and ascendant in L. Therefore $\langle x \rangle$ si $\langle M_1, \ldots, M_k \rangle$ asc L so that $\langle x \rangle$ asc L and $x \in \gamma_6(L)$. □

We call L a *Fitting algebra* if $L = \nu(L)$, a *Baer algebra* if it has characteristic zero and $L = \beta(L)$, and a *Gruenberg algebra* if it has characteristic zero and $L = \gamma(L)$. We denote the respective classes by $\mathfrak{Ft}$, $\mathfrak{B}$, $\mathfrak{Gr}$. It turns out to be convenient to define L to be a *Baer algebra* in characteristic $p > 0$ if $x \in L$ implies $\langle x \rangle$ si L, largely because such an algebra *is* locally nilpotent (see theorem 7.1.5): this does not conflict with the definition in characteristic zero, and we use $\mathfrak{B}$ for the corresponding class without ambiguity. (Whether a similar idea works for $\mathfrak{Gr}$ is not known.)

In characteristic zero we obviously have

$$\nu(L) \leqq \beta(L) \leqq \gamma(L)$$

$$\nu(L) \leqq \rho(L)$$

and

$$\mathfrak{F}\mathfrak{t} \leqq \mathfrak{B} \leqq \mathfrak{G}\mathfrak{r} \leqq \mathrm{L}\mathfrak{N}.$$

In characteristic $p > 0$ all we know is that

$$\nu(L) \leqq \rho(L)$$

and

$$\mathfrak{F}\mathfrak{t} \leqq \mathrm{L}\mathfrak{N}, \ \mathfrak{F}\mathfrak{t} \leqq \mathfrak{B},$$

though using 7.1.5 we can strengthen these last to read

$$\mathfrak{F}\mathfrak{t} \leqq \mathfrak{B} \leqq \mathrm{L}\mathfrak{N}.$$

3. Behaviour under derivations

It is not clear from the definitions whether $\beta(L)$ or $\gamma(L)$ are ideals of L, or whether any of the four radicals might be characteristic. If the field has characteristic $p > 0$ it is well known that the nil radical of a finite-dimensional Lie algebra need not be a characteristic ideal: *a fortiori* the same holds for the Fitting and Hirsch-Plotkin radicals in characteristic p. Since the other two radicals are not defined in characteristic p, we may confine our attention to fields of characteristic 0. Hartley [76] has shown that we now get better behaviour: $\beta(L)$ and $\nu(L)$ are characteristic. However $\rho(L)$ need not be characteristic, and $\gamma(L)$ need not even be an ideal. We proceed to justify these assertions.

THEOREM 3.1 *Suppose L is a Lie algebra over a field of characteristic 0, and that $N \lhd L$, $N \in \mathfrak{N}_c$. Let $d \in \mathrm{Der}(L)$. Then Nd is contained in an ideal I of L such that $I \in \mathfrak{N}_{2c}$.*

Proof: We use formal power series as in chapter 4. We have $N^\uparrow \lhd L^\uparrow$ and $N^\uparrow \in \mathfrak{N}_c$. Thus

$$M = (N^\uparrow)^{\exp(td)} \lhd L^\uparrow$$

and $M \in \mathfrak{N}_c$. Therefore

$$K = M + N^\uparrow \lhd L^\uparrow$$

and $K \in \mathfrak{N}_{2c}$ by theorem 1.2.5. It follows that $K^\downarrow \in \mathfrak{N}_{2c}$ and is an ideal of L.

But if $x \in N$ then K contains x and

$$x+txd+\frac{t^2}{2}xd^2+\dots$$

and hence contains

$$txd+\dots$$

so that $xd \in K^{\downarrow}$. Therefore $Nd \subseteq K^{\downarrow}$ as required. □

COROLLARY 3.2 *If L is a Lie algebra over a field of characteristic 0 then $\nu(L)$ is a characteristic ideal of L.* □

The subscript $2c$ in theorem 3.1 cannot be improved to c, at least for $c = 1$, as is shown by Hartley [76]. Consider the algebra M of zero-triangular 5×5 matrices over $\mathfrak{k}$. Denote elementary matrices by e_{ij}. Now M is nilpotent, and has an ideal L consisting of matrices (a_{ij}) such that $a_{ij} = 0$ unless $i = 1$ or $j = 5$. Then $L^2 = \langle e_{15}\rangle$. Further, L has an ideal N consisting of matrices

$$\alpha e_{12}+\beta e_{15}+\gamma e_{45}$$

$(\alpha, \beta, \gamma \in \mathfrak{k})$ and N is abelian. The inner derivation $d = (e_{23}+e_{34})^*$ is such that Nd contains

$$[e_{12}, e_{23}] = e_{13}$$
$$[e_{34}, e_{45}] = e_{35}.$$

But

$$[e_{13}, e_{35}] = e_{15} \neq 0$$

so that Nd is not contained in an abelian ideal of L.

A similar proof works for the Baer radical:

THEOREM 3.3 *If L is a Lie algebra over a field $\mathfrak{k}$ of characteristic 0 then $\beta(L)$ is a characteristic ideal of L.*

Proof: Let $M \in \mathfrak{F} \cap \mathfrak{N}$ be a subideal of L. Then over $\mathfrak{k}_0 = \mathfrak{k}\langle t\rangle$ we have $M^{\uparrow} \in \mathfrak{F} \cap \mathfrak{N}$ and $M^{\uparrow}$ si $L^{\uparrow}$. Therefore $M^{\uparrow} \leqq \beta(L^{\uparrow})$. Let $d \in \mathrm{Der}(L)$, $\alpha = \exp(td)$. Then since $\beta(L^{\uparrow})$ is automorphism-invariant,

$$M^{\uparrow\alpha} \leqq \beta(L^{\uparrow}).$$

Hence

$$K = \langle M^{\uparrow}, M^{\uparrow\alpha}\rangle \leqq \beta(L^{\uparrow})$$

so that K si $L^{\uparrow}$ and $K \in \mathfrak{N} \cap \mathfrak{F}$ (over $\mathfrak{k}_0$). Then $K^{\downarrow}$ si L and $K^{\downarrow} \in \mathfrak{N}$. But as in theorem 3.1, $Md \subseteq K^{\downarrow} \leqq \beta(L)$. Therefore $\beta(L)$ ch L. □

For the Gruenberg radical there is a weaker result, also due to Hartley (cf. Stewart [213] p. 85).

THEOREM 3.4 *Let L be a Lie algebra over a field $\mathfrak{k}$ of characteristic* 0. *Then $\gamma(L)$ is invariant under every locally finite derivation of L.*

Proof: This time we use the tensor product algebra $L^{\wedge}$. Let d be any locally finite derivation of L. Then $\alpha = \exp(td)$ is an automorphism of $L^{\uparrow}$, and by lemma 4.2.2 it fixes $L^{\wedge}$, so is an automorphism of $L^{\wedge}$.

Now let $(V_\sigma)_{\sigma \leqq \rho}$ be an ascending series in L. Then $(V_\sigma^{\wedge})_{\sigma \leqq \rho}$ is an ascending series in $L^{\wedge}$. This is obvious except at limit ordinals, and follows at limit ordinals because of a previously-mentioned finiteness property of the tensor product: that for any subspace V of L the space $V^{\wedge}$ consists of those elements $x = \sum x_r t^r$ of $V^{\uparrow}$, all of whose coefficients lie in some *finite-dimensional* subspace of V. (See chapter 4 section 1).

Thus if $x \in \gamma(L)$ then $x \in \gamma(L^{\wedge})$. Since α is an automorphism x^{α} also lies in $\gamma(L^{\wedge})$, so that

$$t^{-1}(x - x^{\alpha}) = xd + \tfrac{1}{2}txd^2 + \ldots \in \gamma(L^{\wedge}).$$

Now $L_1 = \{\sum_{i=0}^{\infty} x_i t^i\}$ is a $\mathfrak{k}$-subalgebra of $L^{\wedge}$, and by theorem 2.1(6)

$$xd + \tfrac{1}{2}txd^2 + \ldots \in \gamma(L_1).$$

Also $L_2 = \{\sum_{i=1}^{\infty} x_i t^i\}$ is a $\mathfrak{k}$-ideal of L_1. There is an obvious isomorphism $L_1/L_2 \to L$ which preserves the action of d. From this it follows that $xd \in \gamma(L)$. □

COROLLARY 3.5 *If L is a Lie algebra over a field of characteristic 0 and if L is generated by ascendant* L$\mathfrak{F}$*-subalgebras (and in particular if $L \in$* L$\mathfrak{F}$*) then $\gamma(L)$ is an ideal of L.*

Proof: If x is an element of an ascendant L$\mathfrak{F}$-subalgebra then x^* is a locally finite derivation, as shown after lemma 2.2. □

Of course if $L \in$ L$\mathfrak{N}$ (over a field of characteristic 0) we can prove that $\gamma(L) \lhd L$ using exponentials of inner derivations.

Example 3.6

The following example, due to Hartley [76], shows how far the above results are best-possible.

Let $\mathfrak{k}$ be any field of characteristic 0, and let $P = \mathfrak{k}[t]$ be a polynomial algebra. Considered as an abelian Lie algebra P has derivations

$$x: p \mapsto tp$$

$$y: p \mapsto \frac{\mathrm{d}p}{\mathrm{d}t}.$$

Then $[x, y] = z$ is the identity on P, whilst $[x, z] = [y, z] = 0$. So $\langle x, y, z \rangle = Q \subseteq \mathrm{Der}(P)$ is 3-dimensional and nilpotent of class 2.

Now P is an irreducible Q-module. For if $0 \neq p \in P$ then sufficiently many differentiations map p to a non-zero scalar, from which we can derive all the elements of P using the x-action. Let L be the split extension $P \dotplus Q$. Let $L_1 = P + \langle y \rangle$, $L_2 = P + \langle y \rangle + \langle z \rangle$. Then

$$P \lhd L_1 \lhd L_2 \lhd L.$$

It is easy to see that $L_1 \in \mathrm{L}\mathfrak{N}$, for if P_n comprises those elements of P of degree $\leqq n$, then each finite subset of L_1 lies in some $P_n + \langle y \rangle$, which is nilpotent. Also we have

$$\langle y \rangle \lhd P_0 + \langle y \rangle \lhd \ldots \lhd P_n + \langle y \rangle \lhd \ldots$$

so that

$$\langle y \rangle \operatorname{asc} L_1.$$

Now $L_2 \notin \mathrm{L}\mathfrak{N}$ because for any $k \in \mathbb{N}$,

$$[1, {}_k z] = 1 \neq 0.$$

Therefore $\rho(L_2) = L_1$. Now if $p \in P$ then $[p, z] = p$ so that $p \notin I_L(z)$. So $\langle z \rangle$ is not ascendant in $P + \langle z \rangle$ and so is not ascendant in L_2 or in L. But $d = x^*|_{L_2}$ is a derivation of L_2 mapping y to z. Therefore

$$\rho(L_2)d \nsubseteq \rho(L_2)$$

so that $\rho(L_2)$ is not characteristic in L_2. Also $\gamma(L_2)$ is not characteristic in L_2.

Moreover, $\rho(L)$ cannot contain z, so we must have $\rho(L) = P$. But $\langle y \rangle \operatorname{asc} L$. Hence $\rho(L)$ does not contain every ascendant $\mathrm{L}\mathfrak{N}$-subalgebra of L, in contrast to what happens for groups. Further, $\gamma(L) = L_1$, which is not an ideal of L.

From these examples it follows that the four radicals are in general distinct, except perhaps for $\nu(L)$ and $\beta(L)$. However, we have not yet distinguished between the classes $\mathfrak{Ft}$, $\mathfrak{B}$, $\mathfrak{Gr}$, and $\mathrm{L}\mathfrak{N}$. The next sections take up this problem.

4. Baer and Fitting algebras

The main object of this section will be to construct a Lie algebra which is Baer but not Fitting (cf. [207, 216]). We shall say a little about Gruenberg algebras here, and rather more in section 6.

It is easy to see that Gruenberg algebras need not be Baer, because of:

LEMMA 4.1 *Over a field of characteristic 0, every hypercentral Lie algebra is a Gruenberg algebra.*

Proof: Let L be hypercentral, and let $Z_\alpha = \zeta_\alpha(L)$. Then for any $x \in L$ the series $(Z_\alpha + \langle x \rangle)$ shows that $\langle x \rangle$ asc L. □

Now let X be a vector space with basis $x_0, x_1, x_2, \ldots$ and think of X as an abelian Lie algebra. There is a derivation σ such that

$$x_0\sigma = 0$$

$$x_i\sigma = x_{i-1} \qquad (i > 0),$$

the *downward shift* on X. Form the split extension $L = X \dotplus \langle \sigma \rangle$. Then it is easy to see that

$$\zeta_n(L) = \langle x_0, \ldots, x_{n-1} \rangle$$

for n finite,

$$\zeta_\omega(L) = X$$

$$\zeta_{\omega+1}(L) = L.$$

Therefore L is hypercentral, and hence Gruenberg. But L is not Baer, because $\langle \sigma^L \rangle = L$ so that $\langle \sigma \rangle$ cannot be a subideal of L.

Gruenberg algebras need not be hypercentral. The McLain algebra $\mathscr{L}_{\mathfrak{k}}(\mathbb{Z})$ is generated by abelian ideals (cf. McLain [308]) so is Fitting (and *a fortiori* Baer or Gruenberg if char($\mathfrak{k}$) = 0) but has trivial centre so cannot be hypercentral.

All the radicals behave well under the Mal'cev correspondence. If we use ν, β, γ in the obvious sense for groups, we have:

THEOREM 4.2 *Let G be a locally nilpotent torsion-free group, with completion G^*. Then*

(a) $\mathscr{L}(\nu(G)^*) = \mathscr{L}(\nu(G^*)) = \nu(\mathscr{L}(G^*))$.
(b) $\mathscr{L}(\beta(G)^*) = \mathscr{L}(\beta(G^*)) = \beta(\mathscr{L}(G^*))$.
(c) $\mathscr{L}(\gamma(G)^*) = \mathscr{L}(\gamma(G^*)) = \gamma(\mathscr{L}(G^*))$.

Proof: We prove (a). Let $x \in \nu(G)^*$. By lemma 5.2.2(c) there exists $n \in \mathbb{Z}$ such that $x^n \in \nu(G)$. Thus $x^n \in N \lhd G$ for some nilpotent N. Therefore $x \in \bar{N}$ (completion in G^*). Now $\bar{N} \lhd G^*$ by lemma 5.2.2(a) since $\bar{G}^* = G^*$, and $\bar{N}$ is nilpotent by lemma 5.2.2(d). Thus $x \in \nu(G^*)$, and $\nu(G)^* \leqq \nu(G^*)$.

Now let $y \in \nu(G^*)$. Then $y \in M \lhd G^*$, where M is nilpotent. By lemma 5.2.2(d) and (a), $\bar{M}$ is nilpotent and normal in G^*. By lemma 5.2.2(e) $\bar{M} = \overline{\bar{M} \cap G}$. So for some $m \in \mathbb{Z}$, we have $y^m \in \bar{M} \cap G$. But $\bar{M} \cap G \lhd G$, and is nilpotent. Thus $y^m \in \nu(G)$, so $y \in \nu(G)^*$. Thus $\nu(G^*) \leqq \nu(G)^*$. Combining the two inclusions gives the first part of (a). The second follows from theorem 5.5.2 and corollary 5.5.4(c).

The proofs of (b) and (c) are similar: instead of normal subgroups one uses subnormal or ascendant subgroups. □

COROLLARY 4.3 *Let G be a locally nilpotent torsion-free group with completion G^*. If any one of G, G^*, $\mathscr{L}(G^*)$ is Fitting (resp. Baer, Gruenberg) then so are the other two.* □

In view of corollary 4.3, in order to find a non-Fitting Baer algebra it would suffice to find a non-Fitting torsion-free Baer group. Such a group has been constructed by Roseblade and Stonehewer [337], although its properties are not immediately obvious from its definition. But such a roundabout route turns out not to be necessary: the whole Roseblade-Stonehewer construction can be performed for Lie algebras. Moreover this is desirable, since it gives some information in characteristic p which would not be available through the Mal'cev correspondence.

The Roseblade-Stonehewer algebras

Let $\mathfrak{k}$ be any field (not necessarily of characteristic 0) and let A be an *infinite*-dimensional vector space over $\mathfrak{k}$. Form the *exterior algebra R* generated by A over $\mathfrak{k}$. (First form the *tensor algebra*

$$T = k \oplus A \oplus (A \otimes A) \oplus (A \otimes A \otimes A) \oplus \ldots$$

with multiplication induced by $\otimes$, and let I be the ideal of T generated by all elements $a \otimes a$ ($a \in A$). Put $R = T/I$.)

It is well known and easy to prove that R has the following properties (Chevalley [269]): R is an associative $\mathfrak{k}$-algebra, containing isomorphic copies of $\mathfrak{k}$ and A. Making the obvious identifications, $\mathfrak{k} \cap A = 0$. The algebra R is graded, and the homogeneous elements of degree i are the products of i elements of A when $i > 0$, and the elements of $\mathfrak{k}$ when $i = 0$. Further,

(i) $a\lambda = \lambda a \qquad (a \in A,\ \lambda \in \mathfrak{k})$,
(ii) $a^2 = 0 \qquad (a \in A)$,
(iii) If $x \in R$ is such that $xA = 0$, then $x = 0$.

(If A is finite-dimensional, part (iii) fails.) From (ii) it follows that for all, $a, b \in A$ we have $(a+b)^2 = 0$, so that $ab = -ba$. Hence for any a, b, c, d A we have

$$abc = cab, \quad abcd = cdab. \tag{2}$$

We begin by constructing a Lie algebra J of 2×2 matrices over A (considered as a Lie algebra over $\mathfrak{k}$, and with Lie multiplication $[M, N] = MN - NM$). Let K be the set of all matrices of the form

$$\begin{pmatrix} 0 & 0 \\ \lambda & 0 \end{pmatrix} \qquad (\lambda \in \mathfrak{k})$$

and let H be the set of all matrices of the form

$$\begin{pmatrix} 0 & a \\ 0 & 0 \end{pmatrix} \qquad (a \in A).$$

Clearly H and K are abelian Lie algebras: K is 1-dimensional and H is isomorphic to A as vector space, so is infinite-dimensional. Let $J = \langle H, K \rangle$.

LEMMA 4.4 *The ideal closures $\langle H^J \rangle$ and $\langle K^J \rangle$ are both nilpotent of class 2.*

Proof: Let Z be the subalgebra of J generated by all matrices

$$\begin{pmatrix} ab+c & 0 \\ d & ab-c \end{pmatrix} \qquad (a, b, c, d \in A). \tag{3}$$

Then

$$\left[\begin{pmatrix} ab+c & 0 \\ d & ab-c \end{pmatrix}, \begin{pmatrix} pq+r & 0 \\ s & pq-r \end{pmatrix}\right] = \begin{pmatrix} \alpha & 0 \\ \beta & \gamma \end{pmatrix}$$

for $p, q, r, s \in A$, where

$$\alpha = (ab+c)(pq+r) - (pq+r)(ab+c)$$
$$\beta = d(pq+r) + (ab-c)s - s(ab+c) - (pq-r)d$$
$$\gamma = (ab-c)(pq-r) - (pq-r)(ab-c).$$

Using (2) this reduces to

$$\begin{pmatrix} 2cr & 0 \\ 0 & 2cr \end{pmatrix} \tag{4}$$

which is of the form (3) with $a = 2c$, $b = r$, $c = d = 0$. Thus Z is *spanned* by all matrices of the form (3).

Hence $[Z, H]$ is spanned by all products

$$\left[\begin{pmatrix} ab+c & 0 \\ d & ab-c \end{pmatrix}, \begin{pmatrix} 0 & e \\ 0 & 0 \end{pmatrix}\right]$$

for $a, b, c, d, e \in A$. Using (1) such a product equals

$$\begin{pmatrix} de & 0 \\ 0 & de \end{pmatrix} \tag{5}$$

which lies in Z. So $[Z, H] \subseteq Z$. Similarly $[Z, K] \subseteq Z$. Also $[H, K]$ is generated by all products

$$\left[\begin{pmatrix} 0 & a \\ 0 & 0 \end{pmatrix}, \begin{pmatrix} 0 & 0 \\ \lambda & 0 \end{pmatrix}\right]$$

which works out as

$$\begin{pmatrix} \lambda a & 0 \\ 0 & -\lambda a \end{pmatrix}$$

and is in Z.

Consequently $Z+H$ and $Z+K$ are idealised both by H and by K, and therefore by J. Thus $\langle H^J \rangle \leqq Z+H$, $\langle K^J \rangle \leqq Z+K$. (In fact these are equalities.) To prove the lemma it is sufficient to show that $Z+H$ and $Z+K$ are nilpotent of class 2. Now H is abelian so

$$[Z+H, Z+H] \leqq [Z, Z]+[Z, H].$$

Matrices in $[Z, Z]$ are sums of matrices of the form

$$\begin{pmatrix} pq & 0 \\ 0 & pq \end{pmatrix} \qquad (p, q \in A) \tag{6}$$

and so are matrices in $[Z, H]$, by (4) and (5). Further, anything of the form (6) is in the centre of Z by (5). Therefore $[Z+H, Z+H, Z+H] = 0$ as required.

The calculations for $Z+K$ are much the same. □

Now J acts naturally on the $\mathfrak{k}$-vector space $V = R \times R$, and therefore J can be considered as a Lie algebra of derivations of the abelian Lie algebra V.

Construct the split extension

$$L = V \dotplus J.$$

If $(x, y) \in V$ then

$$\left[(x, y), \begin{pmatrix} 0 & 0 \\ \lambda & 0 \end{pmatrix}\right] = (\lambda y, 0) \tag{7}$$

$$\left[(x, y), \begin{pmatrix} 0 & a \\ 0 & 0 \end{pmatrix}\right] = (0, xa). \tag{8}$$

Let

$$V_1 = \{(0, y) : y \in R\}$$

$$V_2 = \{(x, 0) : x \in R\}.$$

From (7) $C_V(K) = V_2$, and from (8) and (iii) $C_V(H) = V_1$. Now $[V, H] \leqq V_1$ so that $[V, H, H] = 0$. Since $V \lhd L$ and V and H are abelian, it follows that $V+H \in \mathfrak{N}_2$. By lemma 1.3.7

$$H \lhd^2 V+H \lhd^2 V+\langle H^J \rangle \lhd L$$

so that $H \lhd^5 L$. Similarly $K \lhd^5 L$.

From this we can obtain another example where joins of subideals fail to be subideals. We show that $J = I_L(J)$. For if $i \in I_L(J)$ then $i = v+j$ for $v \in V$, $j \in J$. Since $V \lhd L$ and $V \cap J = 0$ we must have $v = 0$, so that $i \in J$. This means that J is the join of two 5-step abelian subideals, one of which is 1-dimensional; yet J is not a subideal, nor even an ascendant subalgebra.

Now $J = \langle H^J \rangle + \langle K^J \rangle$, so by Fitting's theorem $J \in \mathfrak{N}_4$. Further, L is the join of K and $V+H$, both nilpotent subideals of L. But L is not nilpotent, since J is self-idealising. Therefore no analogue of the Derived Join Theorem can hold for nilpotent algebras instead of soluble.

The algebra L is generated by three abelian subideals V, H, and K. So if $\mathrm{char}(\mathfrak{k}) = 0$, L is Baer. We show that L is not Fitting. If L were Fitting then K would lie in a nilpotent ideal, so that $\langle K^L \rangle$ would be nilpotent. But this contains all matrices

$$\left[\begin{pmatrix} 0 & c \\ 0 & 0 \end{pmatrix}, \begin{pmatrix} 0 & 0 \\ 1 & 0 \end{pmatrix}\right] = \begin{pmatrix} c & 0 \\ 0 & -c \end{pmatrix} \qquad (c \in A)$$

and also it contains $\langle K^V \rangle$, which includes all vectors $(\lambda y, 0)$ for $\lambda \in \mathfrak{k}$, $y \in A$. Therefore for any $a, c_1, c_2, \ldots, c_n \in A$ and $n \in \mathbb{N}$; $\langle K^L \rangle^{n+1}$ contains

$$\left[(a, 0), \begin{pmatrix} c_1 & 0 \\ 0 & -c_1 \end{pmatrix}, \ldots, \begin{pmatrix} c_n & 0 \\ 0 & -c_n \end{pmatrix}\right]$$

which is

$$(ac_1c_2\ldots c_n, 0).$$

By (iii) this is not 0 for a suitable choice of $a, c_1, c_2, \ldots, c_n$. Therefore $\langle K^L \rangle$ is not nilpotent.

We have proved:

THEOREM 4.5 *Over any field $\mathfrak{k}$ of characteristic zero there exists a Baer algebra in the class $\mathfrak{A}\mathfrak{N}_4$ which is not a Fitting algebra.* □

In contrast, it may be shown that any Baer algebra in the class $\mathfrak{N}\mathfrak{A}$ *is* Fitting (Stewart [207] p. 142).

The algebra just constructed also answers a question of Hartley [76] p. 260, and shows that if B is a finite-dimensional subideal of a Lie algebra L, there need not exist $I \lhd L$ with $I^n \leqq B \leqq I$ for some $n \in \mathbb{N}$. All we need do is put $B = K$.

The construction of L above does not need a field of characteristic 0. But in characteristic p it is not clear whether it is a Baer algebra under the revised definition. It is certainly not Fitting. In fact we can show that if $\mathfrak{k}$ is a field of characteristic $p > 0$ then L *is* Baer; further we do not need to make any computations. A sketch of the proof runs like this: we can do the whole construction with the field $\mathfrak{k}$ replaced by a commutative ring $\mathfrak{r}$. Let us denote the resulting algebra by $L(\mathfrak{r})$. If $\mathfrak{k}$ is any field we can find a polynomial ring $\mathfrak{r}$ over $\mathbb{Z}$ in a sufficiently vast number of variables to yield a ring epimorphism $\mathfrak{r} \to \mathfrak{k}$. Since $\mathfrak{r}$ is an integral domain it has a field of fractions $\mathfrak{f}$ with $\mathfrak{r} \subseteq \mathfrak{f}$. Further, $\mathfrak{f}$ has characteristic 0. The ring maps induce Lie ring maps

$$L(\mathfrak{f}) \supseteq L(\mathfrak{r}) \to L(\mathfrak{k}).$$

Let x be any element of $L(\mathfrak{k})$. Pull it back to $L(\mathfrak{r})$ and include it in $L(\mathfrak{f})$ to get an element y. Since $\mathfrak{f}$ has characteristic zero, $L(\mathfrak{f})$ is Baer, so that $\langle y \rangle$ si $L(\mathfrak{f})$. Restricting to $L(\mathfrak{r})$ and applying the homomorphism onto $L(\mathfrak{k})$ shows that $\langle x \rangle$ si $L(\mathfrak{k})$. Therefore $L(\mathfrak{k})$ is Baer but not Fitting.

It follows from theorem 4.5 that the Baer and Fitting radicals are in general distinct. However, $\nu(L) \neq 0$ in the example, because $V \subseteq \nu(L)$. No example is known of a Baer algebra with *trivial* Fitting radical.

5. The Levič-Tokarenko theorem

Kovàcs and Neumann, and independently Kargapolov, constructed a locally nilpotent non-Gruenberg group (Kargapolov [296], Robinson [331] p. 28). Since their group is a p-group we cannot use the Mal′cev correspondence to obtain a locally nilpotent non-Gruenberg Lie algebra. However, we can convert their group into a torsion-free group using a theorem of Levič and Tokarenko [145]:

THEOREM 5.1 *Every periodic locally nilpotent group is an epimorphic image of a torsion-free locally nilpotent group.*

Proof: Periodic locally nilpotent groups are uniquely expressible as direct sums of p-groups for distinct primes p (Kuroš [303] p. 229), whence it suffices to prove the theorem for a locally finite p-group P. Form the group algebra $\mathbb{Z}_pP$ and let I be its *augmentation ideal*, spanned by the elements $g-1$ ($g \in P$). Every finite subset of I lies inside the augmentation ideal of $\mathbb{Z}_pQ$, for some finite subgroup Q of P. By Curtis and Reiner [272] p. 189 this is nilpotent, so that I is a locally nilpotent associative algebra over $\mathbb{Z}_p$. Therefore the multiplicative semigroup $\dot{I}$ of I is a locally nilpotent semigroup (with 0). We form the semigroup rings $\mathbb{Z}\dot{I}$ and $\mathbb{Z}_p\dot{I}$ (identifying the zero of $\dot{I}$ with that of the coefficient ring) and these are also locally nilpotent. There are obvious ring homomorphisms

$$\mathbb{Z}\dot{I} \to \mathbb{Z}_p\dot{I} \to I$$

and these induce homomorphisms of adjoint groups

$$(\mathbb{Z}\dot{I})^0 \to (\mathbb{Z}_p\dot{I})^0 \to I^0.$$

Further, the map $g \to g-1$ is a group monomorphism $P \to I^0$.

We claim that $T = (\mathbb{Z}\dot{I})^0$ is a locally nilpotent torsion-free group. For let $t_1, \ldots, t_n \in T$. There exists a finitely generated subring U of $\mathbb{Z}\dot{I}$ containing $t_1, \ldots, t_n$. By local nilpotence, $U^k = 0$ for some $k > 0$. Now every element of the derived group $\gamma_2(T)$ is of the form $(1+u)^{-1}(1+v)^{-1}(1+u)(1+v)$ for $u, v \in U$. This is congruent to 1 modulo U^2. Inductively we see that $t_1, \ldots, t_n$ lie inside a nilpotent subgroup of T. Therefore T is locally nilpotent. Now if $t \in T$ then $t = 1+u$, and $(1+u)^r \equiv 1+ru \pmod{(u^2)}$, so is never 1 for $r \neq 0$. So T is torsion-free.

Identify P with its image in I^0 and co-restrict, giving a subgroup $\mathscr{T}(P)$ of T and a canonical homomorphism

$$\mathscr{T}(P) \to P.$$

Then $\mathscr{T}(P)$ is the desired group. □

We call $\mathscr{T}(P)$ the Levič-Tokarenko *unwrapping* of P. We could take P to be the Kovàcs-Neumann-Kargapolov group: then $\mathscr{T}(P)$ has a non-Gruenberg homomorphic image so is itself non-Gruenberg. So $\mathscr{L}(\mathscr{T}(P)^*)$ would be a non-Gruenberg locally nilpotent Lie algebra over $\mathbb{Q}$. However, this procedure can be simplified a little. Take $\dot{I}$ as above and form the semigroup algebra $\mathbb{Q}\dot{I}$. This is a locally nilpotent associative $\mathbb{Q}$-algebra. Then

$$L = [\mathbb{Q}\dot{I}]$$

is a locally nilpotent Lie algebra over $\mathbb{Q}$. By theorem 5.6.1

$$\mathscr{G}(L) = (\mathbb{Q}\dot{I})^0 \supseteq (\mathbb{Z}\dot{I})^0 \supseteq \mathscr{T}(P).$$

Therefore $\mathscr{G}(L)$ has a subquotient isomorphic to P, which is non-Gruenberg: so L is non-Gruenberg. Thus:

THEOREM 5.2 *Let $\dot{I}$ be the multiplicative semigroup of the augmentation ideal of $\mathbb{Z}_pP$, where P is any non-Gruenberg locally finite p-group. If A is the semigroup algebra $\mathbb{Q}\dot{I}$ then $[A]$ is a locally nilpotent non-Gruenberg Lie algebra over $\mathbb{Q}$.* □

COROLLARY 5.3 *For any field $\mathfrak{k}$ of characteristic 0 there exist locally nilpotent Lie algebras over $\mathfrak{k}$ which are not Gruenberg algebras.*

Proof: Take P to be the Kovàcs-Neumann-Kargapolov group, form A as above, and let $L = \mathfrak{k} \otimes_{\mathbb{Q}} [A]$. It is not hard to check that L is non-Gruenberg and locally nilpotent. □

We can improve on this corollary to obtain locally nilpotent algebras with *trivial* Gruenberg radicals. First we need a characterisation of Gruenberg algebras, due to Simonjan [192].

THEOREM 5.4 *A locally nilpotent Lie algebra over a field of characteristic 0 is a Gruenberg algebra if and only if it has an ascending abelian series.*

Proof: Let L be a Gruenberg algebra. Then there exists $H \operatorname{asc} L$ with H abelian. First we show how to construct and ascending abelian series from H to $\langle H^L \rangle$. There is a series $(H_\alpha)_{\alpha \leqq \sigma}$ with $H_0 = H$, $H_\sigma = L$. Let $H^* = \langle H^L \rangle$,

$H_\alpha^* = \langle H^{H_\alpha}\rangle$ $(0 \leqq \alpha \leqq \sigma)$. By definition $H_\alpha^* \lhd H_\alpha$ so that $H_\alpha^* \lhd H_{\alpha+1}^*$. At limit ordinals λ we have

$$H_\lambda^* = \bigcup_{\alpha<\lambda} H_\alpha^*$$

so that $(H_\alpha^*)_{\alpha\leqq\sigma}$ is an ascending series from H to H^*. We show by induction on β that there exists an ascending abelian series from H_β^* to $H_{\beta+1}^*$. This is obvious for $\beta = 0$, so let $\beta > 0$ and assume the assertion true for all ordinals less than β. Clearly

$$(H_\beta^*)^{H_{\beta+1}} = H_{\beta+1}^*,$$

so

$$H_{\beta+1}^*/H_\beta^* = \sum (H_\beta^* + [H_\beta^*, x_1, \ldots, x_n])/H_\beta^*$$

summed over all sequences of elements $x_i \in H_{\beta+1}$.

Since L is locally nilpotent and the characteristic is 0 we may define $e(x) = \exp(x^*)$ for $x \in L$, and it is an automorphism. By lemma 1.4.9

$$H_{\beta+1}^*/H_\beta^* = \sum (H_\beta^* + H_\beta^{*e(x_1)\ldots e(x_n)})/H_\beta^*.$$

Hence there is an ascending series of ideals between H_β^* and $H_{\beta+1}^*$ of which a typical factor is

$$(H_\beta^{*e} + M)/M$$

where $e = e(x_1)\ldots e(x_n)$ is an automorphism of L, $x_i \in H_{\beta+1}$ for all i, and $M \lhd H_{\beta+1}^*$.

Let $N \lhd H_{\beta+1}^*$. By induction there is an ascending abelian series from 0 to H_β^*. Consider the series obtained from this by adding N to each term. A typical factor is of the form $(Y+N)/(X+N)$ where $X \lhd Y \leqq H_\beta^*$ and Y/X is abelian. It follows that $(Y+N)/(X+N)$ is abelian, and there is an ascending abelian series from N to H_β^*+N. Let $N = M^{e^{-1}}$ and transform by e to get an ascending abelian series from M to $H_\beta^{*e}+M$. This establishes the assertion about $H_{\beta+1}^*/H_\beta^*$. Fitting all the segments together gives an ascending abelian series from 0, through H, to $\langle H^L\rangle$. Now $\langle H^L\rangle \lhd L$, so we may repeat the process on the quotient $L/\langle H^L\rangle$, which is also Gruenberg. Eventually this quotient must be zero, and we have obtained an ascending abelian series for L.

Conversely, let L be locally nilpotent with an ascending abelian series $(V_\alpha)_{\alpha\leqq\sigma}$. Take any element $x \in L$. We will show that $\langle x\rangle$ asc L. First define

$$W_\alpha = \bigcap_{n\in m} V_\alpha^{\exp(nx^*)}.$$

Then $W_0 = 0$, $W_\alpha \lhd W_{\alpha+1}$, and $W_\lambda \geqq \bigcup_{\alpha<\lambda} W_\alpha$ for limit ordinals λ. But if $y \in W_\lambda$ then $\langle y^{\langle x \rangle} \rangle \leqq V_\lambda$, and being finite-dimensional $\langle y^{\langle x \rangle} \rangle \leqq V_\alpha$ for some $\alpha < \lambda$, so that $y \in W_\alpha$. Therefore (W_α) is an ascending series. Also $V_{\alpha+1}^2 \leqq V_\alpha$ so that $W_{\alpha+1}^2 \leqq W_\alpha$, and the series is abelian. Further, all the terms are idealised by x. We now refine it to a series with factors *centralised* by x. Take a typical factor $W = W_{\alpha+1}/W_\alpha$. Then x^* is a nil derivation of W. If we let $K_i = \ker(x^i)$ then

$$0 \leqq K_1 \leqq \ldots \text{ and } \bigcup_{i=1}^{\infty} K_i = W.$$

Further, x centralises each K_{i+1}/K_i.

Let the refined series, centralised by x, be (X_α). Then $(X_\alpha + \langle x \rangle)$ is a series ascending from $\langle x \rangle$ to L, so that $\langle x \rangle \operatorname{asc} L$. Hence L is Gruenberg. □

We now have

THEOREM 5.5 *Over any field $\mathfrak{k}$ of characteristic zero there exists a non-zero locally nilpotent Lie algebra with trivial Gruenberg radical.*

Proof: Let L be a non-Gruenberg locally nilpotent Lie algebra over $\mathfrak{k}$, let $\Gamma = \gamma(L)$. Then $\Gamma \lhd L$ by corollary 3.5. Consider $M = L/\Gamma$. If $K/\Gamma = \gamma(M)$ then K has an ascending abelian series, so is Gruenberg by theorem 5.4. Hence $K \leqq \Gamma$. So M has trivial Gruenberg radical. □

We have now proved that the classes $\mathfrak{Ft}$, $\mathfrak{B}$, $\mathfrak{Gr}$, $\mathrm{L}\mathfrak{N}$, and $\mathfrak{Z}$ are distinct; and the Fitting, Baer, Gruenberg, and Hirsch-Plotkin radicals are generally distinct. The inclusions between the classes can be summed up by the lattice diagram of figure 2.

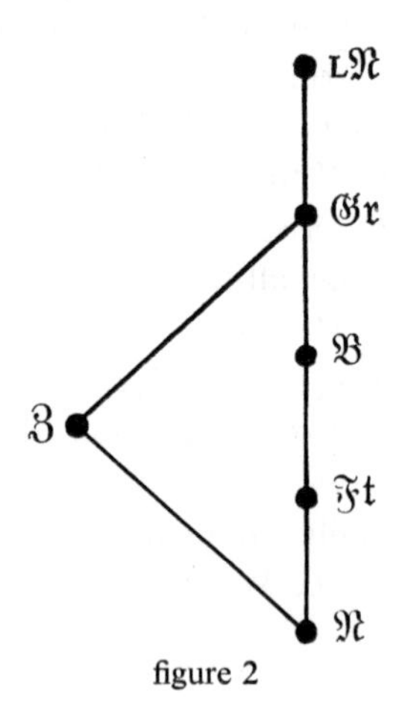

figure 2

The four radicals will find many applications in later chapters, in a variety of different situations.

Numerous other radicals have been considered by Tôgô [230], Tôgô and Kawamoto [231], Vasilesçu [236], and Parfenov [173]. Parfenov defines a Lie algebra L to be *local* if every finitely-generated subalgebra satisfies a non-trivial identity (which includes locally soluble and locally finite algebras) and proves a grand theorem which shows that most radicals are not characteristic:

THEOREM 5.6 *If $\mathfrak{X}$ is any class of local Lie algebras over a field $\mathfrak{k}$, then there exists a Lie algebra L over $\mathfrak{k}$ having a derivation d such that $\rho(L)d$ is not contained in the sum of the $\mathfrak{X}$-ideals of L.*

Proof: See Parfenov [173]. □

COROLLARY 5.7 *Over any field $\mathfrak{k}$ there exist Lie algebras L such that $\rho_{\mathrm{L}\mathfrak{F}}(L)$, $\rho_{(\mathrm{L}\mathfrak{F} \cap \mathrm{LE}\mathfrak{A})}(L)$, and the sum of the $\mathrm{LE}\mathfrak{A}$-ideals of L are not characteristic.* □

However, by using a method similar to that of theorem 3.1 it can be shown that in characteristic 0 the sum of the soluble ideals of L *is* characteristic.

It is an open question whether the sum of the locally soluble ideals of a Lie algebra is always locally soluble.

Chapter 7

Lie algebras in which every subalgebra is a subideal

The observation that in a nilpotent group every subgroup is subnormal led to the question of whether the converse was true: this was problem 22 of the Kuroš and Černikov survey [304]. A negative answer was provided by Heineken and Mohamed [212], who produced a non-nilpotent metabelian p-group with trivial centre in which every proper subgroup is nilpotent and subnormal.

Prior to this counterexample Roseblade considered a related problem in [335]. For positive integers n, r let

$$\mathfrak{U}_{n,r}$$

be the class of groups in which r-generator subgroup is subnormal in at most n-steps. The question was whether for a given n and r we can find a positive integer $\alpha = \alpha(n, r)$ such that

$$\mathfrak{U}_{n,r} \leqq \mathfrak{N}_{\alpha(n,r)}.$$

(In this context $\mathfrak{N}_c$ is the class of groups which are nilpotent of class not exceeding c; $\mathfrak{N}$ and $\mathrm{L}\mathfrak{N}$ the classes of nilpotent and locally nilpotent groups respectively).

For torsion groups the (restricted) wreath product $C \operatorname{wr} D$ of a cyclic group of order 2 and a direct product of a countably infinite number of cyclic groups of order 2 gives an example of a non-nilpotent metabelian 2-group which is in class $\mathfrak{U}_{2r+1,r}$ for every $r = 1, 2, \ldots$.

Roseblade proved that to each n there corresponds positive integers $\alpha_1(n)$ and $\alpha_2(n)$ such that

$$\mathfrak{U}_{n,\alpha_2(n)} \leqq \mathfrak{N}_{\alpha_1(n)}. \tag{1}$$

Our object in this chapter is to consider the Lie-theoretic analogues of these problems. Their development is quite advanced in this case. For instance, by employing the Mal′cev correspondence we are able to show that for torsion-free groups or groups of bounded exponent we can replace the $\alpha_2(n)$ in equation (1) by n (see also Amayo [8]).

We have defined the class of Baer algebras

$$\mathfrak{B}$$

to be the class of Lie algebras L such that if $x \in L$ then $\langle x \rangle$ si L. We define the classes

$$\mathfrak{D}, \mathfrak{E}, \mathfrak{E}_n$$

by: $L \in \mathfrak{D}$ if every subalgebra of L is a subideal; $L \in \mathfrak{E}$ if for every $x, y \in L$ there exists $m = m(x, y)$ such that $[x, {}_m y] = 0$, and $L \in \mathfrak{E}_n$ if for any $x, y \in L$ we have $[x, {}_n y] = 0$. The class $\mathfrak{E}$ is called the class of *Engel algebras* or Lie algebras satisfying the Engel condition; $\mathfrak{E}_n$ is the class of *n-Engel algebras* or Lie algebras satisfying the *n*th *Engel condition*. Let n, r be positive integers and let

$$\mathfrak{D}_{n,r}$$

be the class of Lie algebras in which every $\mathfrak{G}_r$-subalgebra is an n-step subideal. Then

$$\mathfrak{D}_n = \bigcap_{r=1}^{\infty} \mathfrak{D}_{n,r} \tag{2}$$

is the class of Lie algebras in which every subalgebra is an n-step subideal. For let $L \in \mathfrak{D}_{n,r}$ for $r = 1, 2, \ldots$ and let $H \leqq L$. By lemma 2.1.1 we have $L \circ_n H$ as the union of terms of the form $L \circ_n B$ where $B \in \mathfrak{G}$ and $B \leqq H$; by hypothesis $B \lhd^n L$ so $L \circ_n B \leqq B$, whence $L \circ_n H \leqq H$ and $H \lhd^n L$.

It is not hard to show that if $L \in \mathfrak{E}$ and $\langle x \rangle \lhd H \leqq L$, then $[H, x] = 0$. Thus since $\mathfrak{D}_{n,1} \leqq \mathfrak{B} \leqq \mathfrak{E}$ then

$$\mathfrak{D}_{n,1} \leqq \mathfrak{E}_n, \tag{3}$$

which accounts for the importance of Engel conditions in the study of the classes $\mathfrak{D}_{n,r}$.

In fig. 3 below we illustrate the known inclusions between the various classes we have defined. Some of these will be proved in this chapter. All inclusions are strict, in general.

If $\mathfrak{X}$ denotes any of the classes in fig. 3 then

$$\mathfrak{X} = \{\mathrm{s}, \mathrm{Q}\}\mathfrak{X}$$

and if $\mathfrak{X} \neq \mathfrak{B}, \mathfrak{D}$, or $\mathfrak{N}$ then

$$\mathfrak{X} = \{\mathrm{s}, \mathrm{Q}, \mathrm{L}\}\mathfrak{X}.$$

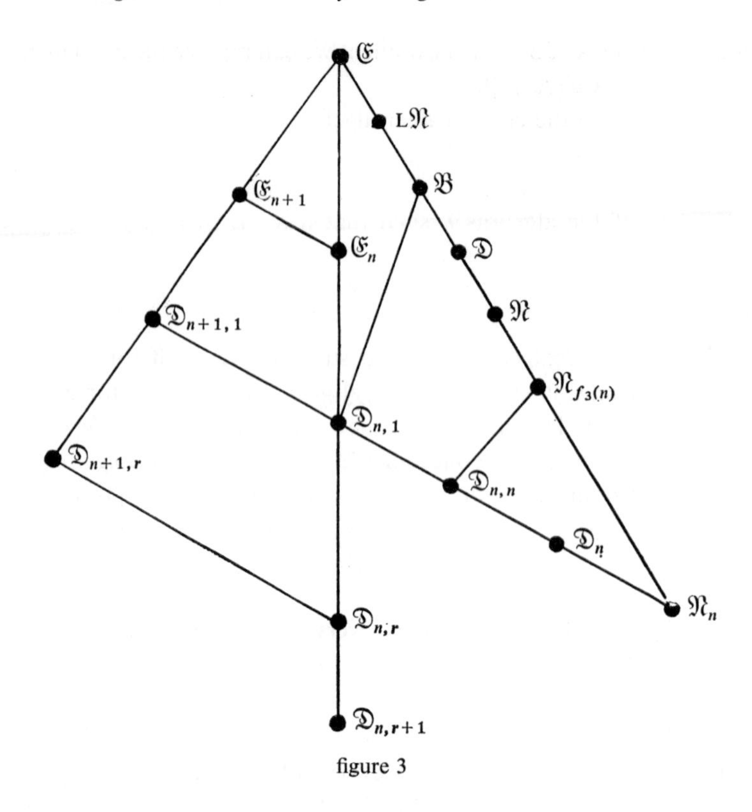

figure 3

In section 1 we prove various preliminary results which we need in later sections. Among these are some sufficiency conditions for nilpotence.

In section 2 we state and prove a key result and deduce some of its immediate applications. We also show that algebras in the class $\mathfrak{D}_{n,n}$ are nilpotent.

Engel conditions are discussed in section 3. Some of these we apply to deduce our main theorem, that algebras in the class $\mathfrak{D}_{n+1,n}$ are nilpotent.

In section 4 our first result is an example to show that over fields of arbitrary characteristic the class $\mathfrak{D}_{n+1,n}$ is the 'largest' nilpotent class among the $\mathfrak{D}_{n,r}$. Then we discuss what happens for fields of characteristic zero.

Finally in section 5 we give an example, due to Unsin [235], of a non-nilpotent metabelian Lie algebra in which every subalgebra is a subideal. This example exists over fields of characteristic > 0. We do not know whether such an example can be found over a field of characteristic zero.

1. Nilpotent subideals

In this section we develop some sufficiency conditions for nilpotence which we need in later sections. We also derive some of their more immediate applications, one of these being that every Baer algebra is locally nilpotent.

PROPOSITION 1.1 *Let H be a subalgebra of a Lie algebra L and let c and d be positive integers. Then*

(a) *If $A \subseteq I_L(H)$ and $[H, {}_dA] \subseteq H^2$ then for each $m \in \mathbb{N}$,*

$$[H^2, {}_{m(2d-1)}A] \subseteq H^{2+m}.$$

(b) *If $H \in \mathfrak{N}_c$, $A \subseteq I_L(H)$ and $[H, {}_dA] \subseteq H^2$ and $[A, {}_mA] \subseteq H$ then*

$$H + \langle A \rangle \in \mathfrak{N}_{\{cd+(c-1)(d-1)+m\}}.$$

(c) *If $H \lhd L$, $H \in \mathfrak{N}_c$ and $L/H^2 \in \mathfrak{N}_d$ then*

$$L \in \mathfrak{N}_{\{cd+(c-1)(d-1)\}}.$$

(d) *If $\mathfrak{X} = \{\mathrm{I}, \mathrm{Q}\}\mathfrak{X}$ and $\mathfrak{X} \cap \mathfrak{A}^2 \leqq \mathfrak{N}_c$ then for every d,*

$$\mathfrak{X} \cap \mathfrak{A}^d \leqq \mathfrak{N}_{((2c-1)^{d-1}+1)/2}.$$

Proof: Assume that (a) is true. Then (c) is an immediate consequence of (a). We obtain (d) from (c) and a simple induction on d.

To prove (a) we need the following result: if A, B, C are subsets of L then

$$[[A, B], {}_nC] \subseteq \textstyle\sum_{i=0}^{n} [[A, {}_iC], [B, {}_{n-i}C]], \tag{4}$$

for every $n \in \mathbb{N}$. Equation (4) follows from the Jacobi identity and induction on n.

Let $e(m) = m(2d-1)$ and apply induction on m, for $m = 0$, there is nothing to prove. Let $m > 0$ and assume that the result (a) is true for all r, $0 \leqq r \leqq m-1$.

$$[H^2, {}_{e(m)}A] \subseteq \textstyle\sum_{i=0}^{e(m)} [U_i, V_i],$$

where $U_i = [H, {}_iA]$ and $V_i = [H, {}_{e(m)-i}A] = U_{e(m)-i}$. Clearly $U_r \subseteq U_i$ (since A idealises H) whenever $i \leqq r$.

If $i < d$, then $e(m) - i > e(m-1) + d - 1$ and so

$$V_i \subseteq [H, {}_dA, {}_{e(m-1)}A] \subseteq [H^2, {}_{e(m-1)}A] \subseteq H^{2+m-1},$$

by induction. But $U_i \subseteq H$ and so $[U_i, V_i] \subseteq H^{2+m}$.

Similarly if $i > e(m-1)+d-1$, then $e(m)-i < d$ and as above we have $U_i \subseteq H^{2+m-1}$ and $V_i \subseteq H$ and so $[U_i, V_i] \subseteq H^{2+m}$. Now for $m = 1$, one of these two situations must occur and so the result is true in this case. So we may assume that $m > 1$.

Suppose that $d \leqq i \leqq e(m-1)+d-1$; then we can find r, $1 \leqq r \leqq m-1$, for which $e(r-1)+d \leqq i \leqq e(r)+d-1$. Then

$$U_i \subseteq [H, {}_dA, {}_{e(r-1)}A] \subseteq [H^2, {}_{e(r-1)}A] \subseteq H^{2+r-1}$$

and, since $e(m)-i > e(m)-e(r)-d+1 = e(m-r-1)+d$,

$$V_i = U_{e(m)-i} \subseteq U_{e(m-r-1)+d} \subseteq H^{2+m-r-1},$$

by induction. Thus $[U_i, V_i] \subseteq H^{2+r-1+2+m-r-1} = H^{2+m}$.

So for any i, $0 \leqq i \leqq e(m)$ we have $[U_i, V_i] \leqq H^{2+m}$ and therefore $[H^2, {}_{e(m)}A] \subseteq H^{2+m}$. This completes our induction on m and proves (a).

(b) Let $B = \langle A \rangle$. Now $[H, {}_dA] \subseteq H^2$ implies that $[H, {}_dB] \subseteq H^2$. We also have $[H, {}_dH+B]$ as a sum of terms of the form $[H, X_1, \ldots, X_d]$ where $X_i = H$ or B for each i. If at least one $X_i = H$, then since $H^2 \lhd I_L(H)$ such a term is contained in H^2; if every $X_i = B$ then the term is $[H, {}_dB] \subseteq H^2$. So we have $[H, {}_dH+B] \subseteq H^2$. Employing (a) and the fact that $H \in \mathfrak{N}_c$ we have

$$[[H, {}_dH+B], {}_{e(c-1)}H+B] \subseteq [H^2, {}_{e(c-1)}H+B] \subseteq H^{2+c-1} = 0.$$

Thus $H \leqq \zeta_{e(c-1)+d}(H+B)$ and the result follows since $[A, {}_mA] \subseteq A$ implies that $B^{m+1} \subseteq H$. We also have

$$e(c-1)+d = (c-1)(2d-1)+d = cd+(c-1)(d-1). \qquad \square$$

We remark that part (c) above is the Lie theoretic analogue of a result of Hall; the bound $cd+(c-1)(d-1)$ is the best possible. For in [340] A. G. R. Stewart constructs for each pair of positive integers c, d, a torsion-free group G having a normal subgroup N with N nilpotent of class c, G/N' nilpotent of class d, and G nilpotent of class precisely $cd+(c-1)(d-1)$. By considering the completions of G and N and employing the Mal'cev correspondence we obtain a Lie algebra L with the same properties.

Next we have a result about metabelian Lie algebras.

LEMMA 1.2 *Let $L \in \mathfrak{A}^2$ and let $x \in L$. If $[L^2 \cap \langle x^L \rangle, {}_nx] \subseteq \langle x \rangle$ then $\langle x^L \rangle \in \mathfrak{N}_m$, where $m = \max\{1, n\}$. In particular if either $[L, {}_nx] \subseteq \langle x \rangle$ or $\langle x \rangle \lhd^n L$ then $\langle x^L \rangle \in \mathfrak{N}_m$, where $m = \max\{1, n-1\}$.*

Proof: Let $A = L^2$ and $B = \langle x^L \rangle$. Clearly

$$B = \langle x \rangle + A \cap B$$

and so if $x \in L^2$ or if $n = 0$, then $B = \langle x \rangle \in \mathfrak{A}$.

Suppose that $n > 0$ and $x \notin L^2$. Then (as $\langle x \rangle$ is 1-dimensional) $[B \cap A, {}_n x] = 0$. A simple induction on r shows that

$$B^{r+1} = [B, {}_r x]$$

and so $B^{n+1} = 0$.

In case $[L, {}_n x] \subseteq \langle x \rangle$ and $x \notin L^2$ then $[L, {}_n x] = 0$. The Jacobi identity and the fact that $L \in \mathfrak{A}^2$ give for all r and i,

$$[x, {}_r L, {}_i x] = [x, L, {}_i x, {}_{r-1} L] = [L, {}_{i+1} x, {}_{r-1} L]$$

and so $[x, {}_r L, {}_{n-1} x] = 0$, for all r. But

$$B \cap A = \textstyle\sum_{r=1}^{\infty} [x, {}_r L]$$

and the result follows from above. If $\langle x \rangle \lhd^n L$, then $[L, {}_n x] \subseteq \langle x \rangle$. □

We restate theorem 1.2.4 in more precise form as follows. *To each pair c, r of positive integers there corresponds a positive integer $f(c, r)$ such that*

$$\mathfrak{N}_c \cap \mathfrak{G}_r \leqq \mathfrak{F}_{f(c,r)}. \tag{5}$$

We note that $f(1, r) = 1$ and that if $c > 1$ and $r > 1$ then

$$r(r+1)/2 \leqq f(c, r) \leqq r(r^{c-1}+1)/2 \tag{6}$$

(see also Cohn [43], p. 296–7).

We know by theorem 1′ of Gruenberg [72] that $\mathfrak{E} \cap \mathrm{E}\mathfrak{A} \leqq \mathrm{L}\mathfrak{N}$. However we require the more precise:

Lemma 1.3 *Let n, m, r be positive integers. Then there exists $f_0(n, m, r)$ such that if $L \in \mathfrak{E}_n \cap \mathfrak{A}^m$ and $L/L^2 \in \mathfrak{G}_r$ then $L \in \mathfrak{N}_{f_0}$. In particular*

$$\mathfrak{E}_n \cap \mathfrak{A}^m \cap \mathfrak{G}_r \leqq \mathfrak{G}_r \cap \mathfrak{N}_{f_0(n,m,r)}.$$

Proof: By induction on m. Assume that $n > 1$. By hypothesis we can find $x_1, \ldots, x_r$ such that $L = X + L^2$, where $X = \langle x_1, \ldots, x_r \rangle$.

If $m \leqq 2$, then by lemma 1.2, $\langle x^L \rangle \in \mathfrak{N}_{n-1}$ for each $x \in L$. Hence $X^L = \sum_{i=1}^{r} \langle x_i^L \rangle \in \mathfrak{N}_{r(n-1)}$ and so $L = X^L + L^2 \in \mathfrak{N}_{r(n-1)+1}$, since $L^2 \in \mathfrak{A}$ for $m \leqq 2$. So put $f_0(n, 2, r) = r(n-1)+1$.

Let $m > 2$ and assume that the result is true for $m-1$. Put $B = L^{(m-2)}$ and $A = L^{(m-1)}$ so that $A = B^2 \in \mathfrak{A}$. Let $c = f_0(n, m-1, r)$. Clearly L/A

satisfies the hypothesis of L and so by induction $L/A \in \mathfrak{N}_c$. Put $H = L^{c+1}$. By lemma 2.1.9 we have $L = X+H$ and so $L/H \in \mathfrak{G}_r \cap \mathfrak{N}_c \leqq \mathfrak{F}_{f(c,r)}$ by equation (5). Hence $B/H \in \mathfrak{F}_k$ where $k \leqq f(c,r)-r$; for $L/L^2 \in \mathfrak{F}_r$ implies that $\dim B/H \leqq \dim L^2/H \leqq \dim L/H - r$ (for we have assumed that L/L^2 is generated by no fewer than r elements). So we can find $b_1, \ldots, b_k \in B$ such that $B = \langle b_1, \ldots, b_k \rangle + H$: and so $B \in \mathfrak{N}_{k(n-1)+1}$, since $H \in \mathfrak{A}$ and $B \in \mathfrak{A}^2 \cap \mathfrak{E}_n$. Therefore by proposition 1.1 we have $L \in \mathfrak{N}_d$ where

$$d = c(nk-k+1)+(c-1)(nk-k).$$

So we define

$$f_0(n,m,r) = f_0(n,m-1,r)((n-1)(f(c,r)-r)+1)+ \\ +(f_0(n,m-1,r)-1)((n-1)(f(c,r)-r)),$$

where $c = f_0(n,m-1,r)$, if $m > 2$. This also holds for $n = 1$. □

COROLLARY 1.4 *Let n, r be positive integers. Then there exists a positive integer $f_1(n,r)$ such that*

$$\mathfrak{G}_r \cap \mathfrak{D}_{n,1} \leqq \mathfrak{G}_r \cap \mathfrak{N}_{f_1(n,r)}.$$

Proof: Let $L = \langle x_1, \ldots, x_r \rangle \in \mathfrak{D}_{n,1}$. By the Derived Join Theorem we can find $m = m(n,r)$ such that $L \in \mathfrak{A}^m$ (since $\langle x_i \rangle \lhd^n L$ for $i = 1, \ldots, r$). By equation (3) we have $\mathfrak{D}_{n,1} \leqq \mathfrak{E}_n$ and so the result follows by lemma 1.3. □

We recall that $\mathfrak{C}$ is the class of Lie algebras L with $L/L^2 \in \mathfrak{G}$. We may also define $\mathfrak{C}_r$ to consist of Lie algebras L with $L/L^2 \in \mathfrak{G}_r$. Then $\mathfrak{C}$ and $\mathfrak{C}_r$ are Q-closed and

$$\mathfrak{C} \cap \mathfrak{N} = \mathfrak{G} \cap \mathfrak{N}; \; \mathfrak{C}_r \cap \mathfrak{N} = \mathfrak{G}_r \cap \mathfrak{N}.$$

Next we have

THEOREM 1.5 *Let H be a subalgebra of a Lie algebra L. Then*

(a) *If $L \in \mathfrak{B} \cap \mathfrak{C} \cap \mathrm{E}\mathfrak{A}$ then $L \in \mathfrak{G} \cap \mathfrak{N}$.*

(b) $\mathfrak{B} \leqq \mathrm{L}\mathfrak{N}$.

(c) *If $x \in H$ implies that $\langle x \rangle$ si L then every $\mathfrak{G}$-subalgebra of H is a subideal of L. In particular L has a unique maximal $\mathfrak{B}$-ideal, the Baer radical of L.*

Proof: (a) Let $L \in \mathfrak{B} \cap \mathfrak{C} \cap \mathfrak{A}^m$ so that $L = \langle x_1, \ldots, x_r \rangle + L^2$ for some $x_i \in L$. We use induction on m to show that $L \in \mathfrak{N}$. Suppose that $m \leqq 2$. Then by lemma 1.2 we have $\langle x^L \rangle \in \mathfrak{N}$ for each $x \in L$. Since also $L^2 \in \mathfrak{A}$ then $L = \sum_{i=1}^r \langle x_i^L \rangle + L^2 \in \mathfrak{N} \cap \mathfrak{C} = \mathfrak{G} \cap \mathfrak{N}$. Let $m > 2$ and suppose that the result is true for $m-1$. Put $B = L^{(m-2)}$ and $A = B^2 = L^{(m-1)}$. Then $L/A \in \mathfrak{B} \cap \mathfrak{C} \cap \mathfrak{A}^{m-1}$ and so by induction $L/A \in \mathfrak{G} \cap \mathfrak{N} = \mathfrak{F} \cap \mathfrak{N}$. Hence $B/A \in \mathfrak{F}$ and

$$B \in \mathfrak{B} \cap \mathfrak{C} \cap \mathfrak{A}^2 \leqq \mathfrak{G} \cap \mathfrak{N} \text{ (from above).}$$

So by proposition 1.1 we have $L \in \mathfrak{N}$; so $L \in \mathfrak{C} \cap \mathfrak{N} = \mathfrak{G} \cap \mathfrak{N}$. This completes our induction on m and proves (a).

(b) By the Derived Join Theorem we have $\mathfrak{B} \leqq \mathrm{LE}\mathfrak{A}$ and so the result follows from (a).

(c) Let $K \leqq H$, $K \in \mathfrak{G}$. If $x \in H$ then $\langle x \rangle \text{ si } L$ and so $\langle x \rangle \text{ si } H$. Thus $H \in \mathfrak{B}$ and from (b) we have $H \in \mathrm{L}\mathfrak{N}$. Hence $K \in \mathfrak{G} \cap \mathfrak{N} = \mathfrak{F} \cap \mathfrak{N}$ and so $K \in \mathfrak{F}_m \cap \mathfrak{N}$ for some m. If $m \leqq 1$, then $K \text{ si } L$, by hypothesis. Let $m > 1$ and suppose that each $\mathfrak{F}_{m-1}$-subalgebra of H is a subideal of L. Now $K > K^2$ and so K has a subalgebra $K_0 \in \mathfrak{F}_{m-1}$. Then $K_0 \text{ si } L$ and we can find $x \in K$ with $K = \langle x \rangle + K_0$. Since $\langle x \rangle \text{ si } L$ then by lemma 2.1.4 $K \text{ si } L$. Hence every $\mathfrak{F}$-subalgebra (and so $\mathfrak{G}$-subalgebra) of H is a subideal of L.

Evidently the union of an ascending chain of $\mathfrak{B}$-ideals of L is a $\mathfrak{B}$-ideal of L. So by Zorn's lemma L has maximal $\mathfrak{B}$-ideals. We now have to show that the sum of a pair of $\mathfrak{B}$-ideals is a $\mathfrak{B}$-ideal. Let M and N be $\mathfrak{B}$-ideals of L. Then $M, N \in \mathrm{L}\mathfrak{N}$. Let $x \in M+N$. We will show that $\langle x \rangle \text{ si } L$ and this will show that $M+N \in \mathfrak{B}$, whence L has a unique $\mathfrak{B}$-ideal. Now we can find $A, B \in \mathfrak{G}$ with $A \leqq M$ and $B \leqq N$ such that $x \in C = \langle A, B \rangle$. From above we have $A, B \in \mathfrak{F} \cap \mathfrak{N}$ and $A \lhd^r M \lhd L$ and $B \lhd^s N \lhd L$, for some r and s. Thus $\langle A^B \rangle \in \mathfrak{G}$ (for $[A, {}_kB] = 0$ for some k) and so $\langle A^B \rangle \text{ si } M$ and $\langle A^B \rangle \in \mathfrak{N}$. Similarly $\langle B^A \rangle \text{ si } N$ and $\langle B^A \rangle \in \mathfrak{N}$ (for $M, N \lhd L$). Hence $C = \langle A^B \rangle + \langle B^A \rangle \in \mathfrak{N}$ and $C \text{ si } L$. So $\langle x \rangle \text{ si } C \text{ si } L$. □

Remark 1. From theorem 1.5(c) we have

$$\mathfrak{B} \overset{\ulcorner}{\underset{\llcorner}{=}} \{\mathrm{S}, \mathrm{Q}, \mathrm{N}_0\}\mathfrak{B}. \tag{7}$$

We have also shown that $L \in \mathfrak{B}$ if and only if every $\mathfrak{G}$-subalgebra of L is a subideal of L.

We now have that the classes $\mathfrak{D}$, $\mathfrak{D}_n$, $\mathfrak{D}_{n,r}$, $\mathfrak{B}$ are contained in the class $\mathrm{L}\mathfrak{N}$ of locally nilpotent algebras. Our next result, due to Hartley [76], is one frequently used in studying classes of $\mathrm{L}\mathfrak{N}$-algebras.

LEMMA 1.6 (*Hartley*) *Let $L \in \mathrm{L}\mathfrak{N}$ and let M be a minimal ideal of L. Then $M \leqq \zeta_1(L)$. In particular if $H \lhd L$, $H \in \mathfrak{F}_m$ and $L \in \mathrm{L}\mathfrak{N}$ then $H \leqq \zeta_m(L)$.*

Proof: Suppose that $[M, L] \neq 0$. Then there exists $a \in L$ and $b \in M$ with $c = [b, a] \neq 0$. Now $c \in M$ and M is a minimal ideal so $M = \langle c^L \rangle$. Hence $b \in \langle c^L \rangle$ and we can find a $\mathfrak{G}$-subalgebra A of L with $b \in \langle c^A \rangle$. Let $A_0 = \langle A, a \rangle$ so that as $L \in \mathrm{L}\mathfrak{N}$ we have $A_0 \in \mathfrak{G} \cap \mathfrak{N}$. Let $B = \langle b^{A_0} \rangle$. Then as $[B, A_0] \lhd A_0$ and $c = [b, a] \in [B, A_0]$ we must have $b \in [B, A_0]$. Hence $B = [B, A_0]$, a contradiction since $A_0 \in \mathfrak{N}$. So $c = 0$ and $[M, L] = 0$. The second result follows by induction on m. □

We have seen that for any n and r we have

$$\mathfrak{D}_{n,r+1} \leqq \mathfrak{D}_{n,r} \leqq \mathfrak{D}_{n+1,r}.$$

It is natural to ask whether given $L \in \mathfrak{D}_{n,r}$ (for $n,r > 1$) we can find non-trivial subalgebras in (say) $\mathfrak{D}_{n-1,r-1}$. A partial answer is provided by the next result, which is often of use in inductive proofs.

LEMMA 1.7 *Let n, r, s be positive integers with $1 \leqq s \leqq r-1$. Suppose that $L \in \mathfrak{D}_{n,r}$ and H is a $\mathfrak{G}_s$-subalgebra of L. Let H_j, $0 \leqq j \leqq n$, be the* jth *ideal closure of H in L. Then*

$$H_j/H_{j+1} \in \mathfrak{D}_{n-j,r-s}$$

for all $j = 0, \ldots, n-1$.

Proof: Suppose that $0 < j < n$ and Y/H_{j+1} is a $\mathfrak{G}_{(r-s)}$-subalgebra of H_j/H_{j+1}. It is enough to show that $Y \lhd^{n-j} H_j$.

Evidently there exists a $\mathfrak{G}_{(r-s)}$-subalgebra X of H_j with $Y = X + H_{j+1}$. Let $K = \langle X, H \rangle$ so that $K \in \mathfrak{G}_r$ and so $K \lhd^n L$. Let K_i be the ith ideal closure of K in L. Since $H \leqq K \leqq H_j$ it follows by proposition 2.1.1 that

$$H_i = H + L \circ_i H \leqq K_i \leqq H_j + L \circ_i H_j.$$

Hence $K_j = H_j$ since $H_j \lhd^j L$. But $K \lhd^{n-j} K_j = H_j$ and $H_{j+1} \lhd H_j$ and so $Y = K + H_{j+1} \lhd^{n-j} H_j$.

Trivially we have $L/H_1 = H_0/H_1 \in \mathfrak{D}_{n,r} \leqq \mathfrak{D}_{n,r-s}$. □

Finally we remark that using corollary 1.4 and the proof of theorem 1.5(c) we can find $f_1^*(n, r)$ such that

$$\mathfrak{D}_{n,1} \leqq \mathfrak{D}_{f_1^*(n,r),r} \tag{8}$$

for all r.

2. The key lemma and some applications

In this section we prove a key result (lemma 2.1) and give some of its immediate applications.

LEMMA 2.1 *Let A, B and H be subalgebras of a Lie algebra L and let $J = \langle A, B, H\rangle$. Suppose that A and B are abelian ideals of J and m is a positive integer such that $A/H \cap A \in \mathfrak{F}_m$ and $B/H \cap B \in \mathfrak{F}_m$. Then*

$$[A, B]/H \cap [A, B] \in \mathfrak{F}_{m-1}.$$

Proof: Evidently $[A, B]$ is an abelian ideal of J and is contained in $A \cap B$. Let $X = [A, B]/H \cap [A, B] = [A, B]/H \cap A \cap [A, B]$. Then

$$(H \cap A + [A, B])/H \cap A \leqq A/H \cap A$$

and

$$(H \cap A + [A, B])/H \cap A \cong [A, B]/H \cap A \cap [A, B] = X.$$

So $\dim X \leqq m$. If $\dim X \nleqq m-1$, then $\dim X = m$ and so $\dim(H \cap A + [A, B])/H \cap A = m$. But $\dim(A/H \cap A) \leqq m$ and so we must have $(H \cap A + [A, B])/H \cap A = A/H \cap A$. Therefore

$$H \cap A + [A, B] = A$$

and by symmetry

$$H \cap B + [A, B] = B.$$

We have $A^2 = B^2 = 0$ and so

$$\begin{aligned}[A, B] &= [H \cap A + [A, B], B] = [H \cap A, B] \\ &= [H \cap A, H \cap B + [A, B]] = [H \cap A, H \cap B] \leqq H.\end{aligned}$$

Thus $\dim X = 0$, a contradiction since $m > 0$. So we must have $\dim X \leqq m-1$. □

As an immediate consequence we have the very useful

LEMMA 2.2 *Let $\{A_i : i \in I\}$ be a family of abelian ideals of a Lie algebra L and let $K = \langle A_i : i \in I\rangle$. Let m be a non-negative integer and let $H \leqq L$ such that $A_i/H \cap A_i \in \mathfrak{F}_m$ for all $i \in I$. Then*

$$K^{(m)} \leqq H.$$

If $K+H \in \mathrm{L}\mathfrak{N}$ *then*

$$H \lhd^{m(m+1)/2} K+H.$$

Proof: If $m = 0$ then each $A_i \leqq H$ and so $K = K^{(0)} \leqq H$. Let $m > 0$ and assume that the result is true for $m-1$. Evidently $K^{(1)} = K^2 = \langle [A_i, A_j] : i, j \in I \rangle$. By lemma 2.1 we have $[A_i, A_j]/H \cap [A_i, A_j] \in \mathfrak{F}_{m-1}$ for all $i, j \in I$. Clearly $[A_i, A_j]$ is an abelian ideal of L for all $i, j \in I$. Hence by induction $(K^2)^{(m-1)} \leqq H$. So $K^{(m)} = (K^2)^{(m-1)} \leqq H$.

By induction we have $H \lhd^{(m-1)m/2} K^2+H$, for $K+H \in \mathrm{L}\mathfrak{N}$ implies that $K^2+H \in \mathrm{L}\mathfrak{N}$. A second induction on r shows that

$$(K+H)^{r+1} \leqq K^2+[K, {}_rH]+H^{r+1}.$$

But for each $i \in I$ we have $A_i/H \cap A_i \lhd (A_i+H)/H \cap A_i$ and so it follows from lemma 1.6 that $A_i/H \cap A_i \leqq \zeta_m((A_i+H)/H \cap A_i)$. This implies that $[A_i, {}_mH] \leqq H \cap A_i$.

Since $K = \sum_{i \in I} A_i$ then $[K, {}_mH] \leqq H$ and so $(K+H)^{m+1} \leqq K^2+H$, whence $K^2+H \lhd^m K+H$ and $H \lhd^{m(m+1)/2} K+H$. □

Remark 2. Lemma 2.1 holds if we replace 'A and B are abelian' by '$A^2 \leqq H$ and $B^2 \leqq H$'. Similarly for lemma 2.2 it is enough to take $A_i^2 \leqq H$ for all $i \in I$.

We recall that if A, V are vector spaces and θ is a linear map from A to V then

$$A/C_A(\theta) \cong A\theta \subseteq V, \text{ where } C_A(\theta) = \ker \theta.$$

In the special case where A is a subspace of a Lie algebra L and θ denotes the adjoint map u^* induced by an element u of L then $A\theta = [A, u] \subseteq [L, u]$ and $C_A(\theta) = C_A(u) = \{a \in A : [a, u] = 0\}$. Thus

$$A/C_A(u) \cong [A, u] \subseteq [L, u], \tag{9}$$

where $A/C_A(u)$ is the quotient space of A by $C_A(u)$ and the isomorphism is linear.

Furthermore $[L, u] = [A, u]$ if and only if $L = A+C_L(u)$.

THEOREM 2.3 *Let n be a positive integer and let X be a subset of a Lie algebra L such that $L = \langle X \rangle$ and*

(i) *if* $x \in X$ *then* $\langle x^L \rangle \in \mathfrak{A}$

(ii) *if* $x_1, \ldots, x_n \in X$ *then* $\langle x_1, \ldots, x_n \rangle \lhd^n L$.

Then $L \in \mathfrak{A}^{f(n,n)}$.

Proof: Evidently $L \in \mathfrak{L}\mathfrak{N}$. Thus if $n = 1$, then by lemma 1.6 we have $\langle x \rangle \lhd L$ and so $[L, x] = 0$ for each $x \in X$. Hence $L^2 = L^{(1)} = 0$.

Suppose that $n > 1$. Let $x_1, \ldots, x_n \in X$ and let $Y = \langle x_1, \ldots, x_n \rangle$. From (i) we have $Y^L = \sum_{i=1}^{n} \langle x_i^L \rangle \in \mathfrak{N}_n$. Hence $Y \in \mathfrak{G}_n \cap \mathfrak{N}_n$ and $Y \in \mathfrak{F}_{f(n,n)}$. Let $u = [x_1, \ldots, x_n]$ and put $H = C_L(u)$. From (ii) we have $Y \lhd^n L$ and so $[L, u] \leqq Y$. Clearly $[L, u] \leqq \zeta_1(Y)$ and so $\dim([L, u]) \leqq m = f(n, n) - 1$ (for otherwise $Y = \zeta_1(Y)$ which implies that $\dim Y \leqq n$ and so $f(n, n) \leqq n$, a contradiction). It follows from (9) that H has codimension m in L; since also L is generated by abelian ideals it follows from lemma 2.2 that $L^{(m)} \leqq H = C_L(u)$ and so $u \in C_L(L^{(m)})$. Thus $\langle u^L \rangle \leqq C_L(L^{(m)})$. But the x_i were arbitrary and so $L^n = \langle [x_1, \ldots, x_n]^L : x_i \in X \rangle \leqq C_L(L^{(m)})$. Now $L^{(m)} \leqq L^{2^m} \leqq L^n$ since $n \leqq 2^m$. Hence $L^{(m+1)} = 0$. □

Remark 3. The Mal'cev correspondence enables us to deduce that if a torsion-free group satisfies the hypothesis of theorem 2.3 then it is soluble of derived length $\leqq f(n, n)$.

Let m be a positive integer and L a Lie algebra.

We define

$$\alpha^m(L) = \{x \in L : \langle x^L \rangle \in \mathfrak{A}^{(m)}\}$$

and we let $\alpha(L) = \alpha^1(L)$. Clearly $\alpha^m(L)$ is invariant under scalar multiplications and Lie multiplications by the elements of L. It is not in general a subspace of L.

Let $n > 1$ and let $L \in \mathfrak{D}_{n,n} \cap \mathfrak{A}^2$. By lemma 1.2 we have $\langle x^L \rangle \in \mathfrak{N}_{n-1}$ for each $x \in L$. Thus if $x_1, \ldots, x_n \in L$ and $X = \langle x_1, \ldots, x_n \rangle$ then $X \in \mathfrak{G}_n \cap \mathfrak{N}_{n(n-1)} \leqq \mathfrak{F}_{f(n(n-1),n)}$. By hypothesis $X \lhd^n L$ and so $[L, x_1, \ldots, x_n] \subseteq X \cap L^2$. Thus since $L^2 \in \mathfrak{A}$ then for any $y_1, \ldots, y_r \in L$ we have

$$\begin{aligned}[][[L, x_1, \ldots, x_n], y_1, \ldots, y_r] &= [L, x_1, y_1, \ldots, y_r, x_2, \ldots, x_n] \\ &\subseteq X \cap L^2,\end{aligned}$$

since $X \lhd^{n-1} X^L$ and $[L, x_1, y_1, \ldots, y_r] \subseteq X^L$. Put $Y = [L, x_1, \ldots, x_n]^L$ and let $m = f((n-1)n, n)$. Then $Y \lhd L$, $Y \leqq X \cap L^2$ and so $Y \in \mathfrak{F}_m$. Since $L \in \mathfrak{L}\mathfrak{N}$ we have $Y \leqq \zeta_m(L)$. But the x_i's were arbitrary and L^{n+1} is generated by terms like Y so $L^{n+1} \leqq \zeta_m(L)$, whence $L = \zeta_{m+n}(L)$. Thus if $n > 1$,

$$\mathfrak{D}_{n,n} \cap \mathfrak{A}^2 \leqq \mathfrak{N}_{n+f(n(n-1),n)-1}.$$

Now $\mathfrak{D}_{n,n}$ is {I, Q}-closed and so by proposition 1.1 we have

LEMMA 2.4 *To each pair of positive integers m, n there corresponds a positive integer $f_2(n, m)$ such that*

$$\mathfrak{D}_{n,n} \cap \mathfrak{A}^m \leqq \mathfrak{N}_{f_2(n,m)}. \qquad \square$$

It is clear that

$$\mathfrak{D}_{1,1} = \mathfrak{D}_1 = \mathfrak{A} = \mathfrak{N}_1.$$

Let $n > 1$ and suppose inductively that $d(r)$ has been defined for all $r < n$ so as to satisfy

$$\mathfrak{D}_{r,r} \leqq \mathfrak{A}^{d(r)}.$$

Define

$$d(n) = (d(1)+\ldots+d(n-1))f(n, n)$$

and let

$$m = d(1)+\ldots+d(n-1).$$

Let $L \in \mathfrak{D}_{n,n}$ and let $X = \langle x \rangle \leqq L$. Then $X \lhd^n L$. Let X_i be the ith ideal closure of X in L. Then for each i, $0 < i < n$, we have (lemma 1.7)

$$X_i/X_{i+1} \in \mathfrak{D}_{n-i,n-1} \leqq \mathfrak{D}_{n-i,n-i} \leqq \mathfrak{A}^{d(n-i)}.$$

Therefore $X_i^{(d(n-i))} \leqq X_{i+1}$ and so $X_1^{(m)} \leqq X$. Now $X_1 = \langle x^L \rangle \lhd L$ implies that $X_1^{(m)} \lhd L$. Thus if $X_1^{(m)} \neq 0$, then $X = \langle x \rangle = X_1^{(m)} \lhd L$ so that $X = X_1$ and $X_1^{(1)} = 0$; this is a contradiction since $m \geqq 1$. So $X_1^{(m)} = 0$. We now have $L = \alpha^m(L)$. For each j, $0 \leqq j \leqq m$, let $L_j = \langle \langle x^L \rangle^{(j)} : x \in L \rangle$. Then $L_j \lhd L$, $L = L_0$ and $L_m = 0$. Clearly $L_{j+1} \leqq L_j$ and $L_j/L_{j+1} = \langle \alpha(L_j/L_{j+1}) \rangle$ for each j, $0 \leqq j \leqq m-1$; and $L_j/L_{j+1} \in \mathfrak{D}_{n,n}$. Therefore by theorem 2.3 we have $L_j^{(f(n,n))} \leqq L_{j+1}$ for each j, $0 \leqq j \leqq m-1$. Thus as $L_0 = L$ and $L_m = 0$ we have (since $d(n) = mf(n, n)$)

$$L^{(d(n))} = 0.$$

Hence $\mathfrak{D}_{n,n} \leqq \mathfrak{A}^{d(n)}$ for each n. Let $f_3(n) = f_2(n, d(n))$. It follows from lemma 2.4 that

THEOREM 2.5 *To each positive integer n there corresponds a positive integer $f_3(n)$ such that*

$$\mathfrak{D}_{n,n} \leqq \mathfrak{N}_{f_3(n)}. \qquad \square$$

Remark 4. A torsion-free group in the class $\mathfrak{U}_{n,n}$ is nilpotent of class $f_3(n)$. This follows by the Mal'cev correspondence.

We now ask for a fixed n, how small r can be to ensure that

$$\mathfrak{D}_{n,r} \leqq \mathfrak{N} \tag{*}$$

We will show in a later section that over a field of characteristic $p > 0$, a necessary condition for (*) is that

$$r > (n-2)/(p-1).$$

We will show that over any field a sufficient condition for (*) to be true is that

$$r > n-2.$$

In other words we show that for every n,

$$\mathfrak{D}_{n+1,n} \leqq \mathfrak{N}.$$

To do this we need some results about Engel algebras.

3. Engel conditions

We will discuss Engel conditions in their own right in chapter 16; but in this chapter we need to develop some of their properties in relation to $\mathfrak{O}$-algebras.

Our first result (lemma 3.1) is due to Higgins [89]. We state it in stronger form, but the proof given there will work for this case as well.

Let L be a Lie algebra defined over a field $\mathfrak{k}$. Suppose that $g(t_1, \ldots, t_r)$ is a polynomial over $\mathfrak{k}$ with zero constant term in the non-commutative indeterminates $t_1, \ldots, t_r$. If $x_i \in L$ and x_i^* is the adjoint map of L induced by x_i we denote by $g(x_1^*, \ldots, x_r^*)$ the element of End(L) obtained from $g(t_1, \ldots, t_r)$ by substituting x_i^* for t_i ($i = 1, \ldots, r$). We define

$$L(g) = L(g(t_1, \ldots, t_r)) =$$
$$= \{a \in L : ag(x_1^*, \ldots, x_r^*) = 0, \text{ for all choices of } x_i \in L\}. \tag{10}$$

Evidently $L(g) = \bigcap_{x_1,\ldots,x_r \in L} \ker(g(x_1^*, \ldots, x_r^*))$.

LEMMA 3.1 (*Higgins*) *Let L be a Lie algebra defined over a field* $\mathfrak{k}$ *and let* $g(t_1, \ldots, t_r)$ *be a polynomial over* $\mathfrak{k}$ *of degree not exceeding n in each of its arguments. Then L(g) is a characteristic ideal of L if either*

(i) $|\mathfrak{k}| > n$ *or*

(ii) $|\mathfrak{k}| > n-1$ *and g is homogeneous in each of those arguments which occur with degree n.* □

(Compare with theorems 14.2.8 and 14.4.1.)

Let n, r be positive integers. We define

$$g_{n,r}(t_1, \ldots, t_r) = t_1^n \ldots t_r^n. \tag{11}$$

Any field has at least two elements. Hence by lemma 3.1, for any Lie algebra L,

$$L(g_{2,r}) \operatorname{ch} L, \quad \text{for all } r. \tag{12}$$

We now have

THEOREM 3.2 *Let L be a Lie algebra defined over a field of characteristic different from 2. Then for any positive integer r,*

$$L(g_{2,r}) \leqq \zeta_{3r}(L).$$

In particular if every $\mathfrak{G}_{r+1}$*-subalgebra of L is nilpotent of class* $\leqq 2r$, *then L is nilpotent of class* $\leqq 3r$.

Proof: We use induction on r. By equation (12) $L(g_{2,1}) \lhd L$. Let $a \in L(g_{2,1})$ and let $x, y, z \in L$. Then

$$0 = [a, x+y, x+y] = [a, x, y]+[a, y, x]+[a, {}_2x]+[a, {}_2y].$$

Therefore

$$[a, [x, y]] = [a, x, y]-[a, y, x] = 2[a, x, y]. \tag{13}$$

Hence

$$[a, [x, y, z]] = 2[a, [x, y], z] = 4[a, x, y, z].$$

But $[a, z] \in L(g_{2,1})$ and by (13), $[a, x, z] = -[a, z, x]$ and so

$$\begin{aligned}[a, [x, y, z]] &= [a, [x, y], z]-[a, z, [x, y]]\\ &= 2[a, x, y, z]-2[a, z, x, y]\\ &= 2[a, x, y, z]+2[a, x, z, y]\\ &= 2[a, x, y, z]-2[a, x, y, z] = 0,\end{aligned}$$

since $[a, x] \in L(g_{2,1})$ and so by (13), $[a, x, y, z] = -[a, x, z, y]$. So we have $4[a, x, y, z] = [a, [x, y, z]] = 0$, whence $[a, x, y, z] = 0$ (L is defined over a field of characteristic not 2). So $a \in \zeta_3(L)$ and $L(g_{2,1}) \leqq \zeta_3(L)$.

Let $r > 1$ and suppose that $L(g_{2,r-1}) \leqq N = \zeta_{3(r-1)}(L)$. Clearly if $a \in L(g_{2,r})$ and $x \in L$ then $[a, {}_2x] \in L(g_{2,r-1}) \leqq N$. Thus

$$(L(g_{2,r})+N)/N \leqq (L/N)(g_{2,1}) \leqq \zeta_3(L/N),$$

from above. Hence $L(g_{2,r}) \leqq \zeta_{3r}(L)$. □

Suppose that L is defined over a field of characteristic p. Put $L_r = L(g_{2,r})$ and let $L_0 = 0$. From equation (13) and the fact that if $a \in L_r$ and $x \in L$ then $[a, {}_2x] \in L_{r-1}$ we deduce that: if $p = 2$, then $[L_r, L^2] \leqq L_{r-1}$ and so $[L_r, {}_rL^2] = 0$; thus

$$L_r \in \mathfrak{N}_r\mathfrak{A} \text{ and } L_1 \in \mathfrak{N}_2. \tag{14}$$

If $p \neq 2$, then $L_r \in \mathfrak{N}_{3r} \leqq \mathfrak{A}^{r+1}$, since $3r+1 \leqq 2^{r+1}$ for all r. So we have

COROLLARY 3.3 *Let L be any Lie algebra. Then* $L(g_{2,1}) \in \mathfrak{N}_3$ *and* $L(g_{2,r}) \in \mathfrak{A}^{r+1}$ *for all positive integers r.* □

Clearly $L \in \mathfrak{E}_n$ if and only if $L = L(g_{n,1})$. So corollary 3.3 includes the well known result that $\mathfrak{E}_2 \leqq \mathfrak{N}_3$.

LEMMA 3.4 *Let L be a Lie algebra defined over a field of characteristic p and let H be an ideal of L such that* $[H, L^2] = 0$. *Then*

(i) *If* $p = 0$ *or* $n \leqq p$ *then* $[H, L(g_{n,1})] \leqq \zeta_{n-1}(L)$.
(ii) *If* $p = 0$ *or* $n < p$ *then* $H \cap L(g_{n,1}) \leqq \zeta_n(L)$.

Proof: Let $a \in H$, $y \in L(g_{n,1})$ and $x \in L$. Then

$$0 = [y, {}_na+x] = [y, a, {}_{n-1}x]+[y, {}_nx] = -[a, y, {}_{n-1}x],$$

since L^2 centralises H. Hence $[H, L(g_{n,1})] \leqq H \cap L(g_{n-1,1})$ and so (i) will follow from (ii) (for if $n \leqq p$ then $n-1 < p$). (ii) Evidently we may assume that $H \leqq L(g_{n,1})$ since $L(g_{n,1}) \lhd L$ by lemma 3.1. Clearly $[H^2, L] \subseteq \subseteq [H, [H, L]] = 0$. Thus if $x \in L$ and d_x is the restriction of the adjoint map x^* to H then d_x is a Lie endomorphism of H; for if $a, b \in H$ then

$$0 = [a, b]d_x = [ad_x, bd_x].$$

Furthermore as L^2 centralises H and $H \leqq L(g_{n,1})$ we have

$$d_xd_y = d_yd_x \text{ and } d_x^n = 0, \tag{15}$$

for all $x, y \in L$. We claim that if $x_1, x_2, \ldots \in L$ and $d_i = d_{x_i}$ then

$$d_1^{n_1} \ldots d_r^{n_r} = 0, \text{ if } n_1 + \ldots + n_r = n. \tag{16}$$

We prove (16) by induction on r. By equation (15) this is true for $r = 1$. Suppose that (16) is true for some r, and all $x_1, \ldots, x_r \in L$. Consider

$$d_0^{n_0} d_1^{n_1} \ldots d_r^{n_r}, \text{ where } n_0 + n_1 + \ldots + n_r = n.$$

Let x_0 and x_1 induce d_0 and d_1 respectively and let $s = n_0 + n_1$. Clearly we may assume that $n_i \neq 0$ for any $i = 0, 1, \ldots, r$. Define for each i, $1 \leqq i \leqq s-1$, $z_i = ix_0 + x_1$ and let d_{z_i} be the endomorphism of H corresponding to z_i. Then $d_{z_i} = id_0 + d_1$ and, since $s + n_2 + \ldots + n_r = n$, the inductive hypothesis on r gives

$$0 = d_{z_i}^s d_2^{n_2} \ldots d_r^{n_r} = (id_0 + d_1)^s d_2^{n_2} \ldots d_r^{n_r}.$$

Using (15) and the inductive hypothesis on r we have

$$iu_1 + \ldots + i^{s-1} u_{s-1} = 0, \tag{17}$$

where for each j, $1 \leqq j \leqq s-1$,

$$u_j = \tbinom{s}{j} d_0^j d_1^{s-j} d_2^{n_2} \ldots d_r^{n_r}$$

and $\binom{s}{j}$ is the binomial coefficient. Thus each u_j is independent of i. If we let i range from 1 to $s-1$ we obtain from (17) a system of $s-1$ equations in $s-1$ variables with matrix $V = (i^j)_{1 \leqq i, j \leqq s-1}$. Clearly the determinant of V is $\det V = (s-1)! \prod_{1 \leqq j < i \leqq s-1} (i-j)$ and so $\det V \neq 0$ if $p = 0$ or $p > n$, since $s \leqq n$. Thus V is non-singular and so (17) has the trivial solution, $u_1 = \ldots = u_{s-1} = 0$. In particular $u_{n_0} = \binom{s}{n_0} d_0^{n_0} d_1^{n_1} \ldots d_r^{n_r} = 0$. But $s \leqq n < p$ or $p = 0$ implies that $\binom{s}{n_0} \neq 0$ and so $d_0^{n_0} d_1^{n_1} \ldots d_r^{n_r} = 0$. Our induction on r is now complete. Hence $d_1 \ldots d_n = 0$ for all d_i and so $H \leqq \zeta_n(L)$. □

THEOREM 3.5 *Let $L \in \mathfrak{A}^m$ be defined over a field of characteristic p. Then*

(a) *If $p = 0$ or $p > n$ then $L(g_{n,r}) \leqq \zeta_{r((n^m-1)/(n-1))}(L)$.*

(b) *If $p = 0$ or $p > n-1$ then $L(g_{n,1}) \in \mathfrak{N}_{(n^m-1)/(n-1)}$.*

Proof: Put $h(n, m) = (n^m - 1)/(n-1)$ (we assume that $n \neq 1$).

(a) Let $L_1 = L(g_{n,1})$. If $m = 1$ there is nothing to prove. Let $m > 1$ and assume inductively that $L^2(g_{n,1}) \leqq \zeta_c(L^2)$ where $c = h(n, m-1)$. Evidently $B = [L_1, L] \leqq L^2(g_{n,1})$. For each i, $0 \leqq i \leqq c$, let $B_i = [B, {}_{c-i}L^2]$. Then $B_c = B$ and $B_0 = 0$ and $[B_i, L^2] \leqq B_{i-1}$. Now $B_0 = 0 \leqq \zeta_0(L)$. Suppose inductively that $B_i \leqq \zeta_{ni}(L) = N$. Apply lemma 3.4 to L/N and $H = (B_{i+1} + N)/N$ to obtain $H \leqq \zeta_n(L/N)$ and so $B_{i+1} \leqq \zeta_{n(i+1)}(L)$. Hence

$B_i \leqq \zeta_{ni}(L)$ for all i. In particular $B = [L_1, L] \leqq \zeta_{nc}(L)$ and so $L_1 \leqq \zeta_{1+nc}(L)$. Clearly $1+nc = h(n, m)$.

So (i) is true for $r = 1$ and for all m. Suppose inductively that (i) is true for $r-1$ and all m. Now if $a \in L(g_{n,r})$ and $x \in L$ then $[a, {}_nx] \in L(g_{n,r-1}) \leqq \leqq \zeta_s(L) = M$, where $s = (r-1)h(n, m)$. Hence $(L(g_{n,r})+M)/M \leqq (L/M)(g_{n,1}) \leqq \leqq \zeta_{h(n,m)}(L/M)$, from above. So $L(g_{n,r}) \leqq \zeta_{rh(n,m)}(L)$ and (i) is established for all r and m.

(b) By induction on m we may assume that $L_1^2 \in \mathfrak{N}_c$ where $L_1 = L(g_{n,1})$ and $c = h(n, m-1)$. For each i, $0 \leqq i \leqq c$, define $H_i = [L_1^2, {}_{c-i}L_1^2]$. Then $H_i \lhd L_1$, $[H_i, L_1^2] \leqq H_{i-1}$, $H_0 = 0$ and $H_c = L_1^2$. Fix i and let $H = H_i/H_{i-1}$. Apply lemma 3.4 to the pair H, L_1/H_{i-1} to obtain (since $L_1/H_{i-1} = = (L_1/H_{i-1})(g_{n,1})$) $[H, L_1/H_{i-1}] \leqq \zeta_{n-1}(L_1/H_{i-1})$.

Hence $[H_i, {}_nL_1] \leqq H_{i-1}$ and so $[H_c, {}_{nc}L_1] \leqq H_0 = 0$, whence

$$L_1 \in \mathfrak{N}_{1+nc}.$$

Clearly $1+nc = 1+nh(n, m-1) = h(n, m) = (n^m-1)/(n-1)$. □

Remark 5. Part (b) of theorem 3.5 extends the result of Higgins [89] theorem 1; for it is easy to see that the proof of lemma 3.4 and hence theorem 3.5 holds for Lie rings (of the appropriate characteristic).

Before we apply the results above it is useful to formalise a general line of approach in deciding whether a class or sequence of classes is a nilpotent class.

Let $\mathfrak{X}_1, \mathfrak{X}_2, \ldots$ be a sequence of {I, Q}-closed classes of Lie algebras. We wish to establish the following hypothesis:

'To each positive integer n there exists a positive integer $h(n)$ such that $\mathfrak{X}_n \leqq \mathfrak{N}_{h(n)}$.'

We consider the following cases.

Case 1. Can we find $c = c(n)$ such that $\mathfrak{X}_n \cap \mathfrak{A}^2 \leqq \mathfrak{N}_c$? If so then by proposition 1.1 for every d we have

$$\mathfrak{X}_n \cap \mathfrak{A}^d \leqq \mathfrak{N}_{((2c-1)^{d-1}+1)/2}.$$

Case 2. Can we find $d = d(n)$ such that $L = \langle \alpha(L) \rangle$ and $L \in \mathfrak{X}_n$ implies that $L \in \mathfrak{A}^d$? If so then for every m, $L = \langle \alpha^m(L) \rangle$ and $L \in \mathfrak{X}_n$ implies that $L \in \mathfrak{A}^{md}$. (If also $L = \alpha^m(L)$ then $L \in \mathfrak{A}^{2+(m-1)d}$).

To see this we define for each j, $0 \leqq j \leqq m$,

$$L_j = \langle H^{(j)} : H = \langle x^L \rangle \text{ for some } x \in \alpha^m(L) \rangle.$$

Then $L_0 = L$, $L_m = 0$, $L_j \lhd L$ and so $L_j/L_{j+1} \in \mathfrak{X}_n$ and $L_j/L_{j+1} = \langle \alpha(L_j/L_{j+1}) \rangle$; hence $L_j/L_{j+1} \in \mathfrak{A}^d$ for all j, $0 \leqq j < m$ and the result follows.

In case $L = \alpha^m(L)$ then clearly $L/L_1 = \alpha(L/L_1)$ and so $L/L_1 \in \mathfrak{E}_2$, whence $L^{(2)} \leqq L_1$; clearly $L_1 \in \mathfrak{A}^{(m-1)d}$ and the required result now follows.

Case 3. What relations exist between the $\mathfrak{X}_n$'s? Suppose that $h(1), \ldots, h(n-1)$ are known. Can we find m depending on n such that

$$L \in \mathfrak{X}_n \text{ implies that } L = \langle \alpha^m(L) \rangle? \tag{18}$$

We will refer to this as *procedure* (18). It is clear that we may perform the various steps in any way we please. First we apply the procedure to the classes $\mathfrak{D}_{n+1,n}$.

We deduce case 1 from the more general

THEOREM 3.6 *Over any field of characteristic zero or $p > m$,*

$$\mathfrak{D}_{mn+1,n} \cap \mathfrak{A}^d \leqq \mathfrak{N}_{f_4(m,n,d)},$$

where $f_4(m, n, d) = \{(2nm+1+2f(n^2m, n))^{d-1}+1\}/2$.

Proof: It is enough to consider the case $L \in \mathfrak{D}_{mn+1,n} \cap \mathfrak{A}^2$. Let $X = \langle x_1, \ldots, x_n \rangle \leqq L$. For each $x \in L$ we have, by lemma 1.2, $\langle x^L \rangle \in \mathfrak{N}_{nm}$ and so $X \in \mathfrak{G}_n \cap \mathfrak{N}_{n^2m} \leqq \mathfrak{F}_{s+1}$, where $s = f(n^2m, n) - 1$. Let $x, y_1, \ldots, y_k \in L$ and let $u = [x, {}_{m+1}x_1, {}_m x_2, \ldots, {}_m x_n]$. Then since $L^2 \in \mathfrak{A}$ we have

$$\begin{aligned}[u, y_1, \ldots, y_k] &= [x, x_1, y_1, \ldots, y_k, {}_m x_1, \ldots, {}_m x_n] \\ &\in [X^L, {}_{nm}X] \subseteq X,\end{aligned}$$

since $X \lhd^{nm+1} L$. Hence $A = \langle u^L \rangle \leqq X$ and so $A \in \mathfrak{F}_s$. But $L \in \mathrm{L}\mathfrak{N}$ and therefore by lemma 1.6, $A \leqq \zeta_s(L)$. Now the x_i were arbitrary and so we have that modulo $\zeta_s(L)$, $[x, {}_{m+1}x_1] \in L(g_{m,n-1}) \cap L^2 = H$, for all $x, x_1 \in L$. Clearly $[H, L^2] = 0$ and $L \in \mathfrak{A}^2$ and so by (the proof of) theorem 3.5, $H \leqq \zeta_{(n-1)m}(L)$. Hence modulo $\zeta_s(L)$ we have $[x, {}_{m+1}x_1] \in \zeta_{(n-1)m}(L)$ for all $x, x_1 \in L$. Thus

$$L/\zeta_{s+(n-1)m}(L) \in \mathfrak{E}_{m+1} \cap \mathfrak{A}^2 \leqq \mathfrak{N}_{m+2},$$

by theorem 3.5(b), whence $L \in \mathfrak{N}_{s+nm+2}$. We have

$$s+nm+2 = f(n^2m, n)+nm+1. \qquad \square$$

By putting $m = 1$ above we obtain the result that over any field,

$$\mathfrak{D}_{n+1,n} \cap \mathfrak{A}^d \leqq \mathfrak{N}_{\{(2n+2f(n^2,n)+1)^{d-1}+1\}/2}.$$

Now suppose that L is a Lie algebra, $a_1, \ldots, a_r \in \alpha(L)$, and $x \in L$ with $[L, {}_m x] = 0$. We claim that

$$\langle a_1, \ldots, a_r, x\rangle \in \mathfrak{N}_{rm+1}. \tag{*}$$

For let $A = \sum_{i=1}^{r} \langle a_i^L \rangle$ so that $A \in \mathfrak{N}_r$ and let $A_j = \zeta_j(A)$. Clearly $A_1 \leqq \zeta_m(A+\langle x\rangle)$, and by induction on j we have $A_j \leqq \zeta_{jm}(A+\langle x\rangle)$ for all j. So $A \leqq \zeta_{rm}(A+\langle x\rangle)$, whence $A+\langle x\rangle \in \mathfrak{N}_{rm+1}$ and the result follows.

Next we turn to case 2. Let $L \in \mathfrak{D}_{n+1,n}$ and suppose that $L = \langle \alpha(L)\rangle$. Let $x \in L$ and let $a_1, \ldots, a_{n-1} \in \alpha(L)$ and let $X = \langle x, a_1, \ldots, a_{n-1}\rangle$ so that by (*), $X \in \mathfrak{N}_{(n-1)(n+1)+1}$. Then $X \in \mathfrak{F}_{f(n^2,n)}$ (by equation (6)). Let $s = f(n^2, n)-n$. Define $u = [a_1, \ldots, a_{n-1}, {}_2 x]$ and let A be an abelian ideal of L. Then $[A, u] \subseteq [A, X^{n+1}] \subseteq [A, {}_{n+1}X] \subseteq X$, since $X \lhd^{n+1} L$. Now $[A, u] \subseteq \zeta_{n-1}(X)$ and so $\dim[A, u] \leqq s$ (for otherwise $\dim(X/\zeta_{n-1}(X)) \leqq f(n^2, n)-s-1 = n-1$ and so $X \in \mathfrak{N}_{2n-2}$, whence $s+1 \leqq \dim X \leqq f(2n-2, n)$, a contradiction). Let $H = C_L(u)$ so that $\dim(A/H \cap A) \leqq s$ (by equation (9), $A/H \cap A \cong [A, u]$). By lemma 2.2, $L^{(s)} \leqq H$. Assume that $n > 1$. Evidently for each j, $L^{(j)} = \langle \alpha(L^{(j)})\rangle$. As the a_i and x were arbitrary we can pick them from $L^{(s)}$. Then $u \in \zeta_1(L^{(s)})$. Furthermore if $B = L^{(s)}$ then B^{n-1} is spanned (*qua* vector space) by elements of the form $[a_1, \ldots, a_{n-1}]$ where the a_i belong to $\alpha(B)$. So modulo $\zeta_1(B)$ we have $B^{n-1} \leqq B(g_{2,1})$ and therefore $(B^{n-1})^{(2)} = 0$, whence $B^{(n+1)} = 0$ and so

$$L \in \mathfrak{A}^{f(n^2,n)+1}. \tag{19}$$

For $n = 1$, we have $L \in \mathfrak{D}_{2,1} \leqq \mathfrak{E}_2 \leqq \mathfrak{A}^2$, so (19) holds for $n = 1$.

Finally we attack the general case. We know that $\mathfrak{D}_{2,1} \leqq \mathfrak{A}^2$. Let $n > 1$ and assume that we have defined $d(r)$ for all r, $1 \leqq r \leqq n-1$, so as to satisfy

$$\mathfrak{D}_{r+1,r} \leqq \mathfrak{A}^{d(r)}. \tag{**}$$

Let $m = d(1)+\ldots+d(n-1)$ and let $L \in \mathfrak{D}_{n+1,n}$. We will show that $L = \alpha^m(L)$ so that by (18) and (19)

$$L \in \mathfrak{A}^{d(n)} \tag{20}$$

where $d(n) = 2+(d(1)+\ldots+d(n-1))(f(n^2, n)+1)$.

Let $X = \langle x\rangle \leqq L$ and let X_i be the ith ideal closure of X in L. Then $X_0 = L$ and $X_n = X$. By lemma 1.7, for each i, $0 < i < n$,

$$X_i/X_{i+1} \in \mathfrak{D}_{n+1-i,n-1} \leqq \mathfrak{D}_{n-i+1,n-i} \leqq \mathfrak{A}^{d(i)},$$

by (**). Thus $X_i^{(d(i))} \leqq X_{i+1}$ and so $X_1^{(m)} \leqq X$. If $X^{(m)} \neq 0$ then $X = X_1^{(m)} \lhd L$; hence $X_1 = X \in \mathfrak{A}$, a contradiction since $m > 0$. Thus $X_1^{(m)} = 0$. Now

$X_1 = \langle x^L \rangle$ and x was an arbitrary element of L, so $L = \alpha^m(L)$. This establishes (20) and so (**) for all r.

Define $f_5(n) = \{(2n+1+2f(n^2, n))^{d(n)-1}+1\}/2$. Then we have

THEOREM 3.7 *To each positive integer n there corresponds a positive integer $f_5(n)$ such that*

$$\mathfrak{D}_{n+1,n} \leqq \mathfrak{N}_{f_5(n)}. \qquad \square$$

Remark 6. We can employ corollary 3.3 to prove that if $L \in \mathfrak{D}_{2n-1,n}$ and $L = \langle \alpha(L) \rangle$ then $L \in \mathfrak{A}^{h_1(n)}$ for some $h_1(n)$. For this we let $X = \langle x_1, \ldots, x_n \rangle \leqq L$ and $u = [x_1, {}_2x_2, \ldots, {}_2x_n]$. By corollary 1.4 we have $X \in \mathfrak{F}_m$ where $m = f_1(2n-1, n)$. Using the argument which led to equation (19) we obtain $L^{(m)} \leqq C_L(u)$. The x_i are arbitrary and so can be chosen from $L^{(m)} = B$. Then $B/\zeta_1(B) = (B/\zeta_1(B))(g_{2,n-1})$ and so by corollary 3.3 we have $B^{(n)} \leqq \zeta_1(B)$. Thus $B^{(n+1)} = 0$ and we can take $h_1(n) = f_1(2n-1, n)+n+1$.

4. A counterexample

In this section we produce an example which shows that the restrictions on characteristics of fields in theorem 3.6 cannot be relaxed. Then we discuss what can be said about the classes $\mathfrak{D}_{n,r}$ $(n > r+1)$ over fields of characteristic zero.

Let $\mathfrak{k}$ be a field of characteristic $p > 0$. By theorem 3.6 for every positive integer n,

$$\mathfrak{D}_{(p-1)n+1,n} \cap \mathfrak{A}^2 \leqq \mathfrak{N}_{np+f(n^2,n)}$$

over the field $\mathfrak{k}$. In view of this it is surprising that we have:

THEOREM 4.1 *Let $\mathfrak{k}$ be a field of characteristic $p > 0$. Then there exists a non-nilpotent metabelian Lie algebra L defined over $\mathfrak{k}$ such that*

$$L \in \bigcap_{n=1}^{\infty} \mathfrak{D}_{(p-1)n+2,n}.$$

Proof: Let B be an abelian Lie algebra over $\mathfrak{k}$ with basis $\{b_i : i = 1, 2, \ldots\}$. Define $U = U(B)$ to be the *universal enveloping algebra* of B (see Jacobson [98], chapter V, p. 151 ff.). Then U is a commutative and associative algebra with unit and is a B-module under the usual action (see also chapter 11). Let V be the submodule of U spanned by b_i^p for $i = 1, \ldots$, so that $V = \sum_{i=1}^{\infty} Ub_i^p$.

Form the quotient module $A = U/V$, consider it as an abelian Lie algebra over $\mathfrak{k}$ and let L be the split extension

$$L = A \dotplus B,$$

where $A \lhd L$ and $A \cap B = 0$.

Evidently $L \in \mathfrak{A}^2$. Let c be any positive integer. Clearly $b_1 b_2 \ldots b_{c+1} + V = [b_1 + V, b_2, \ldots, b_c] \not\equiv 0 \pmod{V}$. Thus $L \notin \mathfrak{N}_c$ for any c and so $L \notin \mathfrak{N}$.

Let $x \in B$ so that $x = \sum \beta_i b_i$ ($\beta_i \in \mathfrak{k}$). Consider x as an element of U. Then

$$x^p = \textstyle\sum_i \beta_i^p b^p \in V,$$

since U is commutative and $\mathfrak{k}$ has characteristic p. Thus if $a \in A$ then $[a, {}_p x] = ax^p = 0$ and so $[A, {}_p x] = 0$. Now let $x_1, \ldots, x_n \in B$ and let $X = \langle x_1, \ldots, x_n \rangle$. Now X is abelian and so for any r,

$$[A, {}_r X] = \textstyle\sum [A, {}_{r_1} x_1, \ldots, {}_{r_n} x_n],$$

where the summation is taken over all n-tuples of non-negative integers $(r_1, \ldots, r_n)$ such that $r_1 + \ldots + r_n = r$. In particular if $r = (n-1)p+1$, then some $r_i > p-1$ in any n-tuple $(r_1, \ldots, r_n)$ and so $[A, {}_{r_1} x_1, \ldots, {}_{r_n} x_n] \subseteq [A, {}_{r_i} x_i, \ldots] = 0$, from above. Hence

$$[A, {}_{(p-1)n+1} X] = 0.$$

It is easy to see that for any r, $(A+X)^{r+1} = [A, {}_r X]$ $(r > 0)$, and so if $r = (p-1)n+1$ then $A + X \in \mathfrak{N}_r$. Clearly $A + X \lhd L$ and so for any subalgebra Y of $A+X$ we have $Y \lhd^r A + X \lhd L$, where $r = (p-1)n+1$. In particular let $a_1, \ldots, a_n \in A$ and let $H = \langle a_1 + x_1, \ldots, a_1 + x_n \rangle$. Then $H \leqq A + X$ and so

$$H \lhd^{(p-1)n+2} L.$$

But every $\mathfrak{G}_n$-subalgebra of L is of the same form as H and so $L \in \mathfrak{D}_{(p-1)n+2,n}$, for every positive integer n. □

We remark that $L \in \mathfrak{D}_{p+1,1} \leqq \mathfrak{E}_{p+1}$. So theorem 4.1 also gives us an example of a non-nilpotent and metabelian Lie algebra over a field a characteristic p which satisfies the $(p+1)$st Engel condition.

Over fields of characteristic zero

Let L be an L$\mathfrak{N}$-algebra over a field of characteristic zero. We know from section 4.2 that if A and B are subsets of L then

$$A^B = \textstyle\sum_{\theta \in \exp(\mathrm{ad}_B(L))} A^\theta,$$

where $\exp(\mathrm{ad}_B(L))$ is the group of automorphisms generated by $\exp(x^*)$ for all $x \in B$. Thus suppose that $H \lhd^m L$ and $H = H_m \lhd H_{m-1} \lhd \ldots \lhd H_1 \lhd H_0 = L$ is the ideal closure series of H in L. Let $0 < i \leqq m-1$. Then

$$H_i = H^{H_{i-1}} = H_{i+1}^{H_{i-1}} = \sum_{\theta \in \exp(\mathrm{ad}_{H_{i-1}}(L))} H_{i+1}^{\theta}.$$

Thus if $H_{i+1} \in \mathfrak{A}^{d_i}$ then

$$H_i = \langle \alpha^{d_i}(H_i) \rangle, \tag{21}$$

for $H_{i+1}^{\theta} \lhd H_i^{\theta} = H_i$ for all $\theta \in \exp(\mathrm{ad}_{H_{i-1}}(L))$, since $H_{i+1} \lhd H_i \lhd H_{i-1}$.

As an application of (21) we see that in applying the procedure (18) to the classes $\mathfrak{D}_{n,r}$, we can omit the last step, namely case 3. For example suppose we know (over fields of characteristic zero) that for some n and r

$$L = \langle \alpha(L) \rangle \text{ and } L \in \mathfrak{D}_{n,r} \text{ implies that } L \in \mathfrak{A}^{d(n,r)}.$$

Then it follows by (18) that $L = \langle \alpha^m(L) \rangle$ and $L \in \mathfrak{D}_{n,r}$ implies that $L \in \mathfrak{A}^{md(n,r)}$. We claim that for this n and r we have

$$\mathfrak{D}_{n,r} \leqq \quad \mathfrak{A}^{(d(n,r))^{n-1}}$$

whence nilpotence (of bounded class) for algebras in this class will follow by theorem 3.6. For let $L \in \mathfrak{D}_{n,r}$ and let $x \in L$ and $H = \langle x \rangle$. Then $H \lhd^n L$. Let H_i denote the ith ideal closure of H in L. Clearly $H = H_n \leqq \zeta_1(H_{n-1})$ and so $H_{n-1} = \zeta_1(H_{n-1}) \in \mathfrak{A}$ (assume that $n > 1$). Suppose inductively that $H_{i+1} \in \mathfrak{A}^{d(n,r)^{n-2-i}}$, where $0 < i < n-1$. Then by equation (21) we have $H_i = \langle \alpha^{d_i}(H_i) \rangle$, where $d_i = d(n,r)^{n-2-i}$. We also have $H_i \in \mathfrak{D}_{n,r}$ and so $H_i \in \mathfrak{A}^{d_{i+1}}$, where $d_{i+1} = d(n,r)^{n-2-i}.d(n,r)$. So we have $H_i \in \mathfrak{A}^{d(n,r)^{n-1-i}}$, for all i, $0 < i < n-1$ $(n > 2)$. We apply the process above again to H_2 and H_1 and get $H_1 \in \mathfrak{A}^m$ where $m = d(n,r)^{n-2}$. Now x was arbitrary and $H_1 = \langle x^L \rangle$ so $L = \langle \alpha^m(L) \rangle$ and hence $L \in \mathfrak{A}^{md(n,r)}$. If $n = 1$ then $L \in \mathfrak{A}$ and so the result holds for $n = 1$ as well. In the case $n = 2$ we have $L = \langle \alpha(L) \rangle \in \mathfrak{A}^{d(n,r)}$. Our claim is now proved for all n and r. We state our result above as

LEMMA 4.2 *Let n, r be positive integers and let $\mathfrak{k}$ be a field of characteristic zero. Let $d(n,r)$ be a positive integer such that if L is a Lie algebra over $\mathfrak{k}$ with $L = \langle \alpha(L) \rangle$ and $L \in \mathfrak{D}_{n,r}$ then $L \in \mathfrak{A}^{d(n,r)}$. Then over the field $\mathfrak{k}$,*

$$\mathfrak{D}_{n,r} \leqq \mathfrak{A}^{d(n,r)^{n-1}} \cap \mathfrak{N}_{\{n(d(n,r)n-1)-1\}/(n-1)}. \qquad \square$$

We note that if L is a Lie algebra then

$$L(g_{n,r}) \in \mathfrak{E}_n^r, \tag{22}$$

if L is defined over a field with at least n elements. This follows by induction on r and by noting that $L(g_{n,r})/L(g_{n,r-1}) = (L/L(g_{n,r-1}))(g_{n,1}) \in \mathfrak{E}_n$. Thus if $\mathfrak{E}_n \leqq \mathfrak{A}^{f_6(n)}$ then

$$L(g_{n,r}) \in \mathfrak{A}^{rf_6(n)}. \tag{23}$$

THEOREM 4.3 *Let n be a positive integer such that over fields of characteristic zero there exists $f_6(n)$ for which $\mathfrak{E}_n \leqq \mathfrak{A}^{f_6(n)}$. Then there exists $f_7(n, r)$ such that over fields of characteristic zero,*

$$\mathfrak{D}_{n(r-1)+1,r} \leqq \mathfrak{N}_{f_7(n,r)},$$

for all positive integers r.

Proof: By lemma 4.2 we need consider only the case

$$L = \langle \alpha(L) \rangle \text{ and } L \in \mathfrak{D}_{n(r-1)+1,r}.$$

Let $X = \langle x_1, \ldots, x_r \rangle \leqq L$ and let $u = [x_1, {}_nx_2, \ldots, {}_nx_r]$. Put $H = C_L(u)$. By corollary 1.4 $X \in \mathfrak{F}_{f_1(n(r-1)+1,r)}$. Put $m = f_1(n(r-1)+1, r)-1$, and assume that $r > 1$. Now let A be an abelian ideal of L. Then $[A, u] \subseteq X$ and so $[A, u] \in \mathfrak{F}_m$. Hence $A/H \cap A \in \mathfrak{F}_m$. Therefore as L is generated by abelian ideals like A we have, by lemma 2.2, that $L^{(m)} \leqq H = C_L(u)$. The x_i were arbitrary and so we deduce that if $B = L^{(m)}$ then

$$B/\zeta_1(B) = (B/\zeta_1(B))(g_{n,r-1}) \in \mathfrak{A}^{(r-1)f_6(n)},$$

by equation (23).

Thus $L \in \mathfrak{A}^{d(n,r)}$, where $d(n, r) = f_1(n(r-1)+1, r)+(r-1)f_6(n)$. The required result now follows by lemma 4.2. The case $r = 1$ is trivial. □

The results of Higgins [89] and Heineken [79] show that we can take

$$f_6(1) = 1,\ f_6(2) = 2,\ f_6(3) = 3,\ f_6(4) = 5. \tag{24}$$

This yields by theorem 4.3 the nilpotence (of bounded class) of algebras (over fields of characteristic zero) in the classes

$$\mathfrak{D}_{2r-1,r},\ \mathfrak{D}_{3r-2,r},\ \mathfrak{D}_{4r-3,r},$$

and so by the Mal'cev correspondence torsion-free groups in any of the classes

$$\mathfrak{U}_{2r-1,r},\ \mathfrak{U}_{3r-2,r},\ \mathfrak{U}_{4r-3,r}$$

are nilpotent. (The bounds are different in the three cases.)

We have not been able to determine whether (for fields of characteristic zero) we can find $f(n)$ such that

$$\mathfrak{D}_{n,1} \leqq \mathfrak{N}_{f(n)}.$$

The result is true for $n \leqq 4$.

5. Unsin's algebras

In this section we give examples, due to Unsin [235], to show that in characteristic $p > 0$ there exist non-nilpotent $\mathfrak{D}$-algebras. The construction is similar to that of Heineken and Mohamed [292] for groups; the proofs are variations on Unsin's. The analogy with the Heineken-Mohamed group is incomplete, as we shall remark later.

We begin by describing a certain associative algebra F over a given field $\mathfrak{k}$ of characteristic $p > 0$. Let $n(i)$ $(i \in \mathbb{N}\backslash\{0\})$ be any sequence of integers $\geqq 1$, and let F have a presentation with generators $a, x_1, x_2, \ldots$ and relations

(i) $a^2 = axa = 0 \quad (x \in F)$
(ii) $x_{i+1}^{p^{n(i+1)}} = x_i + a \quad (i \geqq 1)$
(iii) $x_1^{p^{n(1)}} = (x_1 + a)^{p^{n(1)}} = 0.$

Let A be the ideal of F generated by a. The essential content of (i) is that $A^2 = 0$. From (ii) and (iii) we see that F/A has generators $\bar{x}_i = x_i + A$ subject to relations

$$\bar{x}_{i+1}^{p^{n(i+1)}} = \bar{x}_i$$

$$\bar{x}_1^{p^{n(1)}} = 0.$$

We can exhibit the structure of F/A more explicitly. Let S_m $(m \in \mathbb{N})$ be the semigroup-with-0 having a single generator g subject to the relation

$$g^{p^m} = 0.$$

Then

$$S_m = \{g, g^2, \ldots, g^{p^m - 1}, 0\}$$

is a commutative nilpotent semigroup-with-0. There are embeddings

$$S_{n(1)} \to S_{n(1)+n(2)} \to \ldots \to S_{n(1)+\ldots+n(i)} \to \ldots$$

where the generators g_i map according to

$$g_i \leftrightarrow g_{i+1}^{p^{n(i+1)}}.$$

Let S be the direct limit of this sequence. Then F/A is isomorphic to $\mathfrak{k}S$, the semigroup algebra of S, with the 0 of S identified with that of $\mathfrak{k}$. Let $C = \mathfrak{k}S$.

Lemma 5.1 *C is commutative, and every $x \in C$ generates a nilpotent ideal of C.*

Proof: Each S_i is commutative, so S is, so C is. If $s \in S$ then $s^t = 0$ for some $t \in \mathbb{N}$. Since C is commutative and spanned by S, it follows that if $x \in C$ then $x^u = 0$ for some $u \in \mathbb{N}$. The ideal generated by x is $\mathfrak{k}x + Cx$, and

$$(\mathfrak{k}x+Cx)^u \subseteq \mathfrak{k}x^u + Cx^u = 0. \qquad \square$$

Corollary 5.2 *The Lie algebra $[F]$ is a metabelian Fitting algebra.*

Proof: F has a series $0 \lhd A \lhd F$ with commutative factors, so $[F]$ has a series $0 \lhd [A] \lhd [F]$ with abelian factors. Hence $[F] \in \mathfrak{A}^2$.

Let $x \in F$. Now $x + A$ generates a nilpotent ideal J/A of F/A. Therefore $x \in J$ and $J^t \subseteq A$ for some $t \in \mathbb{N}$. But then $J^{2t} \subseteq A^2 = 0$ so that J is nilpotent. Passing to the Lie algebra, x lies in the nilpotent ideal $[J]$ of $[F]$, whence F is Fitting. $\square$

Now let

$$z_i = x_i + a$$

and set

$$L = \langle z_1, z_2, \ldots \rangle \leqq [F].$$

Then L is also metabelian and Fitting. We claim that if $p > 2$, or if $p = 2$ and $n(1) > 1$, then $L \in \mathfrak{D}$ and $\zeta_1(L) = 0$. Since $L \neq 0$ we will have $L \notin \mathfrak{N}$ as required.

We split the proof into six sections.

I: *Useful formulae*

We begin with some general observations. Let B be any associative algebra over $\mathfrak{k}$, with $x, y \in B$. By induction we have

$$[x, {}_n y] = \textstyle\sum_{i=0}^{n} (-1)^i \binom{n}{i} y^i x y^{n-i}. \tag{1}$$

If $\mathfrak{k}$ has characteristic $p > 0$ it follows that

$$[x, {}_p y] = xy^p - y^p x = [x, y^p]$$

and hence, by another induction,

$$[x, {}_{p^n}y] = [x, y^{p^n}] \qquad (n \in \mathbb{N}). \tag{2}$$

Following Jacobson [98] p. 187 we now show that

$$(x+y)^p \equiv x^p + y^p \qquad (\text{mod } X^2) \tag{3}$$

where X is the *Lie* algebra generated by x and y. To this purpose we introduce an indeterminate t and work in the polynomial algebra $B[t]$. Expanding and collecting coefficients yields

$$(tx+y)^p = t^p x^p + y^p + \sum_{i=1}^{p-1} s_i t^i$$

where the s_i are polynomials in x, y of total degree p. Differentiating this with respect to t gives

$$\sum_{i=0}^{p-1} (tx+y)^i x (tx+y)^{p-i-1} = \sum_{i=1}^{p-1} i s_i t^{i-1},$$

so that by (1)

$$[x, {}_{p-1}(tx+y)] = \sum i s_i t^{i-1}.$$

It is now obvious that each $s_i \in X^2$. If we set $t = 1$ we obtain (3). (The first few s_i are listed in Jacobson [98] p. 187.)

Now let $u_1, \ldots, u_n \in B$ and let X be the Lie algebra generated by the u's. By induction from (3), and using (2), we get

$$(u_1 + \ldots + u_n)^{p^m} \equiv u_1^{p^m} + \ldots + u_n^{p^m} \quad (\text{mod } X^2) \tag{4}$$

for $m, n \in \mathbb{N}$.

We apply these results to the algebra F described above.

From (i) and (2) we get

$$\begin{aligned} z_{i+1}^{p^{n(i+1)}} &= (x_{i+1} + a)^{p^{n(i+1)}} \\ &= x_{i+1}^{p^{n(i+1)}} + [a, {}_{t_i}x_{i+1}] \end{aligned}$$

(where $t_i = p^{n(i+1)} - 1$)

$$= z_i + [a, {}_{t_i}z_{i+1}]$$

so that

$$z_i = z_{i+1}^{p^{n(i+1)}} - [a, {}_{t_i}z_{i+1}]. \tag{5}$$

By induction and (2) we get, for all $i, k \in \mathbb{N}$,

$$\begin{aligned} z_i = z_{i+k}^{p^{n(i+k)+\ldots+n(i+1)}} &- [a, {}_{u_2}z_{i+k}] - \\ &- [a, {}_{u_3}z_{i+k}] - \ldots - [a, {}_{u_{k+1}}z_{i+k}] \end{aligned} \tag{6}$$

where

$$u_j = p^{n(i+k)+\ldots+n(i+1)} - p^{n(i+k)+\ldots+n(i+j)}$$

if $j \leqq k$, and

$$u_{k+1} = p^{n(i+k)+\ldots+n(i+1)} - 1.$$

By (2) we now have

$$z_i^* = (z_{i+k}^*)^{p^{n(i+k)+\ldots+n(i+1)}} - [a, {}_{u_2}z_{i+k}]^* - \\ - [a, {}_{u_3}z_{i+k}]^* - \ldots - [a, {}_{u_{k+1}}z_{i+k}]^*. \tag{7}$$

From (7) it follows that for all $i, k \in \mathbb{N}$,

$$[a, z_i] = [a, {}_u z_{i+k}] \tag{8}$$

where

$$u = p^{n(i+k)+\ldots+n(i+1)}.$$

Also, from (i), (ii), and (2) we have $[a, {}_v x_1] = 0$ where $v = p^{n(1)} - 1$, so that

$$[a, {}_v z_1] = 0. \tag{9}$$

From the known structure of F and (i), (ii), (iii), it is not hard to exhibit a 'canonical' basis for F. Namely, take all elements of the forms

$$a,\ X,\ Xa,\ aX,\ XaY$$

where X and Y are of the form

$$x_1^{\alpha_1} \ldots x_r^{\alpha_r}$$

with $0 \leqq \alpha_i < p^{n(i)}$, and some $\alpha_i \neq 0$. (Terms with $\alpha_\iota = 0$ are deemed to be omitted from the expression.)

II: *Triviality of the centre*

In this section we show that $\zeta_1(L) = 0$.

Define

$$j_1 = p^{n(1)} - 1,$$

$$j_i = p^{n(i)+\ldots+n(2)}(p^{n(1)} - 1) \qquad (i \geqq 2).$$

From (9) and (8) it follows that

$$[a, {}_{j_i} z_i] = 0,$$

whilst use of the 'canonical' basis shows that

$$[a, {}_{j_i-1}z_i] \neq 0.$$

We may now prove:

LEMMA 5.3 *For a given i, the elements* $[a, {}_jz_i]$ $(j < j_i)$ *are linearly independent.*

Proof: Let $\alpha = z_i^*$, $t = j_i$. Then as above $a\alpha^t = 0$, $a\alpha^{t-1} \neq 0$. Suppose that

$$\sum_{j=1}^{t-1} \lambda_j a\alpha^j = 0$$

for $\lambda_j \in \mathfrak{k}$. If we act on this by powers of α and use a reverse induction we get $\lambda_{t-1} = \lambda_{t-2} = \ldots = \lambda_1 = 0$. □

Next we define

$$L_i = \langle z_1, \ldots, z_i, [a, z_1], \ldots, [a, z_i]\rangle.$$

Then we have:

LEMMA 5.4 $\zeta_1(L_i) = \langle [a, {}_{j_i-1}z_i]\rangle$.

Proof: Let $z \in L_i$. By (8) and (9) we can write

$$z = \sum_{j=1}^{j_i-1} \alpha_j[a, {}_jz_i] + \sum_{j=1}^{i-1} \beta_j z_i^{p^{n(i)+\ldots+n(j+1)}} + \beta_i z_i \tag{10}$$

where $\alpha_j, \beta_j, \beta_i \in \mathfrak{k}$. Then $[[a, z_i], z] = 0$, so that

$$\sum_{j=1}^{i-1} \beta_j[a, {}_{w_j}z_i] + \beta_i[a, {}_2z_i] = 0$$

where

$$w_j = p^{n(i)+\ldots+n(j+1)} + 1.$$

Since $p > 2$ or $n_1 > 1$, we have $w_j < j_i$, so by lemma 5.3 $\beta_j = 0$ for $1 \leqq j \leqq i$.

Further, $[z, z_i] = 0$, so by lemma 5.3 again we have $\alpha_j = 0$ for $1 \leqq j < j_i - 1$. But clearly $[a, {}_{j_i-1}z_i] \in \zeta_1(L_i)$. This proves the lemma. □

LEMMA 5.5 *The centre of L is trivial.*

Proof: We have $L = \bigcup_{i\in\mathbb{N}} L_i$, so that $\zeta_1(L) \leqq \bigcap_{i\in\mathbb{N}} \zeta_1(L_i)$ which is 0 by lemma 5.4. □

III: *L is a* $\mathfrak{D}$*-algebra*

Let $U \leqq L$. We must show that U si L. In fact we prove a stronger statement. We note two distinct cases:

(a) U is finitely generated modulo L^2,
(b) U is not.

In case (a) we have

$$U \leqq L^2 + \langle y_1, \ldots, y_r \rangle$$

where the $y_i \in L$ and $r \in \mathbb{N}$. Let $Y_i = \langle y_i^L \rangle$. Since L is Fitting each $Y_i \in \mathfrak{N}$. But now

$$U \leqq L^2 + Y_1 + \ldots Y_r$$

which is nilpotent by theorem 1.2.5 (noting that L^2 is abelian). Hence

$$U \text{ si } L^2 + Y_1 + \ldots + Y_r \lhd L$$

and so U si L.

It remains to discuss case (b). We show that if (b) occurs then U is an *ideal* of L. To this end put

$$I = I_L(U).$$

We wish to show that $I = L$. The argument splits into three parts.

IV: *Proof that* $L = L^2 + I$.

Let V be the associative algebra generated by all u^* for $u \in U$. Clearly *if* $x^* \in V$ *then* $x \in I$.

Choose any $k \in \mathbb{N}$. Since (b) holds we can find $i > k$ such that there is an element

$$y = \alpha_i z_i + \ldots + \alpha_1 z_1 + w \in U$$

where $w \in L^2$, the $\alpha_j \in \mathfrak{k}$, and $\alpha_i \neq 0$. By (7)

$$y^* = \alpha_i z_i^* + \ldots + \alpha_1 z_i^{*p^{n(i)+\ldots+n(2)}} + v^* \tag{11}$$

where $v \in L^2$. Since L is Fitting every inner derivation is nilpotent. By taking p^mth powers of (11) for various m and eliminating, using (4), it follows that

$$z_i^* + v_i^* \in V$$

where $v_i \in L^2$. (Note that $y^* \in V$). Therefore

$$u_i = z_i + v_i \in I.$$

Now $i > k$, so

$$(u_i^*)^{p^{n(i)+\ldots+n(k+1)}} = z_k^* + v_k^* \in V$$

for some $v_k \in L^2$. Therefore

$$u_k = z_k + v_k \in I$$

so that $z_k = u_k - v_k \in I + L^2$. The z_i generate L, from which it follows that $L = I + L^2$.

V: *Proof that* $I \cap L^2 \lhd L$

This is now very easy. For $I \cap L^2$ is idealised by I, and by L^2 since L^2 is abelian. By IV the result follows.

VI: *Proof that* $I = L$

Pick $i \in \mathbb{N}$. We can find $i_1 \in \mathbb{N}$ with

$$v_i = \sum_{k=1}^{k_0} \gamma_k [a, {}_k z_{i_1}] \qquad (\gamma_k \in \mathfrak{k})$$

using (8). Now pick an arbitrary $l > k_0$. There exists $j \geqq \max(i+1, i_1)$ with

$$v_{i+1} = \sum_{k \geqq 1} \delta_k [a, {}_k z_j] \quad (\delta_k \in \mathfrak{k}).$$

We consider the element

$$q = [u_{i+1}, u_i]$$
$$= [z_{i+1}, z_i] + [v_{i+1}, z_i] - [v_i, z_{i+1}]$$

which lies in $I \cap L^2$. By (7), $[z_{i+1}, z_i]$ is a sum of distinct terms of the form

$$[a, {}_k z_{i+1}] \qquad 1 \leqq k \leqq p^{n(i+l)+\dots+n(i+1)}$$

or

$$[a, {}_k z_j] \qquad k \leqq p^{n(j)+\dots+n(i+1)}.$$

Since $k_0 < l$, not all the terms of $[z_{i+1}, z_i]$ appear among those of $[v_i, z_{i+1}]$. Further, $[v_{i+1}, z_i]$ is a sum of terms of the form $[a, {}_g z_j]$, where

$$g = k + p^{n(j)+\dots+n(i+1)},$$

so the summands of $[z_{i+1}, z_i]$ and $[v_{i+1}, z_i]$ are pairwise distinct.

Consequently we can write

$$q = \sum_{k \geqq 1} \varepsilon_k [a, {}_k z_j] \qquad (\varepsilon_k \in \mathfrak{k})$$

where $\varepsilon_{l_0} \neq 0$ for some $l_0 \leqq p^{n(j)+\dots+n(i+1)}$. Since $I \cap L^2 \lhd L$, we have

$$[q, {}_k z_j] \in I \text{ for all } k \in \mathbb{N},$$

and since z_j^* is nilpotent the usual reverse induction yields

$$[a, {}_{l_0}z_j] \in I,$$

and then

$$[a, z_i] = [a, {}_{p^{n(j)+\dots+n(i+1)}}z_j] \in I.$$

But the ideal of L generated by the $[a, z_i]$ is L^2, so that $L^2 \leqq I$. From IV it follows that $L = I$, whence $U \lhd L$.

This proves:

THEOREM 5.6 *The Lie algebras L constructed above, with $p > 2$ or $p = 2$ and $n(1) > 1$, have the following properties:*

(a) *L is a non-zero metabelian Lie algebra.*

(b) *Every subalgebra of L is a subideal.*

(c) *The centre of L is zero so L is not nilpotent.*

(d) *Every subalgebra of L is either nilpotent, and contained in a nilpotent ideal, or is itself an ideal.* □

More information about these algebras may be found in Unsin [235]. In particular he proves that if U is not finitely generated modulo L^2 then $U \geqq L^2$, $U^2 = L^2$, and $\zeta_1(U) = 0$. He also proves that two of the above algebras are isomorphic if and only if they arise from the same choice of sequence $n(1), n(2), \dots$. Thus there are at least $2^{\aleph_0}$ countable-dimensional non-nilpotent $\mathfrak{D}$-algebras over $\mathfrak{k}$.

It is instructive to make a comparison with the Heineken-Mohamed group [292], which has the following properties:

(i) It is metabelian,

(ii) Every proper subgroups is subnormal and nilpotent,

(iii) It has trivial centre.

Condition (ii) fails for Unsin's algebras, in that some proper subalgebras are not nilpotent. In fact it *must* fail, because any Lie algebra satisfying (i) and (ii) is nilpotent. More strongly we have:

PROPOSITION 5.7 *If L is a Lie algebra with $L^2 < L$ and if every proper subalgebra of L is a nilpotent subideal, then L is nilpotent.*

Proof: We can find an ideal T of L of codimension 1, so that $L = T + \langle x \rangle$. By hypothesis T is nilpotent, and $\langle x \rangle$ is a subideal. Now $S = \langle x^L \rangle < L$, so is nilpotent; and then $L = T + S$ is nilpotent by theorem 1.2.5. □

Chapter 8

Chain conditions for subideals

The classes Min-si and Max-si have already made an appearance in chapter 3, where we proved that in characteristic 0 they are coalescent. The object of this chapter is to investigate the structure of algebras in these and related classes. The chapter divides broadly into two parts: the first on minimal conditions, the second on maximal conditions.

1. Classes related to Min-si

The classes Min-si and Min-$\lhd^n$ ($n \in \mathbb{N}$) are related by the series of inclusions

$$\text{Min-}\lhd \supseteq \text{Min-}\lhd^2 \supseteq \text{Min-}\lhd^3 \supseteq \ldots \supseteq \text{Min-si}.$$

There is no very obvious reason why these inclusions should not be strict. But in group theory it is a fact, due to Robinson [327], that the minimal condition for 2-step subnormal subgroups implies the minimal condition for all subnormal subgroups. Following [206], we shall prove that in characteristic 0

$$\text{Min-}\lhd^2 = \text{Min-si}$$

while in characteristic $p > 0$

$$\text{Min-}\lhd^3 = \text{Min-si}.$$

We show in section 3 that this 3 cannot be reduced to 2. Certainly the 2 cannot be improved to 1: the algebra constructed in example 6.3.6 satisfies Min-$\lhd$ (because P is a minimal ideal, being irreducible) but not Min-$\lhd^2$ (since every subspace of P is a 2-step subideal).

The proofs of these results afford interesting applications of the Baer and Hirsch-Plotkin radicals.

LEMMA 1.1 *If L is hypercentral then $L^{(\alpha)} = 0$ for some ordinal α.*

Proof: First we require an analogue of Grün's lemma (Kuroš [303] p. 227). Let K be any Lie algebra such that $\zeta_2(K) > \zeta_1(K)$. We show that $K^{(1)} < K$. For take $a \in \zeta_2(K) \backslash \zeta_1(K)$ and consider the map $a^* : K \to K$. Then a^* is a homomorphism, the image of a^* is contained in $\zeta_1(K)$, but since $a \notin \zeta_1(K)$ this image is not zero. Therefore K has a non-zero abelian homomorphic image, so that $K^{(1)} < K$.

Now let $L \in \mathfrak{Z}$, and let $P = \bigcap_{\beta \geq 0} L^{(\beta)}$. Then $P = L^{(\alpha)}$ for some α, and $P \in \mathfrak{Z}$. Either $P = 0$, or $P = \zeta_1(P)$, or $\zeta_2(P) > \zeta_1(P)$. But in the latter cases $P^{(1)} < P$ so that $L^{(\alpha+1)} < L^{(\alpha)}$, contradicting the definition of P. Hence $P = 0$. □

LEMMA 1.2 *If L is a locally nilpotent Lie algebra satisfying* Min-$\lhd$, *then L is soluble.*

Proof: Let U be the hypercentre of L. If $U \neq L$ then L/U is locally nilpotent. By Min-$\lhd$ we can find a non-zero minimal ideal M/U of L/U; but by lemma 7.1.6 this minimal ideal is central, contrary to the definition of U. Therefore $L = U$ is hypercentral. By lemma 1.1 $L^{(\alpha)} = 0$ for some ordinal α. By Min-$\lhd$ this α can be made finite, so that L is soluble. □

The algebra constructed after lemma 6.4.1 (vector space with a downward shift) is locally nilpotent. It is not hard to show that it satisfies the minimal condition for ideals, for its only ideals are the terms of the upper central series. This shows that in the above lemma we cannot assert that L is finite-dimensional, or that it is nilpotent.

LEMMA 1.3 *If $L \in$ Min-$\lhd^2$ then the Hirsch-Plotkin radical of L is finite-dimensional.*

Proof: $R = \rho(L)$ is locally nilpotent, so by lemma 1.2 it is soluble. Now for any $n \in \mathbb{N}$ we have

$$R^{(n)} \text{ ch } R \lhd L$$

so that $R^{(n)} \lhd L$, and $R^{(n)} \in$ Min-$\lhd$. Thus $R^{(n)}/R^{(n+1)}$ is abelian with Min-$\lhd$, so is finite-dimensional: hence R is finite-dimensional. □

We now come to the main result of this section:

THEOREM 1.4 *If L is a Lie algebra over a field of characteristic 0 satisfying* Min-$\lhd^2$, *then L satisfies* Min-si.

Proof: Assume the contrary. Then there exists M minimal with respect to

$$M \lhd L \text{ and } M \notin \text{Min-si}.$$

Let N be any proper ideal of M. For any integer $i > 0$ $N^i \text{ ch } N \lhd M \lhd L$ so that $N^i \lhd^2 L$. Hence by Min-si $N^\omega = N^c$ for some integer c. By lemma 1.3.2 $N^c \lhd L$. Now N/N^c si L/N^c and is nilpotent, so $N/N^c \leqq \beta(L/N^c)$. But $\beta(L/N^c) \leqq \rho(L/N^c)$, which is finite-dimensional by lemma 1.3. By minimality of M, we have $N^c \in \text{Min-si}$. Since $N/N^c \in \mathfrak{F}$, we have $N \in \text{Min-si}$ by theorem 1.7.4. Hence every proper ideal of M satisfies Min-si. If

$$I_1 > I_2 > \ldots$$

is a properly descending chain of subideals of M, then I_2 si $\langle I_2^M \rangle$ which is a proper ideal of M. But then $I_3, I_4, \ldots$ are subideals of $\langle I_2^M \rangle$, contrary to Min-si.

Therefore the assumption is false, and $L \in \text{Min-si}$. □

In characteristic $p > 0$ we cannot make this use of the Baer radical: the best we are able to prove is:

Proposition 1.5 *If L is a Lie algebra satisfying* Min-$\lhd^3$, *then L satisfies* Min-si.

Proof: Work as in theorem 1.4, except that now we show $N/N^c \in \mathfrak{F}$ directly: $N^i \text{ ch } N \lhd M \lhd L$ so that $N^i \lhd^2 L$. Thus $N^i/N^{i+1} \in \mathfrak{A} \cap \text{Min-}\lhd \leqq \mathfrak{F}$, and $N/N^c \in \mathfrak{F}$. □

In section 3 we show that the 3 is best possible.

2. The structure of algebras in Min-si

It is well known that if a group G has a unique minimal normal subgroup N of finite index, then N has no proper normal subgroups of finite index. For Lie algebras the corresponding statement is false: Hartley's algebra (Example 6.3.6) has a unique ideal P minimal with respect to $L/P \in \mathfrak{F}$, but P has proper ideals of codimension 1, 2, 3, ..., being abelian.

Nonetheless we do have a partial result:

Lemma 2.1 *Suppose I is a subalgebra of L, having finite codimension (as vector space), such that I has no proper ideals of finite codimension. Then I is an ideal of L, and is the unique minimal ideal of finite codimension.*

Proof: Think of L as an I-module. Then I is a submodule, so that L/I (vector space quotient!) is an I-module of finite dimension. If $I \in \mathfrak{F}$ then $I = 0$ and the lemma is true. Otherwise $C_I(L/I)$ is an ideal of I, of finite codimension: for $I/C_I(L/I)$ has L/I as a faithful module. Therefore $C_I(L/I) = I$, so that $[L, I] \leqq I$. Therefore I is an ideal of L. The rest is obvious. □

We define a new class $\mathfrak{T}$ of Lie algebras: $L \in \mathfrak{T}$ if every subideal of L is an ideal. Thus $\mathfrak{T}$ is the class of Lie algebras in which the relation '$\lhd$' is transitive. Robinson [327] proves that every group with the minimal condition for subnormal subgroups is a finite extension of a group in which normality is transitive. We have the following analogue:

THEOREM 2.2 *Let L be a Lie algebra over a field of characteristic 0 satisfying* Min-si. *Then L has an ideal of finite codimension belonging to the class $\mathfrak{T}$.*

Proof: Let F be minimal with respect to F si L and F of finite codimension. Clearly F has no proper ideals of finite codimension, so by lemma 2.1 $F \lhd L$. We show $F \in \mathfrak{T}$. Assume the contrary: then there exists K minimal with respect to K si F, $K \ntriangleleft F$. Now if $K = K^2$ then $K \lhd F$ by lemma 1.3.2. But K^2 si F, so by minimality $K^2 \lhd F$. Then K/K^2 si F/K^2 and is abelian, so $K/K^2 \leqq$ $\leqq B/K^2 = \beta(F/K^2)$. Now $B/K^2 \lhd F/K^2$, and by lemma 1.3 $B/K^2 \in \mathfrak{F}$. Since F has no proper ideals of finite codimension F/K^2 has no proper ideals of finite codimension. By corollary 1.4.3

$$C_{F/K^2}(B/K^2) = F/K^2$$

so that $B/K^2 \leqq \zeta_1(F/K^2)$; and since $K \leqq B$ we have $K \lhd F$ which is a contradiction. Therefore F, which has finite codimension, is a $\mathfrak{T}$-algebra. □

We also have a structure theorem of a different kind:

THEOREM 2.3 *Let L be a Lie algebra over a field of characteristic 0, satisfying* Min-si, *and let F be its unique minimal ideal of finite codimension. Then L has an ascending series of ideals whose factors are either simple or finite-dimensional abelian, and F has an ascending series of ideals whose factors are either infinite-dimensional simple or 1-dimensional.*

Proof: Let $K \in$ Min-si over a field of characteristic 0, and let M be a minimal ideal of K (which necessarily exists). We show that M is either simple or finite-dimensional abelian. If M is not simple it has a non-zero proper ideal I. Then $I = I^c$ for some integer c, and $I^c \lhd L$, so $I^c = 0$. Therefore $R = \rho(L) \neq 0$ and $R \cap M \neq 0$, so $M \leqq R$. By lemma 1.3 R is finite-dimensional and so nilpotent; therefore M is nilpotent. So $M^2 < M$ and $M^2 \lhd L$, therefore $M^2 = 0$ and M is abelian. Since $M \leqq R$ it is finite-dimensional.

We can therefore build an ascending tower of ideals M_α of L by taking $M_{\alpha+1}$ to be a minimal ideal of L/M_α, and taking unions at limit ordinals. This gives the required series for L.

The same procedure gives an ascending series for F. Since F has no ideals of finite codimension, every finite-dimensional factor in the series is central, so can be refined to give 1-dimensional factors. The remaining infinite-dimensional factors must be simple. □

In the group-theoretic case, one proves theorem 2.2 first, with the hypothesis that the group has minimal condition for 2-step subnormal subgroups, and then proves theorem 1.4. A key concept in the proof is the 'Wielandt normaliser' of a group G, defined to be the intersection of the normalisers of the subnormal subgroups of G. It is a characteristic subgroup in which normality is transitive: the problem is to show it has finite index. In the Lie algebra case, everything comes the other way around: we chop off a finite piece at the top and prove that what is left is a $\mathfrak{T}$-algebra.

We can define the *Wielandt idealiser* of a Lie algebra L to be

$$w(L) = \bigcap_{H \operatorname{si} L} I_L(H).$$

Obviously $w(L) \in \mathfrak{T}$ if $w(L) \lhd L$. However, it is not clear whether $w(L)$ is an ideal of L. But we can prove:

THEOREM 2.4 *Let L be a Lie algebra over a field of characteristic zero satisfying* Min-si, *and let F be its unique minimal ideal of finite codimension. Then w(L) contains F, so is of finite codimension.*

Proof: Suppose H si L. Then $H^\omega = H^c \lhd L$ for some $c > 0$. Let $\bar{L} = L/H^c$ and denote images under the natural map $L \to L/H^c$ by bars (the so-called 'bar convention'). Then

$$\bar{H} \leqq \beta(\bar{L}).$$

Now $\beta(\bar{L}) \in \mathfrak{F}$ and is an ideal of L, so $\bar{C} = C_{\bar{L}}(\beta(\bar{L}))$ is an ideal of finite codimension in $\bar{L}$, and therefore contains F. Therefore F centralises H modulo H^c, so that

$$[H, F] \leqq H^c \leqq H$$

and F idealises H. Thus

$$F \leqq \bigcap_{H \operatorname{si} L} I_L(H) = w(L)$$

and the theorem is proved. □

Certain classes of $\mathfrak{T}$-algebras are classified in [205]. In particular, every soluble $\mathfrak{T}$-algebra is either abelian, or the split extension of an abelian algebra by the 1-dimensional algebra of all scalar multiplications. The proofs are straightforward.

This shows that soluble $\mathfrak{T}$-algebras are a subclass of the class $\mathfrak{P}$ of parasoluble Lie algebras, studied by Brazier [33]. We say that L is *parasoluble* of *paraheight* at most n if it has a series of ideals

$$0 = L_0 \leqq L_1 \leqq \ldots \leqq L_n$$

such that L acts on each L_{i+1}/L_i by scalar multiplications. We let $\mathfrak{P}_n$ consist of all Lie algebras with such a series and define

$$\mathfrak{P} = \bigcup_{n=1}^{\infty} \mathfrak{P}_n.$$

(Parasolubility was introduced for groups by Wehrfritz.) It is easy to see that $\mathfrak{P}_n \leqq \mathfrak{N}_n\mathfrak{A}$. Using this, Brazier has proved that $\mathfrak{P}_n$ is L-closed (although it is not a variety). He has also considered similar problems for Lie rings and associative rings.

3. The case of prime characteristic

Following [12] we give examples to show that in characteristic $p > 0$ the condition Min-$\lhd^2$ does not imply Min-si. The genesis of these examples is explained in [12].

Let $\mathfrak{k}$ be any field of characteristic $p > 0$, and for the moment let S be any Lie algebra over $\mathfrak{k}$. For $i = 0, 1, \ldots, p-1$ take vector spaces S_i over $\mathfrak{k}$ linearly isomorphic to S, where

$$\psi_i : S \to S_i$$

is such an isomorphism. For any $x \in S$ define

$$x_i = \begin{cases} x\psi_i & \text{if } 0 \leqq i \leqq p-1 \\ 0 & \text{otherwise.} \end{cases}$$

Let

$$L = S_0 \oplus S_1 \oplus \ldots \oplus S_{p-1}$$

and identify each S_i with the corresponding subspace of L. Define a multiplication on L by

$$[x_i, y_j] = (-1)^j \binom{p-1-i}{j} [x, y]_{i+j} \qquad (x, y \in S)$$

and extend by linearity to the whole of L. We claim that this defines a Lie algebra structure on L.

To prove this we require an identity involving binomial coefficients. In the polynomial algebra $\mathfrak{k}[t]$ we have

$$(1-t)^{p-1} = (1-t)^p/(1-t) = (1-t^p)/(1-t)$$
$$= 1+t+\ldots+t^{p-1}$$

so that

$$(-1)^i\binom{p-1}{i} = 1 = (-1)^j\binom{p-1}{j}$$

for $0 \leqq i, j \leqq p-1$. Hence

$$(-1)^i \frac{(p-1)!}{i!(p-1-i)!} = (-1)^j \frac{(p-1)!}{j!(p-1-j)!}$$

and therefore

$$(-1)^i \frac{(p-1-j)!}{i!} = (-1)^j \frac{(p-1-i)!}{j!}. \qquad (*)$$

Now let $x, y, z \in S$, and $0 \leqq i, j, k \leqq p-1$. First we check antisymmetry:

$$[x_i, x_i] = (-1)^i\binom{p-1-i}{i}[x, x]_{2i} = 0,$$

$$[y_j, x_i] = (-1)^i\binom{p-1-j}{i}[y, x]_{i+j}$$
$$= -(-1)^j\binom{p-1-i}{j}[x, y]_{i+j}$$
$$= -[x_i, y_j]$$

where we have used (*) and the fact that $[y, x]_{i+j} = 0$ if $i+j \geqq p$. Now the Jacobi identity:

$$[[x_i, y_j], z_k] = \lambda_{ijk}[[x, y], z]_{i+j+k}$$

where

$$\lambda_{ijk} = (-1)^j\binom{p-1-i}{j}(-1)^k\binom{p-1-i-j}{k}$$

or is 0 if $i+j+k \geqq p$. So

$$[[x_i, y_j], z_k]+[[y_j, z_k], x_i]+[[z_k, x_i], y_j]$$

$$= (\lambda_{ijk}[[x, y], z]+\lambda_{jki}[[y, z], x]+\lambda_{kij}[[z, x], y])_{i+j+k}.$$

If $i+j+k \geqq p$ this is 0. We show that if $0 \leqq i+j+k \leqq p-1$ then $\lambda_{ijk} = \lambda_{jki} = \lambda_{kij}$, from which the required result follows since the Jacobi identity holds in S. From (*) we have

$$(-1)^{i+k} \frac{(p-1-j)!}{i!} = (-1)^{j+k} \frac{(p-1-i)!}{j!}$$

So that

$$(-1)^{j+k} \frac{(p-1-i)!}{j!(p-1-i-j)!} \frac{(p-1-i-j)!}{k!(p-1-i-j-k)!} =$$

$$= (-1)^{k+i} \frac{(p-1-j)!}{k!(p-1-j-k)!} \frac{(p-1-j-k)!}{i!(p-1-i-j-k)!}$$

which proves that $\lambda_{ijk} = \lambda_{jki}$. Hence also $\lambda_{jki} = \lambda_{kij}$. Therefore L is a Lie algebra.

Now let $N = S_1+\ldots+S_{p-1}$. Then $L/N \cong S$. From the definition of multiplication, we have $[S_i, S_j] \leqq S_{i+j}$, where we let $S_k = 0$ if $k \geqq p$.

Next assume that S is simple. Then $[S_i, S_j] = S_{i+j}$ since $S^2 = S$. Hence

$$N^i = S_i+\ldots+S_{p-1}$$

and N is nilpotent of class exactly $p-1$. We claim that the only ideals of L are L, together with the powers $N, N^2, \ldots, N^p = 0$.

Let M be a proper ideal of L. Let c be the least integer such that $N^c \leqq M$, and suppose for a contradiction that $N^c \neq M$. Then M contains an element

$$x_j^j+x_{j+1}^{j+1}+\ldots+x_{p-1}^{p-1}$$

where the $x^i \in S$, $x^j \neq 0$, and $j < c$. For any $y \in S$, the ideal M contains

$$[y_{c-j-1}, x_j^j+x_{j+1}^{j+1}+\ldots+x_{p-1}^{p-1}]$$

which is congruent, modulo N^c, to $\lambda[y, x^j]_{c-1}$ $(0 \neq \lambda \in \mathfrak{k})$. Hence M contains X_{c-1} where X is the S-submodule of S generated by x^j. Since $x^j \neq 0$, this is the whole of S; but then $N^{c-1} \leqq M$ contradicting the choice of c.

Next we define a derivation δ of L by

$$x_i\delta = x_{i-1} \qquad (x \in S)$$

and note that this *is* a derivation, since

$$\begin{aligned}[x_i\delta, y_j]+[x_i, y_j\delta] &= [x_{i-1}, y_j]+[x_i, y_{j-1}] \\ &= \left((-1)^{j-1}\binom{p-i}{j}+(-1)^j\binom{p-i-1}{j-1}\right)[x, y]_{i+j-1} \\ &= (-1)^j\binom{p-1-i}{j}[x, y]_{i+j-1} \\ &= [x_i, y_j]\delta.\end{aligned}$$

(If either i or $j = 0$ a similar computation applies.)

We now let K be the split extension of L by the 1-dimensional algebra $\langle\delta\rangle$.

We claim that the only ideals of K are $0, L, K$. For let $H \lhd K, H \neq 0, K$. If $H \cap L = 0$ then $[H, L] = 0$ and then $N \lhd K$. But $[N, \delta] = N\delta \nleqq N$. Hence $H \cap L \neq 0$. Since $H \cap L \lhd L$ it follows from the above that $N^{p-1} = S_{p-1} \leqq H$. But then, applying δ several times, we find that $S_i \leqq H$ for $i = p-1, p-2, \ldots, 0$. Therefore $L \leqq H$. Since K/L has dimension 1, $L = H$.

Hence the ideals of K are $0, L, K$. The ideals of L are $L, N, N^2, \ldots, N^p = 0$. So the 2-step subideals of K are precisely $K, L, N, N^2, \ldots, N^p = 0$. These are finite in number, so satisfy not only the descending but also the ascending chain condition: that is, $K \in \text{Min-}\lhd^2 \cap \text{Max-}\lhd^2$.

Finally we choose S to be an *infinite*-dimensional simple algebra. Then $K \notin \text{Min-si}$. In fact, N^{p-1} is an infinite-dimensional abelian 2-step subideal of K, and hence contains an infinite properly descending chain of 3-step subideals.

Even when S has finite dimension, it is still true that L is a minimal ideal of K, so is *characteristically simple*, that is, has no characteristic ideals other than itself and 0. It is easy to see that the finite-dimensional characteristically simple Lie algebras over a field of characteristic zero are precisely the simple or abelian algebras. The construction*) shows that in characteristic $p > 0$ other algebras may occur; with any simple algebra as top section, and a nilpotent algebra underneath.

Further, both K and L are *uniserial*, that is, have a unique chief series.

---

*) Indeed, Block (Determination of the differentiably simple rings with a minimal ideal, Ann. of Math. 90 (1969) 433–459) has classified, in terms of simple algebras, all characteristically simple Lie algebras containing a minimal ideal. They are either finite-dimensional abelian, simple, or of the form SG where S is a simple algebra over a field of characteristic p and G is a finite elementary abelian p-group. Our example above corresponds to G being cyclic of order p. Block's main theorem, in fact, shows that similar results hold for any kind of non-associative algebra.

By further specialising the choice of S we get algebras K with extra properties. For example, if S is locally finite (e.g. the algebra of all trace-zero transformations of finite rank of an infinite-dimensional vector space) then it is easy to see that K is locally finite. So in characteristic $p > 0$

$$\mathrm{L}\mathfrak{F} \cap \text{Min-}\lhd^2 \cap \text{Max-}\lhd^2 \nleqq \text{Min-si}.$$

However, theorem 2.2 remains true in characteristic $p > 0$, although the argument needed to prove it is slightly more complicated.

THEOREM 3.1 *Let L be a Lie algebra over a field $\mathfrak{k}$ of characteristic $p > 0$, satisfying* Min-si. *Then L has an ideal of finite codimension belonging to the class $\mathfrak{T}$.*

Proof: Let F be minimal with respect to $F \text{ si } L$, $L/F \in \mathfrak{F}$. By lemma 2.1 $F \lhd L$, and F has no proper ideals of finite codimension. To prove that $F \in \mathfrak{T}$ it is sufficient to show that

$$H \lhd K \lhd F \Rightarrow H \lhd F.$$

Choose E minimal with respect to $E \text{ si } K$, $K/E \in \mathfrak{F}$ and D minimal with respect to $D \text{ si } H$, $H/D \in \mathfrak{F}$. By lemma 2.1 D and E have no proper ideals of finite codimension. Hence $D^2 = D$, $E^2 = E$, and both D and E are ideals of L. Now $E \cap H$ is an ideal of finite codimension in H, so $D \leqq E \cap H$. Further, $K \lhd F$. Since K/E is a finite-dimensional F-module it is centralised by F. But $E \leqq E+H \leqq K$, and therefore $E+H \lhd F$. Also $(E+H)/E$ is abelian, so that $H/(E \cap H)$ is abelian. Let $X = (E+H)/(E \cap H)$. This splits as the direct sum of $E/(E \cap H)$ and $H/(E \cap H)$, so that

$$E/(E \cap H) \leqq C_X(H/(E \cap H)).$$

Further $H/(E \cap H)$ is abelian, so

$$H/(E \cap H) \leqq C_X(H/(E \cap H)).$$

Therefore $[H, E+H] \leqq E \cap H$. Now H/D is a finite-dimensional K-module, so is centralised by E. Therefore $[H, E] \leqq D$. It follows that

$$[H, E+H, E+H] \leqq D+H^3 \leqq D+[E \cap H, H] \leqq D$$

and so

$$H/D \leqq Z/D = \zeta_2((E+H)/D).$$

Now

$$Z/D \text{ ch } (E+H)/D \lhd F/D$$

so that $Z \lhd F$. But then Z/D is nilpotent and satisfies Min-si, so is finite-dimensional; hence it is centralised by F. Since $D \leqq H \leqq Z$ we have $[H, F] \leqq D \leqq H$, so that $H \lhd F$. Therefore $H \lhd^2 F$ implies $H \lhd F$, so that $F \in \mathfrak{T}$. □

4. Examples of algebras with Min-si

Most results in this section will be quoted without proof, since the constructions involved require a lot of tedious computation.

Trivial instances of Lie algebras satisfying Min-si are $\mathfrak{F}$-algebras, simple algebras, or more generally algebras having a finite series with factors simple or finite-dimensional. The first non-trivial examples may be found in [206], where it is shown that there exist algebras in Min-si having an ascending series with simple factors whose type is an arbitrary ordinal. In this example all the simple factors are isomorphic; and the construction is similar to one of Robinson [327].

A more natural class of algebras with Min-si may be found in [212]. Let $\mathfrak{k}$ be any field; let c be any infinite cardinal with successor c^+. Select a vector space V of dimension c over $\mathfrak{k}$. For any infinite cardinal $d \leqq c^+$ define $E(c, d)$ to be the set of linear maps $\alpha : V \to V$ such that the image of α has dimension $< d$. This is an associative $\mathfrak{k}$-algebra, so we may form the Lie algebra $L(c, d) = [E(c, d)]$.

We let $L = L(c, c^+)$, and denote by S the set of scalar multiplications $v \mapsto \lambda v$ ($\lambda \in \mathfrak{k}$, $v \in V$). We put $F = L(c, \aleph_0)$, which is the set of linear transformations of finite rank. If $\alpha \in F$ we can define the trace of α (see e.g. [206] p. 306), and we set $T = \{\alpha \in F : \text{trace } \alpha = 0\}$. We then have:

THEOREM 4.1 *Let $L = L(c, c^+)$. Then the ideals of L are precisely the following:*

(a) $L(c, d)$ *for* $\aleph_0 \leqq d \leqq c^+$,
(b) $L(c, d) + S$ *for* $\aleph_0 \leqq d \leqq c$,
(c) S,
(d) $\{0\}$,
(e) *Any subspace X of L such that* $T \leqq X \leqq F + S$.

The lattice of ideals of L is as shown in figure 4.

Proof: See [212]. □

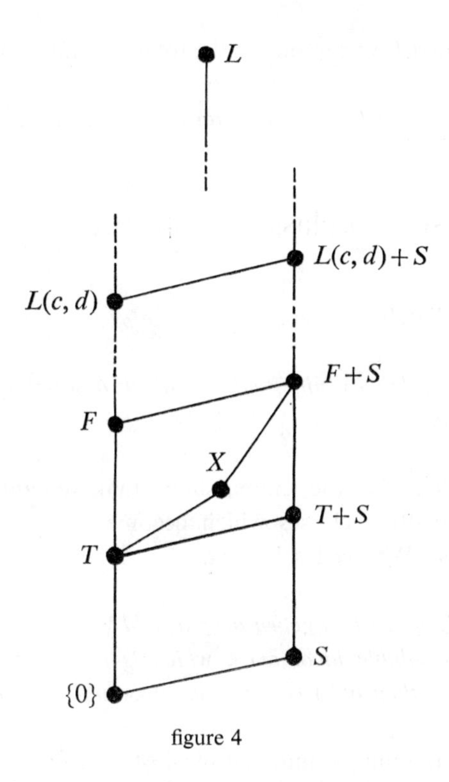

figure 4

The main line of the proof is this: we can easily find the ideals of $E(c, c^+)$, and they are just the $E(c, d)$ for $\aleph_0 \leqq d \leqq c^+$. We use the following theorem of Herstein [82]:

THEOREM 4.2 *If A is a simple associative $\mathfrak{k}$-algebra and $L = [A]$, then*

$$L^2/(L^2 \cap \zeta_1(L))$$

is a simple Lie algebra, unless A has dimension 4 over its centre which is a field of characteristic 2. □

By computing derived algebras this implies that $L(c, d^+)/L(c, d)$ is simple for $d < c$. Further computations of centralisers and various *ad hoc* arguments lead to theorem 4.1.

It is clear from theorem 4.1 that L has no ideals of finite codimension. Since cardinals are well-ordered, L satisfies Min-si. By taking c sufficiently large we get an arbitrarily long ascending series of ideals of L. The presence of 1-dimensional factors, such as S or F/T, shows that these cannot be omitted from the conclusion of theorem 2.3.

Using theorem 4.1 we can obtain the following interesting result (cf. [212]):

PROPOSITION 4.3 *Every Lie algebra can be embedded in a simple Lie algebra.* □

5. Min-si in special classes of algebras

Lemma 1.3 immediately implies:

PROPOSITION 5.1 *A locally nilpotent Lie algebra satisfying* Min-$\lhd^2$ *is finite-dimensional nilpotent.* □

The example quoted after the lemma shows that we cannot relax Min-$\lhd^2$ to Min-$\lhd$. There are other classes which become finite-dimensional under the influence of Min-si. We need a lemma.

LEMMA 5.2 *Let L be a Lie algebra over a field $\mathfrak{k}$ satisfying* Min-si. *Then L has a unique maximal soluble ideal $\sigma(L)$, which is finite-dimensional. If further $\mathfrak{k}$ has characteristic 0 then $\sigma(L)$ contains every soluble subideal of L.*

Proof: Let F be the unique minimal ideal of L of finite codimension, and let $f = \dim(L/F)$. Suppose S is a soluble ideal of L. Then $S \in \mathrm{E}\mathfrak{A} \cap \text{Min-si} \leqq \mathfrak{F}$. The usual centraliser argument shows that $F \cap S \leqq \zeta_1(F)$, which is finite-dimensional by Min-si. Let $z = \dim(\zeta_1(F))$. Then

$$\dim(S) = \dim(F \cap S) + \dim(S+F/F) \leqq z + f.$$

Therefore the sum of all soluble ideals of L is in fact the sum of finitely many soluble ideals, so is soluble by lemma 1.2.6(c). This defines $\sigma(L)$.

Now suppose char$(\mathfrak{k}) = 0$. Let S be a soluble subideal. Define $B_1 = \beta(L)$, $B_{i+1}/B_i = \beta(L/B_i)$. The B_i are soluble ideals of L by lemma 1.2, so each $B_i \leqq \sigma(L)$. But an easy induction on derived length shows that $S \leqq B_i$ for some i, so $S \leqq \sigma(L)$. □

Suppose now that $\mathfrak{W}$ is the class of Lie algebras such that every non-trivial homomorphic image has a non-trivial abelian subideal, and $\mathfrak{W}^*$ is the class of Lie algebras such that every non-trivial homomorphic image has a non-trivial abelian ideal. Then we have a result of [209]:

THEOREM 5.3 *For fields of characteristic 0,*

$$\mathfrak{W} \cap \text{Min-si} = \mathrm{E}\mathfrak{A} \cap \mathfrak{F}.$$

For arbitrary fields,

$$\mathfrak{W}^* \cap \text{Min-si} = \text{E}\mathfrak{A} \cap \mathfrak{F}.$$

Proof: If L belongs to $\mathfrak{W} \cap$ Min-si (over characteristic 0 fields) or $\mathfrak{W}^* \cap$ Min-si, we have $L = \sigma(L)$. Therefore L is finite-dimensional soluble. The converse inclusions are obvious. □

It is not hard to find alternative characterisations of the classes $\mathfrak{W}$, $\mathfrak{W}^*$. Obviously $\mathfrak{W}^*$ is the class of Lie algebras possessing an ascending abelian series of ideals, and is analogous to the SI^*-groups of Kuroš [303] p. 183. Algebras in $\mathfrak{W}$ are analogues of the subsoluble groups of Baer [260], shown by Phillips and Combrink [321] to be the same as SJ^*-groups: the Lie analogue of an SJ^*-group is an algebra with an ascending abelian series of subideals, and an argument similar to that in [321] shows that these are precisely the algebras in $\mathfrak{W}$.

A useful corollary of theorem 5.3 follows from:

Lemma 5.4 *A minimal ideal of a locally soluble Lie algebra is abelian.*

Proof: Let N be a minimal ideal of $L \in \text{LE}\mathfrak{A}$ and suppose N is not abelian. Then there exist $a, b \in N$ such that $[a, b] = c \neq 0$. By minimality $N = \langle c^L \rangle$ so there exist $x_1, \dots, x_n \in L$ such that $a, b \in \langle c, x_1, \dots, x_n \rangle = H$, say. Clearly H is soluble. Now $C = \langle c^H \rangle \lhd H$, and $a, b \in C$, so $c = [a, b] \in C^2 \text{ ch } C \lhd H$, so $c \in C^2 \lhd H$, and $C = C^2$. But $C \leqq H$ so C is soluble, which is a contradiction. Therefore N is abelian as claimed. □

Corollary 5.5 *Locally soluble (or even poly-locally-soluble) Lie algebras satisfying* Min-si *are finite-dimensional and soluble.*

Proof: Clearly all we need prove is that $\text{LE}\mathfrak{A} \cap \text{Min-si} \leqq \mathfrak{F}$. But by lemma 5.4, inductively, $\text{LE}\mathfrak{A} \cap \text{Min-si} \leqq \mathfrak{W}^*$. Theorem 5.3 finishes the proof. □

6. Max-si in special classes of algebras

Now we turn our attention to Max-si. It seems to be a very hard problem whether anything like theorem 1.4 holds; and the analogous problem for groups is still open. Minimal conditions tend to be stronger than maximal conditions, so probably no such theorem is true. But by restricting our attention to less general results, we can prove some interesting theorems. Most of what follows is taken from [209].

LEMMA 6.1 *A soluble Lie algebra satisfying* Max-$\lhd$ *is finitely generated.*

Proof: Let $L \in \mathfrak{A}^d \cap \text{Max-}\lhd$. We show by induction on d that $L \in \mathfrak{G}$. This is clear if $d = 1$. Let $A = L^{(d-1)}$. Then L/A is finitely generated by induction, and A is abelian. There exists a finitely generated subgroup H of L such that $L = A+H$ (let H be generated by an A-transversal). By Max-$\lhd$ there exist $a_1, \ldots, a_n \in A$ such that

$$A = \langle a_1^L \rangle + \ldots + \langle a_n^L \rangle.$$

But if $a \in A$, $h \in H$, then $[a_i, a+h] = [a_i, h]$, whence

$$H+A = H+\langle a_1^H \rangle + \ldots + \langle a_n^H \rangle$$

$$= \langle H, a_1, \ldots, a_n \rangle$$

which is finitely generated. □

Various partial converses to this result will be proved in chapter 11.

Soluble Lie algebras with Max-$\lhd$ need not be finite-dimensional, as is shown by the Hartley algebra (constructed in example 6.3.6). Since the ideal P is an irreducible L-module, and $L/P \in \mathfrak{F}$, it follows that $L \in \text{Max-}\lhd \cap \text{Min-}\lhd$. Now L is soluble, but has infinite dimension, which contrasts with the well-known fact that a soluble group with maximal and minimal condition for normal subgroups is finite. In chapter 11 we will see that everything is better behaved over fields of characteristic $p > 0$.

Now we turn to locally soluble classes. We know very little about $\text{L}\acute{\text{E}}\mathfrak{A} \cap \text{Max-}\lhd$, but we can prove a theorem analogous to one of McLain [313] which shows that $\text{L}\mathfrak{N}^k \cap \text{Max-}\lhd \leqq \mathfrak{G}$ for any $k \in \mathbb{N}$, so that in particular $\text{L}\mathfrak{N} \cap \text{Max-}\lhd = \mathfrak{N} \cap \mathfrak{F}$. First we need several lemmas.

LEMMA 6.2 *Let* $H \lhd L \in \text{L}\acute{\text{E}}\mathfrak{A} \cap \text{Max-}\lhd$. *Then* $H = 0$ *or* $H^2 < H$.

Proof: If $H \neq 0$ we can find K maximal with respect to $K \lhd L$, $K < H$. Then H/K is a minimal ideal of L/K, so by lemma 4.4 H/K is abelian. Therefore $H^2 < H$. □

Let $\mathfrak{Y}$ be any class of Lie algebras. Define the $\mathfrak{Y}$-*residual* of a Lie algebra L to be

$$\lambda_{\mathfrak{Y}}(L) = \bigcap \{N : N \lhd L \text{ and } L/N \in \mathfrak{Y}\}.$$

LEMMA 6.3 *If* $L \in \mathrm{LE}\mathfrak{A} \cap \text{Max-}\lhd$ *and* $L_k = \lambda_{\mathfrak{N}^k}(L)$, *then* L/L_k *is soluble.*

Proof: We use induction on k. If $k = 0$ the result is trivial. We assume $L/L_k \in \mathrm{E}\mathfrak{A}$ and prove the same for L/L_{k+1}. Now $L/L_k^2 \in \mathrm{E}\mathfrak{A} \cap \text{Max-}\lhd \leqq \mathfrak{G}$ by lemma 6.1, so by taking a transversal we can find a finitely generated subalgebra H such that $L = H + L_k^2$. Now $H \in \mathfrak{A}^d$ for some d. Suppose $Q \lhd L$ is such that $L/Q \in \mathfrak{N}^{k+1}$. Then there exists $P \lhd L$ such that $Q \leqq P$, $L/P \in \mathfrak{N}^k$, $P/Q \in \mathfrak{N}$. By definition $P \geqq L_k$, so $P^2 \geqq L_k^2$ and $L = H + P^2$. By lemma 2.1.9 we have $L = H + P^n$ for all n. Since $P/Q \in \mathfrak{N}$ it follows that $L = H + Q$. Therefore $L/Q = H + Q/Q \cong H/(H \cap Q) \in \mathfrak{A}^d$. Now L_{k+1} is the intersection of all such Q, so that $L/L_{k+1} \in \mathrm{R}\mathfrak{A}^d$. But $\mathfrak{A}^d$ is a variety, so $L/L_{k+1} \in \mathfrak{A}^d$. □

COROLLARY 6.4 *If* $L \in \mathrm{L}\mathfrak{N}^k \cap \text{Max-}\lhd$ *then* $L/L_k \in \mathfrak{N}^k$ *and so is the unique minimal ideal with quotient in* $\mathfrak{N}^k$.

Proof: $L/L_k \in \mathrm{E}\mathfrak{A} \cap \text{Max-}\lhd$ by lemma 6.3, so is finitely generated by lemma 5.1. Therefore $L/L_k \in \mathfrak{G} \cap \mathrm{L}\mathfrak{N}^k \leqq \mathfrak{N}^k$. □

THEOREM 6.5 $\mathrm{L}\mathfrak{N}^k \cap \text{Max-}\lhd \leqq \mathfrak{G} \cap \mathfrak{N}^k$.

Proof: Let $L \in \mathrm{L}\mathfrak{N}^k \in \text{Max-}\lhd$. It is sufficient to show $L \in \mathfrak{G}$. Let $L_k = \lambda_{\mathfrak{N}^k}(L)$ as before, and suppose if possible that $L_k \neq 0$. Then by lemma 6.2 $L_k^2 < L_k$. So L/L_k^2 is soluble, so by Max-$\lhd$ is finitely generated, so by the fact that $L \in \mathrm{L}\mathfrak{N}^k$, $L/L_k^2 \in \mathfrak{N}^k$. This contradicts the definition of L_k. Therefore $L_k = 0$, so L is soluble, so finitely generated. □

In [209] this theorem and the Mal'cev correspondence are used to obtain results on chain conditions in Mal'cev groups, some of which were found previously by Gluškov [277].

7. Examples of algebras satisfying Max-si

A large class of non-trivial Lie algebras satisfying Max-si can be constructed as tensor products $A \otimes_{\mathfrak{k}} B$, where A is a suitable commutative associative $\mathfrak{k}$-algebra and B a suitable Lie algebra. This type of construction is discussed in [217].

Another type of example is available for fields of characteristic 0. Let $\mathfrak{k}$ be such a field. Let W be a Lie algebras with basis $\{w_1, w_2, \ldots\}$ and multiplication

$$[w_i, w_j] = (i-j)w_{i+j}.$$

It is easy to check that the Jacobi identity holds. Let I_n be the subspace of W spanned by all w_i with $i \geqq n$, for $n = 1, 2, \ldots$. Clearly I_n is an ideal of L, of finite codimension.

THEOREM 7.1 *The algebra W constructed above is infinite-dimensional, residually nilpotent, and satisfies* Max-si. *Further, every subideal of W contains some I_n and is of finite codimension.*

Proof: Obviously dim W is infinite. It is not hard to see that $W^n = I_n$, so $W^\omega = 0$ and W is residually nilpotent.

Let $S \neq 0$ be a subspace of W idealised by some I_n. We claim that S contains some I_m. For let

$$0 \neq x = \sum x_i w_i \in S.$$

Then S contains the element $[x, w_n] = z$ of I_n. Either $x = \lambda w_n$ $(\lambda \in \mathfrak{k})$ or $z \neq 0$. In the latter case

$$z = \sum_{j \in J} z_j w_j \tag{*}$$

where J is a subset of $\{j \in \mathbb{N} : j \geqq n\}$ chosen to make all $z_j \neq 0$ for $j \in J$. Choose an element of S having the form (*) with $|J|$ as small as possible. Rename this element z. Now if we pick $k \in J$, we have

$$S \ni [z, w_k] = \sum_{j \in J} z_j(j-k)w_{j+k}$$

and since the term involving $z_k(k-k)$ is zero, the sum on the right hand side has fewer terms. By minimality of $|J|$ it must be 0. But this means that $z = z_k w_k$ since the other terms do not cancel.

Thus we have shown that S contains some w_r. But then S contains all $[w_r, w_i]$ and therefore contains I_{2r+1}, as claimed.

Now suppose $0 \neq T \lhd^n W$. We prove by induction on n that T contains some I_m. The above argument deals with the case $n = 1$. Otherwise $T \lhd U \lhd^{n-1} W$. Therefore U contains some I_r; this I_r idealises T; so T contains some I_s.

It follows that T is of finite codimension. Since every non-zero subideal of W is of finite codimension, we have $W \in$ Max-si. $\square$

This algebra W has another interesting property: the join of any two subideals of W is a subideal. For if H, K si W we can find r such that H and K contain I_r. Modulo I_r we have W nilpotent, so that $\langle H, K \rangle$ si W.

Chapter 9

Chain conditions on ascendant abelian subalgebras

In chapter 8 we considered chain conditions on the subideals of a Lie algebra and then applied the results to various classes of Lie algebras. Here we look at the effect of imposing maximal and minimal conditions on the ascendant abelian subalgebras. This is the Lie-theoretic analogue of the work of Robinson [328] for groups and our account is based on chapter 10 of Amayo [2]. We state our results in a very general fashion: inevitably they will overlap some of the results obtained in chapter 8.

Let Δ be any of the relations $\leqq$, $\lhd$, $\lhd^{\sigma}$ (σ is an ordinal), si, asc. Let $\mathfrak{X}$ be a class of Lie algebras.

Define the class

$\acute{\mathrm{E}}(\Delta)\mathfrak{X}$

to consist of all Lie algebras L having an ascending $\mathfrak{X}$-series $(L_\alpha)_{\alpha \leqq \sigma}$ such that $L_\alpha \,\Delta\, L$ for all $\alpha \leqq \sigma$. (Clearly $\acute{\mathrm{E}}(\leqq)\mathfrak{X} = \acute{\mathrm{E}}(\mathrm{asc})\mathfrak{X} = \acute{\mathrm{E}}\mathfrak{X}$. Note that, $\acute{\mathrm{E}}(\lhd^{\sigma})$ and $\acute{\mathrm{E}}(\lhd)$ are not closure operations.)

Let L be a Lie algebra. If $H \in \mathfrak{X}$ and $H \,\Delta\, L$ we say that H is a *$\Delta\mathfrak{X}$-subalgebra* of L.

We write

$L \in \Delta\mathfrak{X}$-Max

(or say that L satisfies $\Delta\mathfrak{X}$-Max) if every $\Delta\mathfrak{X}$-subalgebra H of L satisfies Max (i.e. if $H \in \mathfrak{X}$ and $H \,\Delta\, L$ then $H \in$ Max). We write

$L \in$ Max-$\Delta\mathfrak{X}$

(or say that L satisfies Max-$\Delta\mathfrak{X}$) if L has no infinite strictly ascending chains of $\Delta\mathfrak{X}$-subalgebras (i.e. L satisfies the maximal condition on its $\Delta\mathfrak{X}$-subalgebras).

Similarly we define the classes

$\Delta\mathfrak{X}$-Min and Min-$\Delta\mathfrak{X}$.

If $\mathfrak{X}$ happens to be the class of all Lie algebras we write Max-Δ for Max-$\Delta\mathfrak{X}$ and Δ-Max for $\Delta\mathfrak{X}$-Max and so on. We also abbreviate si and asc by s and a respectively so that we have Max-s$\mathfrak{X}$ for Max-si$\mathfrak{X}$, a$\mathfrak{X}$-Max for asc$\mathfrak{X}$-Max and so on. Finally we write

$$L \in \Delta\mathfrak{X}\text{-Fin}$$

to mean that every $\Delta\mathfrak{X}$-subalgebra of L is finite-dimensional. We have seen that

$$\text{Max} \cap \text{E}\mathfrak{A} = \text{Min} \cap \text{E}\mathfrak{A} = \mathfrak{F} \cap \text{E}\mathfrak{A}.$$

Thus if $\mathfrak{X}$ is any of the classes E$\mathfrak{A}$, $\mathfrak{N}$, $\mathfrak{A}$, then

$$\Delta\mathfrak{X}\text{-Max} = \Delta\mathfrak{X}\text{-Min} = \Delta\mathfrak{X}\text{-Fin}.$$

We will study the conditions we have defined in the cases where $\mathfrak{X}$ is one of classes,

$$\mathfrak{A},\ \mathfrak{N},\ \text{E}\mathfrak{A}.$$

For a Lie algebra L we have defined the Fitting radical $\nu(L)$ as the join of its nilpotent ideals and $\sigma(L)$ as the join of its soluble ideals. For fields of characteristic zero, both these radicals are characteristic ideals of L (see theorem 6.3.1 and corollary 6.3.2).

1. Maximal conditions

LEMMA 1.1 *Let R be the join of a collection of $\mathfrak{F} \cap \text{E}\mathfrak{A}$-ideals of a Lie algebra L and suppose that R is infinite-dimensional. Then R contains an infinite-dimensional $\mathfrak{Z}$-ideal H of L with $H = \zeta_\omega(H)$.*

Proof: Evidently we can find a strictly ascending chain

$$0 = R_0 < R_1 < R_2 < \dots$$

of $\mathfrak{F} \cap \text{E}\mathfrak{A}$-ideals of L such that $R_i \leqq R$ for all i. Clearly we can refine this chain to one of the form

$$0 = N_0 < N_1 < N_2 < \dots$$

where each N_j is an $\mathfrak{F} \cap \text{E}\mathfrak{A}$-ideal of L and $N_{j+1}/N_j \in \mathfrak{A}$. Let $N = \bigcup_{j=0}^{\infty} N_j$ so that $N \lhd L$ and N is infinite-dimensional. Put $H_1 = N_1 \lhd L$; so $H_1 \in \mathfrak{F} \cap \mathfrak{A}$. Suppose that we have defined H_i such that $H_i \in \mathfrak{F} \cap \mathfrak{N}$, $H_i \lhd L$ and $H_i \leqq N$. Since $H_i \in \mathfrak{F}$ we have $N/C_N(H_i) \in \mathfrak{F}$ and so $C_N(H_i) \notin \mathfrak{F}$, whence

$C_N(H_i) \nleqq H_i$. Thus we can find n_i minimal with respect to

$$C_{n_i} = C_N(H_i) \cap N_{n_i} \nleqq H_i$$

(for $C_N(H_i) = \bigcup N_j \cap C_N(H_i)$ and $H_i \in \mathfrak{F}$). Clearly $n_i \neq 0$ and

$$C_{n_i}^2 \leqq C_N(H_i) \cap N_{n_i - 1} \leqq H_i$$

(for $N_{n_i}^2 \leqq N_{n_i - 1}$ and by the minimality of n_i) and so $C_{n_i}^3 = 0$. Define $H_{i+1} = C_{n_i} + H_i$. Then $H_{i+1} \in \mathfrak{F} \cap \mathfrak{N}$ and $H_{i+1} \leqq N$. Furthermore $C_N(H_i) = C_L(H_i) \cap N \lhd L$ (for $H_i \lhd L$ implies that $C_L(H_i) \lhd L$, and $N \lhd L$) and $N_{n_i} \lhd L$ and so $C_{n_i} = C_N(H_i) \cap N_{n_i} \lhd L$. Therefore $H_{i+1} \lhd L$ and $H_i < H_{i+1}$. Proceeding inductively we define H_i for all i so that $H_i \lhd L$, $H_i \in \mathfrak{F} \cap \mathfrak{N}$ and $H_i < H_{i+1} \leqq N$. Evidently $H_{i+1}^2 \leqq H_i$ and H_{i+1} is centralised by $C_{n_{i+k}}$ for all $k \geqq 1$. Now let

$$H = \bigcup_{i=1}^{\infty} H_i,$$

so that H is infinite-dimensional and $H \lhd L$. Clearly

$$H = \langle H_{i+1}, C_{n_{i+1}}, C_{n_{i+2}}, \ldots \rangle$$

for all $i = 1, 2, \ldots$. Therefore

$$[H, H_{i+1}] \subseteq H_{i+1}^2 \leqq H_i,$$

whence $H_i \leqq \zeta_i(H)$ for all i (for $H_1 \leqq \zeta_1(H)$) and $H = \zeta_\omega(H)$. Finally $H \leqq N \leqq R$ and the lemma is proved. □

Remark 1. Evidently the conclusion of lemma 1.1 holds if we merely require that $R \notin \mathfrak{F} \cap \mathrm{E}\mathfrak{A}$ (for then either $R \notin \mathfrak{F}$ or $R \notin \mathrm{E}\mathfrak{A}$, which also implies that $R \notin \mathfrak{F}$). Further if we insist that every nilpotent ideal of L contained in R be finite-dimensional then we get $\zeta_i(H) < \zeta_{i+1}(H)$ for all i.

We have seen that a minimal ideal of a $\mathrm{L}\mathfrak{N}$-algebra is central (lemma 7.1.6) and that a minimal ideal of a $\mathrm{LE}\mathfrak{A}$-algebra is abelian (lemma 8.4.4). Now we have

LEMMA 1.2 (a) *A maximal abelian ideal of a* $\mathfrak{Z}$*-algebra is self-centralising.*

(b) *If L is any Lie algebra then* $\sigma(L) \in \acute{\mathrm{E}}(\lhd)\mathfrak{A}$.

(c) *Let $H \lhd L$ and let Δ be any of the relations* $\leqq$, $\lhd$, $\lhd^\alpha$, si, asc. *If* $L \in \acute{\mathrm{E}}(\Delta)\mathfrak{A}$ *and if H contains every* $\Delta\mathfrak{N}_2$*-subalgebra of L then* $C_L(H) \leqq H$.

(d) *Let $H \lhd L$ such that H contains every* $\mathfrak{N}_2$*-ideal of L. If* $K = \langle I : I \lhd L \text{ and } I \in \mathrm{E}\mathfrak{A} \cup \mathfrak{Z} \rangle$, *then*

$$K \cap C_L(H) = \zeta_1(H) \leqq \zeta_1(H \cap K).$$

In particular for any Lie algebra L, $C_{\sigma(L)}(\nu(L)) = \zeta_1(\nu(L))$.

(e) *Let* Δ *be one of the relations* si, asc *and let* $K \, \Delta \, L$ *with* $K \in \acute{\mathrm{E}}(\Delta)\mathfrak{A}$. *If* $H \lhd L$ *and* H *contains every* $\Delta\mathfrak{N}_2$*-subalgebra of* L *then*

$$K \cap C_L(H) \leqq C_K(H \cap K) = \zeta_1(H \cap K).$$

Proof: (a) Let $L \in \mathfrak{Z}$ and let $\{Z_\alpha : 0 \leqq \alpha \leqq \tau\}$ be its upper central series. Let A be a maximal abelian ideal of L and let $C = C_L(A)$. Then $C \lhd L$ and $A \leqq C$. If $C \neq A$ then we can find an ordinal α minimal with respect to $C \cap \zeta_\alpha(L) \nleqq A$. Clearly α is not a limit ordinal. Let $x \in C \cap \zeta_\alpha(L)$ with $x \notin A$. If $B = \langle x \rangle + A$ then $B^2 = 0$ and

$$[B, L] \subseteq [C \cap \zeta_\alpha(L) + A, L] \subseteq C \cap \zeta_{\alpha-1}(L) + A \subseteq A,$$

and so $B \lhd L$, which contradicts the maximality of A. So $C = A$.

(b) Evidently if $M \in \acute{\mathrm{E}}(\lhd)\mathfrak{A}$ and $N \leqq M$ then $N \in \acute{\mathrm{E}}(\lhd)\mathfrak{A}$. Now $\sigma(L) \leqq K = \langle I : I \lhd L \text{ and } I \in \mathfrak{Z} \cup \mathrm{E}\mathfrak{A} \rangle$, so it is enough to show that $K \in \acute{\mathrm{E}}(\lhd)\mathfrak{A}$. Assume $K \neq 0$. Define $K_0 = 0$. If τ is a limit ordinal and K_α has been defined for all ordinals $\alpha < \tau$, put $K_\tau = \bigcup_{\alpha<\tau} K_\alpha$. Suppose that K_α has been defined so that $K_\alpha \lhd L$ and $K_\alpha < K$. Then there is $I \lhd L$, such that $I \in \mathfrak{Z} \cup \mathrm{E}\mathfrak{A}$ and $I \nleqq K$. If $I \in \mathfrak{Z}$, let β be minimal with respect to $\zeta_\beta(I) = I_\beta \nleqq K_\alpha$. Then β is not a limit ordinal so $\beta - 1$ exists and $\zeta_{\beta-1}(I) \leqq K_\alpha$. Put $K_{\alpha+1} = K_\alpha + I_\beta$, so that $K_{\alpha+1}^2 \leqq K_\alpha + I_\beta^2 \leqq K_\alpha + \zeta_{\beta-1}(I) \leqq K_\alpha$, since $K_\alpha \lhd L$. Furthermore $\zeta_\beta(I)$ ch $I \lhd L$, so $\zeta_\beta(I) \lhd L$ and $K_{\alpha+1} \lhd L$. If on the other hand $I \in \mathrm{E}\mathfrak{A}$ then $I^{(d)} = 0$ for some d. Let n be maximal with respect to $I^{(n)} \nleqq K_\alpha$. Then $I^{(n+1)} \leqq K_\alpha$. In this case put $K_{\alpha+1} = K_\alpha + I^{(n)}$, so that $K_{\alpha+1}^2 \leqq K_\alpha + I^{(n+1)} \leqq K_\alpha$. Further $I^{(n)}$ ch $I \lhd L$ so that $I^{(n)} \lhd L$ and $K_{\alpha+1} \lhd L$. So in any case we see that if $K_\alpha \leqq K$, $K_\alpha \lhd L$ and $K_\alpha \neq K$, then we can define $K_{\alpha+1}$ with $K_\alpha < K_{\alpha+1} \leqq K_\alpha$, $K_{\alpha+1} \lhd L$ and $K_{\alpha+1}^2 \leqq K_\alpha$. Evidently at limit ordinals τ, if $K_\alpha \lhd L$ for all $\alpha < \tau$, then $K_\tau = \bigcup_{\alpha<\tau} K_\alpha \lhd L$. Now by set-theoretic considerations we can find an ordinal μ so that $K = K_\mu$. Hence K has an ascending series of ideals of L, $\{K_\alpha : 0 \leqq \alpha \leqq \mu\}$, with abelian factors. In particular $K \in \acute{\mathrm{E}}(\lhd)\mathfrak{A}$ and so $\sigma(L) \in \acute{\mathrm{E}}(\lhd)\mathfrak{A}$.

(c) The result is obvious if Δ is $\leqq$, for then $H = L$. Let Δ be one of $\lhd^\lambda$ (λ an ordinal > 0), si, asc, and let $\{L_\alpha : 0 \leqq \alpha \leqq \tau\}$ be an ascending abelian series of L with $L_\alpha \, \Delta \, L$ for all $\alpha \leqq \tau$. Let $C = C_L(H)$. Then $C \lhd L$ and so $C \cap L_\alpha \, \Delta \, L$ for all $\alpha \leqq \tau$. Suppose that $C \nleqq H$. Then we can find α minimal with respect to $C \cap L_\alpha \nleqq H$. Evidently $\alpha > 0$ and α is not a limit ordinal so that $\alpha - 1$ exists. Now $(C \cap L_\alpha)^2 \leqq C \cap (L_\alpha^2) \leqq C \cap L_{\alpha-1} \leqq H$, by the minimality of α, and so $(C \cap L_\alpha)^3 \leqq [H, C] = 0$, whence $C \cap L_\alpha \in \mathfrak{N}_2$ and $C \cap L_\alpha \leqq H$, a contradiction (for $C \cap L_\alpha \, \Delta \, L$). Thus $C \leqq H$.

(d) Clearly $C = K \cap C_L(H) \lhd L$. First we show that $C \leqq H$ (and hence $C \leqq \zeta_1(H)$). By the proof of (b) above K has an ascending abelian series $\{K_\alpha : 0 \leqq \alpha \leqq \mu\}$ such that $K_\alpha \lhd L$ for all $\alpha \leqq \mu$. Suppose that $C \nleqq H$. Then we can find an ordinal α minimal with respect to $C \cap K_\alpha \nleqq H$. Then α is not zero or a limit ordinal and so $\alpha - 1$ exists. Since $K_\alpha^2 \leqq K_{\alpha-1}$ then $(C \cap K_\alpha)^2 \leqq$ $\leqq C \cap K_{\alpha-1} \leqq H$, whence $C \cap K_\alpha \in \mathfrak{N}_2$ and $C \cap K_\alpha \lhd L$ (for $C \lhd L$ and $K_\alpha \lhd L$) and so $C \cap K_\alpha \leqq H$, a contradiction. Therefore $C \leqq H$ and so $C \leqq \zeta_1(H)$. But $\zeta_1(H)$ is an abelian ideal of L ($\zeta_1(H) \operatorname{ch} H \lhd L$) so that $\zeta_1(H) \leqq K \cap C_L(H) = C$. Thus $C = \zeta_1(H)$. Evidently $C \leqq C_K(H \cap K) \cap H =$ $= C_{H \cap K}(H \cap K) = \zeta_1(H \cap K)$. In particular $R = \sigma(L) \leqq K$ and so if we put $H = \nu(L)$, then $C_R(H) = R \cap C_L(H) \leqq \zeta_1(H \cap K) = \zeta_1(H)$ since $H \leqq R \leqq K$. Conversely $\zeta_1(H) \leqq R \cap C_L(H) = C_R(H)$.

(e) Trivially $C_L(H) \leqq C_L(H \cap K)$ and so $K \cap C_L(H) \leqq C_K(H \cap K)$. Let $N \in \mathfrak{N}_2$ and $N \,\Delta\, K$. Since $K \,\Delta\, L$, then $N \,\Delta\, L$ (for 'si' and 'asc' are transitive) and so $N \leqq H$, whence $N \leqq H \cap K$ and $H \cap K$ contains every $\Delta\mathfrak{N}_2$-subalgebra of K. Further $H \cap K \lhd K$ and so by (c) we have $C_K(H \cap K) \leqq H \cap K$. □

There are several results that we can deduce from lemmas 1.1 and 1.2. Of these the most important for our purposes are:

COROLLARY 1.3 *Let L be a Lie algebra. Then*

(a) *$\nu(L) \in \mathfrak{F} \cap \mathfrak{N}$ if and only if $\sigma(L) \in \mathfrak{F} \cap \mathrm{E}\mathfrak{A}$.*

(b) *If $L \in \text{Max-}\lhd\, \mathfrak{A}$ and A is an abelian ideal of L then to each $\mathfrak{Z}$-ideal K of L there corresponds an integer n such that $A \cap K = A \cap \zeta_n(K)$.*

(c) *If $L \in \mathfrak{Z} \cap \text{Max-}\lhd\, \mathfrak{A}$ then $L \in \mathfrak{F} \cap \mathfrak{N}$.*

(d) *If $L \in \text{Max-}\lhd^2\, \mathfrak{A}$ then $\sigma(L) \in \mathfrak{F} \cap \mathrm{E}\mathfrak{A}$.*

Proof: (a) Let $H = \nu(L) \in \mathfrak{F} \cap \mathfrak{N}$ and let $R = \sigma(L)$. Then $R/C_R(H) \in \mathfrak{F}$ and by lemma 1.2(d) we have $C_R(H) = \zeta_1(H) \in \mathfrak{F}$ and so $R \in \mathfrak{F}$. But R is a sum of soluble ideals, so $R \in \mathfrak{F} \cap \mathrm{E}\mathfrak{A}$. The converse is trivial as $\nu(L) \in \mathrm{L}\mathfrak{N}$, and $\nu(L) \leqq \sigma(L)$.

(b) For each α, $\zeta_\alpha(K) \operatorname{ch} K \lhd L$ and so $\zeta_\alpha(K) \lhd L$, whence $A \cap \zeta_\alpha(K)$ is an abelian ideal of L. Since $L \in \text{Max-}\lhd\, \mathfrak{A}$ then we can find n such that $A \cap \zeta_n(K) = A \cap \zeta_{n+1}(K) = \ldots$, and so $A \cap \zeta_\omega(K) = A \cap \zeta_n(K)$. If $A \cap \zeta_\alpha(K) \leqq$ $\leqq A \cap \zeta_n(K)$, then $[A \cap \zeta_{\alpha+1}(K), K] \leqq A \cap \zeta_\alpha(K) \leqq A \cap \zeta_n(K)$ and so $A \cap \zeta_{\alpha+1}(K) \leqq A \cap \zeta_{n+1}(K) = A \cap \zeta_n(K)$. If λ is a limit ordinal and $A \cap \zeta_\alpha(K) \leqq$ $\leqq A \cap \zeta_n(K)$ for all $\alpha < \lambda$ then $A \cap \zeta_\lambda(K) = \bigcup_{\alpha<\lambda} A \cap \zeta_\alpha(K) \leqq A \cap \zeta_n(K)$. Thus $A \cap \zeta_\alpha(K) \leqq A \cap \zeta_n(K)$ for all α, whence $A \cap K = A \cap \zeta_n(K)$.

(c) Let A be a maximal abelian ideal of L. By (b) we have $A = A \cap L =$ $= A \cap \zeta_n(L)$ for some n. Let $X/A \cap \zeta_i(L) \leqq A \cap \zeta_{i+1}(L)/A \cap \zeta_i(L)$ so that X is an abelian ideal of L (for $[X, L] \subseteq A \cap \zeta_i(L) \leqq X$). Thus $A \cap \zeta_{i+1}(L)/A \cap \zeta_i(L) \in$

$\in \text{Max} \cap \mathfrak{A} = \mathfrak{F} \cap \mathfrak{A}$, whence $A = A \cap \zeta_n(L) \in \mathfrak{F}$. Therefore $L/C_L(A) \in \mathfrak{F}$ and since $C_L(A) = A$ (lemma 1.2(a)) then $L \in \mathfrak{F} \cap \mathfrak{Z} = \mathfrak{F} \cap \mathfrak{N}$.

(d) If $L \in \text{Max-}\lhd^2 \mathfrak{A}$ then every ideal of L satisfies Max-$\lhd$ $\mathfrak{A}$. In particular if N is a nilpotent ideal of L then $N \in \mathfrak{N} \cap \text{Max-}\lhd \mathfrak{A} \leqq \mathfrak{F} \cap \mathfrak{N}$, by (c). Thus $H = \nu(L)$ is a sum of $\mathfrak{F} \cap \mathfrak{N}$-ideals of L. If $H \notin \mathfrak{F}$, then by lemma 1.1 H contains an infinite-dimensional $\mathfrak{Z}$-ideal K of L. But $K \in \mathfrak{Z} \cap \text{Max-}\lhd \mathfrak{A}$ and so $K \in \mathfrak{F} \cap \mathfrak{N}$, a contradiction. Thus $H \in \mathfrak{F}$ and so $H \in \mathfrak{F} \cap \mathfrak{N}$ (H is a sum of nilpotent ideals of L), whence by (a) we have $\sigma(L) \in \mathfrak{F} \cap \text{E}\mathfrak{A}$. □

Let Δ be one of the relations $\lhd^\alpha$ (α an ordinal > 0), si, asc and let $\mathfrak{X}$ be a class of Lie algebras. For a Lie algebra L we define

$$J_{\Delta\mathfrak{X}}(L) = \langle H : H \,\Delta\, L,\ H \in \mathfrak{X} \rangle.$$

Thus $\nu(L) = J_{\lhd\mathfrak{N}}(L)$ and $\sigma(L) = J_{\lhd\text{E}\mathfrak{A}}(L)$.

We note that if $H \text{ ch } K \,\Delta\, L$ then $H \,\Delta\, L$. In particular if $K \,\Delta\, L$ then $\zeta_\beta(K) \,\Delta\, L$ for all ordinals β. Hence if $L \in \text{Max-}\Delta\mathfrak{N}$ then every $\Delta\mathfrak{Z}$-subalgebra is a $\Delta\mathfrak{N}$-subalgebra. Evidently

$$\text{Max-}\Delta\, \text{E}\mathfrak{A} \leqq \text{Max-}\Delta\, \mathfrak{N} = \text{Max-}\Delta\, \mathfrak{Z} \leqq \text{Max-}\Delta\, \mathfrak{A}.$$

Let α, β, λ be ordinals such that $\beta + \lambda \leqq \alpha$. Clearly

$$K \lhd^\lambda L \in \text{Max-}\lhd^\alpha \mathfrak{X} \text{ implies that } K \in \text{Max-}\lhd^\beta \mathfrak{X}.$$

THEOREM 1.4 *Let $1 < \alpha < \omega$ and let $L \in \text{Max-}\lhd^\alpha \mathfrak{A}$ over a field of characteristic zero. Then*

(a) $J_{\lhd^{\alpha-1}\text{E}\mathfrak{A}}(L) = \sigma(L) \in \mathfrak{F} \cap \text{E}\mathfrak{A}$ *and*
(b) $J_{\lhd^{\alpha-1}\mathfrak{Z}}(L) = \nu(L) \in \mathfrak{F} \cap \mathfrak{N}$.

Proof: By induction on α. The case $\alpha = 2$ follows from corollary 1.3(c, d). Suppose that $\alpha > 2$ and the result is true for $\alpha - 1$. Let $H \in \text{E}\mathfrak{A}$ such that $H \lhd^{\alpha-1} L$ and let $H_1 = H^L$. Then $H \lhd^{\alpha-2} H_1 \in \text{Max-}\lhd^{\alpha-1} \mathfrak{A}$ and so $H \leqq \sigma(H_1) \text{ ch } H_1$ and $\sigma(H_1) \in \mathfrak{F} \cap \text{E}\mathfrak{A}$. Thus $H \leqq \sigma(H_1) \leqq \sigma(L)$ since $\sigma(H_1) \lhd L$. Further $L \in \text{Max-}\lhd^\alpha \mathfrak{A} \leqq \text{Max-}\lhd^2 \mathfrak{A}$ so that $\sigma(L) \in \mathfrak{F} \cap \text{E}\mathfrak{A}$. This proves (a), and (b) follows similarly. □

We remark that if Δ is any one of $\lhd^\alpha$, si, asc then

$$\nu(L) \leqq J_{\Delta\mathfrak{N}}(L) \leqq J_{\Delta\mathfrak{Z}}(L) \text{ and } \sigma(L) \leqq J_{\Delta\text{E}\mathfrak{A}}(L).$$

For most of this chapter we will be concerned with deriving conditions under which

$$J_{\Delta\mathfrak{Z}}(L) = \nu(L) \text{ and } J_{\Delta\text{E}\mathfrak{A}}(L) = \sigma(L).$$

For the first equality our remarks above show that it is enough to prove that $J_{\Delta\mathfrak{N}}(L) \leqq \nu(L)$ in the case where $\nu(L) \in \mathfrak{F}$. Before proceeding we need a result like lemma 1.2(d) which makes things considerably easier.

LEMMA 1.5 *Let Δ be one of the relations $\lhd^{\alpha}$,* si, asc *and let $H \lhd L$ such that H contains every $\Delta\mathfrak{N}_2$-subalgebra of L. Suppose that K is the union of an ascending chain*

$$0 = K_0 \leqq K_1 \leqq \dots$$

of subalgebras of L such that for each non-limit ordinal β, $K_\beta \, \Delta \, L$, and $K_\beta \in \mathrm{E}\mathfrak{A} \cup \mathfrak{Z}$ or $K_{\beta-1} \lhd K_\beta$ and $K_\beta/K_{\beta-1} \in \mathrm{E}\mathfrak{A} \cup \mathfrak{Z}$. Then

$$C_L(H) \cap K \leqq H.$$

In particular if $H \in \mathfrak{F}$ then $K \in \mathfrak{F} \cap \mathrm{E}\mathfrak{A}$.

Proof: Let $C = C_L(H)$ so that $C \lhd L$. Clearly if $I \, \Delta \, L$ then $C \cap I \, \Delta \, L$. Suppose that $C \cap K \nleqq H$. Then there exists an ordinal β minimal with respect to $C \cap K_\beta \nleqq H$. Then β is not a limit ordinal and so $\beta - 1$ exists. We have two cases: either $K_\beta \in \mathrm{E}\mathfrak{A} \cup \mathfrak{Z}$ or $K_{\beta-1} \lhd K_\beta$ and $K_\beta/K_{\beta-1} \in \mathrm{E}\mathfrak{A} \cup \mathfrak{Z}$. We also have $C \cap K_{\beta-1} \leqq H$. Suppose then that $K_{\beta-1} \lhd K_\beta$ and $K_\beta/K_{\beta-1} \in \mathrm{E}\mathfrak{A} \cup \mathfrak{Z}$. Put $U = K_\beta$ and $V = K_{\beta-1}$. If $U/V \in \mathrm{E}\mathfrak{A}$ then $U^{(d)} \leqq V$ for some d and so $C \cap U^{(d)} \leqq H$. Pick i maximal with respect to $C \cap U^{(i)} \nleqq H$. Then $C \cap U^{(i+1)} \leqq H$ and so $C \cap U^{(i)} \in \mathfrak{N}_2$. Clearly $U^{(i)}$ ch $U \, \Delta \, L$ implies $U^{(i)} \, \Delta \, L$ and so $C \cap U^{(i)} \, \Delta \, L$, whence $C \cap U^{(i)} \leqq H$, a contradiction. Now suppose that $U/V \in \mathfrak{Z}$ and let $U_\lambda/V = \zeta_\lambda(U/V)$. Then $U_0 = V$ and $C \cap U_0 \leqq V$. Let μ be minimal with respect to $C \cap U_\mu \nleqq H$. Then μ is not a limit ordinal (for at limit ordinals γ, $U_\gamma = \bigcup_{\lambda<\gamma} U_\lambda$) so $\mu - 1$ exists and we have $C \cap U_{\mu-1} \leqq H$. But $C \lhd L$ and $[U_\mu, U] \leqq U_{\mu-1}$ so $[C \cap U_\mu, U] \leqq C \cap U_{\mu-1} \leqq H$. This implies that if $S/H = \zeta_1((U+H)/H)$ then $C \cap U_\mu \leqq C \cap S$. Now $S^2 \leqq H$ so $C \cap S \in \mathfrak{N}_2$. Furthermore $U \, \Delta \, L$ implies that $U+H/H \, \Delta \, L/H$ and so $S/H \, \Delta \, L/H$ (for S/H ch $(U+H)/H$), whence $S \, \Delta \, L$ and $C \cap S$ is a $\Delta\mathfrak{N}_2$-subalgebra of L. Hence $C \cap U_\mu \leqq C \cap S \leqq H$, a contradiction. If on the other hand we have the case $U = K_\beta \in \mathrm{E}\mathfrak{A} \cup \mathfrak{Z}$, then arguing in the same was as above we arrive at a contradiction. Hence $C \cap K \leqq H$.

Finally if $H \in \mathfrak{F}$ then $L/C_L(H) \in \mathfrak{F}$ so $K/C \cap K \in \mathfrak{F}$ and hence $K \in \mathfrak{F}$. That $K \in \mathrm{E}\mathfrak{A}$ follows from its definition. □

COROLLARY 1.6 *Let Δ be one of the relations $\lhd^{\alpha}$,* si, asc *and let L be a Lie algebra over a field of characteristic zero. Suppose that $H \lhd L$ and H contains* *every $\Delta\mathfrak{N}_2$-subalgebra of L. If $H \in \mathfrak{F}$ then $\sigma(L) \in \mathfrak{F} \cap \mathrm{E}\mathfrak{A}$ and $\sigma(L)$ contains every $\Delta\,\mathrm{E}\mathfrak{A}$-subalgebra of L.*

Proof: We consider the case $\Delta = \lhd^{\alpha}$ (the others are much easier). Let $R \in \mathrm{E}\mathfrak{A}$ with $R \lhd^{\alpha} L$ so that we have a series

$$R = R_0 \lhd R_1 \lhd \dots R_{\alpha} = L.$$

Evidently for each ordinal β we have $S_{\beta} = \sigma(R_{\beta}) \lhd^{\alpha} L$. Suppose that $1 \leqq \beta < \alpha$. (If $\alpha = 1$, we have $R \leqq \sigma(L)$ so we may assume that $\alpha > 1$.) Clearly $J \lhd S_{\beta}$ implies that $J \lhd^{\alpha} L$ (for S_{β} ch $R_{\beta} \lhd R_{\beta+1}$ gives $J \lhd S_{\beta} \lhd$ $\lhd R_{\beta+1} \lhd \dots R_{\alpha} = L$). Furthermore $S_{\beta} = \sigma(R_{\beta}) \in \acute{\mathrm{E}}(\lhd)\mathfrak{A}$ and so by lemma 1.5 (putting $S_{\beta} = K$) we have $S_{\beta} \in \mathfrak{F} \cap \mathrm{E}\mathfrak{A}$ and therefore

$$S_{\beta} \leqq S_{\beta+1}. \tag{*}$$

It is not hard to see that $S_{\alpha} = \sigma(R_{\alpha}) = \sigma(L)$ has an ascending series of ideals of L with abelian factors. Therefore by lemma 1.5, $S_{\alpha} \in \mathfrak{F} \cap \mathrm{E}\mathfrak{A}$ (this also proves the required result for the case $\alpha = 1$). Hence

$$S_{\beta} \in \mathfrak{F} \cap \mathrm{E}\mathfrak{A} \text{ and } S_{\beta} \lhd^{\alpha} L \text{ for all } \beta,\ 1 \leqq \beta \leqq \alpha.$$

We claim that if $1 \leqq \lambda \leqq \alpha$, then for any pair of ordinals β, γ with $1 \leqq \beta \leqq$ $\leqq \gamma \leqq \lambda$, we have

$$S_{\beta} \leqq S_{\gamma}. \tag{**}$$

We use induction on λ. If $\lambda = 1$, the result is trivial. Let $\lambda > 1$ and assume the result is true for all $\mu < \lambda$. If $\lambda - 1$ exists, then by (*) we have $S_{\lambda-1} \leqq S_{\lambda}$, and so (**) holds for λ. If λ is a limit ordinal then $K = \bigcup_{\mu<\lambda} S_{\mu}$ satisfies the hypothesis of lemma 1.5 and so $K \in \mathfrak{F} \cap \mathrm{E}\mathfrak{A}$. Furthermore $K \lhd R_{\lambda} = \bigcup_{\mu<\lambda} R_{\mu}$ (for if $\beta, \mu < \lambda$, then $[S_{\beta}, R_{\mu}] \subseteq [S_{\gamma}, R_{\gamma}] \subseteq S_{\gamma}$, where $\gamma = \max\{\beta, \mu\}$) and so $K \leqq S_{\lambda}$. Hence (**) holds for λ as well. This completes the proof of (**). In particular

$$R \leqq S_1 \leqq S_{\alpha} = \sigma(L) \in \mathfrak{F} \cap \mathrm{E}\mathfrak{A}. \qquad \square$$

Our main use for corollary 1.6 is the following result:

COROLLARY 1.7 *Let L be a Lie algebra over a field of characteristic zero and let Δ be one of the relations $\lhd^{\alpha}$,* si, asc. *If $J_{\Delta\mathfrak{N}}(L) = \nu(L) \in \mathfrak{F} \cap \mathfrak{N}$ then $J_{\Delta\mathrm{E}\mathfrak{A}}(L) = \sigma(L)$ and $\sigma(L) \in \mathfrak{F} \cap \mathrm{E}\mathfrak{A}$.* $\square$

THEOREM 1.8 *Let α be an ordinal > 1 and let $L \in \text{Max-}\lhd^{\alpha}\mathfrak{N}$ over a field of characteristic zero. Then*

(a) $J_{\lhd^{\alpha}\mathfrak{Z}}(L) = \nu(L) \in \mathfrak{F} \cap \mathfrak{N}$ *and*
(b) $J_{\lhd^{\alpha}\mathrm{E}\mathfrak{A}}(L) = \sigma(L) \in \mathfrak{F} \cap \mathrm{E}\mathfrak{A}$.

Proof: By corollary 1.7 it is enough to prove (a) and by our previous remarks it is enough to show that $J_{\lhd^{\alpha}\mathfrak{N}}(L) \leqq \nu(L) \in \mathfrak{F} \cap \mathfrak{N}$. Now $\alpha \geqq 2$ and so by corollary 1.3 we have $H = \nu(L) \in \mathfrak{F} \cap \mathfrak{N}$. Evidently every $\lhd^{\alpha}$ $\mathfrak{N}$-subalgebra of L is contained in a maximal one. So let N be a maximal $\lhd^{\alpha}$ $\mathfrak{N}$-subalgebra of L. Then we have a series

$$N = N_0 \lhd N_1 \lhd \ldots N_{\alpha} = L.$$

We claim that if $1 \leqq \beta \leqq \alpha$ then $N \lhd N_{\beta}$. We use induction on β. For $\beta = 1$ this is trivial. If β is a limit ordinal and $N \lhd N_{\lambda}$ for all $\lambda < \beta$, then $N \lhd N_{\beta}$, since $N_{\beta} = \bigcup_{\lambda<\beta} N_{\lambda}$. Suppose that $\beta - 1$ exists and $N \lhd N_{\beta-1} \lhd N_{\beta}$. Now N is a maximal nilpotent ideal of $N_{\beta-1}$; for $\beta - 1 \geqq 1$ and so if $M \lhd N_{\beta-1}$ then $M \lhd^{\alpha} L$ so that if $N \leqq M \in \mathfrak{N}$ then by the maximality of N we must have $N = M$. Hence $N = \nu(N_{\beta-1})$ ch $N_{\beta-1} \lhd N_{\beta}$ and so $N \lhd N_{\beta}$. This establishes our claim. In particular $N \lhd N_{\alpha} = L$, and so $N \leqq H$ and $N = H$. □

We note that $\lhd^2 \mathfrak{A}\text{-Fin} \leqq \text{Max-}\lhd^2 \mathfrak{A}$ (the reverse inequality is false by section 8.3) since for any positive integers n and c, the union of any ascending chain of $\lhd^n \mathfrak{N}_c$-subalgebras of a Lie algebra is also a $\lhd^n \mathfrak{N}_c$-subalgebra of that algebra. Thus if $L \in \lhd^2 \mathfrak{A}\text{-Fin}$ then $H = \nu(L) \in \mathfrak{F} \cap \mathfrak{N}$, by corollary 1.3. Now let A be an abelian 2-step subideal of L. Then $A \in \mathfrak{F}$. By corollary 4.1.5 if L is defined over a field of characteristic zero then $A^L \in \mathrm{L}\mathfrak{N}$. Hence $A \leqq \zeta_m(A^L)$ for some m (lemma 7.1.6). But $\zeta_m(A^L)$ is a nilpotent ideal of L, so

$$A \leqq \zeta_m(A^L) \leqq H = \nu(L).$$

Let M be a Lie algebra over a field of characteristic zero and let n, c be positive integers. If A is a maximal $\lhd^n \mathfrak{N}_c$-subalgebra of M then $A^{\uparrow}$ is a maximal $\lhd^n \mathfrak{N}_c$-subalgebra of $M^{\uparrow}$. For by lemma 4.1.1, $A^{\uparrow} \in \mathfrak{N}_c$ and $A^{\uparrow} \lhd^n M^{\uparrow}$. If $A^{\uparrow} \leqq K \leqq M^{\uparrow}$ and $K \in \mathfrak{N}_c$, $K \lhd^n M^{\uparrow}$ then $A \leqq K^{\downarrow} \lhd^n M$ and $K^{\downarrow} \in \mathfrak{N}_c$. Hence $A = K^{\downarrow}$ and so by lemma 4.1.2(f), $A^{\uparrow} = K$.

THEOREM 1.9 *Let α be an ordinal > 1 and let $L \in \lhd^{\alpha} \mathfrak{A}\text{-Fin}$ over a field of characteristic zero. Then*

(a) $J_{\lhd^{\alpha}\mathfrak{Z}}(L) = \nu(L) \in \mathfrak{F} \cap \mathfrak{N}$ *and*
(b) $J_{\lhd^{\alpha}\mathrm{E}\mathfrak{A}}(L) = \sigma(L) \in \mathfrak{F} \cap \mathrm{E}\mathfrak{A}$.

Proof: We will prove by induction on α that $J_{\lhd^\alpha \mathfrak{N}}(L) \leqq \nu(L)$ and $\nu(L) \in \mathfrak{F} \cap \mathfrak{N}$. This will give $L \in \text{Max-}\lhd^\alpha \mathfrak{N}$ and the required result will follow by theorem 1.8.

The inductive step from α to $\alpha+1$ is trivial (see proof of theorem 1.4). So we need consider only two cases.

Case 1. $\alpha = 2$. Let $H = \nu(L)$. Then $H \in \mathfrak{F} \cap \mathfrak{N}$ and H contains every $\lhd^2 \mathfrak{A}$-subalgebra of L, by our remarks above. Let $c > 1$ and assume inductively that H contains every $\lhd^2 \mathfrak{N}_{c-1}$-subalgebra of L. Now let $N \in \mathfrak{N}_c$ with $N \lhd^2 L$ and let $K = N^L$. Define $P = \langle M^L : M \in \mathfrak{N}_{c-1}, M \lhd K\rangle$. Then $P \leqq H \cap K \in \mathfrak{F}$ and so $P \leqq \zeta_r(K)$ for some finite r (lemma 7.1.6). We will show that $K \in \mathfrak{N}_{2r}$, so that as $K \lhd L$ we have $N \leqq K \leqq H$.

By lemma 4.1.1(g) we have

$$P^\dagger \leqq \zeta_r(K^\dagger).$$

Let $Y = \langle N^{\dagger\theta} : \theta \in \exp(tL)\rangle$. Then $Y \leqq K^\dagger$ and $Y^\downarrow = K$, by corollary 4.1.4. For any $\theta \in \exp(tL)$ we have

$$(N^{\dagger\theta})^2 = (N^{\dagger 2})^\theta \leqq P^{\dagger\theta} = P^\dagger,$$

since $P \lhd L$. Let $\theta_1, \ldots, \theta_{r+1} \in \exp(tL)$ and define for each i, $1 \leqq i \leqq r+1$, $N_i = N^{\dagger\theta_i}$. Then $N_i \lhd K^\dagger$ and $N_i^2 \leqq P^\dagger \leqq \zeta_r(K^\dagger)$. Thus if $U = [N_1, \ldots, N_{r+1}]$ then $U \lhd K^\dagger$ and

$$U^2 \leqq \textstyle\sum_\tau [N_1, \ldots, N_{r+1}, N_{\tau(1)}, \ldots, N_{\tau(r+1)}],$$

where the summation is taken over all permutations τ of $1, \ldots, r+1$. Since for any such permutation we have $U = [N_1, \ldots, N_{r+1}] \leqq N_{\tau(1)}$, then

$$U^2 \leqq \textstyle\sum_\tau [N_{\tau(1)}^2, N_{\tau(2)}, \ldots, N_{\tau(r+1)}] = 0,$$

since $N_{\tau(1)}^2 \leqq \zeta_r(K^\dagger)$. Thus U is an abelian ideal of $K^\dagger$. Now let A be a maximal abelian ideal of K. Then $A \leqq P \leqq \zeta_r(K)$. Therefore $[A^\dagger, U] = 0$ and so $A^\dagger + U$ is an abelian ideal of $K^\dagger$. By our remarks above $A^\dagger$ is a maximal abelian ideal of $K^\dagger$, so $A^\dagger + U = A^\dagger$, whence $U \leqq A^\dagger \leqq \zeta_r(K^\dagger)$. Evidently Y^{r+1} is a sum of terms like U so $Y^{r+1} \leqq \zeta_r(K^\dagger)$ and hence $Y \in \mathfrak{N}_{2r}$. Thus $Y^\downarrow = K \in \mathfrak{N}_{2r}$ and so $N \leqq K \leqq H = \nu(L)$. This completes our induction on c and so H contains every $\lhd^2 \mathfrak{N}$-subalgebra of L. This proves case 1.

Case 2. α is a limit ordinal and for each ordinal $\beta < \alpha$, if $M \in \lhd^\beta \mathfrak{A}\text{-Fin}$ then $J_{\lhd^\beta \mathfrak{N}}(M) \leqq \nu(M) \in \mathfrak{F} \cap \mathfrak{N}$.

Let $N \lhd^\alpha L$ with $N \in \mathfrak{N}$. Then we have an ascending series

$$N = N_0 \lhd N_1 \lhd \ldots N_\alpha = \textstyle\bigcup_{\beta<\alpha} N_\beta = L.$$

For each β, $1 \leqq \beta < \alpha$, $N_\beta \in \lhd^\beta \mathfrak{A}$-Fin. We also have $N_1 \in \lhd^2 \mathfrak{A}$-Fin, since $1+\alpha = \alpha$. So by induction we have

$$N \leqq V_\beta = \nu(N_\beta) \in \mathfrak{F} \cap \mathfrak{N} \text{ and } V_\beta \leqq V_\gamma,$$

for all β, γ with $1 \leqq \beta \leqq \gamma < \alpha$. We also have $V_\beta \operatorname{ch} N_\beta \lhd N_{\beta+1}$, so $V_\beta \lhd N_{\beta+1}$ and $V_\beta \lhd^\alpha L$. Put $V = \bigcup_{\beta<\alpha} V_\beta$. Then $V \lhd L$, since $L = N_\alpha = \bigcup_{\beta<\alpha} N_\beta$.

Now $H = \nu(L) \in \mathfrak{F} \cap \mathfrak{N}$ (since $L \in \lhd^2 \mathfrak{A}$-Fin) and every $\lhd^\alpha \mathfrak{N}$-subalgebra of L is finite-dimensional. Clearly for each finite i, $\zeta_i(V) \lhd L$, so $\zeta_i(V) \leqq H \in \mathfrak{F}$. Thus we can find n such that $K = \zeta_n(V) = \zeta_{n+1}(V) = \ldots$. Thus $K \in \mathfrak{F} \cap \mathfrak{N}$ and $\zeta_1(V/K) = 0$. We can find $\lambda < \alpha$ with $K \leqq V_\lambda$, since $V = \bigcup_{\beta<\alpha} V_\beta$. For each $\beta \geqq \lambda$ define S_β by $S_\beta/K = \zeta_1(V_\beta/K)$. Since $V_\beta/K \lhd N_{\beta+1}/K$ then $S_\beta/K \lhd N_{\beta+1}/K$ and so $S_\beta \lhd N_{\beta+1}$. Further if $\lambda \leqq \beta \leqq \gamma < \alpha$ then $[N_{\beta+1}, S_\gamma] \leqq S_\gamma$ and

$$[S_\beta, S_\gamma] \subseteq [V_\beta, S_\gamma] \subseteq [V_\gamma, S_\gamma] \leqq K. \tag{*}$$

Define $S = \langle S_\beta : \lambda \leqq \beta < \alpha \rangle$. Then for each $\beta \geqq \lambda$,

$$\langle N_\beta, S \rangle = \langle N_\beta, S_\gamma : \beta+1 \leqq \gamma < \alpha \rangle, \text{ and so}$$
$$\langle N_\beta, S \rangle \lhd \langle N_{\beta+1}, S \rangle = \langle N_{\beta+1}, S_\gamma : \beta+2 \leqq \gamma < \alpha \rangle.$$

So we have an ascending series

$$S \lhd \langle N_\lambda, S \rangle \lhd \langle N_{\lambda+1}, S \rangle \lhd \ldots \langle N_\alpha, S \rangle = L$$

of type α from S to L (note that N_λ idealises every S_β). By (*) S/K is abelian and so $S \in \mathfrak{N}$ (for $K \leqq S \cap \zeta_n(V)$ and $S \leqq V$). Therefore $S \in \mathfrak{F} \cap \mathfrak{N}$ and so we can find μ, with $\lambda \leqq \mu < \alpha$ such that $S \leqq V_\mu$. Thus $S_\mu \geqq S_{\mu+1} \geqq \ldots$, whence for some γ with $\lambda \leqq \gamma < \alpha$, we have $S_\gamma = S_{\gamma+1} = \ldots$ and $S_\gamma/K \leqq \zeta_1(V/K) = 0$. Hence $S_\gamma = K$ and $K = V_\gamma = V_{\gamma+1} = \ldots = V \leqq H$ (for if $\beta \geqq \gamma$ and $K < V_\beta$, then as $V_\beta \in \mathfrak{N}$ we have $S_\beta/K \neq 0$ and so $S_\beta \neq K$, a contradiction). So $N \leqq H$ and case 2 is proved. □

If α is an infinite ordinal then $1+\alpha = \alpha$. Thus if A is a $\lhd^\alpha \mathfrak{A}$-subalgebra of a Lie algebra L then so is every subalgebra of A. Hence over fields of characteristic zero,

$$\text{Max-}\lhd^\alpha \mathfrak{A} \leqq \lhd^\alpha \mathfrak{A}\text{-Fin} \leqq \text{Max-}\lhd^\alpha \mathrm{E}\mathfrak{A} \cap \lhd^\alpha \mathrm{E}\mathfrak{A}\text{-Fin},$$

by theorem 1.9. Using theorems 1.4, 1.8 and 1.9 it is not hard to deduce the following:

COROLLARY 1.10 *Over any field of characteristic zero the classes occurring in* (a), (b), (c), (d), *respectively coincide.*

(a): *α is an ordinal* > 1: Max-$\lhd^{\alpha}\mathfrak{N}$, Max-$\lhd^{\alpha}\mathrm{E}\mathfrak{A}$
$\lhd^{\alpha}\mathfrak{A}$-Fin, $\lhd^{\alpha}\mathfrak{N}$-Fin, $\lhd^{\alpha}\mathrm{E}\mathfrak{A}$-Fin.

(b): *α is an infinite ordinal:* Max-$\lhd^{\alpha}\mathfrak{A}$, Max-$\lhd^{\alpha}\mathfrak{N}$, Max-$\lhd^{\alpha}\mathrm{E}\mathfrak{A}$
$\lhd^{\alpha}\mathfrak{A}$-Fin, $\lhd^{\alpha}\mathfrak{N}$-Fin, $\lhd^{\alpha}\mathrm{E}\mathfrak{A}$-Fin.

(c): Max-s$\mathfrak{A}$, Max-s$\mathfrak{N}$, Max-s$\mathrm{E}\mathfrak{A}$
s$\mathfrak{A}$-Fin, s$\mathfrak{N}$-Fin, s$\mathrm{E}\mathfrak{A}$-Fin.

(d): Max-a$\mathfrak{A}$, Max-a$\mathfrak{N}$, Max-a$\mathrm{E}\mathfrak{A}$
a$\mathfrak{A}$-Fin, a$\mathfrak{N}$-Fin, a$\mathrm{E}\mathfrak{A}$-Fin. □

COROLLARY 1.11 *Let $\mathfrak{X}$ be a class of Lie algebras over a field of characteristic zero. If $\mathfrak{A} \leqq \mathfrak{X} \leqq \mathrm{E}\mathfrak{A}$ then*

$$\text{Max-s}\mathfrak{X} = \bigcap_{n=1}^{\infty} \text{Max-}\lhd^{n}\mathfrak{X}$$

and

$$\text{s}\mathfrak{X}\text{-Max} = \text{s}\mathfrak{X}\text{-Min} = \text{s}\mathfrak{X}\text{-Fin} = \bigcap_{n=1}^{\infty} \lhd^{n}\mathfrak{X}\text{-Fin}.$$

Proof: Let $L \in \bigcap_{n=1}^{\infty}$ Max-$\lhd^{n}\mathfrak{X}$. Then $L \in$ Max-$\lhd^{n+1}\mathfrak{X} \leqq \lhd^{n}\mathfrak{A}$-Fin, for each n. Hence by theorem 1.9, $\sigma(L) \in \mathfrak{F} \cap \mathrm{E}\mathfrak{A}$ and contains every n-step subideal of L. Hence $\sigma(L)$ contains every $\mathfrak{X}$-subideal of L and so $L \in$ Max-s$\mathfrak{X}$. Thus Max-s$\mathfrak{X} \leqq \bigcap_{n=1}^{\infty}$ Max-$\lhd^{n}\mathfrak{X} \leqq$ Max-s$\mathfrak{X}$. The second half follows similarly. □

It is worth remarking that we do have:

PROPOSITION 1.12 *For any class $\mathfrak{X}$ of Lie algebras,*

$$\text{Max-a}\mathfrak{X} = \bigcap_{\alpha>0} \text{Max-}\lhd^{\alpha}\mathfrak{X} \text{ and } \text{Min-a}\mathfrak{X} = \bigcap_{\alpha>0} \text{Min-}\lhd^{\alpha}\mathfrak{X}.$$

Proof: Let $L \in \bigcap_{\alpha>0}$ Max-$\lhd^{\alpha}\mathfrak{X}$ and let α be an ordinal of cardinality $> \dim L$. On set-theoretic grounds, if H asc L then $H \lhd^{\alpha} L$. Therefore as $L \in$ Max-$\lhd^{\alpha}\mathfrak{X}$ we have $L \in$ Max-a$\mathfrak{X}$. Similarly Min-a$\mathfrak{X} \leqq \bigcap_{\alpha>0}$ Min-$\lhd^{\alpha}\mathfrak{X}$. □

2. Minimal conditions

LEMMA 2.1 *Let $L \in$ Min-$\lhd^{2}\mathfrak{A}$. Then $\nu(L) \in \mathfrak{F} \cap \mathfrak{N}$ and $\sigma(L) \in \mathfrak{F} \cap \mathrm{E}\mathfrak{A}$. In particular $\acute{\mathrm{E}}(\lhd)\mathfrak{A} \cap$ Min-$\lhd^{2}\mathfrak{A} \leqq \mathfrak{F} \cap \mathrm{E}\mathfrak{A}$.*

Proof: Let A be an abelian ideal of L. Then $A \in \text{Min-}\lhd\,\mathfrak{A}$ and so $A \in \text{Min} \cap \mathfrak{A} \leqq \mathfrak{F} \cap \mathfrak{A}$.

Now let N be a nilpotent ideal of L and let A be an abelian ideal of N. For each i let $A_i = A \cap \zeta_i(N)$. Then $A_0 = 0$ and $A_c = A$ (if $N \in \mathfrak{N}_c$). Now A_{i+1}/A_i is a central factor of N and $N \in \text{Min-}\lhd\,\mathfrak{A}$. Hence $A_{i+1}/A_i \in \text{Min}$ (for if $X/A_i \leqq A_{i+1}/A_i$ then $X \lhd N$ and $X^2 = 0$) and so $A_{i+1}/A_i \in \text{Min} \cap \mathfrak{A} \leqq \mathfrak{F} \cap \mathfrak{A}$. Therefore $A \in \mathfrak{F}$. Thus $N \in \mathfrak{N} \cap \lhd\,\mathfrak{A}\text{-Fin} \leqq \mathfrak{F} \cap \mathfrak{N}$, by corollary 1.3.

Suppose that K is a $\mathfrak{Z}$-ideal of L and $B = \zeta_\omega(K)$. Then $B \lhd L$ and $B \in \text{Min-}\lhd\,\mathfrak{A}$. Let A be a maximal abelian ideal of B. For each positive integer i define $C_i = A \cap C_B(\zeta_i(K))$. Then $C_i \lhd B$ and $A = C_1 \geqq C_2 \geqq \dots$. By $\text{Min-}\lhd\,\mathfrak{A}$ this chain is stationary after finitely many steps. So we can find m such that $C_m = C_{m+1} = \dots$ and so $C_m \leqq A \cap \zeta_1(B)$ (for $B = \bigcup_{i=0}^{\infty} \zeta_i(K)$). Now $\zeta_1(B)$ is an abelian ideal of L and so is finite-dimensional. Thus $C_m \in \mathfrak{F}$. Furthermore $\zeta_m(K)$ is a nilpotent ideal of L and so is in $\mathfrak{F}$. Hence $A/C_m \in \mathfrak{F}$ (for $B/C_B(\zeta_m(K)) \in \mathfrak{F}$) and $A \in \mathfrak{F}$. But $C_B(A) = A$ so we have $B/A \in \mathfrak{F}$ and $A \in \mathfrak{F}$, whence $B \in \mathfrak{F}$. Since $B = \bigcup_{i=0}^{\infty} \zeta_i(K)$ we can find a finite n such that $B = \zeta_n(K) = \zeta_{n+1}(K) = \dots$. Therefore $K = \zeta_n(K) \in \mathfrak{N}$ and so $K \in \mathfrak{F}$.

Now let $H = \nu(L)$. Then H is the join of $\mathfrak{F} \cap \mathfrak{N}$-ideals of L. If $H \notin \mathfrak{F}$ then by lemma 1.1 H contains an infinite-dimensional $\mathfrak{Z}$-ideal of L, which is impossible from above. Thus $H \in \mathfrak{F}$ and so $H \in \mathfrak{F} \cap \mathfrak{N}$. By corollary 1.3 $\sigma(L) \in \mathfrak{F} \cap \text{E}\mathfrak{A}$. By lemma 1.5 it follows that $\acute{\text{E}}(\lhd)\mathfrak{A} \cap \text{Min-}\lhd^2\,\mathfrak{A} \leqq \mathfrak{F} \cap \text{E}\mathfrak{A}$. □

If α is an infinite ordinal then $n+\alpha = \alpha$ for every positive integer n. So if $H \lhd^n K \lhd^\alpha L$ then $H \lhd^\alpha L$. We can now prove an analogue of corollary 1.10 which is true for all fields.

COROLLARY 2.2 *Let Δ be one of the relations* si, asc, $\lhd^\alpha$ (α *an infinite ordinal*). *Then the classes*

$\text{Min-}\Delta\mathfrak{A}$, $\text{Min-}\Delta\mathfrak{N}$, $\text{Min-}\Delta\text{E}\mathfrak{A}$,
$\Delta\mathfrak{A}\text{-Fin}$, $\Delta\mathfrak{N}\text{-Fin}$, $\Delta\text{E}\mathfrak{A}\text{-Fin}$,

coincide.

Proof: Let $L \in \text{Min-}\Delta\mathfrak{A} \cup \Delta\mathfrak{A}\text{-Fin}$ and let K be a $\Delta\text{E}\mathfrak{A}$-subalgebra of L. If $H \text{ si } K$ then $H \,\Delta\, L$ and so $K \in (\text{Min-}\lhd^2\,\mathfrak{A} \cup \lhd^2\,\mathfrak{A}\text{-Fin}) \cap \text{E}\mathfrak{A}$. Therefore by lemma 2.1 (for $\lhd^2\,\mathfrak{A}\text{-Fin} \leqq \text{Min-}\lhd^2\,\mathfrak{A}$) $K \in \mathfrak{F} \cap \text{E}\mathfrak{A}$. Hence $L \in \Delta\text{E}\mathfrak{A}\text{-Fin} \leqq \text{Min-}\Delta\text{E}\mathfrak{A}$. We also have

$$\text{Min-}\Delta\text{E}\mathfrak{A} \leqq \text{Min-}\Delta\mathfrak{N} \leqq \text{Min-}\Delta\mathfrak{A}$$

and

$$\Delta\text{E}\mathfrak{A}\text{-Fin} \leqq \Delta\mathfrak{N}\text{-Fin} \leqq \Delta\mathfrak{A}\text{-Fin}.$$ □

Suppose that $\mathfrak{X}$ is an I-closed class of Lie algebras and $L \in \bigcap_{n=1}^{\infty}$ Min-$\lhd^n \mathfrak{X}$. Let H si L with $H \in \mathfrak{X}$. Then $H \lhd^n L$ for some n. If $K \lhd^3 H$ then $K \in \mathfrak{X}$ and $K \lhd^{n+3} L \in$ Min-$\lhd^{n+3} \mathfrak{X}$. Hence $H \in$ Min-$\lhd^3$ = Min-si (proposition 8.1.5) and so $H \in$ Min-s$\mathfrak{X}$, whence $L \in$ Min-s$\mathfrak{X}$.

Let $\mathfrak{A} \leqq \mathfrak{X} \leqq \mathrm{E}\mathfrak{A}$ and let $L \in \bigcap_{n=1}^{\infty}$ Min-$\lhd^n \mathfrak{X}$. Suppose that $H \lhd^n L$ and $H \in \mathfrak{X}$. If $K \in \mathfrak{A}$ and $K \lhd^2 H$ then $K \lhd^{n+2} L$ and, since $L \in$ Min-$\lhd^{n+2} \mathfrak{X} \leqq$ $\leqq$ Min-$\lhd^{n+2} \mathfrak{A}$, then $H \in$ Min-$\lhd^2 \mathfrak{A} \cap \mathrm{E}\mathfrak{A}$ so $H \in \mathfrak{F} \cap \mathrm{E}\mathfrak{A} \cap \mathfrak{X}$. Thus $H \in$ Min-s$\mathfrak{X}$. So $L \in$ Min-s$\mathfrak{X}$.

So we have proved:

PROPOSITION 2.3 *Let $\mathfrak{X}$ be a class of Lie algebras.*

(a) *If $\mathfrak{X}$ is* I-*closed then* Min-s$\mathfrak{X} = \bigcap_{n=1}^{\infty}$ Min-$\lhd^n \mathfrak{X}$.

(b) *If $\mathfrak{A} \leqq \mathfrak{X} \leqq \mathrm{E}\mathfrak{A}$ then* Min-s$\mathfrak{X} = \bigcap_{n=1}^{\infty}$ Min-$\lhd^n \mathfrak{X}$
$= \mathrm{s}\mathfrak{A}$-Fin $= \bigcap_{n=1}^{\infty} \lhd^n \mathfrak{A}$-Fin. □

Now suppose that $L \in$ Min-$\lhd^2 \mathfrak{A}$ and $A \lhd M \lhd L$ such that $A^2 = 0$ and $M \in \mathrm{L}\mathfrak{N}$. Then $A \leqq \nu(L)$ and so $A \in \mathfrak{F}$. (1)

For $M \in$ Min-$\lhd \mathfrak{A}$ and so $A \in$ Min-M (the minimal condition for ideals of M contained in A). Thus we can find an ascending series

$$0 = A_0 \lhd A_1 \lhd \dots A_\lambda = A \text{ (for some ordinal } \lambda)$$

of ideals of M such that each factor $A_{\alpha+1}/A_\alpha$ is a chief factor of M, and so central in M (by lemma 7.1.6). By induction we have $A \leqq \zeta_\lambda(M)$. But $\zeta_\lambda(M)$ is a $\mathfrak{Z}$-ideal of L and so is finite dimensional and nilpotent and contained in $\nu(L)$ (by proof of lemma 2.1). Hence $A \leqq \nu(L)$ and $A \in \mathfrak{F}$.

In particular for fields of characteristic zero,

$$\text{Min-}\lhd^2 \mathfrak{A} \leqq \lhd^2 \mathfrak{A}\text{-Fin} \leqq \text{Min-}\lhd^2 \mathfrak{A}, \qquad (*)$$

and for each positive integer $n > 1$,

$$\text{Min-}\lhd^n \mathfrak{A} \leqq \lhd^n \mathfrak{A}\text{-Fin} \leqq \text{Min-}\lhd^n \mathfrak{A}. \qquad (**)$$

For if $L \in$ Min-$\lhd^2 \mathfrak{A}$ over a field of characteristic zero and A is a $\lhd^2 \mathfrak{A}$-subalgebra of L then $A_1 = \langle A^L \rangle = \nu(A_1) \in \mathrm{L}\mathfrak{N}$. By (1) we have $A \in \mathfrak{F}$ and so $L \in \lhd^2 \mathfrak{A}$-Fin. This proves (*) and (**) follows by induction on n.

From (**), corollary 2.2 and theorem 1.9 it is easy to deduce:

THEOREM 2.4 *Let α be an ordinal > 1 and let $L \in \text{Min-}\lhd^{\alpha}\mathfrak{A}$ over a field of characteristic zero. Then*

(a) $J_{\lhd^{\alpha}3}(L) = \nu(L) \in \mathfrak{F} \cap \mathfrak{N}$ *and*
(b) $J_{\lhd^{\alpha}\mathrm{E}\mathfrak{A}}(L) = \sigma(L) \in \mathfrak{F} \cap \mathrm{E}\mathfrak{A}$. □

From this there follows immediately:

COROLLARY 2.5 *If α is an ordinal > 1 then over fields of characteristic zero the classes*

$$\text{Min-}\lhd^{\alpha}\mathfrak{A},\ \text{Min-}\lhd^{\alpha}\mathfrak{N},\ \text{Min-}\lhd^{\alpha}\mathrm{E}\mathfrak{A},$$
$$\lhd^{\alpha}\mathfrak{A}\text{-Fin},\ \lhd^{\alpha}\mathfrak{N}\text{-Fin},\ \lhd^{\alpha}\mathrm{E}\mathfrak{A}\text{-Fin},$$

coincide. □

The Exceptional Case $\alpha = 1$

(i) The Lie algebra L of example 6.3.6 is such that $L \notin \mathfrak{F}$ and $L \in \text{Min-}\lhd \cap \text{Max-}\lhd \cap \mathrm{E}\mathfrak{A}$. Hence theorems 1.4, 1.8, 1.9 and 2.4 all fail for the case $\alpha = 1$.

(ii) A trivial induction on derived lengths gives

$$\lhd\mathfrak{N}_2\text{-Fin} \leqq \lhd\mathrm{E}\mathfrak{A}\text{-Fin}.$$

(iii) We also have $\lhd\mathfrak{N}_2\text{-Fin} \leqq \text{Max-}\lhd\mathrm{E}\mathfrak{A}$.
For let $L \in \lhd\mathfrak{N}_2$-Fin and let $R = \sigma(L)$ and let N be a maximal $\mathfrak{N}_2$-ideal of L. Then $N \in \mathfrak{F}$. Now R has an ascending series of ideals $\{R_\beta : \beta \leqq \lambda\}$ of L with abelian factors. Let $C = C_R(N)$. Then $R/C \in \mathfrak{F}$. If $C \nleqq N$ then there exists β

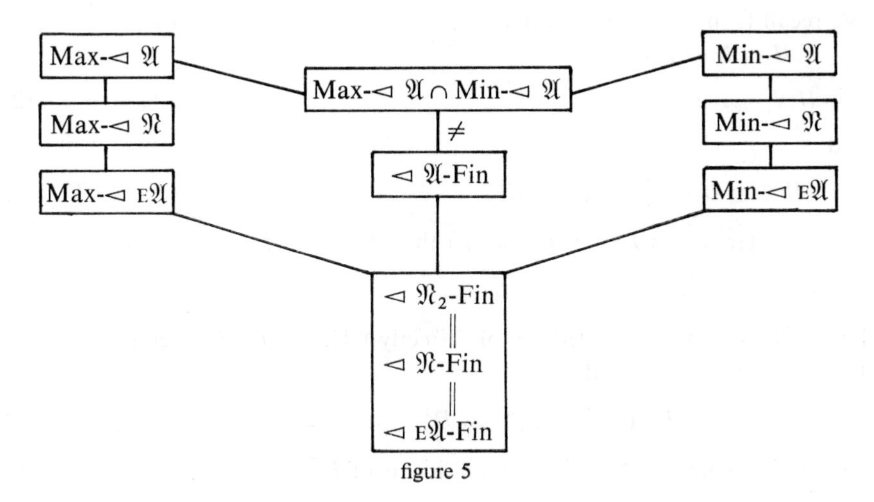

figure 5

minimal with respect to $C \cap R_\beta \nleqq N$. Then β is not a limit ordinal so $\beta - 1$ exists and $C \cap R_{\beta-1} \leqq N$. Now $R_\beta^2 \leqq R_{\beta-1}$ and so $C \cap R + N \in \mathfrak{N}_2$. Further $C \cap R_\beta \lhd L$ and so $C \cap R_\beta + N \lhd L$, whence $C \cap R_\beta \leqq N$, a contradiction. Thus $C \leqq N \in \mathfrak{F}$ and so $R \in \mathfrak{F} \cap \mathrm{E}\mathfrak{A}$.

(iv) The Lie algebra L constructed after lemma 6.4.1 is such that $L \in$ Min-$\lhd$ but $L \notin$ Max-$\lhd$ $\mathfrak{A}$.

(v) We also have Max-$\lhd$ $\nleqq$ Min-$\lhd$ $\mathfrak{A}$.
For let L be the split extension $L = \langle x_1, x_2, \ldots \rangle \dotplus \langle d \rangle$, where $[x_i, x_j] = 0$ for all i, j and $x_i d = x_{i+1}$ for all i. Then every ideal of L has finite codimension in L but $\langle x_2, \ldots \rangle > \langle x_3, \ldots \rangle \ldots$ is a strictly descending chain of abelian ideals of L.

The situation is illustrated by the lattice of figure 5. There are no other inclusions, except possibly $\lhd$ $\mathfrak{N}_2$-Fin $\geqq$ $\lhd$ $\mathfrak{A}$-Fin, or

$$\lhd\ \mathfrak{A}\text{-Fin} \geqq \text{Max-}\lhd\ \mathfrak{N} \cup \text{Min-}\lhd\ \mathfrak{N},$$

or between the classes in the first column or third column.

3. Applications

Before discussing some of the applications of our results so far we consider a new radical for Lie algebras.

Let L be a Lie algebra and let $x_1, \ldots, x_n \in L$ and $H \leqq L$. Define

$$H(x_1, \ldots, x_n) = H + \sum [H, x_i] + \sum_{1 \leqq i < j \leqq n} [H, x_i, x_j] + \ldots + [H, x_1, \ldots, x_n].$$

We recall from proposition 2.1.10 that

if $H \lhd K \lhd L$ then

$$H(x_1, \ldots, x_n) \lhd K \tag{2}$$

and

$$\langle H^L \rangle = H^L = \sum_{n=0}^{\infty} \sum_{x_i \in L} H(x_1, \ldots, x_n). \tag{3}$$

In particular if $I \lhd L$ and $d \in \mathrm{Der}(L)$ then $I \lhd L \lhd L \dotplus \mathrm{Der}(L)$ so

$$I + Id = I(d) \lhd L, \tag{4}$$

by (2). (It is also easy to deduce this directly.) Thus if $I \in \mathfrak{F}_m$ then $I + Id \in \mathfrak{F}_{2m}$. From (2) and (3) we deduce:

$$\text{If } H \lhd K \lhd L \text{ and } H \in \mathfrak{F}_m \text{ then } H^L \in \acute{\mathrm{E}}(\lhd)\mathfrak{F}_m, \tag{5}$$

$$\text{and } H^L \text{ is a sum of finite-dimensional ideals of } K. \tag{6}$$

For (6) we note that $H(x_1, \ldots, x_n) \in \mathfrak{F}_{rm}$ where $r = 2^n$. Define $H_0 = H$. If H_α has been defined for some ordinal α and $H \neq H^L$ then we can find $x_1, \ldots, x_n \in L$ such that $H(x_1, \ldots, x_n) \nleqq H_\alpha$; and so let n be minimal with respect to this. Then $H(y_1, \ldots, y_k) \subseteq H$ for all $y_1, \ldots, y_k \in L$ with $k \leqq n-1$. Define $H_{\alpha+1} = H_\alpha + H(x_1, \ldots, x_n) \lhd H^L$. Clearly $H_{\alpha+1} = H_\alpha + [H, x_1, \ldots, x_n]$, and so $H_{\alpha+1}/H_\alpha \in \mathfrak{F}_m$. If α is a limit ordinal and H_β has been defined for all $\beta < \alpha$, let $H_\alpha = \bigcup_{\beta<\alpha} H_\beta$. By set-theoretic considerations we must have $H_\lambda = H^L$ for some ordinal λ. Thus $H^L \in \acute{\mathrm{E}}(\lhd)\mathfrak{F}_m$.

(Note: $H(x_1, \ldots, x_n) = H(x_1, \ldots, x_{n-1}) + [H(x_1, \ldots, x_{n-1}), x_n]$

$= H(x_1, \ldots, x_{n-1})(x_n)$

$= [H, x_1, \ldots, x_n] + \sum_{r=1}^{n-1} \sum_{1 \leqq i_1 < \ldots < i_r \leqq n} H(x_{i_1}, \ldots, x_{i_r})$.)

Let L be a Lie algebra. We define the $\mathfrak{F}$-radical $\rho_{\mathfrak{F}}(L)$ by

$$\rho_{\mathfrak{F}}(L) = \langle H : H \in \mathfrak{F} \text{ and } H \lhd L \rangle. \tag{7}$$

From (4)–(7) we have:

LEMMA 3.1 *Let L be a Lie algebra. Then*

(a) *$\rho_{\mathfrak{F}}(L)$ is a characteristic ideal of L.*

(b) *$\rho_{\mathfrak{F}}(L) \in \acute{\mathrm{E}}(\lhd)\mathfrak{F}$ and has an ascending series of ideals of L whose factors are finite-dimensional.* □

THEOREM 3.2 *Let α be a non-zero ordinal.*

(a) *If $L \in \text{Max-}\lhd^\alpha \mathfrak{F}$ then $J_{\lhd^\alpha \mathfrak{F}}(L) = \rho_{\mathfrak{F}}(L) \in \mathfrak{F}$.*

(b) *If $\alpha > 1$ and $L \in \text{Min-}\lhd^\alpha$ then $J_{\lhd^\alpha \mathfrak{F}}(L) = \rho_{\mathfrak{F}}(L) \in \mathfrak{F}$.*

Proof: By transfinite induction on α. Let $P = \rho_{\mathfrak{F}}(L)$.

(a) For $\alpha = 1$ the result is immediate by (7). Assume that the result holds for α and let $L \in \text{Max-}\lhd^{\alpha+1}\mathfrak{F}$. Let $H \lhd^{\alpha+1} L$ with $H \in \mathfrak{F}$ and let $H_1 = H^L$. Then $H_1 \in \text{Max-}\lhd^\alpha \mathfrak{F}$ and $H \lhd^\alpha H_1$ so $H \leqq P_1 = \rho_{\mathfrak{F}}(H_1) \in \mathfrak{F}$. By lemma 3.1 $P_1 \text{ ch } H_1$ so $P_1 \lhd L$, whence $H \leqq P_1 \leqq P$.

Let α be a limit ordinal and assume that the result is true for all β with $1 \leqq \beta < \alpha$. Let $H \in \mathfrak{F}$ and $H \lhd^\alpha L$ so that we have an ascending series

$$H = H_0 \lhd H_1 \lhd \ldots H_\alpha = L$$

from H to L. For each β, $1 \leqq \beta < \alpha$, $H_\beta \in \text{Max-}\lhd^\beta \mathfrak{F}$. Hence by the inductive hypothesis $H \leqq P_\beta$ for all $\beta > 0$. Now $P_\beta \text{ ch } H_\beta \lhd H_{\beta+1}$ so $P_\beta \lhd H_{\beta+1} \lhd \lhd H_{\beta+2} \lhd \ldots H_\alpha = L$, whence $P_\beta \lhd^\alpha L$. Thus if H is chosen to be a maximal

$\lhd^\alpha \mathfrak{F}$-subalgebra of L (which exists since $L \in \text{Max-}\lhd^\alpha \mathfrak{F}$) then $H = P_\beta$ for all $\beta > 0$ with $\beta < \alpha$. Hence $H \lhd \bigcup_{\beta<\alpha} H_\beta = H_\alpha = L$ and so $H \leqq P$, whence $H = P \in \mathfrak{F}$. But every $\lhd^\alpha \mathfrak{F}$-subalgebra is contained in a maximal one and so the result follows. This proves (a).

(b) The inductive step from α to $\alpha+1$ follows as above. So we need consider only two cases.

Case 1. $\alpha = 2$.

Let $I \lhd L$ and let $J = \rho_{\mathfrak{F}}(J)$. Then $J \operatorname{ch} I \lhd L$ so $J \lhd L$. Now if $H \in \mathfrak{F}$ and $H \lhd J$ then $J/C_J(H) \in \mathfrak{F}$ and so $K \leqq C_J(H)$, where $K = \bigcap \{N : N \lhd J$ and $J/N \in \mathfrak{F}\}$. Hence $K \leqq \zeta_1(J) \leqq \nu(L)$. By lemma 2.1, $\nu(L) \in \mathfrak{F}$ so $K \in \mathfrak{F}$. But $J \in \text{Min-}\lhd$ so that $J/K \in \mathfrak{F}$ and so $J \in \mathfrak{F}$, whence $J \leqq P$.

If we put $I = L$ we get $J = P \in \mathfrak{F}$. If $H \in \mathfrak{F}$ and $H \lhd^2 L$ then by (6) we have $I = H^L = \rho_{\mathfrak{F}}(I) \leqq P$. So $H \leqq P$ and case 1 is proved.

Case 2. α is a limit ordinal and the result holds for all β with $1 < \beta < \alpha$. Then $P \in \mathfrak{F}$ since $\text{Min-}\lhd^\alpha \leqq \text{Min-}\lhd^2$. Let H be a $\lhd^\alpha \mathfrak{F}$-subalgebra of L so that we have an ascending series

$$H = H_0 \lhd H_1 \lhd \dots H_\alpha = L.$$

Let $P_\beta = \rho_{\mathfrak{F}}(H_\beta)$ for $2 \leqq \beta < \alpha$. Now $H_\beta \in \text{Min-}\lhd^\beta$ and so for all β, γ with $2 \leqq \beta \leqq \gamma < \alpha$ we have by induction,

$$H \leqq P_\beta \leqq P_\gamma \in \mathfrak{F} \text{ and } P_\beta \lhd^\alpha L$$

(for $P_\beta \operatorname{ch} H_\beta \lhd H_{\beta+1}$ so $P_\beta \lhd H_{\beta+1} \lhd \dots$, whence $P_\beta \lhd^\gamma H_\gamma$ for all $\gamma \geqq \beta$). By lemma 1.3.4 we have $P_\beta^\omega \lhd L$ and so $P_\beta^\omega \leqq P$ for all $\beta < \alpha$. Further for each β we can find a positive integer c with $P_\beta^\omega = P_\beta^c$. Let $K = \langle P_\beta^\omega : \beta < \alpha \rangle$. So $K \leqq P$ and $K \in \mathfrak{F}$. Hence we can find $\gamma < \alpha$ such that $K = P_\gamma^\omega = P_\gamma^n$ for some positive integer n. We also have $K \lhd L$ and $L/K \in \text{Min-}\lhd^\alpha$. Let $J = \bigcup_{\beta<\alpha} P_\beta$. Then $J \lhd \bigcup_{\beta<\alpha} H_\beta = H_\alpha = L$ and J/K is the union of an ascending chain of $\lhd^\alpha (\mathfrak{F} \cap \mathfrak{N})$-subalgebras of L/K.

We claim that $J/K \in \mathfrak{Z}$; so that $J/K \leqq \nu(L/K) \in \mathfrak{F}$ by lemma 2.1 and so $J \in \mathfrak{F}$ and $H \leqq J \leqq P$.

Suppose that the claim is false. Now $\zeta_\omega(J/K) \leqq \nu(L/K) \in \mathfrak{F}$ and so we can find a finite i for which $\zeta_\omega(J/K) = \zeta_i(J/K) = M/K$. Since $M \in \mathfrak{F}$ we have for some $\lambda > 0$, $M \leqq P_\lambda$. For each $\beta \geqq \lambda$ we define S_β by $S_\beta/M = \zeta_1(P_\beta/M)$ (for $P_\beta/M \in \mathfrak{N}$). We let $S = \langle S_\beta : \lambda \leqq \beta < \alpha \rangle$. Arguing as in the proof of theorem 1.9 we get that $S/M \lhd L/M$ and so $S/M \in \mathfrak{F}$ and $S/M \leqq \zeta_1(J/M) = 0$, a contradiction. So our claim is true and case 2 is proved. □

COROLLARY 3.3

(a) $\text{Max-s}\mathfrak{F} = \bigcap_{n=1}^{\infty} \text{Max-}\lhd^{n}\mathfrak{F}$.

(b) *Let Δ be one of the relations* si, asc.

If $L \in \text{Min-}\Delta \cup \text{Max-}\Delta\mathfrak{F}$ *then* $\rho_{\mathfrak{F}}(L) \in \mathfrak{F}$ *and* $\rho_{\mathfrak{F}}(L)$ *contains every* $\Delta\mathfrak{F}$*-subalgebra of* L. □

We are now ready for the applications mentioned before. Evidently over any field we have

$$\text{Max-s}\mathfrak{A} \leqq \text{Min-s}\mathfrak{A} = \text{s}\mathfrak{A}\text{-Fin}. \tag{8}$$

First we observe that if $L \in \acute{\text{E}}\mathfrak{A}$ and $H \leqq L$ then $H \in \acute{\text{E}}\mathfrak{A}$. Furthermore $\mathfrak{F} \cap \acute{\text{E}}\mathfrak{A} \leqq \mathfrak{F} \cap \text{E}\mathfrak{A}$ and so if $L \in \acute{\text{E}}\mathfrak{A}$ then $\rho_{\mathfrak{F}}(L)$ is a sum of $\mathfrak{F} \cap \text{E}\mathfrak{A}$-ideals of L.

Let Δ be one of the relations $\lhd^{\alpha}$ $(\alpha > 0)$, si, asc. Following the remarks after corollary 2.5 it is not hard to show that

$$\Delta\mathfrak{N}_2\text{-Fin} = \Delta\mathfrak{N}\text{-Fin} = \Delta\mathfrak{Z}\text{-Fin} = \Delta\,\text{E}\mathfrak{A}\text{-Fin}. \tag{9}$$

(it is enough to show this in the case $\Delta = \lhd$; and this case was established by showing that if $L \in \lhd\, \mathfrak{N}_2$-Fin, then $\sigma(L) \in \mathfrak{F}$.)

THEOREM 3.4 *Let Δ be one of the relations* $\lhd^{\alpha}$, si, asc. *Then*

$$\acute{\text{E}}(\Delta)(\text{E}\mathfrak{A} \cup \mathfrak{Z}) \cap \Delta\mathfrak{N}_2\text{-Fin} \leqq \mathfrak{F} \cap \text{E}\mathfrak{A}.$$

Proof: Let $L \in \mathfrak{X} = \acute{\text{E}}(\Delta)(\text{E}\mathfrak{A} \cup \mathfrak{Z}) \cap \Delta\, \mathfrak{N}_2$-Fin. Then $L \in \Delta\, \text{E}\mathfrak{A}$-Fin, so $\sigma(L) \in \mathfrak{F} \cap \text{E}\mathfrak{A}$, whence $L/\sigma(L) \in \mathfrak{X}$. So we may without loss of generality put $\sigma(L) = 0$ and prove that $L = 0$. Now if $H \in \mathfrak{F}$ and H asc L then $H^{\omega} = H^{c}$ for some c and $H^{\omega} \lhd L$, whence $H^{c} = 0$ (for $H \in \mathfrak{F}$ implies that $H \in \acute{\text{E}}\mathfrak{A} \cap \mathfrak{F} \leqq \mathfrak{F} \cap \text{E}\mathfrak{A}$), as $\sigma(L) = 0$.

Suppose that $L \neq 0$. We consider the case $\Delta = \lhd^{\alpha}$ $(\alpha > 0)$ as the others follow similarly. Now L has an ascending series (for some ordinal $\lambda > 0$)

$$0 = L_0 \lhd L_1 \lhd \dots L_{\lambda} = L$$

such that for each $\beta < \lambda$, $L_{\beta}\, \Delta\, L$ and $L_{\beta+1}/L_{\beta} \in \text{E}\mathfrak{A} \cup \mathfrak{Z}$. In particular $L_1 \in \text{E}\mathfrak{A} \cup \mathfrak{Z}$ and $L_1\, \Delta\, L$ (even if $\lambda = 1$) and so $L_1 \in \mathfrak{F} \cap \text{E}\mathfrak{A}$ (by (9)), whence $L_1 \in \mathfrak{F} \cap \mathfrak{N}$ from above. Let $N = L_1 \lhd^{\alpha} L$, so we have an ascending series (possibly finite)

$$N = N_0 \lhd N_1 \lhd \dots N_{\alpha} = L.$$

For each β define $P_{\beta} = \rho_{\mathfrak{F}}(N_{\beta})$. From above we have that if H asc L and $H \in \mathfrak{F}$ then $H \in \mathfrak{N}$ (as $\sigma(L) = 0$) so P_{β} is a sum of $\mathfrak{F} \cap \mathfrak{N}$-ideals and so

$P_\beta = \zeta_\omega(P_\beta)$ (by lemma 7.1.6). Now P_β ch $N_\beta \lhd N_{\beta+1} \lhd \dots$ and so $P_\beta \lhd N_{\beta+1}$ (in case $\beta+1 \leqq \alpha$) and $P_\beta \lhd^\mu N_\mu$ for all $\mu \geqq \beta$. By (9) we have $L \in \lhd^\alpha \mathfrak{Z}$-Fin, so $P_\beta \in \mathfrak{F} \cap \mathfrak{N}$. Therefore $P_\beta \in \mathfrak{F} \cap \mathfrak{N}$ for all $\beta \leqq \alpha$,

$$P_\beta \leqq P_{\beta+1} \quad (\text{if } \beta+1 \leqq \alpha) \tag{*}$$

and

$$P_\beta \lhd N_{\beta+1},\ P_\beta \lhd^\mu N_\mu \quad (\text{if } \mu \geqq \beta). \tag{**}$$

We claim that:

$$\text{if } \beta \leqq \gamma \leqq \mu \leqq \alpha, \text{ then } P_\beta \leqq P_\gamma. \tag{!}$$

Assume that (!) is true. Then $N \neq 0$ and $N = P_0 \leqq P_\alpha = 0$, (since $\rho_{\mathfrak{F}}(L) \leqq \sigma(L) = 0$), a contradiction. Hence $L = 0$, so that $L = \sigma(L) \in \mathfrak{F} \cap \mathrm{E}\mathfrak{A}$ and the theorem is true.

We prove (!) by induction on μ. For $\mu = 0$ the result is trivial. If it holds for μ, then by (*) it is true for $\mu+1$ as well. So let μ be a limit ordinal and assume that (!) holds for all $\gamma < \mu$. So if $\beta \leqq \gamma < \mu$ then $P_\beta \leqq P_\gamma$. Define $H = \bigcup_{\beta<\mu} P_\beta$. Then $H \lhd N_\mu = \bigcup_{\beta<\mu} N_\mu$. By (**) we have $P_\beta \lhd^\mu N_\mu$ for all $\beta < \mu$. Furthermore $P_\beta \in \mathfrak{F} \cap \mathfrak{N}$ for all β. In particular $K = P_\mu \leqq N_\gamma$ for some $\gamma < \mu$, whence $P_\mu \leqq P_\gamma$. We note that $N_\mu \in \lhd^\mu \mathfrak{N}_2$-Fin and so $\sigma(N_\mu) \in \mathfrak{F} \cap \mathrm{E}\mathfrak{A}$, whence $\sigma(N_\mu) \leqq P_\mu = K$. Suppose that $H \neq K$. Then employing the argument of the proof of case 2 of theorem 1.9 we can show that for some $\beta \geqq \gamma$, $0 \neq \zeta_1(P_\beta/K) \leqq M/K = \zeta_1(H/K)$. But then $M \lhd N_\mu$ and $M \in \mathrm{E}\mathfrak{A}$ so $M \leqq \sigma(N_\mu) \leqq K$, a contradiction. Hence $H = K$ and so (!) holds for μ as well. Thus (!) is true for all $\mu \leqq \alpha$. □

We remark that if Δ is one of the relations $\lhd^\alpha$ (α an infinite ordinal), si, asc then

$$\Delta\mathfrak{N}_2\text{-Fin} = \text{Min-}\Delta\mathfrak{A} = \Delta\mathfrak{A}\text{-Fin} \geqq \text{Max-}\Delta\mathfrak{A}.$$

We also have for each positive integer n,

$$\text{Max-}\lhd^{n+1}\mathfrak{A} \cup \text{Min-}\lhd^{n+1}\mathfrak{A} \leqq \lhd^n\mathfrak{N}_2\text{-Fin}.$$

Since a chief factor of a $\mathrm{LE}\mathfrak{A}$-algebra is abelian then it follows that $\mathrm{LE}\mathfrak{A} \cap \text{Min-}\lhd \leqq \acute{\mathrm{E}}(\lhd)\mathfrak{A}$. So we have

Corollary 3.5 $\mathrm{LE}\mathfrak{A} \cap \text{Min-}\lhd \cap \lhd\, \mathfrak{N}_2\text{-Fin} \leqq \mathfrak{F} \cap \mathrm{E}\mathfrak{A}$. □

Suppose that $L \in \mathrm{LE}\mathfrak{A} \cap \text{Min-s}\mathfrak{A}$. Let H si L with $H \in \mathrm{E}\mathfrak{A}$. Then we have $H \lhd^n L$ for some n. Put $R = \sigma(L) \in \mathfrak{F} \cap \mathrm{E}\mathfrak{A}$. We have a series

$$H = H_n \lhd H_{n-1} \lhd \dots \lhd H_0 = L.$$

Let $P_i = \rho_{\mathfrak{F}}(H_i)$. Since each $H_i \in \text{Min-s}\mathfrak{A}$, we have $\sigma(H_i) \in \mathfrak{F}$. But each $H_i \in \text{LE}\mathfrak{A}$, so $P_i \leqq \sigma(H_i) \leqq P_i$. Therefore, since $P_i \text{ ch } H_i$, we have

$$H = P_n \lhd P_{n-1} \lhd \ldots \lhd P_1 \lhd P_0 = \sigma(L).$$

So $\sigma(L)$ contains every $\text{sE}\mathfrak{A}$-subalgebra of L.

Corollary 3.6

(a) $\text{J}_s\text{E}\mathfrak{A} \cap \text{Min-s}\mathfrak{A} \leqq \mathfrak{F} \cap \text{E}\mathfrak{A}$.
(b) $\mathfrak{B} \cap \text{Min-s}\mathfrak{A} \leqq \mathfrak{F} \cap \mathfrak{N}$.
(c) $\acute{\text{E}}\mathfrak{A} \cap \text{Min-a}\mathfrak{A} \leqq \mathfrak{F} \cap \text{E}\mathfrak{A}$.

Proof: (a) If $L \in \text{J}_s\text{E}\mathfrak{A}$ then L is generated by soluble subideals and so by the Derived Join Theorem $L \in \text{LE}\mathfrak{A}$; from above each soluble subideal of L lies in $\sigma(L) \in \mathfrak{F} \cap \text{E}\mathfrak{A}$.

(b) Clearly $\mathfrak{B} \leqq \text{J}_s\text{E}\mathfrak{A}$ and by theorem 7.1.5 $\mathfrak{B} \leqq \text{L}\mathfrak{N}$. The result now follows from (a).

(c) Let $L \in \acute{\text{E}}\mathfrak{A} \cap \text{Min-a}\mathfrak{A}$ and let α be an ordinal of cardinality $> \dim L$. Clearly if $H \text{ asc } L$ then $H \lhd^{\alpha} L$. Further we have $\text{Min-a}\mathfrak{A} = \text{a}\mathfrak{N}_2\text{-Fin}$. So $L \in \acute{\text{E}}(\lhd^{\alpha})\mathfrak{A} \cap \lhd^{\alpha}\mathfrak{N}_2\text{-Fin}$ and by theorem 3.4 $L \in \mathfrak{F} \cap \text{E}\mathfrak{A}$. □

Chapter 10

Existence theorems for abelian subalgebras

A famous problem in group theory, credited to Schmidt [339], asks whether every infinite group possesses an infinite abelian subgroup. It is now known that the answer is negative: Novikov and Adjan [317] proved that the Burnside groups of odd exponent $\geqq 4381$ are infinite, but that every abelian subgroup is finite cyclic. (This exponent has since been improved to 697; see Adjan [255]). On the other hand the answer is affirmative for locally finite groups (Hall and Kulatilaka [289], Kargapolov [299]). Strunkov [345] has extended these results to 'binary finite' groups, all of whose 2-generator subgroups are finite. And Kulatilaka [301] has shown that for certain generalised soluble classes one may place further restrictions on the nature of the desired abelian subgroup, such as requiring that it be subnormal.

We can raise the analogous question for Lie algebras: *under what circumstances does an infinite-dimensional Lie algebra possess an infinite-dimensional abelian subalgebra?* Unlike the group-theoretic case, it is easy to see that some extra hypotheses are required. For if L is a free Lie algebra on 2 or more generators then L is infinite-dimensional; on the other hand every subalgebra of L is free (Witt [249]) so every abelian subalgebra has dimension $\leqq 1$.

For generalised soluble classes we need only recast some of the results of the previous chapter. This is done in the first section. In the second we prove an analogue of the Hall-Kulatilaka-Kargapolov theorem, following [211]. In the final section we give examples, other than free algebras, to show that infinite-dimensional abelian subalgebras need not always exist. These examples allow us to avoid quoting the theorem about subalgebras of free Lie algebras: they have several other properties which make them interesting in their own right.

1. Generalised soluble classes

The statements in the next theorem all follow without difficulty from theorems 9.1.3 and 9.1.9. Alternatively one may consult [211]. Collectively they give Lie algebra analogues of the results of Kulatilaka [301].

THEOREM 1.1 (a) *Every infinite-dimensional hypercentral Lie algebra has an infinite-dimensional abelian ideal.*

(b) *Every infinite-dimensional* $\acute{E}(\lhd)\mathfrak{A}$*-algebra (and in particular every infinite-dimensional Fitting algebra) has an infinite-dimensional abelian 2-step subideal.*

(c) *Every infinite-dimensional* $\acute{E}(\mathrm{si})\mathfrak{A}$*-algebra has an infinite-dimensional abelian subideal.*

(d) *Every infinite-dimensional Baer algebra has an infinite-dimensional abelian subideal.*

(e) *Every infinite-dimensional* $\acute{E}\mathfrak{A}$*-algebra (and in particular every infinite-dimensional Gruenberg algebra over a field of characteristic 0) has an infinite-dimensional ascendant abelian subalgebra.* □

The existence of a locally nilpotent algebra with trivial Gruenberg radical (theorem 6.5.5) shows that in (e) we cannot replace 'Gruenberg' by 'locally nilpotent'. However, we will show that every infinite-dimensional locally nilpotent algebra has an infinite-dimensional abelian subalgebra. We need a slight generalisation of this for the next section, and it is convenient to introduce two new classes of algebras.

$\mathfrak{Q}$: $L \in \mathfrak{Q}$ if $L \in \mathfrak{F}$ or if L has an infinite-dimensional abelian subalgebra.
$\mathfrak{R}$: $L \in \mathfrak{R}$ if $L \in \mathfrak{F}$ or if there exists $x \in L$, $x \neq 0$, with $C_L(x) \notin \mathfrak{F}$.

LEMMA 1.2 *Let* $\mathfrak{X} = \mathrm{Q}\mathrm{s}\mathfrak{X}$ *be a class of Lie algebras. Then* $\mathfrak{X} \leqq \mathfrak{Q}$ *if and only if* $\mathfrak{X} \leqq \mathfrak{R}$.

Proof: Obviously $\mathfrak{Q} \leqq \mathfrak{R}$ so the implication one way is clear. For the opposite implication, consider the set $\mathscr{C}$ of all finite-dimensional abelian subalgebras A of L for which $C_L(A) \notin \mathfrak{F}$. Since $0 \in \mathscr{C}$ it follows that $\mathscr{C}$ is not empty.

Suppose $A \in \mathscr{C}$. Then $A \lhd C = C_L(A)$ and $C/A \notin \mathfrak{F}$. But $C/A \in \mathrm{Q}\mathrm{s}\mathfrak{X} = \mathfrak{X} \leqq \mathfrak{R}$, so there exists $x \in C \backslash A$ such that

$$D/A = C_{C/A}(A+x) \notin \mathfrak{F}.$$

Now $[D, x] \leqq A$. Let $A_1 = A + \langle x \rangle$, which is finite-dimensional abelian, and an ideal of C. Then

$$C_1 = C_L(A_1) = C_D(x).$$

Consider $x^*|_D : D \to [D, x]$. The kernel of this is C_1, and its image is contained in A so is finite-dimensional. Thus $\dim D/C_1$ is finite, so that $C_1 \notin \mathfrak{F}$. Thus $A_1 \in \mathscr{C}$.

But $A < A_1$. So $\mathscr{C}$ has no maximal element. We can therefore find a strictly increasing chain

$$A_1 < A_2 < \ldots$$

of elements of $\mathscr{C}$. The union of this chain is infinite-dimensional and abelian; so $L \in \mathfrak{Q}$ as claimed. □

THEOREM 1.3 *Every infinite-dimensional locally nilpotent Lie algebra has an infinite-dimensional abelian subalgebra.*

Proof: We have to show that $\mathrm{L}\mathfrak{N} \leqq \mathfrak{Q}$. It is sufficient to show that $\mathrm{L}\mathfrak{N} \leqq \mathfrak{N}$, by lemma 1.2. Suppose for a contradiction that $L \in \mathrm{L}\mathfrak{N} \backslash \mathfrak{N}$.

First we show that if F is an $\mathfrak{F}$-subalgebra of L then there exists another $\mathfrak{F}$-subalgebra F^* of L such that

$$C_L(F^*) < C_L(F).$$

Since $L \notin \mathfrak{N}$, we have $C_L(F) \in \mathfrak{F}$. Pick $0 \neq c \in C_L(F)$. Then $C_L(c) \in \mathfrak{F}$, so we can find $x \in L \backslash C_L(c)$. Let $F^* = \langle F, x \rangle$ which is finite-dimensional since $\mathrm{L}\mathfrak{N} \leqq \mathrm{L}\mathfrak{F}$. Then $C_L(F^*) = C_L(F) \cap C_L(x) \subseteq C_L(F) \backslash \langle c \rangle$.

Hence we can construct some $\mathfrak{F}$-subalgebra K of L such that $C_L(K) = 0$. But $L \in \mathrm{L}\mathfrak{N}$ so $K \in \mathfrak{N}$, and $0 \neq \zeta_1(K) \leqq C_L(K)$. This is a contradiction and proves the theorem. □

THEOREM 1.4 *Every infinite-dimensional locally soluble Lie algebra over a field of characteristic 0 has an infinite-dimensional abelian subalgebra.*

Proof: Let L be infinite-dimensional locally soluble. If L is not locally finite then L has an infinite-dimensional soluble subalgebra H, which has an infinite-dimensional abelian subalgebra by 1.1(b). If L is locally finite then by Jacobson [98] p. 51 L is locally $\mathfrak{N}\mathfrak{A}$, and hence it follows easily that $L^2 \in \mathrm{L}\mathfrak{N}$ (compare lemma 13.3.10). If $L^2 \in \mathfrak{F}$ then again L is soluble and the theorem follows: but if $L^2 \notin \mathfrak{F}$ we can apply theorem 1.3. □

In this context it is worth noting a theorem of Zaĭcev [349]: *every infinite group of derived length s has a proper subgroup of derived length s.* Now let $\mathfrak{k}$ be a field of characteristic 0. The Hartley algebra (6.3.6) has derived length 3. But it is not hard to see that every proper subalgebra has derived length at most 2. Hence Zaĭcev's theorem has no Lie analogue in characteristic 0.

2. Locally finite algebras

We prove the following theorem of [211], which is the characteristic 0 analogue of the Hall-Kulatilaka-Kargapolov theorem:

THEOREM 2.1 *Every infinite-dimensional locally finite Lie algebra over a field of characteristic 0 has an infinite-dimensional abelian subalgebra.*

COROLLARY 2.2 *Over a field of characteristic 0, a locally finite Lie algebra satisfies* Min *if and only if it is finite-dimensional.* □

(The corollary is, perhaps, a Lie analogue of the theorem of Kegel and Wehrfritz [300] that a locally finite group with the minimal condition for subgroups is abelian-by-finite.)

Before proving the theorem we need a lemma about finite-dimensional Lie algebras, which is given as an exercise in Jacobson [98] p. 149 (ex. 3). The reader is referred to Jacobson for the traditional finite-dimensional terminology which we use.

LEMMA 2.3 *Let L, L^* be finite-dimensional semisimple algebras over a field of characteristic 0, such that $L \leqq L^*$. Let H be a Cartan subalgebra of L. Then there exists a Cartan subalgebra H^* of L^* with $H \leqq H^*$.*

Proof: L^* is an L-module in a natural fashion. The complete reducibility theorem (Jacobson [98] p. 79) says that L^* is a direct sum of irreducible L-modules. Each of these may be thought of as an H-module; and by Samelson [183] decomposes as a sum of 1-dimensional H-submodules. Hence

$$L^* = V_1 \oplus \ldots \oplus V_t$$

where each V_i is a 1-dimensional H-module. Thus if $v \in V_i$, $h \in H$, we have

$$[v, h] = \lambda_i(h)v$$

where λ_i is a linear map $H \to \mathfrak{k}$. Collect together those V_i for which λ_i is equal to a given μ, and let their sum be W_μ. Then

$$L^* = W_0 \oplus W_{\mu_1} \oplus \ldots \oplus W_{\mu_r}$$

and clearly W_μ is the weight space for H with weight μ. As in Jacobson [98] p. 64,

$$\begin{aligned}[W_\lambda, W_\mu] &\subseteq W_{\lambda+\mu} \text{ if } \lambda+\mu \text{ is a weight} \\ &= 0 \qquad \text{otherwise.}\end{aligned}$$

Thus W_0 is a subalgebra of L^*. Now H is abelian (Jacobson [98] p. 110) and $H \leqq W_0$. By definition of W_0, $H \leqq \zeta_1(W_0)$. Let H^*/H be a Cartan subalgebra for W_0/H. We claim that H^* is a Cartan subalgebra of L^*.

Certainly H^* is nilpotent. We show it is self-idealising. Suppose $x \in L^*$ idealises H^*. Then

$$x = x_0 + x_{\mu_1} + \ldots + x_{\mu_r}$$

where $x_\mu \in W_\mu$. Let $h \in H$. Then $[x, h] \in H^* \leqq W_0$. But

$$[x, h] = \mu_1(h)x_{\mu_1} + \ldots + \mu_r(h)x_{\mu_r}$$

This belongs to W_0 for all h if and only if $x_{\mu_1} = \ldots = x_{\mu_r} = 0$. Therefore $x \in W_0$. But $x+H$ idealises H^*/H, so $x \in H^*$. This proves that $H^* = I_{L^*}(H^*)$, so that H^* is a Cartan subalgebra of L. By construction $H \leqq H^*$. □

Proof of theorem 2.1

Let L be infinite-dimensional locally finite over the field $\mathfrak{k}$ of characteristic 0. We wish to construct an infinite-dimensional abelian subalgebra of L. Without loss of generality

$$L = \bigcup_{n=1}^{\infty} L_i$$

where $L_1 < L_2 < \ldots$ and each $L_i \in \mathfrak{F}$. Let R_i be the soluble radical of L_i. Since $R_i \lhd L_i$ it follows that $R = \sum_{i=1}^{\infty} R_i$ is locally soluble. Either it is infinite-dimensional, and we can apply theorem 1.4 to obtain an infinite-dimensional abelian subalgebra, or R is finite-dimensional. In the latter case dim R_i is bounded. By Jacobson [98] p. 91 there exist semisimple Levi factors S_i so that

$$L_i = R_i \oplus S_i,$$

and by [98] p. 93 we may choose these so that $S_i \leqq S_{i+1}$. It follows that *the dimensions of the* S_i *are unbounded.*

Now pick Cartan subalgebras C_i of S_i, and use lemma 2.3 to arrange that $C_i \leqq C_{i+1}$. Each C_i is abelian ([98] p. 110), so

$$C = \bigcup_{i=1}^{\infty} C_i$$

is abelian. Either dim C is infinite, and the theorem is proved, or dim C is finite, equal to some integer c. Then for each i we have

$$\dim C_i \leqq c.$$

Consider now the general situation of a semisimple Lie algebra S over $\mathfrak{k}$, of finite dimension s, having a Cartan subalgebra H of dimension h. Let $\mathfrak{k}^*$ be the algebraic closure of $\mathfrak{k}$, and let

$$S^* = \mathfrak{k}^* \otimes_{\mathfrak{k}} S$$

$$H^* = \mathfrak{k}^* \otimes_{\mathfrak{k}} H.$$

Now S^* is a semisimple $\mathfrak{k}^*$-algebra ([98] p. 70) and H^* is a Cartan subalgebra of S^*. Further,

$$\dim_{\mathfrak{k}^*}(S^*) = s$$

$$\dim_{\mathfrak{k}^*}(H^*) = h.$$

Now S^* is a direct sum

$$S^* = J_1 \oplus \ldots \oplus J_m$$

of simple algebras over $\mathfrak{k}^*$ ([98] p. 71). If we let H_i be a Cartan subalgebra of J_i then

$$H_1 \oplus \ldots \oplus H_m$$

is a Cartan subalgebra of S^*. The conjugacy of Cartan subalgebras ([98] p. 273) implies that

$$H^* = H_1 \oplus \ldots \oplus H_m$$

so that

$$h = h_1 + \ldots + h_m$$

where $h_i = \dim_{\mathfrak{k}^*}(H_i)$. Each $h_i > 0$, so we have $m \leqq h$. An inspection of the standard list of classical simple Lie algebras ([98] p. 272) shows that if $\dim_{\mathfrak{k}^*} J_i = j_i$ then

$$j_i \leqq 2h_i^2 + h_i + 112$$

(the 112 deals with the five exceptional algebras). Then we can find a bound for s: for instance,

$$s = j_1 + \ldots + j_m$$
$$\leqq 2(h_1^2 + \ldots + h_m^2) + (h_1 + \ldots + h_m) + 112h$$
$$\leqq 2h^2 + 113h.$$

But now, in our original situation, we must have

$$\dim_{\mathfrak{k}} S_i \leqq 2c^2 + 113c$$

contrary to the unboundedness of $\dim_{\mathfrak{k}} S_i$. □

This theorem is applied in [211] to prove the existence of an infinite-dimensional commutative subalgebra in any locally finite associative infinite-dimensional $\mathfrak{k}$-algebra when $\mathrm{char}(\mathfrak{k}) = 0$. It has since been shown by Laffey that the restriction on the characteristic and the requirement of local finiteness are superfluous in the associative case. The proofs are purely ring-theoretic.

3. Generalisations of Witt algebras

Let $\mathfrak{k}$ be any field, and let G be a subgroup of the additive group $\mathfrak{k}^+$. Define a Lie algebra $\mathscr{W}_G$ having a basis $\{w_g : g \in G\}$ in bijective correspondence with G, subject to the Lie multiplication

$$[w_g, w_h] = (g-h)w_{g+h}. \qquad (1)$$

The Jacobi identity is easily verified. These algebras are natural generalisations of the *Witt algebra* (cf. Jacobson [98] p. 196) which itself arises from certain infinite-dimensional algebras studied by Cartan [35] vol. 2 part 2 p. 567.

The algebras $\mathscr{W}_G$ are almost always simple:

THEOREM 3.1 *If $\mathfrak{k}$ has characteristic $\neq 2$ and $|G| > 1$ then $\mathscr{W}_G$ is a simple non-abelian Lie algebra. If $\mathfrak{k}$ has characteristic 2 and $|G| > 2$ then $\mathscr{W}_G^2$ is a simple non-abelian Lie algebra.*

Proof: Assume $\mathfrak{k}$ has characteristic $\neq 2$, and $|G| > 1$. Then any non-zero ideal I of $\mathscr{W}_G$ contains an element

$$0 \neq i = \textstyle\sum_{g \in J} \lambda_g w_g$$

where $J \in G$ and $\lambda_g \in \mathfrak{k}$. We may assume $\lambda_g \neq 0$ for $g \in J$, and choose i in such a way that $|J|$ is as small as possible. For any $h \in J$ the ideal I contains

$$[i, w_h] = \sum_{g \in J} \lambda_g [w_g, w_h] = \sum_{g \in J'} \lambda_g (g-h) w_{g+h}$$

where $J' = J \backslash \{h\}$. By minimality $|J| = 1$, and I contains some w_k $(k \in G)$. Then for any $g \in G$, I contains

$$(g-k)^{-1}[w_g, w_k] = w_{g+k}$$

provided $g \neq k$. Hence I contains all w_f where $2k \neq f \in G$. Since $\mathrm{char}(\mathfrak{k}) \neq 2$, it follows that $3k \neq k$ and $-k \neq k$, and neither $3k$ nor $-k$ is equal to $2k$, so that I contains

$$(4k)^{-1}[w_{3k}, w_{-k}] = w_{2k}.$$

(If our chosen $k = 0$ we choose another one from the w_f and work on this instead.) Therefore $I = \mathscr{W}_G$ and $\mathscr{W}_G$ is simple. It is manifestly non-abelian.

If $\mathfrak{k}$ has characteristic 2 and $|G| > 2$ similar arguments show that

$$\mathscr{W}_G^2 = \sum_{g \neq 0} \langle w_g \rangle$$

is simple. □

For finite G (which can occur only if $\mathfrak{k}$ has prime characteristic) the algebras $\mathscr{W}_G$ are all known types of finite-dimensional simple algebras (Seligman [188]).

Now we locate the abelian subalgebras of $\mathscr{W}_G$ when $\mathfrak{k}$ has characteristic 0. It is convenient to find all the finite-dimensional subalgebras, and then to pin down the abelian ones by using the local finiteness of abelian algebras. From now on $\mathfrak{k}$ *has characteristic 0 and G is a fixed subgroup of* $\mathfrak{k}^+$ *with* $|G| > 1$ (*and therefore infinite*).

Any torsion-free abelian group can be linearly ordered (Neumann [314]) so we may equip G with a linear ordering $<$. For any element

$$0 \neq x = \sum x_g w_g$$

of $\mathscr{W}_G$ (where $x_g \in \mathfrak{k}$ and the summation is over all $g \in G$) we define

$$\max(x) = \max\{g: x_g \neq 0\}$$
$$\min(x) = \min\{g: x_g \neq 0\}$$

and for any subset $S \neq \{0\}$ of $\mathscr{W}_G$ we define

$$\max(S) = \{\max(s): s \in S\}$$
$$\min(S) = \{\min(s): s \in S\}.$$

We say that a subset U of G is *closed* if $u, v \in U$ and $u \neq v$ implies $u+v \in U$.

LEMMA 3.2 *If V is a Lie subalgebra of $\mathscr{W}_G$ then* $\max(V)$ *and* $\min(V)$ *are closed subsets of G.*

Proof: Let $g \neq h$ be elements of $\max(V)$. Then there exist $s, t \in V$ such that $\max(s) = g$, $\max(t) = h$. From (1) it is clear that

$$\max([s, t]) = g+h$$

and $\max(V)$ is closed. Dually we can deal with $\min(V)$. □

LEMMA 3.3 *Any non-empty finite closed subset of G takes one of the following forms:*

(a) $\{g\}$ $(g \in G)$,
(b) $\{0, g\}$ $(g \in G)$,
(c) $\{-g, 0, g\}$ $(g \in G)$.

Proof: If S is closed and contains two distinct positive elements g, h we may assume $g > h$; and then S contains $h+g, h+2g, \ldots$ and S is infinite. Dually S contains at most one negative element. If S does not contain both positive and negative elements it is of form (a) or (b). Otherwise either $S = \{h, g\}$ or $S = \{h, 0, g\}$ where $h < 0 < g$. In either case $h+g \in S$, and is not equal to h or g: therefore $h+g = 0$ and S is of type (c). □

THEOREM 3.4 *If $\mathfrak{k}$ has characteristic 0 and $0 \neq G \leqq \mathfrak{k}^+$, then the finite-dimensional subalgebras of $\mathscr{W}_G$ are precisely*

(a) *The 1-dimensional subalgebras $\langle x \rangle$ for $x \in \mathscr{W}_G$,*
(b) *The subalgebras of 3-dimensional algebras spanned by $\{w_{-g}, w_0, w_g\}$ for $0 \neq g \in G$.*

Proof: Let S be a finite-dimensional subalgebra of $\mathscr{W}_G$. Then S lies inside a subspace of $\mathscr{W}_G$ spanned by a finite set of elements w_g, so $\max(S)$ and $\min(S)$ are finite sets. By lemma 3.2 they are closed, so must be of types (a), (b), or (c) as in lemma 3.3. Suppose $\max(S)$ contains $g > 0$. Let $y \in S$ be such that $\max(y) = g$, and take any $x \in S$. No basis element w_h with $0 < h \neq g$ can occur in x with non-zero coefficient, for if so then $\max(x) = g$ also, and some linear combination z of x and y will have $\max(z) \geqq h, \neq g$, which is a contradiction. Similarly we may deal with negative components. Hence there exist elements $h < 0 < g$ of G such that S contains only elements of the form

$$\alpha w_h + \beta w_0 + \gamma w_g \tag{2}$$

for $\alpha, \beta, \gamma \in \mathfrak{k}$.

Suppose that dim $S > 1$, and assume for a contradiction that $h+g \neq 0$. Then S contains two linearly independent elements of the form (2), and hence contains non-zero elements of the form $\alpha w_h + \beta w_0$ and $\gamma w_0 + \delta w_g$ ($\alpha, \beta, \gamma, \delta \in \mathfrak{k}$). It contains the product of these elements, which has a term

$$\alpha\delta(h-g)w_{g+h}.$$

Since $g+h \neq 0, g, h$ we must have $\alpha\delta = 0$. Without loss of generality $\alpha = 0$ and S contains w_0. Then S contains w_g unless $\delta = 0$, and hence w_h and w_{g+h}, a contradiction. Therefore $\delta = 0$ and S is spanned by w_0 and some element $\lambda w_h + \mu w_g$ ($\lambda, \mu \in \mathfrak{k}$). But then

$$[\lambda w_h + \mu w_g, w_0] = \lambda h w_h + \mu g w_g$$

which cannot be in S since $g \neq h$: another contradiction.

Therefore $g+h = 0$ and S is contained in the subspace spanned by $\{w_{-g}, w_0, w_g\}$ for some $g \neq 0$. But

$$\begin{aligned} [w_{-g}, w_0] &= -g w_{-g} \\ [w_g, w_0] &= g w_g \\ [w_{-g}, w_g] &= -2g w_0 \end{aligned} \tag{3}$$

so this subspace is a subalgebra. □

COROLLARY 3.5 *With the above hypotheses,*

(a) *Every abelian subalgebra of* $\mathscr{W}_G$ *is of dimension* $\leqq 1$.

(b) *Every nilpotent subalgebra of* $\mathscr{W}_G$ *is of dimension* $\leqq 1$.

(c) *Every soluble subalgebra of* $\mathscr{W}_G$ *is of dimension* $\leqq 2$.

(d) *Every finite-dimensional subalgebra of* $\mathscr{W}_G$ *is of dimension* $\leqq 3$.

(e) *The* 3-*dimensional subalgebras of* $\mathscr{W}_G$ *are all isomorphic to the split simple algebra*

$$\langle e, f, h \colon [e, h] = 2e,\ [f, h] = -2f,\ [e, f] = h \rangle.$$

Proof: Everything follows from (e). This is a consequence of (3): define an isomorphism by

$$\begin{aligned} h &\leftrightarrow 2g^{-1}w_0 \\ e &\leftrightarrow g^{-1}w_g \\ f &\leftrightarrow g^{-1}w_{-g} \end{aligned}$$

□

Thus the $\mathscr{W}_G$, and their subalgebras, form a wide class of infinite-dimensional algebras without infinite-dimensional abelian subalgebras, and show that simple algebras can have this property. The algebra of chapter 8 section 6 is a subalgebra of $\mathscr{W}_{\mathbb{Z}}$, so residually nilpotent algebras satisfying Max-si may have the property.

To see just how wide a class this is we must solve the isomorphism problem for the $\mathscr{W}_G$: under what conditions do we have $\mathscr{W}_G \cong \mathscr{W}_H$ for two subgroups G, H of $\mathfrak{k}^+$?

LEMMA 3.6 *The intersection of the 3-dimensional subalgebras of $\mathscr{W}_G$ is $\langle w_0 \rangle$.* □

Let X_g ($g \in G$) be the subspace $\langle w_g \rangle$. Then lemma 3.6 characterises X_0 in an isomorphism-invariant fashion. We can also pick out the X_g (as a set rather than as individuals):

LEMMA 3.7 *$\mathscr{W}_G$, considered as an X_0-module, decomposes uniquely into a direct sum of 1-dimensional X_0-submodules, which are precisely the X_g ($g \in G$).*

Proof: Let M be a 1-dimensional X_0-submodule, and let

$$0 \neq m = \sum m_g w_g \in M$$

for $m_g \in \mathfrak{k}$, $g \in G$. Then for any nonzero $x \in X_0$ we must have

$$[m, x] = \lambda m \quad (\lambda \in \mathfrak{k}).$$

Take $x = w_0$: then we must have

$$\sum \lambda m_g w_g = \sum g m_g w_g$$

so that $m_g = 0$ except for one value of g. □

Let H be another subgroup of $\mathfrak{k}^+$ and let

$$\varphi : \mathscr{W}_G \to \mathscr{W}_H$$

be a Lie isomorphism. From lemmas 3.6 and 3.7 it follows that we must have

$$\varphi(w_g) = \lambda_g w_{g'}$$

where $0 \neq \lambda_g \in \mathfrak{k}$, and $' : G \to H$ is a bijection. The condition for φ to be an isomorphism is then

$$(g'-h')\lambda_g \lambda_h w_{g'+h'} = (g-h)\lambda_{g+h} w_{(g+h)'}$$

so that

$$(g+h)' = g'+h' \tag{4}$$

$$(g'-h')\lambda_g \lambda_h = (g-h)\lambda_{g+h}. \tag{5}$$

By (4) the map $'$ is an isomorphism, so in particular $0' = 0$. Putting $h = 0$ in (5) we get

$$g'\lambda_g\lambda_0 = g\lambda_g$$

and since $\lambda_g \neq 0$,

$$g'g^{-1} = \lambda_0^{-1}$$

for all $g \in G$. If we let $\pi = \lambda_0^{-1}$ then

$$g' = \pi g \tag{6}$$

for all $g \in G$. From (5)

$$\pi(g-h)\lambda_g\lambda_h = (g-h)\lambda_{g+h}$$

so provided $g \neq h$ then

$$\lambda_g\lambda_h = \pi\lambda_{g+h}. \tag{7}$$

But then $\lambda_g\lambda_{-g} = \pi\lambda_0 = 1$ and $\lambda_{2g}\lambda_{-g} = \pi\lambda_g$, so that $\lambda_{2g} = \pi\lambda_g\lambda_g$ and (7) holds for all $g, h \in G$. If we put

$$\mu_g = \pi\lambda_g$$

then μ is a homomorphism $G \to \mathfrak{k}^*$, the multiplicative group of nonzero elements of $\mathfrak{k}$. Thus we have:

PROPOSITION 3.8 *Let $\mathfrak{k}$ be a field of characteristic 0, and let G and H be subgroups of $\mathfrak{k}^+$. Then a map $\varphi : \mathscr{W}_G \to \mathscr{W}_H$ is a Lie isomorphism if and only if there is a non-zero element $\pi \in \mathfrak{k}$ and a homomorphism $\mu : G \to \mathfrak{k}^*$ such that for all $g \in G$*

$$\varphi(w_g) = \pi^{-1}\mu_g w_{\pi g}.$$

In particular, $\mathscr{W}_G \cong \mathscr{W}_H$ if and only if $H = \pi G$ for some $\pi \in \mathfrak{k}^*$, in which case G and H are isomorphic as groups. □

Let us say that subgroups G, H of $\mathfrak{k}^+$ are *projectively equivalent* if $H = \pi G$ for some $\pi \in \mathfrak{k}^*$. Then we have the:

THEOREM 3.9 *For any field $\mathfrak{k}$ of characteristic 0 there are at least as many isomorphism classes of simple Lie algebras of dimension $\leqq |\mathfrak{k}|$ over $\mathfrak{k}$ as there are projective equivalence classes of subgroups of $\mathfrak{k}^+$ (and a fortiori as there are isomorphism classes of subgroups of $\mathfrak{k}^+$).* □

COROLLARY 3.10 *Let $\mathfrak{k}$ be a field of characteristic 0 with $|\mathfrak{k}| = c$. Then there exist at least 2^c isomorphism classes of simple Lie algebras of dimension $\leqq c$ over $\mathfrak{k}$.*

Proof: The additive group $\mathfrak{k}^+$ is a vector space over $\mathbb{Q}$. If $c > \aleph_0$ its dimension is c, and it has at least 2^c subspaces (one for each subset of a basis). Each projective equivalence class contains at most c subgroups, so there are at least 2^c equivalence classes. If $c = \aleph_0$ we use instead the fact (Kuroš [302] p. 210, Fuchs [275] p. 149) that $\mathbb{Q}^+$ has $2^{\aleph_0}$ non-isomorphic subgroups. □

From proposition 3.8 we can compute the automorphism group of $\mathscr{W}_G$ by letting $G = H$. It is a split extension of Hom$(G, \mathfrak{k}^*)$ by the subgroup of $\mathfrak{k}^*$ of those π such that $\pi G = G$, with the obvious action: in particular it is metabelian. So the $\mathscr{W}_G$ afford examples of large simple Lie algebras with soluble automorphism group.

We might also consider the derivation algebra of $\mathscr{W}_G$ when $\mathfrak{k}$ has characteristic 0. It can be shown that every derivation of $\mathscr{W}_G$ can be written in the form

$$\delta + \delta'$$

where δ is inner, and where

$$w_g \delta' = \lambda_g w_g$$

for a homomorphism $\lambda: G \to \mathfrak{k}^+$. In particular, Der$(\mathscr{W}_G)$/Inn$(\mathscr{W}_G)$ is abelian.

The simple structure of the generalised Witt algebras makes them a useful testing-ground for conjectures, especially about infinite-dimensional simple algebras.

If $\mathfrak{k}$ has prime characteristic most of the above results break down. Theorem 3.4 goes badly wrong, in that $\mathscr{W}_G$ is always locally finite for prime characteristic. Further, $\mathscr{W}_G$ (if infinite-dimensional) now has an infinite-dimensional abelian subalgebra, generated by all elements

$$\sum_{v \in V} v$$

for finite subgroups V of G.

Chapter 11

Finiteness conditions for soluble Lie algebras

We have seen (lemma 8.5.1) that a soluble Lie algebra satisfying the maximal condition for ideals is finitely generated. In the first section of this chapter we show that although the converse is not in general true certain interesting classes of finitely generated soluble Lie algebras do satisfy such a maximal condition: specifically the class $\mathfrak{G} \cap \mathfrak{A}(\mathfrak{F} \cap \mathrm{E}\mathfrak{A})$, which includes in particular $\mathfrak{G} \cap \mathfrak{A}\mathfrak{N}$. In fact we prove a little more and study the class $\mathfrak{G} \cap \mathfrak{A}\mathfrak{F}$. The methods used are module-theoretic, and follow those employed by Hall [283] in which he proves that every finitely generated group which is an extension of an abelian group by a polycyclic group satisfies the maximal condition for normal subgroups. To use these methods we need to recall several standard facts and prove some less standard ones concerning the universal enveloping algebra of a finite-dimensional Lie algebra.

Next we consider soluble algebras with the double chain condition (Max-$\triangleleft$ $\cap$ Min-$\triangleleft$). For fields of characteristic $p > 0$ these must be finite-dimensional, which corresponds to the situation for groups; but for characteristic zero other possibilities arise.

The third section follows more work of Hall [286] in which it is proved that every finitely generated abelian-by-nilpotent group is residually finite. Again there are two cases for Lie algebras, depending on the characteristic. In characteristic $p > 0$ we prove a little more, namely that $\mathfrak{G} \cap \mathfrak{A}\mathfrak{F}$-algebras are residually finite. In characteristic zero the best we can prove is that finitely generated metabelian algebras are residually finite. Similar results have been found independently by Bahturin [20].

Lastly we turn to the concept of stuntedness, introduced by Lennox and Roseblade [306]. They prove that if G is a finitely generated abelian-by-nilpotent group, then the nilpotency class of the subgroups of G is bounded by some integer. We prove similar results, with the usual bifurcation: in characteristic $p > 0$ we can deal with the class $\mathfrak{G} \cap \mathfrak{A}\mathfrak{F}$, in characteristic zero only with $\mathfrak{G} \cap \mathfrak{A}^2$. The reason for this bifurcation lies in the representation theory of finite-dimensional Lie algebras, which for our purposes behaves better in characteristic $p > 0$.

1. The maximal condition for ideals

Universal enveloping algebras

We sketch briefly some of the properties of the universal enveloping algebra $U = U(L)$ of a Lie algebra L, as given in Jacobson [98] pp. 151ff.

With respect to an ordered basis $\{u_j : j \in J\}$ of L, every element of U is a linear combination of *standard monomials*

$$u_{j_1} \ldots u_{j_r},$$

where $r \geqq 0$ and $j_1 \leqq \ldots \leqq j_r$. If L is defined over a field $\mathfrak{k}$ then the standard monomials form a basis for U over $\mathfrak{k}$. We may (and do) identify L with the subspace of U spanned by the monomials of weight 1, namely u_j $(j \in J)$. The inclusion map $L \to U$ is a Lie homomorphism $L \to [U]$, so that if $x, y \in L$ then

$$[x, y] = xy - yx. \tag{1}$$

If M is any L-module then M has a natural U-module structure defined by

$$a(u_{j_1} \ldots u_{j_r}) = (\ldots(au_{j_1})\ldots u_{j_r}), \tag{2}$$

for $a \in M$. Conversely if M is a U-module then M becomes an L-module by restriction; for if $x, y \in L$ then $[x, y] = xy - yx$ in U and so for $a \in M$,

$$a[x, y] = a(xy - yx) = (ax)y - (ay)x.$$

This correspondence between L-modules and U-modules preserves submodules (see Jacobson [98] p. 160).

For a given Lie algebra L, $U(L)$ has an (associative) ideal Z spanned by the standard monomials of weight $\geqq 1$. If L is defined over a field $\mathfrak{k}$ then

$$U(L) = \mathfrak{k} \oplus Z.$$

We recall that Max (Max-$\lhd$) is the class of Lie algebras satisfying the maximal condition for subalgebras (ideals). For a Lie algebra L and an L-module M we have defined $M \in$ Max-L to mean that M satisfies the maximal condition for L-submodules; similarly we define $M \in$ Max-U, where $U = U(L)$. From our remarks above we see that $M \in$ Max-U if and only if $M \in$ Max-L.

We define the class Max-u by

$L \in$ Max-u

if and only if $U(L)$ satisfies the maximal condition for right ideals (i.e. $U(L)$ is a right noetherian algebra).

LEMMA 1.1 *If R is a right noetherian ring and M is a finitely generated right R-module then M satisfies* Max-R (i.e. *M is a noetherian module*).

Proof: We can find $a_1, \ldots, a_n \in M$ such that $M = a_1R + \ldots + a_nR$. Let $R^n = R \times \ldots \times R$ (n times). Then R^n is also a right noetherian ring. The map θ: $R^n \to M$ defined by

$$(x_1, \ldots, x_n) \mapsto a_1x_1 + \ldots + a_nx_n$$

is an R-module epimorphism from R^n onto M, whence M is a noetherian R-module since R^n is a noetherian R-module. □

PROPOSITION 1.2 $\mathfrak{F} \leqq$ Max-u $\leqq$ Max.

Proof: The inequality $\mathfrak{F} \leqq$ Max-u is theorem 6 of Jacobson [98] p. 166–167. For the second inequality we first prove: Let L be a Lie algebra, $U = U(L)$ and H, K subalgebras of L.

$$\text{If } H < K \text{ then } HU < KU. \tag{3}$$

Let $\{h_i : i \in I_1\}$ be an ordered basis for H; extend by $\{k_i : i \in I_2\}$ to an ordered basis for K and then extend $\{h_i : i \in I_1\} \cup \{k_i : i \in I_2\}$ by $\{f_i : i \in I_3\}$ to an ordered basis for L in such a way that $h_i < k_j < f_m$ for any $i \in I_1$, $j \in I_2$ and $m \in I_3$. The right ideal HU is spanned by monomials of the form

$$x_{r,s} = h_{i_1} \ldots h_{i_r} u_{j_1} \ldots u_{j_s} \quad (r \geqq 1,\ s \geqq 0),$$

where u may be an h, k or f. We wish to express such monomials as linear combinations of standard monomials. Evidently we may assume that $h_{i_1} \ldots h_{i_r}$ and $u_{j_1} \ldots u_{j_s}$ are already in standard form.

We claim that $x_{r,s}$ is a linear combination of standard monomials of the form $h_i u_{n_1} \ldots u_{n_q}$ (where $q > 0$ and u may be an h, k or f). For this we use induction on s. If $s = 0$, the result is trivial. Let $s > 0$ and assume the result for $s-1$. If $h_{i_r} \leqq u_{j_1}$ then $x_{r,s}$ is already in standard form and of the form required. If $u_{j_1} < h_{i_r}$ then u must be an h (by our ordering above). So

$$x_{r,s} = (h_{i_1} \ldots h_{i_r} h_{j_1}) u_{j_2} \ldots u_{j_s}.$$

Now $h_{i_1} \ldots h_{i_r} h_{j_1}$ can be written as a linear combination of standard monomials each of weight at least 1 and involving only h's. Hence $x_{r,s}$ is a linear combination of monomials of the form $x_{r',s-1}$; but each of these (by induction) is a linear combination of standard monomials of the required form and so $x_{r,s}$ is expressible in the required form. This completes our induction on s and proves our claim. Hence HU is spanned (*qua* vector space) by the standard monomials

$$h_i u_{j_1} \ldots u_{j_s} \quad (s \geqq 0)$$

where each u_j is an h, k or f. Now the standard monomials form a basis for U and so are independent. In particular $k_j \notin HU$. But U has an identity so $k_j \in KU$. Since $HU \leqq KU$, we have $HU < KU$ and (3) is proved.

(In view of the basis for HU above it is not hard to show that if $H, K \leqq L$ and $H \neq K$ then $HU \neq KU$.)

Now suppose that $L \in$ Max-u and $L \notin$ Max. Then we can find an infinite strictly ascending chain of subalgebras,

$$H_1 < H_2 < \ldots,$$

whence we have a strictly ascending chain of right ideals of U,

$$H_1 U < H_2 U < \ldots,$$

a contradiction. Hence $L \in$ Max and so Max-$u \leqq$ Max. □

We have seen that $\mathrm{E}\mathfrak{A} \cap \mathrm{Max} \leqq \mathfrak{F} \cap \mathrm{E}\mathfrak{A}$. Thus we have the following analogue of a theorem of Hall [283] p. 421:

COROLLARY 1.3 $\mathfrak{F} \cap \mathrm{E}\mathfrak{A} = \text{Max-}u \cap \mathrm{E}\mathfrak{A} = \mathrm{Max} \cap \mathrm{E}\mathfrak{A}$. □

We know of no example of a Lie algebra satisfying the maximal condition for subalgebras which is not finite-dimensional. We are unable to sharpen the inequalities

$$\mathfrak{F} \leqq \text{Max-}u \leqq \text{Max}.$$

Let L be a Lie algebra, $U = U(L)$ and let Z be the unique maximal ideal of U spanned by the monomials of weight $\geqq 1$.

$$\text{If } A \leqq L \text{ and } A^2 = 0 \text{ then } L \cap ZA = 0. \qquad (4)$$

To prove (4) it is enough to show that ZA is spanned by certain standard monomials each of weight not less than two. Pick an ordered basis $\{a_i : i \in I_1\}$ for A and extend this by $\{b_i : i \in I_2\}$ to an ordered basis for L in such a way that $b_j < a_i$ for all $i \in I_1$ and $j \in I_2$. Evidently ZA is spanned by monomials of the form $u_{j_1}\ldots u_{j_r}a_i$, where $r \geqq 1$ and $u_{j_1}\ldots u_{j_r}$ is a standard monomial in the a's and b's. Let k be minimal with respect to $u_{j_k} \nleqq a_i$. Then $u_{j_m} > a_i$ for $m = k, \ldots, r$ and so each u_{j_m} is an a_{j_m} (by our ordering above) for $m = k, \ldots, r$. But for any $x, y \in A$ we have $0 = [x, y] = xy - yx$ so that $xy = yx$ in U. Hence

$$u_{j_1}\ldots u_{j_r}a_i = u_{j_1}\ldots u_{j_{k-1}}a_i a_{j_k}\ldots a_{j_m},$$

a standard monomial. Hence ZA is spanned by the standard monomials of the form $u_{i_1}\ldots u_{i_r}a_i$, where $r \geqq 1$ and each of these has weight at least two. This proves (4).

(*Note:* Similarly $L \cap AZ = 0$; if $A^2 \neq 0$ then $A \cap ZA \neq 0$.)

LEMMA 1.4 *Let L be a Lie algebra and let $U = U(L)$.*

(a) *If $I \lhd L$ then $UIU = IU = UI$.*
(b) *If $L = \langle x_1, \ldots, x_n \rangle \in \mathfrak{G}$ then $Z = x_1U + \ldots + x_nU$.*

Proof: (a) Evidently $UI, IU \subseteq UIU$. Since U has an identity then $I \subseteq IU$. If $a \in L$ and $b \in I$ then in U, $ab = [a, b] + ba \in IU$. Hence $LI \subseteq IU$ so $LIU \subseteq IU$. By induction it follows that for any $a_1\ldots a_r \in L$ we have $(a_1\ldots a_r)IU \subseteq IU$ and so $UIU \subseteq IU$, whence $UIU = IU$. Similarly we show that $UI = UIU$.

(b) Clearly $T = x_1U + \ldots + x_nU \subseteq Z$ and $Z = LU$. Now each element of L is a linear combination of Lie products of the form $[x_{j_1}, \ldots, x_{j_r}]$ where $\{j_1, \ldots, j_r\} \subseteq \{1, \ldots, n\}$. For $r = 1$, this product is in T. Suppose $r > 1$ and the result holds for $r-1$. Then $y = [x_{j_1}, \ldots, x_{j_{r-1}}] \in T$ and

$$[y, x_r] = yx_r - x_r y \in TU + x_rU \subseteq T.$$

Hence $L \subseteq T$ and so $Z = LU \subseteq TU \subseteq T$. So $Z = T$. □

(*Note:* if $\{x_j : j \in J\}$ is a set of generators for L then

$$Z = \textstyle\sum_{j \in J} x_jU = \sum_{j \in J} Ux_j.)$$

From Jacobson [98] p. 159–162, Theorem 6, corollary 2 we have:

$$\text{If } A \lhd L \text{ then } U(L/A) \cong U/UAU \tag{5}$$

and the isomorphism θ is given by

$$\theta: (x_1 + A)\ldots(x_r + A) \leftrightarrow x_1\ldots x_r + UAU, \tag{6}$$

for $x_1, \ldots, x_r \in L$.

LEMMA 1.5 *Let L be a Lie algebra and let $U = U(L)$.*

(a) *If $A \lhd L$ and $A^2 = 0$ then $A \cong (A+ZAU)/ZAU$ as U/AU-modules under the canonical map $a \leftrightarrow a+ZAU$ $(a \in A)$.*

(b) *If also $L \in \mathfrak{G}$ and $U(L/A)$ is right noetherian then A is a noetherian U/AU-module and hence a noetherian $U(L/A)$-module.*

Proof: (a) By lemma 1.4 we have $ZAU = ZUA = ZA$, since $ZU = Z$. Hence $A \cap ZAU = A \cap ZA \subseteq L \cap ZA = 0$, by (4). Thus the map $a \leftrightarrow a+ZAU$ has trivial kernel. Now A is an L-module under right Lie multiplication and so A is a U-module under the action defined by equation (2), and hence a U/AU-module since $A^2 = [A, A] = 0$. Furthermore Z is a U-module under juxtaposition, so $(Z+ZAU)/ZAU$ is a U/AU-module in the obvious way. Now $AAU \subseteq ZAU$ and $AU = UA = (\mathfrak{k}+Z)A = A+ZA$ and therefore $(A+ZAU)/ZAU$ is a U/AU-submodule of $(Z+ZAU)/ZAU$. Pick any $x_1, \ldots, x_r \in L$. For the U/AU-module action on A we have for $a \in A$,

$$a(x_1 \ldots x_r + AU) = [a, x_1, \ldots, x_r];$$

and for the U/AU-module action on $(A+ZAU)/ZAU$ we have

$$(a+ZAU)(x_1 \ldots x_r + AU) = ax_1 \ldots x_r + ZAU.$$

An induction on r shows that

$$[a, x_1, \ldots, x_r] + ZAU = ax_1 \ldots x_r + ZAU.$$

Hence the map $a \leftrightarrow a+ZAU$ of A onto $(A+ZAU)/ZAU$ is an isomorphism of U/AU-modules.

(b) By lemma 1.4 if $L \in \mathfrak{G}$ then Z is a finitely generated U-module and so $(Z+ZAU)/ZAU$ is a finitely generated U/AU-module. By (5) and lemma 1.4 we have $U(L/A) \cong U/UAU = U/AU$, so U/AU is also right noetherian as $U(L/A)$ is right noetherian. Hence by lemma 1.1 we have that $(Z+ZAU)/ZAU$ is a noetherian U/AU-module so $(A+ZAU)/ZAU$ is a noetherian U/AU-submodule, and therefore by (a), A is a noetherian U/AU-module. Now A is an L/A module under the action $a(x+A) = [a, x]$ and so A is a $U(L/A)$-module under the action defined by (2),

$$\begin{aligned} a((x_1+A)\ldots(x_r+A)) &= (\ldots a(x_1+A)\ldots(x_r+A)) = [a, x_1, \ldots, x_r] \\ &= a(x_1 \ldots x_r + AU). \end{aligned}$$

Thus the $U(L/A)$-module action on A is the same as the U/UAU-module action on A (for $UAU = AU$) and agrees with the isomorphism given by equation (6). Hence A is a noetherian $U(L/A)$-module so that $A \in$ Max-$U(L/A)$ and so $A \in$ Max-L/A, whence $A \in$ Max-L. □

THEOREM 1.6 $\mathfrak{G} \cap \mathfrak{A}\text{Max-}u \leqq \text{Max-}\lhd$.

Proof: Let $L \in \mathfrak{G} \cap \mathfrak{A}\text{Max-}u$. Then L has an abelian ideal A such that $L/A \in \text{Max-}u$. Hence $U(L/A)$ is right noetherian and so by lemma 1.5, $A \in \text{Max-}L$. But $L/A \in \text{Max-}u \leqq \text{Max} \leqq \text{Max-}\lhd$, by proposition 1.2. Hence $L \in \text{Max-}\lhd$. □

It is not hard to show that if $A = \langle a_1, \ldots, a_r \rangle^L \lhd L$ and $A^2 = 0$ and $L/A \in \mathfrak{G}$ then $L \in \mathfrak{G}$.

In chapter 9 we defined the class $\text{Max-}\lhd\,\mathfrak{A}$ of Lie algebras L with the maximal condition on abelian ideals. Evidently it is the class of Lie algebras L in which every abelian ideal is a finitely generated L-module. We also have

$$\text{Max-}\lhd \leqq \text{Max-}\lhd\,\mathfrak{A}.$$

By proposition 1.2 we have $\text{Max-}u \leqq \text{Max} \leqq \mathfrak{G}$.

THEOREM 1.7 *Let $\mathfrak{X}$ be a class of Lie algebras.*

(a) *If* $\mathfrak{X} \leqq \text{Max-}u$ *then*

$$\mathfrak{G} \cap \mathfrak{A}\mathfrak{X} = \text{Max-}\lhd \cap \mathfrak{A}\mathfrak{X} = \text{Max-}\lhd\,\mathfrak{A} \cap \mathfrak{A}\mathfrak{X}.$$

(b) $\mathfrak{G} \cap \mathfrak{A}\mathfrak{X} = \text{Max-}\lhd \cap \mathfrak{A}(\mathfrak{G} \cap \mathfrak{X}) = \text{Max-}\lhd\,\mathfrak{A} \cap \mathfrak{A}(\mathfrak{G} \cap \mathfrak{X})$ *if and only if* $\mathfrak{G} \cap \mathfrak{X} \leqq \text{Max-}u$.

(c) *If* $\mathfrak{G} \cap \mathfrak{X} \neq 0$ *then* $\mathfrak{G} \cap \mathfrak{A}\mathfrak{X} \nleqq \text{Max-}\lhd^2\mathfrak{A}$.

Proof: (a) By theorem 1.6 we have

$$\mathfrak{G} \cap \mathfrak{A}\mathfrak{X} \leqq \text{Max-}\lhd \cap \mathfrak{A}\mathfrak{X} \leqq \text{Max-}\lhd\,\mathfrak{A} \cap \mathfrak{A}\mathfrak{X}.$$

Suppose that $L \in \text{Max-}\lhd\,\mathfrak{A} \cap \mathfrak{A}\mathfrak{X}$. Then L has an abelian ideal A such that $L/A \in \mathfrak{X} \leqq \text{Max-}u \leqq \text{Max} \leqq \mathfrak{G}$. Furthermore A is a finitely generated L-module and so by our remarks above we have $L \in \mathfrak{G}$ and so $L \in \mathfrak{G} \cap \mathfrak{A}\mathfrak{X}$. This proves (a).

(b) We note that for any class $\mathfrak{X}$, $\mathfrak{G} \cap \mathfrak{A}\mathfrak{X} = \mathfrak{G} \cap \mathfrak{A}(\mathfrak{G} \cap \mathfrak{X})$. Hence if $\mathfrak{G} \cap \mathfrak{X} \leqq \text{Max-}u$ then the result follows from (a). Conversely suppose that $\mathfrak{G} \cap \mathfrak{A}\mathfrak{X} \leqq \text{Max-}\lhd\,\mathfrak{A}$. Let $L = \langle x_1, \ldots, x_n \rangle \in \mathfrak{G} \cap \mathfrak{X}$ be defined over a field $\mathfrak{k}$. Let $B = \mathfrak{k}b$ be a 1-dimensional space over $\mathfrak{k}$ and form $A = B \otimes_{\mathfrak{k}} U(L)$. Then A is spanned by elements of the form

$$b \otimes (u_1 \ldots u_m) \qquad (m > 0,\ u_i \in L).$$

Make A into an L-module by defining for $u \in L$,

$$(b \otimes u_1 \ldots u_m)u = b \otimes u_1 \ldots u_m u.$$

Clearly A is a cyclic L-module and $A \cong U(L)$ as L-modules under the map $u_1 \ldots u_m \leftrightarrow b \otimes u_1 \ldots u_m$. Now consider A as an abelian Lie algebra and form the split extension

$$K = A \dotplus L.$$

Then A is a cyclic L-module and so a cyclic K-module, whence $K \in \mathfrak{G} \cap \mathfrak{A}(\mathfrak{G} \cap \mathfrak{X})$. If $K \in \text{Max-}\lhd\, \mathfrak{A}$ then A is a noetherian K-module and so A is a noetherian L-module ($K/A \cong L$). Therefore $U(L)$ is a noetherian L-module and so $L \in \text{Max-}u$. So $\mathfrak{G} \cap \mathfrak{X} \leq \text{Max-}u$. Thus $\mathfrak{G} \cap \mathfrak{A}\mathfrak{X} \leq \text{Max-}\lhd\, \mathfrak{A} \Leftrightarrow \mathfrak{G} \cap \mathfrak{X} \leq \text{Max-}u$.

(c) Let $L \in \mathfrak{G} \cap \mathfrak{X}$ and form K as above. If $K \in \text{Max-}\lhd^2 \mathfrak{A}$ then $A \in \mathfrak{A} \cap \text{Max-}\lhd$ and so $A \in \mathfrak{F}$, whence $U(L)$ is a finite-dimensional space, which is impossible unless $L = 0$. □

The construction of A as a cyclic L-module in the proof of theorem 1.7(b) holds even if $L \notin \mathfrak{G}$.

From the construction we deduce the following statement: $L \in \text{Max-}u$ *if and only if every cyclic L-module is noetherian.*

Part (c) of theorem 1.7 shows that from $L \in \mathfrak{G} \cap \mathfrak{A}\text{Max-}u$, we cannot deduce that $L \in \lhd \mathfrak{A}\text{-Fin}$ or any effectively stronger conclusion than that given by theorem 1.6. This applies particularly to the case

$$L \in \mathfrak{G} \cap \mathfrak{A}^2 = \mathfrak{G} \cap \mathfrak{A}(\mathfrak{A} \cap \mathfrak{F}),$$

whence we deduce that $\mathfrak{G} \cap \mathfrak{A}^3 = \mathfrak{G} \cap \mathfrak{A}(\mathfrak{G} \cap \mathfrak{A}(\mathfrak{A} \cap \mathfrak{F})) \not\leq \text{Max-}\lhd$ (by theorem 1.7(b) since $\text{Max-}u \cap \text{E}\mathfrak{A} \leq \mathfrak{F}$).

The class Max-$\lhd$ is Q-closed and

$$\text{Max-}\lhd \cap \mathfrak{N} \leq \mathfrak{G} \cap \mathfrak{N} \leq \mathfrak{F} \leq \text{Max-}u.$$

From this and theorem 1.7 it follows readily that:

Corollary 1.8 *Let $L \in \mathfrak{A}\mathfrak{F} \cup \mathfrak{A}\mathfrak{N} \cup \mathfrak{A}^2$. Then L is finitely generated if and only if L satisfies the maximal condition for ideals.* □

Finally we give an explicit example to illustrate that $\mathfrak{G} \cap \mathfrak{A}^3 \not\leq \text{Max-}\lhd$. A Lie algebra L is said to be *centre-by-metabelian* if $[L^{(2)}, L] = 0$. In particular, L then belongs to $\mathfrak{A}^3$.

Proposition 1.9 *Over any field there exists a 2-generator centre-by-metabelian Lie algebra whose centre has infinite dimension and which does not satisfy the maximal condition for ideals.*

Proof: Let $\mathfrak{k}$ be a field and let M be a vector space over $\mathfrak{k}$ with basis $\{a_1, a_3, a_5, \ldots; b_0, b_1, b_2, \ldots\}$. Make M a Lie algebra by defining:

$$[a_i, a_j] = 0 = [a_i, b_k] \text{ for all } i, j, k:$$

$$[b_k, b_m] = \begin{cases} 0 & \text{if } k+m \text{ is even} \\ a_{k+m} & \text{if } k \text{ is odd and } m \text{ is even} \\ -a_{k+m} & \text{if } k \text{ is even and } m \text{ is odd.} \end{cases}$$

Extend the product by linearity to the whole of M. Let d be the linear transformation of M defined by

$$b_k d = b_{k+1}; \ a_i d = 0 \text{ for all } i, k.$$

From the equations above we see that

$$[b_{k+1}, b_m] + [b_k, b_{m+1}] = 0 \text{ for all } k, m.$$

Hence d is a derivation of M. Let $D = \mathfrak{k}d$ and form the split extension

$$L = M \dotplus D.$$

Clearly $L = \langle b_0, d \rangle \in \mathfrak{G}_2$. Furthermore $A = \langle a_1, a_3, \ldots \rangle \leqq \zeta_1(L)$ and $A \notin \mathfrak{F}$. We also have $L^2 \leqq M$ and so $L^{(2)} \leqq M^2 = A \leqq \zeta_1(L)$. Finally $a_{2n+1} = \pm [[b_0, {}_n d], [b_0, {}_{n+1} d]] \in L^{(2)}$ if $n > 0$. Since $\zeta_1(L) \notin \mathfrak{F}$ then $L \notin \text{Max-}\lhd$. □

P.M. Neumann has informed us that free centre-by-metabelian Lie algebras (i.e. Lie algebras of the form $F/[F^{(2)}, F]$ where F is a free Lie algebra) on at least two generators have infinite-dimensional centres. So they provide further examples for proposition 1.9.

2. The double chain condition

In this section we consider the class $\text{Max-}\lhd \cap \text{Min-}\lhd$ of Lie algebras with the double chain condition. Example 6.3.6 shows that over fields of characteristic zero, $\text{Max-}\lhd \cap \text{Min-}\lhd \cap \mathfrak{A}\mathfrak{N}_2 \nleqq \mathfrak{F}$. One of our results here shows that this cannot happen over a field of non-zero characteristic.

First we observe that $\text{Max-}\lhd \cap \text{Min-}\lhd$ *is the class of Lie algebras which have a finite chief series.* Trivially if $L \in \text{Max-}\lhd \cap \text{Min-}\lhd$ then L has a finite chief series. Conversely if L has a finite chief series

$$0 = L_0 \lhd L_1 \lhd \ldots \lhd L_n = L,$$

then each chief factor $L_{i+1}/L_i \in \text{Max-}L \cap \text{Min-}L$, whence (by theorem 1.7.4) each $L_i \in \text{Max-}L \cap \text{Min-}L$ and so $L \in \text{Max-}L \cap \text{Min-}L$. Hence

$$L \in \text{Max-}\lhd \cap \text{Min-}\lhd.$$

In particular if every irreducible submodule of L is finite-dimensional then every chief factor of L is finite-dimensional; hence $L \in \text{Max-}\lhd \cap \text{Min-}\lhd$ implies that $L \in \mathfrak{F}$.

For a Lie algebra L we have defined (L) to be the class Lie algebras of zero dimension or isomorphic to L.

THEOREM 2.1 *If* $L \in \text{Max-}\lhd \cap \text{Min-}\lhd$, *the following conditions are equivalent:*

(a) $\mathfrak{N}(L) \cap \text{Max-}\lhd \cap \text{Min-}\lhd \leqq \mathfrak{F}$.

(b) *Every irreducible L-module is finite-dimensional.*

Proof: (a) $\Rightarrow$ (b): Let A be an irreducible L-module. Consider A as an abelian Lie algebra and form the split extension $K = A \dotplus L$. Then $K/A \cong L \in \text{Max-}\lhd \cap$ $\cap \text{Min-}\lhd$. Furthermore $A \in \text{Min-}L \cap \text{Max-}L$ so $A \in \text{Min-}K/A \cap \text{Max-}K/A$, whence $K \in \mathfrak{A}(L) \cap \text{Max-}\lhd \cap \text{Min-}\lhd \leqq \mathfrak{F}$. So A is finite dimensional.

(b) $\Rightarrow$ (a): (even if $L \notin \text{Max-}\lhd \cap \text{Min-}\lhd$). Let $K \in \mathfrak{N}(L) \cap \text{Max-}\lhd \cap \text{Min-}\lhd$. Then K has a nilpotent ideal N such that $K/N \in (L)$. We also have $K/N \in \text{Max-}\lhd \cap \text{Min-}\lhd$. If $K/N = 0$ then $K \in \mathfrak{N} \cap \text{Max-}\lhd \leqq \mathfrak{G} \cap \mathfrak{N} \leqq \mathfrak{F}$. If $K/N \neq 0$ then $K/N \cong L$ and so every irreducible K/N-module is finite dimensional. Now if A/B is a chief factor of K then $[A, N] \leqq B$ (for otherwise $A = [A, N]+B = [A, {}_rN]+B$ for all r, a contradiction since $N \lhd K$ and $N \in \mathfrak{N}$) and so A/B is an irreducible K/N-module. Thus $A/B \in \mathfrak{F}$, whence $K \in \mathfrak{F}$. □

It is worth remarking that if L is a Lie algebra and every irreducible L-module is finite-dimensional then

(a) $(L) \cap \text{Min-}\lhd \leqq \acute{E}(\lhd)\mathfrak{F} \cap \mathfrak{Z}\mathfrak{F}$ and

(b) $(L) \cap \text{Max-}\lhd \leqq \grave{E}(\lhd)\mathfrak{F} \cap (\grave{E}(\lhd)(\mathfrak{A} \cap \mathfrak{F}))\mathfrak{F}$.

Next we note a result of Curtis [48] theorem 5.1, p. 952–953, which is proved as a by-product of corollary 3.6.

THEOREM 2.2 (*Curtis*) *Let L be a finite-dimensional Lie algebra over a field of non-zero characteristic. Then every irreducible L-module is finite dimensional.* □

COROLLARY 2.3 *Over any field of non-zero characteristic,*

$$\acute{E}(\lhd)(\mathfrak{A} \cup \mathfrak{F}) \cap \text{Max-}\lhd \cap \text{Min-}\lhd \leqq \mathfrak{F}.$$

Proof: Let $L \in \acute{E}(\lhd)(\mathfrak{A} \cup \mathfrak{F}) \cap \text{Max-}\lhd \cap \text{Min-}\lhd$. Then L has an ascending chain of ideals with abelian or finite-dimensional factors. By Max-$\lhd$ this chain must be of finite length:

$$0 = L_0 \lhd L_1 \lhd \dots \lhd L_n = L$$

and each $L_i \lhd L$. If $n = 1$ then $L \in (\mathfrak{F} \cup \mathfrak{A}) \cap \text{Max-}\lhd \leqq \mathfrak{F}$. Inductively assume that $L/L_1 \in \mathfrak{F}$ (for L/L_1 satisfies the hypothesis for L and has chain of length $n-1$). If $L_1 \in \mathfrak{F}$ then $L \in \mathfrak{F}$. If $L_1 \in \mathfrak{A}$ then $L \in \mathfrak{A}(L/L_1) \leqq \mathfrak{F}$, by theorem 2.1 since by theorem 2.2 every irreducible module of L/L_1 is finite-dimensional. □

LEMMA 2.4 *Let $L \in \mathfrak{F} \cap \mathfrak{A}$. Then every irreducible L-module is finite-dimensional. In particular*

$$\mathfrak{N}\mathfrak{A} \cap \text{Max-}\lhd \cap \text{Min-}\lhd \leqq \mathfrak{F}.$$

Proof: Let $L \in \mathfrak{F} \cap \mathfrak{A}$ over a field $\mathfrak{k}$ and let $U = U(L)$. Then U is a commutative noetherian algebra with unit and

$$U \cong \mathfrak{k}[t_1, \dots, t_n],$$

(if $\dim L = n$), where $\mathfrak{k}[t_1, \dots, t_n]$ is the polynomial ring in the indeterminates $t_1, \dots, t_n$. Suppose that A is an irreducible L-module. Then A is an irreducible U-module and for some $a \in A$ we have $A = aU$. Let P be the kernel of the map $u \to au$ of U onto A. Then P is a maximal ideal of U since $A \cong U/P$ as U-modules. By the Hilbert Nullstellensatz (Zariski and Samuel [351] p. 165) U/P is a finite algebraic extension of $\mathfrak{k}$ (for U/P is a field and is a finitely generated as a $\mathfrak{k}$-algebra). Hence $A \cong U/P$ is finite-dimensional.

If $K \in \mathfrak{N}\mathfrak{A} \cap \text{Max-}\lhd \cap \text{Min-}\lhd \leqq \mathfrak{N}(\mathfrak{F} \cap \mathfrak{A}) \cap \text{Max-}\lhd \cap \text{Min-}\lhd$ then K has a nilpotent ideal N such that $K/N \in \mathfrak{F} \cap \mathfrak{A}$. Thus $K \in \mathfrak{N}(K/N) \leqq \mathfrak{F}$, by theorem 2.1 since every irreducible K/N-module is finite-dimensional from the above. □

Our next result shows just how typical example 6.3.6 is of the situation in fields of characteristic zero.

THEOREM 2.5 *Let $L \in \mathfrak{N} \cap \mathfrak{F}$ over a field of characteristic zero. Then the following conditions are equivalent:*

(a) $L \in \mathfrak{A} \cap \mathfrak{F}$
(b) *Every irreducible L-module is finite-dimensional.*

Proof: That (a) implies (b) follows by lemma 2.4. Suppose that (b) holds but $L \notin \mathfrak{A}$. Let $N \lhd L$ maximal with respect to $L/N \notin \mathfrak{A}$. Put $K = L/N$. Then every proper quotient of K is abelian and $K \in \mathfrak{N}_2$. Further $\zeta_1(K) = K^2$ and has dimension one (for otherwise $K^2 = 0$). Let $Z = \langle z \rangle = \zeta_1(K)$. It is not hard to show that if $0 \neq x \in K$ then $C_K(x) \lhd K$ and $K/C_K(x)$ has dimension 1 if $x \notin Z$. Thus let $x_1 \in K$, $x_1 \notin Z$. Then there exists y_1 such that $[x_1, y_1] = z$. Put $K_1 = \langle x_1, y_1, z \rangle$. Then $K/(C_K(x_1) \cap C_K(y_1))$ is two-dimensional. If $K_1 < K$ pick $x_2 \in C_K(x_1) \cap C_K(y_1)$ with $x_2 \notin Z$. Since $C_K(x_2)$ contains K_1 and has codimension 1 in K we can find y_2 with $[x_2, y_2] = z$ and

$$y_2 \in C_K(x_1) \cap C_K(y_1)$$

(for $K = \langle x_1 \rangle + \langle y_1 \rangle + C_K(x_1) \cap C_K(y_1)$). Let $K_2 = \langle x_1, x_2, y_1, y_2, z \rangle$. Suppose that K_{n-1} has been defined with

$$K_{n-1} = \langle x_1, \ldots, x_{n-1}, y_1, \ldots, y_{n-1}, z \rangle$$

such that $K/C_K(K_{n-1})$ has dimension $2(n-1)$. If $K_{n-1} < K$ then proceeding as above we can find $x_n, y_n \in C_K(K_{n-1})$ with $[x_n, y_n] = z$. So we put $K_n = \langle K_{n-1}, x_n, y_n \rangle$ and find that $K/C_K(K_n)$ has dimension $2n$ and $K = \langle x_1, \ldots, x_n \rangle + \langle y_1, \ldots, y_n \rangle + C_K(K_n)$; and $[x_i, x_j] = 0 = [y_i, y_j]$ for all i, j; $[x_i, y_j] = \delta_{ij} z$ (where $\delta_{ij} = 0$ if $i \neq j$ and $\delta_{ii} = 1$) for all $i, j = 1, \ldots, n$.

Now $K \in \mathfrak{F}$ and so $K = K_n = \langle x_1, \ldots, x_n, y_1, \ldots, y_n, z \rangle$ for some n. Suppose that L is defined over the field $\mathfrak{k}$.

Let M be the polynomial ring $\mathfrak{k}[t_1, \ldots, t_n]$ and let $u_1, \ldots, u_n, v_1, \ldots, v_n$ be linear transformations of M defined as follows: for $f = f(t_1, \ldots, t_n) \in M$,

$$fu_i = t_i f$$

$$fv_i = \partial f / \partial t_i.$$

Consider M as an abelian Lie algebra and let

$$S = \langle u_1, \ldots, u_n, v_1, \ldots, v_n \rangle \leqq \mathrm{Der}(M).$$

Clearly

$$[u_i, u_j] = 0 = [v_i, v_j] \text{ for all } i, j;$$

$$[u_i, v_j] = \delta_{ij} e; \ [u_i, e] = [v_i, e] = 0,$$

where e is the identity transformation of M. Evidently $S \in \mathfrak{N}_2$ and $S \cong K$ under the map: $u_i \leftrightarrow x_i$, $v_i \leftrightarrow y_i$ (and so $e \leftrightarrow z$). So we make M a K-module via the isomorphism $K \cong S$. It is not hard to see that M is an irreducible S-module and so an irreducible K-module.

Now make M into an L-module by defining for each $a \in L$ and $f \in M$, $fa = f(a+N)$. Then M is an infinite dimensional and irreducible L-module, a contradiction. Hence $L \in \mathfrak{A} \cap \mathfrak{F}$. □

The construction of the quotient K of L holds even in fields of characteristic $p > 0$. However the module M has proper submodules generated by polynomials of the form $\lambda_1 t_1^{k_1 p} + \ldots + \lambda_n t_n^{k_n p}$, where $k_1, \ldots, k_n$ are non-negative integers and $\lambda_1, \ldots, \lambda_n \in$ field; which is as it should be by theorem 2.2.

We conjecture that '$L \in \mathfrak{N} \cap \mathfrak{F}$' in theorem 2.5 can be replaced by '$L \in \mathfrak{A} \cap \mathfrak{F}$'.

Finally it is not hard to see from theorem 2.5 that if $L \in$ Max-$\lhd \cap$ Min-$\lhd$ and if every irreducible L-module is finite-dimensional (over a field of characteristic zero) then $L^2 = L^3$. If our conjecture above is true this may be replaced by $L^2 = L^{(2)}$.

3. Residual finiteness

In this section we discuss the question of residual finiteness in finitely generated soluble algebras, along the lines of Hall [286]. It is necessary to begin with some ring-theoretic results concerning a special kind of ring. We introduce some convenient but non-standard terminology: an associative $\mathfrak{k}$-algebra U is a *Curtis algebra* if there exist finitely many elements $x_1, \ldots, x_n$ in the centre of U such that U is a finitely generated R-module, where $R = \mathfrak{k}[x_1, \ldots, x_n]$. Our interest in these stems from the fact (Jacobson [98] p. 189 lemma 4 and p. 204 lemma 5) that if L is a finite-dimensional Lie algebra over a field $\mathfrak{k}$ of prime characteristic, then the universal enveloping algebra $U(L)$ is a Curtis algebra. Such algebras were investigated by Curtis [48] (hence the name) and we shall need some of his results.

Note further that if L is finite-dimensional abelian over a field of characteristic zero, then $U(L)$ trivially is a Curtis algebra.

Let R be any ring. Recall that a *prime* (*right*) *ideal* of R is a (right) ideal P such that if A and B are any two (right) ideals with $AB \subseteq P$, then either $A \subseteq P$ or $B \subseteq P$. If R happens to be commutative then equivalently if $r, s \in R$ and $rs \in P$ then $r \in P$ or $s \in P$. Let A be an R-module and B a submodule of A. For $a \in A$ we define

$$\mathrm{Ann}_R(a+B) = \{r \in R : ar \in B\}.$$

In case $B = 0$ we write $\mathrm{Ann}_R(a)$. We also define

$$\mathrm{Ann}_R(A) = \{r \in R : Ar = 0\}.$$

We observe that $\mathrm{Ann}_R(a+B)$ is always a right ideal of R and $\mathrm{Ann}_R(A)$ is a two-sided ideal of R.

Our first result is an extension of a well known result in commutative rings (see for instance Lang [305] p. 142–156). For the applications we have in mind we must refer to the proof of this lemma as well as to its statement.

LEMMA 3.1 *Let R be a commutative and noetherian ring with unit and suppose that R is a subring of a ring U.*

If $U = u_1R+\ldots+u_mR$ is a finite extension of R then U is right noetherian.

If also A is a finitely generated U-module then A is a noetherian R-module and there is a finite sequence of R-submodules

$$0 = A_0 \subset A_1 \subset \ldots \subset A_n = A,$$

such that each factor module A_i/A_{i-1} $(i > 0)$ is cyclic and

$$A_i/A_{i-1} \cong_R R/P_i \qquad (i = 1, \ldots, n)$$

for some prime ideal P_i of R.

Proof: Consider U as a right R-module. Then by lemma 1.1 U is a noetherian R-module and so U is right noetherian.

Suppose that $A = b_1U+\ldots+b_kU$. Then clearly $A = \sum_{j=1}^{k}\sum_{i=1}^{m} b_ju_iR$, a finitely generated R-module and so by lemma 1.1 A is a noetherian R-module.

Define $A_0 = 0$. Suppose inductively that $i > 0$ and $A_0, \ldots, A_{i-1}$ have been defined so as to satisfy the required conditions, and $A_{i-1} < A$.

Case 1. If $A_{i-1}U \subseteq A_{i-1}$. Let P_i be a maximal element of the set of ideals $\{\mathrm{Ann}_R(a+A_{i-1}) : a \in A-A_{i-1}\}$ of R and let a_i be such that $P_i = \mathrm{Ann}_R(a_i+A_{i-1})$ $(a_i \notin A_{i-1})$. Since R is commutative then $P_i \subseteq \mathrm{Ann}_R(a_ir+A_{i-1})$ for all $r \in R$. Hence if $a_ir \notin A_{i-1}$ then by the maximality of P_i we have $P_i = \mathrm{Ann}_R(a_ir+A_{i-1})$ and so P_i is a prime ideal of R. Set $A_i/A_{i-1} = (a_iR+A_{i-1})/A_{i-1}$. Clearly (for R has a unit) the map $r \mapsto a_ir+A_{i-1}$ is an R-module homomorphism of R onto A_i/A_{i-1} with kernel P_i. So $R/P_i \cong_R A_i/A_{i-1}$.

Case 2. If $A_{i-1}U \nsubseteq A_{i-1}$. In this case let P_i be a maximal element in the set of ideals $\{\mathrm{Ann}_R(au+A_{i-1}) : a \in A_{i-1},\ u \in U,\ au \notin A_{i-1}\}$ of R: and pick $b_i \in A_{i-1}$ and $u \in U$ such that $b_iu \notin A_{i-1}$ and $P_i = \mathrm{Ann}_R(b_iu+A_{i-1})$. Set $A_i/A_{i-1} = (b_iuR+A_{i-1})/A_{i-1}$. As in case 1 we have that P_i is a prime ideal of R, $A_i/A_{i-1} \cong_R R/P_i$ and if $b_iur \notin A_{i-1}$ then $P_i = \mathrm{Ann}_R(b_iur+A_{i-1})$.

So in either case we have defined A_i such that A_i/A_{i-1} satisfies the required conditions. Now A is a noetherian R-module and so the sequence $A_0, A_1, \ldots$ must stop at some finite n. Hence by our construction we must have $A = A_n$. □

COROLLARY 3.2 *With the notation and hypothesis of lemma* 3.1 *suppose in addition that R is contained in the centre of U. Let M be a U-submodule of A which intersects every non-trivial U-submodule of A non-trivially and let* $P = \mathrm{Ann}_U(M)$. *Then*

$$P \cap R \subseteq P_i$$

for $i = 1, \ldots, n$.

Proof: Put $P_0 = R$. Then $P \cap R \subseteq P_0$ and $A_0(P \cap R)^0 = 0$. Let $i > 0$ and assume inductively that $P \cap R \subseteq P_0 \cap \ldots \cap P_{i-1}$. Then $A_{i-1}(P \cap R)^{i-1} = 0$.

Case 1. $A_{i-1}U \subseteq A_{i-1}$. Then $A_i U P_i = A_i P_i U \subseteq A_{i-1}U \subseteq A_{i-1}$, since$_i$ $P_i \subseteq R \subseteq$ centre of U. Thus P_i (by its maximality and the fact that $A_i U = a_i U + A_{i-1}$) is the annihilator in R of every non-zero element of $A_i U/A_{i-1}$. Suppose that $(P \cap R)^{i-1} \subseteq P_i$. Then if $i = 1$ we have $1 \in P_i$ and so $R \subseteq P_i$ whence $P \cap R \subseteq P_i$; and if $i > 1$ then since P_i is a prime ideal we have $P \cap R \subseteq P_i$. Suppose then that $(P \cap R)^{i-1} \nsubseteq P_i$ and let $x \in (P \cap R)^{i-1}$ with $x \notin P_i$. Multiplication by x induces a homomorphism of U-modules: $A_i U \to A_i Ux = a_i Ux$, and the kernel contains A_{i-1}. Clearly $a_i Ux = a_i xU$ is a U-submodule of A and so for some $v \in U$ have $a_i vx \neq 0$ and $a_i vx \in M$. If $a_i v \in A_{i-1}$ then $a_i vx = 0$ $(A_{i-1}x = 0)$, a contradiction. Thus $a_i v \notin A_{i-1}$. Now $a_i vxP = 0$. Hence $x(P \cap R) \subseteq \mathrm{Ann}_R(a_i v) \subseteq \mathrm{Ann}_R(a_i v + A_{i-1}) = P_i$ and since P_i is a prime ideal of R and $x \notin P_i$ then $P \cap R \subseteq P_i$. (Note: as $x \notin P_i$ we cannot have $a_i Ux = 0$ for this would imply that $x \in \mathrm{Ann}_R(a_i) \subseteq \mathrm{Ann}_R(a_i + A_{i-1}) = P_i$, a contradiction; and so by definition M intersects $a_i Ux$ non-trivially.)

Case 2. $A_{i-1}U \nsubseteq A_{i-1}$. From the proof of lemma 3.1 we see that $A_i = b_i uR + A_{i-1}$ for some $b_i \in A_{i-1}$ and $u \in U$. Hence $A_i(P \cap R)^{i-1} = (b_i(P \cap R)^{i-1})uR + A_{i-1}(P \cap R)^{i-1} = 0$, since $A_{i-1}(P \cap R)^{i-1} = 0$. Hence $(P \cap R)^{i-1} \subseteq \mathrm{Ann}_R(A_i/A_{i-1}) = P_i$, and so as above we have $P \cap R \subseteq P_i$.

Thus in either case $P \cap R \subseteq P_i$ and our inductive step is proved. Hence $P \cap R \subseteq P_i$ for all i. □

We say that a ring R is a *finite extension* of a subring S if R is finitely generated as a (right) S-module.

For the next lemma we need a simple ring-theoretic result, which occurs as lemma 3.1 in Curtis [48]. Suppose that U is a ring and R is a subring of the centre of U and R contains a unit, 1. If U is a finite extension of R then every element x of U satisfies an equation of the form

$$x^n + r_1 x^{n-1} + \ldots + r_n = 0 \quad (r_i \in R). \tag{7}$$

More generally we consider a ring U in which (7) holds for each $x \in U$ (where $n = n(x)$). This is equivalent to saying that the polynomial ring $R[x]$ be finite over R for each $x \in U$.

LEMMA 3.3 (*Curtis*) *Let R be a subring (with unit) of the centre of a ring U, and suppose that for each $x \in U$, the polynomial ring $R[x]$ is a finite extension of R. If M is an irreducible U-module and $P = \mathrm{Ann}_U(M)$ then $R/P \cap R$ is a field and $P \cap R$ is a maximal ideal of R.*

Proof: Let $0 \neq a \in M$ and let $P_1 = \mathrm{Ann}_U(a)$. Then $P \subseteq P_1$ and as $M = aU$ and $aU(P_1 \cap R) = a(P_1 \cap R)U = 0$ we have $P_1 \cap R \subseteq P$; thus $P \cap R = P_1 \cap R$. Now let $r \in R$ and suppose that $ar \neq 0$. Then $P_1 \subseteq \mathrm{Ann}_U(ar)$. Conversely if $x \in \mathrm{Ann}_U(ar)$ and if $x \notin P_1$ then $arx = axr = 0$ so $r \in \mathrm{Ann}_R(ax) \subseteq P \cap R$, from above. Hence as $P \cap R \subseteq P_1$ we have $ar = 0$, a contradiction. Therefore $x \in P_1$ and $P_1 = \mathrm{Ann}_U(ar)$ for all $r \in R - P_1$. If $r \in R - P_1$ then $ar \neq 0$ so $M = arU$. Thus there exists $u \in U$ such that $a = aru$. Hence $0 = aP_1 = aruP_1$ so that $uP_1 \subseteq \mathrm{Ann}_U(ar) = P_1$. Furthermore we have $0 = aru - a = a(ru - 1)$ and so $ru = 1 + v$ for some $v \in P_1$. Now $R[u]$ is a finite extension of R so for some $n = n(u)$ we can find $r_1, \ldots, r_n \in R$ such that

$$u^n + r_1 u^{n-1} + \ldots + r_n = 0.$$

Hence multiplying through by r^{n-1} we have

$$u(ru)^{n-1} + r_1(ru)^{n-1} + \ldots + r_n r^{n-1} = 0.$$

We have $ru = 1 + v$ so for each i, $(ru)^i = 1 + v_i$, where $v_i \in P_1$. Thus

$$u = -uv_{n-1} - r_1(1 + v_{n-1}) - r_2 r(1 + v_{n-2}) - \ldots - r_n r^{n-1}$$

and so $u = s + w$ for some $s \in R$ and $w \in P_1$, since $uv_{n-1} \in P_1$ and $RP_1 \subseteq P_1$. Thus $a = aru = ars + arw = ars$, since $arP_1 = 0$. So $0 = ars - a = a(rs - 1)$, whence $sr - 1 \in P_1 \cap R = P \cap R$. Therefore every non-zero element of $R/P \cap R$ has an inverse and so $R/P \cap R$ is a field and $P \cap R$ is a maximal ideal of R. □

Let R be an associative $\mathfrak{k}$-algebra (or L a Lie algebra over $\mathfrak{k}$) and let A be an R-module (L-module). We say that A is *residually finite* if there exists a family (B_α) of submodules of A such that A/B_α is finite-dimensional over $\mathfrak{k}$ and $\bigcap_\alpha B_\alpha = 0$. Equivalently to every $a \neq 0$ of A we can find a submodule B_a with $a \notin B_a$ and A/B_a finite-dimensional.

Note that by the Hilbert Nullstellensatz if $R = \mathfrak{k}[x_1, \ldots, x_n]$ is a polynomial algebra in finitely many variables, and if P is a maximal ideal in R, then R/P is finite-dimensional.

We now have a crucial result:

PROPOSITION 3.4 *Let U be a Curtis algebra, and A a finitely generated U-module. Then A is residually finite.*

Proof: We have finitely many elements $x_1, \dots, x_n$ in the centre of U, and U is finitely generated as a $\mathfrak{k}[x_1, \dots, x_n]$-module. Let $R = \mathfrak{k}[x_1, \dots, x_n]$. Then R is noetherian, so A is a noetherian R-module (lemma 3.1). Define B_a to be a submodule of A maximal with respect to $a \notin B_a$ (by Zorn). Let $B = A/B_a$. Then every non-trivial submodule of B contains $a + B_a$ and so $M = (aU + B_a)/B_a$ is an irreducible U-module, contained in every submodule of B. Let $P =$ $= \mathrm{Ann}_U(M)$. By lemma 3.3 $P \cap R$ is a maximal ideal of R and so by our remark above $R/P \cap R$ is finite dimensional over $\mathfrak{k}$. By lemma 3.1 and corollary 3.2 we have a finite sequence of submodules

$$0 = B_0 \subset B_1 \subset \dots \subset B_m = B$$

such that each factor $B_i/B_{i-1} \cong R/P_i$ for some prime ideal P_i of R. We also have $P \cap R \subseteq P_i$ for each i, so B_i/B_{i-1} is finite-dimensional. Hence B is finite-dimensional. □

For the sake of generality we define a class Max-*cu* of Lie algebras. We say that $L \in$ Max-*cu* if and only if $U(L)$ is a Curtis algebra. Clearly Max-*cu* is Q-closed, and Max-*cu* $\leqq$ Max-*u*. We can now prove:

THEOREM 3.5 *Let L be a Lie algebra and A an ideal of L such that $L/A \in \mathfrak{F}$, $L/C_L(A) \in$* Max-*cu, and A is finitely generated as an L-module. Then L is residually finite.*

In particular if A is an abelian ideal of a finitely generated Lie algebra L such that $L/A \in \mathfrak{F}$ and $L/C_L(A) \in$ Max-*cu, then L is residually finite. Hence*

$$\mathfrak{G} \cap \mathfrak{A}(\mathfrak{F} \cap \text{Max-}cu) \leqq \mathrm{R}\mathfrak{F}.$$

Proof: A is finitely generated as a module for $U(L/C_L(A))$ which is a Curtis algebra, so A is a residually finite L-module. Since the L-submodules of A are ideals of L, and L/A is finite-dimensional, it follows that L is residually finite. □

COROLLARY 3.6 *Over any field of characteristic $p > 0$,*

$$\mathfrak{G} \cap \mathfrak{A}\mathfrak{F} \leqq \mathrm{R}\mathfrak{F}.$$

Over any field of characteristic zero,

$$\mathfrak{G} \cap \mathfrak{A}^2 \leqq \mathrm{R}\mathfrak{F}.$$

Proof: By Jacobson [98] pp. 189, 204 we know that in characteristic p, $\mathfrak{F} \leqq$ Max-*cu*; and it is obvious that $\mathfrak{A} \cap \mathfrak{F} \leqq$ Max-*cu*. □

The first of these results is due to McInerney [163], who proved it using properties of ideals with 'centralising sets of generators' (cf. McConnell [159]), and independently to Bahturin [20].

It is now easy to see that Max-*cu* $\leqq$ R$\mathfrak{F}$. We do not know whether Max-*cu* contains algebras of infinite dimension. If not, the apparent generality of theorem 3.5 over its corollary is spurious.

Let us say that a module is *monolithic* if it has a non-trivial submodule, the *monolith*, contained in every non-trivial submodule. Then theorem 3.4 implies that every finitely generated monolithic module for a Curtis algebra is finite-dimensional, and in particular proves the theorem of Curtis (theorem 11.2.2) quoted above.

In the characteristic zero case we can say more.

For a positive integer c we define the class $\mathfrak{K}_c$ by

$$L \in \mathfrak{K}_c \text{ if } [L^2, L^c] = 0.$$

Suppose that $L \in \mathfrak{G} \cap \mathfrak{K}_c$ and let $A = L^c$. Then $L \in \mathfrak{G} \cap \mathfrak{A}\mathfrak{N}$ and so $L \in$ Max-$\lhd$ by corollary 1.8. Hence A is a finitely generated L-module. By definition $L^2 \leqq C_L(A)$ so $L/C_L(A) \in \mathfrak{G} \cap \mathfrak{A} = \mathfrak{F} \cap \mathfrak{A}$. Now let $0 \neq x \in L$. If $x \notin A$ we have $L/A \in \mathfrak{G} \cap \mathfrak{N}_c \leqq \mathfrak{F}$. If $x \in A$ then by proposition 3.4 we have some ideal B_a of L such that $A/B_a \in \mathfrak{F}$ and $a \notin B_a$. But $L/A \in \mathfrak{F}$ so $L/B_a \in \mathfrak{F}$. Thus to each $0 \neq x \in L$ there corresponds an ideal N of L with $x \notin N$ and $L/N \in \mathfrak{F}$. Hence $L \in$ R$\mathfrak{F}$.

So we have proved:

THEOREM 3.7 $\mathfrak{G} \cap \mathfrak{K}_c \leqq$ R$\mathfrak{F}$, *for every positive integer c.* □

In the next section we shall need a slightly more general class. Suppose we define a new class Max-*pu* by saying that $L \in$ Max-*pu* if there exists a subring R of the centre of $U(L)$ such that $1 \in R$, R is noetherian and $U(L)$ is a finitely generated R-module. Then we have

$$\mathfrak{F} \cap \mathfrak{A} \leqq \text{Max-}cu \leqq \text{Max-}pu \leqq \text{Max-}u.$$

Furthermore Max-*pu* is Q-closed. It is not hard to show that theorem 3.5 holds if we replace '$L/C_L(A) \in$ Max-*cu*' by '$L/C_L(A) \in$ Max-*pu* and every irreducible $L/C_L(A)$-module is finite-dimensional'.

The construction of theorem 2.5 enables us to prove:

THEOREM 3.8 *Let $L \in \mathfrak{N} \cap \mathfrak{F}$ over a field of characteristic zero. Then the following conditions are equivalent:*

(a) $L \in \mathfrak{A} \cap \mathfrak{F}$
(b) *Every irreducible L-module is finite dimensional*
(c) $\mathfrak{G} \cap \mathfrak{A}(L) \leqq \mathrm{R}\mathfrak{F}$. □

Thus we see that over fields of characteristic zero, $\mathfrak{F} \cap \mathfrak{N} \nleqq$ Max-*cu* and so Max-*cu* < Max-*u*.

We end this section with an example, showing that $\mathfrak{G} \cap \mathfrak{N}\mathfrak{A} \nleqq \mathrm{R}\mathfrak{F}$. Say that a Lie algebra L is *monolithic* if it is a monolithic L-module under the adjoint action.

PROPOSITION 3.9 $\mathfrak{G} \cap \text{Max-}\lhd \cap \mathfrak{N}_2\mathfrak{A} \nleqq \mathrm{R}\mathfrak{F}$.

Proof: Let $\mathfrak{k}$ be any field and let N be a vector space over $\mathfrak{k}$ with basis $\{z, x_i, y_i : i = 0, 1, 2, \ldots\}$. Define a bilinear product [,] on N as follows:

$$[x_i, z] = [x_i, x_j] = [y_i, z] = [y_i, y_j] = [z, z] = 0$$

for all i, j; and

$$[x_i, y_j] = -[y_j, x_i] = \delta_{ij} z.$$

Evidently this makes N a Lie algebra over $\mathfrak{k}$ and $N \in \mathfrak{N}_2$ with $\zeta_1(N) = \mathfrak{k}z$.

Let $x_{-1} = y_{-1} = 0$. Define a linear transformation d of N by

$$\begin{aligned} d\colon\ & z \mapsto 0 \\ & x_i \mapsto x_{i-1} + x_i + x_{i+1} \\ & y_i \mapsto -y_{i-1} - y_i - y_{i+1}, \end{aligned}$$

for all $i = 0, 1, \ldots$. It is not hard to see that for all $i, j = 0, 1, 2, \ldots$ we have

$$[x_{i-1} + x_i + x_{i+1}, y_j] + [x_i, -y_{j-1} - y_j - y_{j+1}] = 0,$$

whence d is a derivation of N.

Form the split extension

$$L = N \dotplus \langle d \rangle.$$

Clearly $L \in \mathfrak{N}_2\mathfrak{A}$ and $L = \langle x_0, y_0, d \rangle$. Furthermore $\zeta_1(L) = \mathfrak{k}z$ and $L/\mathfrak{k}z \in \mathfrak{G} \cap \mathfrak{A}^2 \leqq \text{Max-}\lhd$, by corollary 1.8 and so $L \in \text{Max-}\lhd$.

Finally suppose that $0 \neq I \lhd L$. Let $0 \neq a + \lambda d \in I$ $(a \in N)$. If $\lambda = 0$ then $0 \neq a \in N \cap I$. Assume that $\lambda \neq 0$. If $a \in \mathfrak{k}z$ then $0 \neq \lambda(x_0 + x_1) =$

$= [x_0, a+\lambda d] \in N \cap I$ whilst if $a \notin \mathfrak{k}z$ then $0 \neq ad = [a+\lambda d, d] \in N \cap I$. In any case $N \cap I \neq 0$, whence $z \in N \cap I \leqq I$ and L is monolithic. But $L \notin \mathfrak{F}$ so $L \notin \mathrm{R}\mathfrak{F}$. □

4. Stuntedness

Let α be an ordinal. A Lie algebra L is said to be *(centrally) stunted of height* α if α is the least ordinal such that for any subalgebra H of L, $\zeta_*(H) = \zeta_\alpha(H)$. We call α the *stunted height* of L and write

$$\alpha = \mathrm{sth}(L).$$

Evidently if $H \leqq L$ then $\mathrm{sth}(H) \leqq \mathrm{sth}(L)$ and set theoretic considerations show that for any Lie algebra L, $\mathrm{sth}(L)$ exists. Our object here is to find more precise bounds for $\mathrm{sth}(L)$ for Lie algebras L in certain classes of soluble Lie algebras. We will show that if $L \in \mathfrak{G} \cap \mathfrak{A}^2$ then $\mathrm{sth}(L) < \omega$. However example 6.3.6 shows that over fields of characteristic zero we may have $L \in \mathfrak{G} \cap \mathfrak{A}\mathfrak{N}_2$ and $\mathrm{sth}(L) > \omega$.

Let X be a Lie algebra and let A be an X-module. If $H \leqq X$ we define the annihilator $A_1(H)$ of H in A by

$$A_1(H) = \{a \in A : aH = 0\}.$$

Let $A_0(H) = 0$ and define inductively $A_{\alpha+1}(H)/A_\alpha(H)$ to be $(A/A_\alpha(H))_1(H)$ (for $A_\alpha(H)$ is an H-module so $A/A_\alpha(H)$ is an H-module); at limit ordinals μ we let $A_\mu(H) = \bigcup_{\alpha<\mu} A_\alpha(H)$. We define $A_*(H) = \bigcup_\beta A_\beta(H)$, where the union is taken over all ordinals. We define the *stunted height* $\mathrm{sth}_X(A)$ *(with respect to* X*) of* A to be the least ordinal α such that $A_*(H) = A_\alpha(H)$ for all subalgebras H of X: equivalently $A_\beta(H) = A_\alpha(H)$ for all ordinals $\beta \geqq \alpha$. In this case we also say that *the pair* (A, X) *is stunted of height* α. Evidently if we put $A = X$ then we have $\mathrm{sth}_X(X) = \mathrm{sth}(X)$ and for $H \leqq X$ we have $X_1(H) = C_X(H)$.

Let X be a Lie algebra, A an X-module and B a submodule of A. Then A/B is an X-module and if Y is any ideal of X such that $AY \subseteq B$ then A/B has an obvious X/Y-module structure and for each ordinal α, $(A/B)_\alpha(H) = (A/B)_\alpha(H+Y/Y)$ for every subalgebra H of X. In particular

$$\mathrm{sth}_X(A/B) = \mathrm{sth}_{X/Y}(A/B). \tag{8}$$

Evidently if $H \leqq X$ then for each ordinal α, $B_\alpha(H) = B \cap A_\alpha(H)$ and so

$B_*(H) = B \cap A_*(H)$. Furthermore

$$A_\alpha(H)/B_\alpha(H) \cong_H (A_\alpha(H)+B)/B \subseteq (A/B)_\alpha(H), \tag{9}$$

and so

$$A_*(H)/B_*(H) \cong_H (A_*(H)+B)/B \subseteq (A/B)_*(H). \tag{10}$$

LEMMA 4.1 *Let L be a Lie algebra, A an L-module and B a submodule of A. Suppose that $I \lhd L$ with $AI \subseteq B$. If the pair $(A/B, L/I)$ is stunted of height α and if $H \leqq L$ such that $B_*(H) = B_\beta(H)$ then $A_*(H) = A_{\beta+\alpha}(H)$.*

Proof: By equation (8) and the given hypothesis we have $(A/B)_*(H) = (A/B)_\alpha(H)$. But by (10) $A_*(H)/B_*(H)$ is H-isomorphic to an H-submodule of $(A/B)_*(H) = (A/B)_\alpha(H)$, whence there is an ascending series of type α from $B_*(H)$ to $A_*(H)$ with factors annihilated by H. Since also $B_*(H) = B_\beta(H) = B \cap A_\beta(H) \subseteq A_\beta(H)$, then $A_*(H) \subseteq A_{\beta+\alpha}(H)$. □

By an obvious induction and a similar argument we have:

COROLLARY 4.2 *Let X be a Lie algebra and let A be an X-module with a finite sequence of submodules*

$$0 = A_0 \subset A_1 \subset \ldots \subset A_n = A$$

such that for each i the pair $(A_i/A_{i-1}, X)$ is stunted of height α_i. Then the pair (A, X) is stunted of height $\leqq \alpha_1+\ldots+\alpha_n$. □

Suppose that $X \in \mathfrak{F} \cap \mathfrak{A}$ and $U = U(X)$ is the universal algebra of X. Let B be an X-module which is U-isomorphic to U/P for some prime ideal P of U. Now U is commutative and noetherian and so the annihilator of every non-zero element of B is P. Thus if $H \leqq X$ and $B_*(H) \neq 0$ then $B_1(H) \neq 0$ and H annihilates $B_1(H)$ so $H \subseteq P$. Therefore $BH \subseteq BP = 0$, whence $B_*(H) \subseteq B = B_1(H)$ so $B_*(H) = B_1(H)$. Hence the pair (B, X) is stunted of height $\leqq 1$.

Now suppose that A is a finitely generated X-module. Then by lemma 3.1 A has a finite sequence of submodules (from 0 to A) with each factor isomorphic to a quotient of U by a prime ideal. So by corollary 4.2 and our result above we have:

PROPOSITION 4.3 *If A is a finitely generated module for a Lie algebra L and if $L/C_L(A) \in \mathfrak{F} \cap \mathfrak{A}$ then*

$$\mathrm{sth}_L(A) < \omega.$$ □

THEOREM 4.4 (a) *If* $L \in \mathfrak{N}\mathfrak{A} \cap \text{Max-}\lhd$ *then* L *is centrally stunted of finite height.*

(b) *If* $L \in \mathfrak{G} \cap \mathfrak{K}_c$ *then* L *is centrally stunted of finite height.*

(c) *Every finitely generated metabelian Lie algebra is centrally stunted of finite height.*

Proof: Let $A = L$ and let $B = L^2$ and define for each i, $B_i = (L^2)^i$. Now $L^2 \in \mathfrak{N}_c$ for some c, so $B_{c+1} = 0$. By Max-$\lhd$ $A, B, \ldots, B_c$ are all finitely generated L-modules. Clearly $A/B, B/B_2, \ldots, B_c/B_{c+1}$ are all L/L^2-modules and $L/L^2 \in \mathfrak{F} \cap \mathfrak{A}$. Therefore by proposition 4.3 $\text{sth}_{L/L^2}(A/B) < \omega$, $\text{sth}_{L/L^2}(B/B_2) < \omega, \ldots, \text{sth}_{L/L^2}(B_c/B_{c+1}) < \omega$, whence each pair $(A/B, L), \ldots, (B_c/B_{c+1}, L)$ is stunted of finite height. Hence the pair (A, L) is stunted of finite height (corollary 4.2) so $\text{sth}(L) = \text{sth}_L(A) < \omega$. This proves (a).

If $L \in \mathfrak{G} \cap \mathfrak{K}_c$ then $L \in \text{Max-}\lhd$ (corollary 1.8) and $L \in \mathfrak{N}\mathfrak{A}$, since $[L^2, L^c] = 0$. Hence (b) and (c) follow from above. □

Evidently if L is stunted of finite height h say, then every locally nilpotent subalgebra of L is nilpotent of class $\leqq h$. In particular L has a unique maximal nilpotent ideal. Consider the Lie algebra constructed in the proof of theorem 2.5. Then it is clear that we can prove:

THEOREM 4.5 *Let* $L \in \mathfrak{N} \cap \mathfrak{F}$ *over a field of characteristic zero. Then the following conditions are equivalent:*

(a) $L \in \mathfrak{F} \cap \mathfrak{A}$

(b) *Every irreducible L-module is finite dimensional*

(c) *If* $K \in \mathfrak{G} \cap \mathfrak{A}(L)$ *then* $\text{sth}(K) < \omega$

(d) *If* A *is a finitely generated L-module then* $\text{sth}_L(A) < \omega$.

Proof: We know that (a) $\Leftrightarrow$ (b) and it is easy to see that (c) $\Leftrightarrow$ (d). By the construction of theorem 2.5 (c) $\Rightarrow$ (b), and from above (b) $\Rightarrow$ (a) $\Rightarrow$ (c). □

In [306] Lennox and Roseblade prove that every finitely generated abelian-by-nilpotent group is centrally stunted of finite height and deduce the same conclusion for nilpotent-by-nilpotent groups with the maximal condition for normal subgroups. Now theorem 4.5 shows that these results are not true for Lie algebras defined over a field of characteristic zero. However, in the spirit of the preceeding sections, the results will hold for Lie algebras defined over fields of non-zero characteristic.

Let R be a ring with unit which is noetherian and contained in the centre of a ring U. Suppose that $U = u_1R + \ldots + u_nR$ is finitely generated as an R-module. Let $u \in U$. Then for each i we can find $r_{ij} \in R$ such that $u_iu = \sum_{j=1}^{n} u_j r_{ij}$

and so

$$0 = -u_1 r_{i1} - \ldots + u_i(u - r_{ii}) - u_{i+1} r_{i,i+1} - \ldots - u_n r_{in}. \qquad (*)$$

From (*) we deduce, since $R[u]$ is a commutative ring, that the determinant $d = \det(Iu - D)$ (where I is the identity $n \times n$ matrix and D is the matrix with entries r_{ij}) is such that

$$Ud = 0.$$

But U is a faithful module over $R[u]$ (trivially since $1 \in U$) and so

$$0 = d = u^n + r_{1u}{}^{n-1} + \ldots + r_{n-1} u + r_n,$$

an equation of type (7) with the $r_j \in R$ (note that $r_n = (-1)^n \det(D)$) and of degree n.

We recall that $L \in$ Max-*pu* if $U = U(L)$ is finitely generated as a module over a noetherian subring R (with unit) of the centre of U. Thus if $L \in$ Max-*pu* then we can find a positive integer $n = n(L)$ such that every $u \in U$ satisfies an equation of the form

$$u^n + r_1 u^{n-1} + \ldots + r_n = 0 \qquad (r_j \in R).$$

If A is an L-module and $x \in L$ we write $A_\alpha(x)$ and $A_*(x)$ for $A_\alpha(\langle x \rangle)$ and $A_*(\langle x \rangle)$ respectively (α is an ordinal).

LEMMA 4.6 *Let $L \in$ Max-pu, let $U = U(L)$ and let R be a noetherian ring in the centre of U such that $1 \in R$ and U is finitely generated as an R-module. Let $A = aU$ be a cyclic L-module and suppose that if $P = \mathrm{Ann}_R(a)$ then $P = \mathrm{Ann}_R(au)$ for any $u \in U$ such that $au \neq 0$. Then there exists $n = n(L)$ such that $A_*(x) = A_n(x)$ for every $x \in L$.*

Proof: From our remarks above we can find a positive integer $n = n(L)$ such that each x in L satisfies an equation of the form $x^n + r_1 x^{n-1} + \ldots + r_n = 0$, where the $r_j \in R$. If every $r_j \in P$ then $Ax^n = 0$ so $A_*(x) \subseteq A = A_n(x)$. Otherwise let j be maximal with respect to $r_j \notin P$. Then $1 \leqq j \leqq n$. Suppose that there exists $b \in A_{n-j+1}(x) - A_{n-j}(x)$. From the equation above we have (since $r_{j+1}, \ldots, r_n \in P$),

$$0 = bx^n + br_1 x^{n-1} + \ldots + br_{j-1} x^{n-j+1} + br_j x^{n-j}$$

and so $br_j x^{n-j} = bx^{n-j} r_j = 0$. This implies that $r_j \in P$, a contradiction (for $bx^{n-j} \neq 0$ and $bx^{n-j} = au$ for some $u \in U$ and by hypothesis $P = \mathrm{Ann}_R(au)$). Hence $A_{n-j+1}(x) = A_{n-j}(x)$, and so $A_*(x) = A_{n-j}(x) \subseteq A_n(x)$. □

LEMMA 4.7 *Let L be a Lie algebra and A an L-module. Let n be a positive integer such that for every $x \in L$, $A_*(x) = A_n(x)$. Then to each positive integer m there corresponds a positive integer $f(m) = n^m$ such that if $H \leqq L$ and $H \in \mathfrak{F}_m$ then $A_*(H) = A_{f(m)}(H)$.*

Proof: By induction on m. Put $f(1) = n$. Suppose that $m > 1$ and $f(k) = n^k$ has been defined for all $k \leqq m-1$ so as to satisfy the given conditions. Let $H \leqq L$ with $\dim(H) = m$. Suppose that H is simple. Then $H = H^i$ for all i. Hence if $b \in A_2(H)$ then $bH = bH^2 \subseteq (bH)H = 0$ so $b \in A_1(H)$, whence $A_2(H) = A_1(H)$ and $A_*(H) = A_1(H) \subseteq A_{f(m)}(H)$. If H is not simple then it has a proper non-zero ideal K. Let $k = \dim(K)$ so that $m-k = \dim(H/K)$. Evidently $A_*(H) \subseteq A_*(K) = A_{f(k)}(K)$, by induction. Let $B = A_*(H)$ and define $B_0 = B$ and $B_{i+1} = B_iK$ for $i = 0, 1, \ldots, f(k)-1$. Then $B_{f(k)} = 0$, each B_i is an H-module (induct on i and use the fact that $K \lhd H$) and each factor $C_i = B_i/B_{i+1}$ is an H/K-module. Let $y = x+K \in H/K$. We have $A_*(H) \subseteq A_*(x) = A_n(x)$ and so $A_*(H)x^n = 0$, whence $C_iy^n = 0$ and $(C_i)_*(y) = (C_i)_n(y)$ for every $y \in H/K$. Thus C_i and H/K satisfy the hypothesis of A and L and, since $\dim(H/K) = m-k$, then by the inductive hypothesis the pair $(C_i, H/K)$ is stunted of height $\leqq f(m-k)$ and therefore the pair (C_i, H) is stunted of height $\leqq f(m-k)$ (since $K \leqq C_H(C_i)$). Thus by corollary 4.2 the pair (B, H) is stunted of height $\leqq f(k).f(m-k) = n^k.n^{m-k} = = n^m = f(m)$. But $B = A_*(H)$ and so $B = B_*(H) = B_{f(m)}(H)$, whence $A_*(H) = A_{f(m)}(H)$ and our inductive step is proved. □

(The result of lemma 4.7 can be obtained by proving that if $H \in \mathfrak{F}_m$, A is an H-module and $A_*(x) = A_n(x)$ for all $x \in H$ then the pair (A, H) is stunted of height $\leqq f(m) = n^m$.)

A similar result holds if we allow n to be any ordinal, finite or infinite.

Consider the cyclic module $A = aU$ described in lemma 4.6. Then it follows from lemma 4.7 that if $H \leqq L$ and $H \in \mathfrak{F}_m$ then the pair (A, H) is stunted of height $f(m) = n^m$ or, equivalently $A_*(H) = A_{f(m)}(H)$.

With this remark we can now prove:

PROPOSITION 4.8 *Let L be a Lie algebra and A a finitely generated L-module such that $L/C_L(A) \in$ Max-pu. Then there exists a positive integer $n = n(L)$ such that if m is any positive integer and $H \leqq L$ with $(H+C_L(A))/C_L(A) \in \mathfrak{F}_m$, the pair (A, H) is stunted of height $\leqq h(m) = k.n^m$; $k = k(A) < \omega$.*

In particular if $L/C_L(A) \in$ Max-pu$\cap\mathfrak{F}$ then the pair (A, L) is stunted of finite height.

Proof: Let $X = L/C_L(A)$. Evidently the stunted height of the pair (A, H) is the same as those of the pairs $(A, H+C_L(A))$ and $(A, (H+C_L(A))/C_L(A))$ so without loss of generality we may assume that H contains $C_L(A)$. Furthermore A is a finitely generated X-module (same number of generators as it has as an L-module) so we may further assume that $L \in$ Max-*pu*.

Let $U = U(L) = u_1R+\ldots+u_nR$, where R is noetherian, contains the identity and is contained in the centre of U. Now A is a finitely generated $U(L)$-module and so a noetherian R-module. By lemma 3.1 there is a certain finite sequence of R-submodules of A,

$$0 = A_0 \subset A_1 \subset \ldots \subset A_k = A$$

whose factors are cyclic and R-isomorphic to quotients of R by prime ideals. For a given $i > 0$, if $A_{i-1}U_\mu \subseteq A_{i-1}$ then $A_i = a_iR+A_{i-1}$ where a_i is so chosen that $\mathrm{Ann}_R(a_i+A_{i-1})$ is maximal in the set $\{\mathrm{Ann}_R(a+A_{i-1})\colon a \in A\backslash A_{i-1}\}$. In this case since R is in the centre of U we see that if $P_i = \mathrm{Ann}_R(a_i+A_{i-1})$ then $P = \mathrm{Ann}_R(a_iu+A_{i-1})$ for every $u \in U$ such that $a_iu \notin A_{i-1}$. Thus $B_i = A_iU/A_{i-1}$ is a cyclic U-module and P_i is the annihilator in R of every non-zero element of B_i. Therefore by our remarks above the pair (B_i, H) is stunted of height $\leqq n^m$, whenever $H \leqq L$ and $H \in \mathfrak{F}_m$. On the other hand if $A_{i-1}U \nsubseteq A_{i-1}$ then $A_i = b_iuR+A_{i-1}$, for some $b_i \in A_{i-1}$ and $u \in U$. So $A_iU = A_{i-1}U$.

Now let $H \leqq L$ with $H \in \mathfrak{F}_m$. We claim that for every $i = 0, 1, \ldots, k$, the pair (A_iU, H) is stunted of height $\leqq i.n^m$. For this we use induction on i. If $i = 0$ this is trivial since $A_0 = 0$ so $A_0U = 0$. Let $i > 0$ and assume that our claim holds for $i-1$. If $A_{i-1}U \subseteq A_{i-1}$ then by the argument above the pair $(A_iU/A_{i-1}, H)$ is stunted of height $\leqq n^m$. We also have $A_{i-1}U = A_{i-1}$ so that by induction the pair (A_{i-1}, H) is stunted of height $\leqq (i-1).n^m$ and so by corollary 4.2 the pair (A_iU, H) is stunted of height $\leqq n^m+(i-1)n^m = i.n^m$. If $A_{i-1}U \nsubseteq A_{i-1}$ then we know that $A_iU = A_{i-1}U$ so that the pair (A_iU, H) is stunted of height $\leqq (i-1).n^m \leqq i.n^m$. Thus our claim holds for i and in particular for all i. Now $A = AU = A_kU$, so the pair (A, H) is stunted of height $\leqq k.n^m$. In particular if $L \in \mathfrak{F}$ then $L \in \mathfrak{F}_m$ for some m, so the pair (A, L) is stunted of height $\leqq k.n^m$. □

THEOREM 4.9 *If* $L \in \mathfrak{N}(\text{Max-}pu \cap \mathfrak{F}) \cap \text{Max-}\lhd$ *then* L *is centrally stunted of finite height. In particular every* $\mathfrak{G} \cap \mathfrak{A}(\mathfrak{F} \cap \text{Max-}pu)$*-algebra is centrally stunted of finite height.*

Proof: Let $N \lhd L$ with $N \in \mathfrak{N}_c$ for some c and $L/N \in \text{Max-}pu \cap \mathfrak{F}$. Now N^i is a finitely generated L-module for each i, so N^i/N^{i+1} is a finitely generated

L/N-module so by proposition 4.8 each pair $(N^i/N^{i+1}, L/N)$ is stunted of finite height, whence $(N^i/N^{i+1}, L)$ is stunted of finite height. By corollary 4.2 the pair (N, L) is stunted of finite height. But clearly the pair $(L/N, L/N)$ is stunted of finite height so $(L/N, L)$ is stunted of finite height and thus by corollary 4.2 the pair (L, L) is stunted of finite height. Thus $\mathrm{sth}(L) < \omega$.

For the second part we know that $\mathfrak{G} \cap \mathfrak{A}\mathfrak{F} \leqq \text{Max-}\lhd$. □

We note that the statement, 'the pair (L, L) is stunted of height α' is stronger than '$\mathrm{sth}(L) = \alpha$'.

In view of theorems 4.5 and 4.9 we see that over fields of characteristic zero, $\mathfrak{F} \cap \mathfrak{N} \nleqq \text{Max-}pu$ and so $\text{Max-}pu < \text{Max-}u$.

On the other hand over any field of characteristic not zero we have by the results of Jacobson [98], $\mathfrak{F} \leqq \text{Max-}cu \leqq \text{Max-}pu \leqq \text{Max-}u$. So we have:

Corollary 4.10 *Let L be a Lie algebra over a field of characteristic not zero. If* $L \in (\mathfrak{G} \cap \mathfrak{A}\mathfrak{F}) \cup (\mathfrak{N}\mathfrak{F} \cap \text{Max-}\lhd) \cup (\mathfrak{N}^2 \cap \text{Max-}\lhd)$, *then L is centrally stunted of finite height.*

Proof: Immediate from theorem 4.9. □

Now we consider briefly Lie algebras which are stunted of infinite height. First we have:

Proposition 4.11 *Let X be a Lie algebra and A an X-module. If* $X/C_X(A) \in \mathfrak{F}$ *then the pair* (A, X) *is stunted of height* $\leqq \omega$.

Proof: Let $H \leqq X$. Then $H/C_H(A) \in \mathfrak{F}$. By transfinite induction on α it is easy to show that $A_\alpha(H/C_H(A)) \subseteq A_\omega(H/C_H(A))$ for all ordinals α and so $A_*(H) = A_*(H/C_H(A)) = A_\omega(H/C_H(A)) = A_\omega(H)$ (see the argument of theorem 9.21 of Amayo [2]). □

Theorem 4.12 *Let K be an ideal of a Lie algebra L and suppose that K is stunted of height* λ.

(a) *If* $L/K \in \mathfrak{F} \cap \mathfrak{N}_c$ *then L is stunted of height* $\leqq \omega\lambda + c$.

(b) *If* $L/K \in \mathfrak{F}_m$ *then L is stunted of height* $\leqq \omega\lambda + m$.

In particular if $L \in \mathfrak{N}_c(\mathfrak{F} \cap \mathfrak{A})$ *then L is centrally stunted of height* $\leqq \omega c + 1$.

Proof: We prove (a) and (b) together. Now if $H \leqq L$ then $H \cap K$ is stunted of height $\leqq \lambda$, and $H/H \cap K \in \mathfrak{F} \cap \mathfrak{N}_c \cup \mathfrak{F}_m$. So without loss of generality it is enough to prove that $\zeta_*(L) = \zeta_{\omega\lambda+1}(L)$ or $\zeta_{\omega\lambda+m}(L)$ as the case may be.

Let $B_\alpha = \zeta_\alpha(K)$ and $D_\alpha = \zeta_\alpha(L)$ for all ordinals α and let $D = \zeta_*(L)$. We use transfinite induction on α to show that

$$B_\alpha \cap D \leqq D_{\omega\alpha} \text{ for all } \alpha. \qquad (*)$$

(*) is trivially true for $\alpha = 0$. Suppose that (*) holds for some α. Let $U = B_{\alpha+1}$ and $V = B_\alpha$. Then U and V are L-modules and $V_*(L) = V \cap D \leqq V \cap D_{\omega\alpha} = V_{\omega\alpha}(L)$. Now K annihilates U/V and $L/K \in \mathfrak{F}$ so by proposition 4.11 the pair $(U/V, L/K)$ is stunted of height $\leqq \omega$. Thus the pair $(U/V, L)$ is stunted of height $\leqq \omega$. By lemma 4.1 we have $U_*(L) = U_{\omega\alpha+\omega}(L) = U_{\omega(\alpha+1)}(L) = U \cap D_{\omega(\alpha+1)}$. But $U_*(L) = U \cap D$ and so (*) holds for $\alpha+1$. Suppose that μ is a limit ordinal and (*) holds for all $\alpha < \mu$. Then

$$\begin{aligned} B_\mu \cap D &= \bigcup\nolimits_{\alpha<\mu} B_\alpha \cap D \\ &\leqq \bigcup\nolimits_{\alpha<\mu} D_{\omega\alpha} = D_{\omega\mu}, \end{aligned}$$

since $\omega\mu$ is a limit ordinal and $\omega\mu = \lim_{\beta<\omega\mu}\beta = \lim_{\beta<\mu}\omega\beta$. Thus (*) holds for μ. This completes our induction and proves (*). In particular since K is stunted of height $\leqq \lambda$,

$$K \cap D \leqq \zeta_*(K) \cap D \leqq \zeta_\lambda(K) \cap D \leqq D_{\omega\lambda}.$$

If $L/K \in \mathfrak{F} \cap \mathfrak{N}_c$ then $[D, {}_cL] \leqq D \cap K \leqq D_{\omega\lambda}$, whence $D \leqq D_{\omega\lambda+c}$.

If $L/K \in \mathfrak{F}_m$ then $(D+K)/K \in \mathfrak{F}_m$ so $D/D \cap K \in \mathfrak{F}_m$, whence $D/D_{\omega\lambda} \in \mathfrak{F}_m$ and $D_{\omega\lambda+m+1} = D_{\omega\lambda+m}$; therefore $D = D_{\omega\lambda+m}$.

If $L \in \mathfrak{N}_c(\mathfrak{F} \cap \mathfrak{A})$ then L has an ideal K such that $K \in \mathfrak{N}_c$ and $L/K \in \mathfrak{F} \cap \mathfrak{A}$. The result follows from (a). □

In view of the bounds determined in proposition 4.11 and theorem 4.12 it is interesting to note that we do have:

THEOREM 4.13 (a) *Over any field there exists a Lie algebra* $L \in \mathfrak{N}_c(\mathfrak{F} \cap \mathfrak{A})$ *such that* $L \in \mathfrak{Z}$ *and* $\zeta_{\omega c}(L) < \zeta_{\omega c+1}(L) = L$.

(b) *Over any field there exists a* $\mathfrak{Z}$*-algebra* L *with an ideal* K *such that* K *is stunted of height* ω, L/K *has dimension one and* $\zeta_{\omega^2}(L) < \zeta_{\omega^2+1}(L) = L$.

Proof: (a) Let F be the free Lie algebra (see Jacobson [98], pp. 167ff.) on $x_1, x_2, \ldots$. The map

$$\begin{aligned} \theta\colon\ & x_1 \mapsto 0 \\ & x_{i+1} \mapsto x_i \text{ for } i = 1, 2, \ldots, \end{aligned}$$

extends to a derivation d of F in a natural way. By induction on n it is not hard to show that

$$[x_{i_1}, \ldots, x_{i_s}]d^n = 0 \text{ if } \textstyle\sum_{k=1}^s i_k \leqq n.$$

But if $j > 1$ then clearly

$$x = [x_{n+1}, \ldots, x_{n+j}]d^n = [x_1, x_{n+2}, \ldots, x_{n+j}]+y,$$

where y is a linear combination of left normed commutators of the form

$$w(i_1, \ldots, i_j) = [x_{n+1-i_1}, \ldots, x_{n+j-i_j}]$$

with $i_1 < n$ and $i_1 + \ldots + i_j = n$. Thus each $w(i_1, \ldots, i_j)$ involves an x_m (for instance x_{n+1-i_1}) with $1 < m < n+2$. Since F is free then the map

$$\phi: x_m \mapsto 0 \quad (1 < m < n+2)$$
$$x_m \mapsto x_m \quad \text{(otherwise)}$$

extends to a Lie endomorphism of F. Now F^{n+1} is invariant under the endomorphism ϕ and $x\phi = [x_1, x_{n+2}, \ldots, x_{n+j}] \notin F^{n+1}$ (for $y\phi = 0$) and so $x \notin F^{n+1}$. Thus $x \neq 0 \bmod F^{n+1}$.

Let c be a positive integer. Since F^{c+1} is invariant under d we may form the split extension

$$L = F/F^{c+1} \dotplus \langle d \rangle.$$

Let $A = F/F^{c+1}$. Pick $i \leqq c$ and let $B = F^i/F^{i+1}$. Now B is a $\langle d \rangle$-module and from our remarks above it is easy to see that $B = B_\omega(d) \neq B_m(d)$ for any m (for $[x_{m+2}, {}_{i-1}x_1]+F^{i+1} \in B_{m+1}(d) \backslash B_m(d)$). From this we deduce by induction that $F^i/F^{c+1} = (F^i/F^{c+1})_{\omega(c-i+1)}(L)$ and so $A = A_{\omega c}(L)$ whence $A = \zeta_{\omega c}(L) < \zeta_{\omega c+1}(L) = L$.

(b) For each c let $F_c = F/F^{c+1}$. Let K be the direct sum $K = \mathrm{Dr}_{c=1}^{\infty} F_c$ and let d act on K, componentwise. Then d is a derivation of K so we may form the split extension

$$L = K \dotplus \langle d \rangle.$$

Evidently by (a) we have $F_c \leqq \zeta_{\omega c}(L)$ and $F_{c+1} \nleqq \zeta_{\omega c}(L)$. Furthermore $d \notin \zeta_{\omega c}(L)$ for any c. Thus

$$K \leqq \bigcup_{c=1}^{\infty} \zeta_{\omega c}(L) = \zeta_{\omega^2}(L) < L$$

and $L = \zeta_{\omega^2+1}(L)$. Finally K is a sum of nilpotent ideals and trivially $F_c \leqq \zeta_c(K)$, $F_c \nleqq \zeta_{c-1}(K)$ for all c. Hence $K = \zeta_\omega(K) > \zeta_c(K)$ for all c. □

The modules B described in the proof of part (a) of theorem 4.13 show that the bound ω mentioned in proposition 4.11 is the best possible. From theorem 4.9, theorem 4.13 and the fact that $\mathfrak{F} \cap \mathfrak{A} \leqq$ Max-*pu*, we see that over any field a $\mathfrak{N}(\mathfrak{F} \cap \mathfrak{A})$-algebra cannot in general be embedded in a $\mathfrak{N}(\mathfrak{F} \cap \mathfrak{A})$-algebra satisfying the maximal condition for ideals, even when the obvious restriction on the dimensions is satisfied.

Chapter 12

Frattini theory

1. The Frattini subalgebra

We continue the study, by ring-theoretic methods, of the structure of finitely generated soluble Lie algebras; and we consider in this vein the work on Frattini subgroups of Hall [287]. By analogy with group theory we make the following definition: the *Frattini subalgebra* $F(L)$ of a Lie algebra L is the intersection of its maximal (proper) subalgebras. If there are no such algebras, $F(L) = L$. The Frattini subalgebra of a finite-dimensional Lie algebra has been studied by Marshall [157], Barnes [21], and Towers [233]. The infinite-dimensional case has been studied by McInerney [163] and Towers [232].

We begin by noting some more or less obvious facts.

PROPOSITION 1.1 *For any Lie algebra L,*

$$F(L) \leqq L^2.$$

Proof: Subalgebras of codimension 1 are maximal, and L^2 is an intersection of ideals of codimension 1. □

An element $x \in L$ is a *nongenerator* if whenever $L = \langle X, x \rangle$ for $X \subseteq L$ it follows that $L = \langle X \rangle$. Exactly as for groups we have:

PROPOSITION 1.2 *For any Lie algebra L the Frattini subalgebra is the set of nongenerators of L.*

Proof: Let x be a nongenerator, and suppose $x \notin F(L)$. Then $F(L) \neq L$, and $x \notin M$ for some maximal subalgebra M of L. Now $L = \langle M, x \rangle$, whence $L = M$, a contradiction.

Now suppose $x \in F(L)$. We show that x is a nongenerator. Let $X \subseteq L$ be such that $L = \langle X, x \rangle$. If $x \in X$ then $L = \langle X \rangle$. If not there exists M

maximal subject to $\langle X\rangle \leqq M$, $x \notin M$. If M' is a subalgebra $> M$ then $x \in M'$ so $M' = L$. Therefore M is maximal, so $x \in M$, another contradiction. □

In general a Lie algebra may not possess any maximal subalgebras. For an example, first note:

LEMMA 1.3 *Every maximal subalgebra of a locally nilpotent Lie algebra is an ideal, and has codimension 1.*

Proof: Let L be locally nilpotent, with M maximal in L. Suppose that M is not an ideal of L. Then $L^2 \nleqq M$, and we may choose $x \in L^2 \backslash M$. Then

$$x = \sum_{i=1}^{n} [y_i, z_i]$$

for $y_i, z_i \in L$. Now $L = \langle M, x\rangle$ by maximality. Hence x, the y_i, and the z_i are contained in a subalgebra of L of the form

$$H = \langle x, a_1, \ldots, a_m\rangle$$

where $a_1, \ldots, a_m \in M$. Let $A = \langle a_1, \ldots, a_m\rangle \leqq M$. Then $x \notin A$. By local nilpotence, $H \in \mathfrak{N} \cap \mathfrak{F}$. Take K to be a subalgebra of H, maximal with respect to $A \leqq K$, $x \notin K$. As in proposition 1.2 K is maximal in H. Further, $x \in H^2$. Therefore $H = H^2 + K$ so by lemma 2.1.9 $H = K$, a contradiction.

Therefore $M \lhd L$. Then L/M is simple and locally nilpotent, hence 1-dimensional. □

COROLLARY 1.4 *If L is locally nilpotent then $F(L) = L^2$.*

Proof: By proposition 1.1 it is sufficient to show that $L^2 \leqq F(L)$. But if M is maximal in L, then by lemma 1.3 $L^2 \leqq M$. □

Thus a locally nilpotent Lie algebra has no maximal subalgebras if and only if it is perfect. The McLain algebra $\mathscr{L}_{\mathfrak{k}}(\mathbb{Z})$ is locally nilpotent and perfect.

On the other hand proposition 1.2 immediately implies:

PROPOSITION 1.5 *If $L \neq 0$ is finitely generated then $F(L) < L$.*

Proof: $F(L) = L$ if and only if every element of L is a nongenerator. If $L = \langle x_1, \ldots, x_n\rangle$ it follows that $L = \langle \varnothing \rangle = 0$. □

COROLLARY 1.6 If $F(L)$ is finitely generated and H is a subalgebra of L such that $L = F(L) + H$, then $L = H$. □

In group theory the Frattini subgroup is obviously characteristic. For Lie algebras it need not even be an ideal, as is shown by an example of Barnes [21].

Let

$$L = \langle a, b, c\colon [a, b] = c, [b, c] = a, [c, a] = b\rangle$$

over the field $\mathbb{Z}_2$. Then L is simple, but $F(L) = \langle a+b+c\rangle$.

However, $F(L)$ is characteristic in L if L is finite-dimensional over a field of characteristic zero (Towers [233]) and is an ideal of L if L is finite-dimensional soluble over an arbitrary field (Barnes and Gastineau-Hills [22]).

We therefore define the *Frattini ideal* $\Phi(L)$ to be the largest ideal of L contained in $F(L)$. Clearly we may replace $F(L)$ by $\Phi(L)$ in propositions 1.4, 1.5, 1.6.

2. Soluble algebras: preliminary reductions

Let Ω be a totally ordered set, and let L be a Lie algebra having a series $(\Lambda_\sigma, V_\sigma)_{\sigma\in\Omega}$ of type Ω (section 1.8). If each Λ_σ and each V_σ is an ideal of L we refer to $(\Lambda_\sigma, V_\sigma)_{\sigma\in\Omega}$ as an *ideal series*. A *chief series* is an ideal series which cannot be refined: no ideal of L lies strictly between some Λ_σ and the corresponding V_σ. As in Robinson [331] pp. 16–17 it follows easily that any ideal series refines to a chief series. A *chief factor* of L is a pair (H, K) of ideals of L, such that no ideal of L lies strictly between H and K, and such that $H > K$. By the usual refinement argument using Zassenhaus's lemma it follows that the chief factors are exactly the pairs $(\Lambda_\sigma, V_\sigma)$ coming from chief series of L. By abuse of language we speak of the chief factor H/K. Every chief factor is an irreducible L-module.

The usual refinement argument (Zassenhaus [352] p. 57) shows that any two *finite* series of ideals of L have isomorphic refinements, the isomorphisms being L-module isomorphisms. This result does not extend to arbitrary chief series. For example, let $\mathfrak{k} = \mathbb{Q}$, let $A = \mathbb{Q}[t]$ be a polynomial algebra; take isomorphic copies A_1 and A_2 of A, thought of as abelian algebras, and form the split extension

$$L = A_1 \dotplus A_2$$

where $A_1 \lhd L$ and the A_2-action is defined by $[p, q] = pq$ for $p \in A_1$, $q \in A_2$. Then the ideals of L inside A_1 are just the associative algebra ideals of $A = \mathbb{Q}[t]$. We can use the descending series of associative ideals

$$\mathbb{Q}[t] \geqq (t) \geqq (t^2) \geqq \ldots \geqq (t^n) \geqq \ldots$$

to construct a chief series from A_1 down to 0, and also from L down to A_1.

All the factors of this series have dimension 1. On the other hand, $\mathbb{Q}[t]/(t^2+1)$ is a chief factor of dimension 2. It can therefore be included in a chief series with no refinement isomorphic to the original series.

Suppose now that H/K is a chief factor of L, and let I be an ideal of L. The two series

$$0 \leqq H \leqq K \leqq L$$

$$0 \leqq I \leqq L$$

have L-isomorphic refinements. Thus either H/K is L-isomorphic to a chief factor H'/K' where $H' > K' \geqq I$, or to one where $I \geqq H' > K'$. We then say that H/K is (respectively) *above* or *below* I. This observation often helps to simplify proofs.

The object of this chapter is to study, for certain classes of Lie algebras, relations between the Frattini ideal, the Fitting and Hirsch-Plotkin radicals, and two other ideals which we now define.

Let

$$\psi(L) = \bigcap C_L(H/K)$$

where the intersection is taken over all chief factors H/K of L. Define $\tilde{\nu}(L)$ by

$$\tilde{\nu}(L)/\Phi(L) = \nu(L/\Phi(L)).$$

We say that L has *good Frattini structure* if

$$\nu(L) = \rho(L) = \psi(L) = \tilde{\nu}(L)$$

and if all four ideals are nilpotent. From this it follows that $\Phi(L)$ is nilpotent, being a subalgebra of $\tilde{\nu}(L)$.

It is a well-known fact that, with analogous definitions, finite groups have good Frattini structure; and Hall [287] proves that the same is true of finitely generated metanilpotent (i.e. nilpotent-by-nilpotent) groups. We shall show that a Lie algebra L has good Frattini structure in the following cases:

(i) $L \in \mathfrak{G} \cap \mathfrak{N}^2$ over a field of characteristic $p > 0$,
(ii) $L \in \mathfrak{G} \cap \mathfrak{N}\mathfrak{A}$
(iii) $L \in \mathfrak{G} \cap \mathfrak{N}^2 \cap \text{Min-}\lhd$.

Cases (ii) and (iii) are proved by Towers [232] pp. 71–83, and our proof for case (i) follows the same lines. McInerney [163] has studied analogous problems for Lie rings.

In the present section we reduce the problem to a special case: in the next section we solve the resulting problem.

It is very easy to deal with the Fitting radical:

LEMMA 2.1 *If $L \in \mathfrak{G} \cap \mathfrak{N}(\mathfrak{A}\mathfrak{F})$ then $\nu(L)$ is nilpotent.*

Proof: There exists $N \lhd L$ such that $N \in \mathfrak{N}$ and $L/N \in \mathfrak{G} \cap \mathfrak{A}\mathfrak{F}$. By theorem 11.1.6 L/N satisfies Max-$\lhd$, so we can choose an ideal M of L maximal with respect to $M \geqq N$, $M \in \mathfrak{N}$. Clearly $M = \nu(L)$, so $\nu(L) \in \mathfrak{N}$. □

It should be noted that $\mathfrak{G} \cap \mathfrak{N}^2 \leqq \mathfrak{G} \cap \mathfrak{N}\mathfrak{F} \leqq \mathfrak{G} \cap \mathfrak{N}(\mathfrak{A}\mathfrak{F})$.

LEMMA 2.2 *If $L \in \mathfrak{G} \cap \mathfrak{N}^2$ and $N = \nu(L)$, then $N^2 \leqq \Phi(L)$.*

Proof: Let M be a maximal subalgebra of L and suppose that $N^2 \nleqq M$. Since N is nilpotent there exists $k \geqq 2$ maximal with respect to $N^k \nleqq M$. Thus $L = N^k + M$. But then

$$N^2 = N \cap L^2 = (N^k + N \cap M)^2 \leqq N^{k+1} + N \cap M^2 \leqq M,$$

a contradiction. Hence $N^2 \leqq F(L)$, and since it is an ideal, $N^2 \leqq \Phi(L)$. □

LEMMA 2.3 *If $L \in \mathfrak{G} \cap \mathfrak{N}^2$ and $I = \nu(L)$, then $\nu(L/I^2) = \nu(L)/I^2$.*

Proof: We know that $\nu(L/I^2) = B/I^2$ is nilpotent by lemma 2.1. By proposition 7.1.1(c) B is nilpotent, so $B \leqq I$. Hence $B = I = \nu(L)$. □

Now we turn to the Hirsch-Plotkin radical.

PROPOSITION 2.4 *If $L \in \mathfrak{G} \cap \mathfrak{N}^2$ then $\rho(L) \leqq \psi(L)$.*

Proof: Let A/B be a chief factor of L. We wish to show that $\rho(L)$ centralises A/B, and to do this we may work modulo B or equivalently assume that A is a minimal ideal of L and $B = 0$. Let $R = \rho(L)$, $K = \nu(L) \leqq R$. Then $L/K \in \mathfrak{N} \cap \mathfrak{F}$. Therefore $R = \langle K, t_1, \ldots, t_r \rangle$ for finite r, the t_i being elements of R. Since A is abelian $A \leqq K \leqq R$. Since K is nilpotent $A \cap \zeta_1(K) \neq 0$, so by minimality $A \leqq \zeta_1(K)$. Let $0 \neq a \in A$. Then $N = \langle a, t_1, \ldots, t_r \rangle \leqq R$ so is nilpotent, and therefore $A \cap \zeta_1(N) \neq 0$. Thus there exists $c \in A$ such that $[N, c] = 0$. But $[K, c] = 0$ so $[R, c] = 0$. Therefore $A \cap \zeta_1(R) \neq 0$ and so $A \leqq \zeta_1(R)$. Thus $[A, R] = 0$ as required. □

We may now make a reduction in the problem.

PROPOSITION 2.5 *Let $L \in \mathfrak{G} \cap \mathfrak{N}^2$, $I = \nu(L)$. If L/I^2 has good Frattini structure, then L has good Frattini structure.*

Proof: We have $I \in \mathfrak{N}$ by lemma 2.1. By hypothesis,

$$\nu(L/I^2) = \rho(L/I^2) = \psi(L/I^2) = \tilde{\nu}(L/I^2) \in \mathfrak{N}.$$

By lemma 2.3,

$$\nu(L)/I^2 \in \mathfrak{N}$$

so that

$$\nu(L) \in \mathfrak{N}.$$

Further $\rho(L) \leqq \psi(L)$ by lemma 2.4, and clearly $\psi(L)+I^2/I^2 \leqq \psi(L/I^2)$, so that $\psi(L) \leqq \nu(L)$ and so

$$\rho(L) = \psi(L) = \nu(L).$$

Finally $I^2 \leqq \Phi(L)$ by lemma 2.2, which implies that $\Phi(L/I^2) = \Phi(L)/I^2$, whence $\tilde{\nu}(L/I^2) = \tilde{\nu}(L)/I^2$, so that

$$\tilde{\nu}(L) = \nu(L). \qquad \square$$

This enables us to concentrate on L/I^2, which lies in the class $\mathfrak{G} \cap \mathfrak{A}\mathfrak{N}$. Furthermore, its Fitting radical is I/I^2 which is *abelian*.

Following Towers [232] we define, for any $b \in L$, the *Engel subalgebra*

$$E_L(b) = \{x \in L\colon [x, {}_r b] = 0 \text{ for some } r\}.$$

(This is often called the *null-component* of b.) By Leibniz's rule for derivations it follows that $E_L(b)$ is a subalgebra. The next result is a version of Fitting's lemma in linear algebra:

LEMMA 2.6 *Let $b \in L$ be such that*

$$[L, {}_n b] = [L, {}_{n+1} b]$$

for some integer n. Then

$$L = E_L(b) + [L, {}_n b].$$

Proof: Let $x \in L$. Then $[L, {}_n b] = [L, {}_{2n} b]$ so there exists $y \in L$ such that

$$[x, {}_n b] = [y, {}_{2n} b].$$

Therefore

$$\begin{aligned} x &= (x - [y, {}_n b]) + [y, {}_n b] \\ &\in E_L(b) + [L, {}_n b]. \end{aligned} \qquad \square$$

We now depart from the group-theoretic line of argument, and look at what happens for monolithic algebras.

THEOREM 2.7 *Let $L \in \mathfrak{G} \cap \mathfrak{N}^2$ be monolithic, with monolith A. Then either*

(a) *$L/A \in \mathfrak{N}$ and $\Phi(L) = 0$ or*
(b) *$\nu(L/A) = \nu(L)/A$.*

Proof: Let $\nu(L/A) = N/A$. If $N \leqq C_L(A)$ then case (b) holds. So we may assume $N \nleqq C_L(A)$. If $\nu(L) = 0$ then $L = 0$ and the result is trivial. Therefore A, being the monolith, is contained in $\nu(L)$. Therefore $A \leqq \zeta_1(\nu(L))$ and $\nu(L) \leqq C_L(A)$. Hence $L/C_L(A) \in \mathfrak{N}$, and we can find a nonzero element of $(N+C_L(A))/C_L(A)$ central in $L/C_L(A)$, say $b+C_L(A)$. Then $b \in N \backslash C_L(A)$, and $[L, b] \leqq C_L(A)$. Let $D = \langle b \rangle + C_L(A)$ which is an ideal of L. Since $[D, A] \neq 0$ we have $[D, A] = A$, so $[b, A] = A$. There exists n such that $[L, {}_n b] = A$. Hence by lemma 2.6

$$L = E_L(b) + [L, {}_n b] = E_L(b) + A.$$

Now since $[D, A] \neq 0$, we have $\zeta_1(D) = 0$. Also,

$$\begin{aligned} C_L(A) &= (E_L(b)+A) \cap C_L(A) \\ &= (E_L(b) \cap C_L(A)) + A. \qquad (*) \end{aligned}$$

Now $E_L(b) \cap C_L(A)$ is idealised by $E_L(b)$ and by A, so is an ideal of L. If it is non-zero we have $A \leqq E_L(b)$. But if $a \in A$ then $[a, {}_t b] \neq 0$, $[a, {}_{t+1} b] = 0$, for some t. So

$$[[a, {}_t b], D] = [[a, {}_t b], b + C_L(A)] = 0$$

and therefore $[a, {}_t b] \in \zeta_1(D) = 0$, a contradiction. Therefore $E_L(b) \cap C_L(A) = 0$. From (*) we have $C_L(A) = A$. Therefore $A = \nu(L)$, and L/A is nilpotent.

Further, since $L = A \dotplus E_L(b)$ (vector space direct sum) $E_L(b)$ is a maximal subalgebra, and $\Phi(L) \leqq E_L(b)$. But if $\Phi(L) \neq 0$ then $A \leqq \Phi(L)$, another contradiction. Thus $\Phi(L) = 0$. □

The major obstacle between us and the proof that finitely generated metanilpotent Lie algebras L have good Frattini structure is the proof that $\psi(L) = \nu(L)$. To avoid an immediate confrontation we introduce the property

$$\psi(L/K) = \nu(L/K) \textit{ for all ideals } K \textit{ of } L. \qquad (\Delta)$$

THEOREM 2.8 *If $L \in \mathfrak{G} \cap \mathfrak{A}\mathfrak{N}$ has property (Δ) then*

$$\tilde{\nu}(L) = \nu(L).$$

Proof: Assume the contrary. Then there exists an ideal I of L maximal with respect to $\tilde{\nu}(L/I) \neq \nu(L/I)$. Replacing L by L/I we may assume that:

$$\tilde{\nu}(L/J) = \nu(L/J) \text{ for all } 0 \neq J \lhd L. \tag{**}$$

By property (Δ) there is some chief factor A/B of L not centralised by $\tilde{\nu}(L)$. If $B \neq 0$ then A/B is a chief factor of L/B, so is centralised by $\nu(L/B) = \tilde{\nu}(L/B)$ by (**), and hence is centralised by $\tilde{\nu}(L)$. Therefore $B = 0$ and A is a minimal ideal of L. If there is an ideal C such that $A \cap C = 0$ then $A \cong_L A+C/C$ and a similar argument applies. Hence L is monolithic with monolith A. If $\Phi(L) = 0$ we are home. Otherwise by theorem 2.7 $\nu(L/A) = \nu(L)/A$. Since $A \leqq \Phi(L)$ it is also clear that $\Phi(L/A) = \Phi(L)/A$. But then, since $\tilde{\nu}(L/A) = \nu(L/A)$, we have $\tilde{\nu}(L) = \nu(L)$, a contradiction. □

Corollary 2.9 *If $L \in \mathfrak{G} \cap \mathfrak{A}\mathfrak{N}$ has property (Δ) then L has good Frattini structure.*

Proof: We have $\nu(L) \leqq \rho(L) \leqq \psi(L)$. By (Δ) we have $\psi(L) \leqq \nu(L)$. So $\nu(L) = \rho(L) = \psi(L)$ and clearly these are nilpotent. By theorem 2.8 $\tilde{\nu}(L) = \nu(L)$. □

There is one case where (Δ) trivially holds:

Proposition 2.10 *If L has a finite chief series, i.e. a series*

$$0 \leqq L_0 \leqq L_1 \leqq \ldots \leqq L_n = L$$

with each L_{i+1}/L_i a chief factor of L, then L has property (Δ).

Proof: If $K \lhd L$ then L/K has a finite chief series. Since $\psi(L/K)$ centralises the factors, we have $\psi(L/K) \in \mathfrak{N}$. Thus $\nu(L/K) \leqq \psi(L/K) \leqq \nu(L/K)$. □

3. Proof of the main theorem

Let U be an associative $\mathfrak{k}$-algebra, with centre Z. If M is a U-module we define a *chief factor* of M to be a module of the form H/K, where H and K are submodules of M and H/K is irreducible. Let

$$\psi_U(M)$$

be the intersection of the annihilators in U of all chief factors H/K of M. We say that U has the *chief annihilator property* if whenever M is a finitely generated U-module and $z \in Z \cap \psi_U(M)$ there exists an integer n such that $Mz^n = 0$.

The importance of this for us lies in

LEMMA 3.1 *Let $L \in \mathfrak{G} \cap \mathfrak{N}^2$ and suppose that $A = \nu(L)$. Suppose that the universal enveloping algebra $U(L/A)$ has the chief annihilator property. Then $\nu(L) = \psi(L)$.*

Proof: By lemma 2.3 we may assume A abelian. We know that $\nu(L) \leqq \psi(L)$. If these are not equal we can find an element $a+A$ of the centre of L/A, where $a \in \psi(L)\backslash A$. Now let $z = a+A \in U(L/A)$. If Z is the centre of $U(L/A)$ we have $z \in Z$. Since $a \in \psi(L)$ it follows that $z \in \psi_U(A)$. Now A is a finitely generated $U(L/A)$-module by lemma 11.1.5(b). By the chief annihilator property $Az^n = 0$. Therefore $\langle A, a \rangle$ is a nilpotent ideal of L, contrary to the definition of A. □

COROLLARY 3.2 *Let $L \in \mathfrak{G} \cap \mathfrak{N}^2$. Suppose that for every ideal K of L the universal enveloping algebra*

$$U((L/K)/\nu(L/K))$$

has the chief annihilator property. Then L has good Frattini structure.

Proof: By lemma 3.1 L has property (Δ), whence the result follows from lemma 2.1 and 2.3. □

Conversely, suppose we can find a finite-dimensional nilpotent Lie algebra N such that $U(N)$ does not have the chief annihilator property (with the element z relevant to this lying in the centre of N). Then the split extension

$$L = M \dotplus N$$

lies in $\mathfrak{G} \cap \mathfrak{A}\mathfrak{N}$. Further $z \in \psi(L)$. But $z \notin \nu(L)$ since $[M, {}_n z] \neq 0$ for the relevant module M.

This shows that the chief annihilator property is the crucial point in the theory. Hall [287] shows that if $\mathfrak{k}$ is an absolutely algebraic field of prime characteristic and G is a finitely generated nilpotent group, the group algebra $\mathfrak{k}G$ has the chief annihilator property. Towers [232] applies Hall's method to a polynomial algebra to obtain his theorem. However, more is possible:

THEOREM 3.3 *Let L be a finite-dimensional Lie algebra over $\mathfrak{k}$ and suppose that either*

(i) *$\mathfrak{k}$ has characteristic $p > 0$,*
(ii) *L is abelian.*

Then the universal enveloping algebra U of L has the chief annihilator property.

Proof: Let M be a finitely generated U-module, and let $z \in Z \cap \psi_U(M)$ where Z is the centre of U. We must show that $Mz^n = 0$ for some n. Now M is a Noetherian U-module. If $Mz^n \neq 0$ for all n, we can choose a submodule I of M maximal with respect to $Mz^n \nleqq I$ for all n. Replacing M by M/I we can assume that $Mz^n \leqq J$ for every non-zero submodule J of M.

Let $J = \{m \in M : mz = 0\}$. If $J \neq 0$ then $Mz^n \leqq J$ for some n, and then $Mz^{n+1} = 0$. Therefore $J = 0$. The map

$$\beta\colon M \to M$$

$$m \mapsto mz$$

is a module monomorphism (since z is central). We can therefore find a sequence

$$M = M_0 \xrightarrow{\beta_0} M_1 \xrightarrow{\beta_1} M_2 \xrightarrow{\beta_2} \dots$$

of U-modules, all isomorphic to M, as follows: let $m \mapsto m_k$ be a U-isomorphism $M \to M_k$. Define

$$\beta_k(m_k) = (mz)_{k+1}.$$

Let $\overline{M}$ be the direct limit of this sequence. We may then assume that

$$M = M_0 \leqq M_1 \leqq \dots \leqq \overline{M}$$

and that each M_i is a U-submodule of $\overline{M}$. Let Y be an indeterminate, and make $\overline{M}$ into a $U[Y]$-module by letting Y act as β^{-1}, where

$$\beta\colon \overline{M} \to \overline{M}$$

$$m \mapsto mz$$

is now an automorphism. Clearly $\overline{M}$ is a finitely generated $U[Y]$-module. There is a $U[Y]$-submodule K maximal with respect to $M \nleqq K$. If L is larger than K then $M \leqq L$, so $L = MU[Y] = \overline{M}$. Thus K is maximal in $\overline{M}$. Now $U[Y]$ is the universal enveloping algebra of a direct sum $L \oplus T$ where T is 1-dimensional Lie algebra. Then $\overline{M}/K$, being an irreducible $U[Y]$-module, is finite dimensional: this follows in case (i) by Curtis's theorem, and in case (ii) by the Hilbert Nullstellensatz, in the usual manner.*

Let $K_0 = M \cap K$. Then K_0 is a U-module, and $M/K_0 \cong M + K/K$ which is finite-dimensional. Now $K_0 \neq M$, and we can take $K_1 \leqq M$ maximal with respect to $K_0 \leqq K_1$. Since $z \in \psi_U(M)$ we have $Mz \leqq K_1$. Further, $K_0 z \leqq K$. Hence the endomorphism induced by β on M/K_0 is not an automorphism.

---

*) In the example given on page 702 of [11] it is tacitly assumed that the field has characteristic 0.

By finiteness of the dimension there exists $m \in M$ such that $mz \in K_0$ but $m \notin K_0$, so that $mz \in K$ but $m \notin K$. But $\beta(mz) = m$, so that $m \in K$ since K is a $U[Y]$-module. This is a contradiction, and the theorem follows. □

THEOREM 3.4 *Suppose that either*

(i) $L \in \mathfrak{G} \cap \mathfrak{N}^2$ *over a field of characteristic* $p > 0$, *or*
(ii) $L \in \mathfrak{G} \cap \mathfrak{N}\mathfrak{A}$.

Then L has good Frattini structure.

Proof: This follows from theorem 3.3 and corollary 3.2. □
As in Hall [287] we have the following corollary:

COROLLARY 3.5 *Suppose that* $L \in \mathfrak{G} \cap \mathfrak{N}^{k+1}$ *over a field of characteristic* $p > 0$, *where k is an integer* $\geqq 1$. *Then* $\Phi(L) \in \mathfrak{N}^k$. □

4. Nilpotency criteria

In this section we give some criteria whereby certain finitely generated soluble algebras may be proved nilpotent. They are largely due to Towers [232].

THEOREM 4.1 *Let L have good Frattini structure. Then* $\Phi(L) = L^2$ *if and only if L is nilpotent.*

Proof: If L has good Frattini structure and $\Phi(L) = L^2$ then $L = \tilde{\nu}(L) = \nu(L)$ is nilpotent. The converse follows by corollary 1.4. □

COROLLARY 4.2 *If L is soluble and has good Frattini structure then L is nilpotent if and only if every maximal subalgebra is an ideal.* □

It is known (Robinson [330]) that a finitely generated soluble (even hyper-abelian) group, all of whose finite homomorphic images are nilpotent, is itself nilpotent. However, the Hartley algebra has all its finite-dimensional homomorphic images nilpotent, but is not itself nilpotent, so the analogous result is false even for the class $\mathfrak{G} \cap \mathfrak{A}\mathfrak{N}$ (in characteristic zero). However, we have two positive results.

THEOREM 4.3 *Let L be a finitely generated soluble Lie algebra over a field of characteristic $p > 0$, and suppose that every finite-dimensional homomorphic image of L is nilpotent. Then L is nilpotent.*

Proof: We use induction on the derived length of L, the theorem being trivial if this is 1. Thus if $L^{(d)} = 0$ then every finite-dimensional homomorphic image of $L/L^{(d-1)}$ is a homomorphic image of L, so nilpotent; and by induction $L/L^{(d-1)}$ is nilpotent. So we may assume that $L \in \mathfrak{G} \cap \mathfrak{A}\mathfrak{N}$. Now we may use the fact that L has good Frattini structure. We show that every maximal subalgebra M of L is an ideal, and the result will follow from corollary 4.2.

We can find an integer n such that L^n is abelian; and $L/L^n \in \mathfrak{G} \cap \mathfrak{N}$ so is finite-dimensional. If $L^n \leqq M$ the result is immediate, so we may assume that $L = L^n + M$. But now

$$\begin{aligned}[L^n \cap M, L^n + M] &\leqq [L^n, L^n] + [L^n \cap M, M] \\ &\leqq L^n \cap M\end{aligned}$$

since L^n is an abelian ideal. Therefore $L^n \cap M \lhd L$. We show that if C is an ideal of L such that

$$L^n \cap M < C \leqq L^n$$

then $C = L^n$. For if $C \leqq M$ then $C \leqq L^n \cap M$. So $L = C + M$, and

$$L^n = L^n \cap (L^n + M) = L^n \cap (C + M) = C + (L^n \cap M) = C.$$

Therefore $L^n/(L^n \cap M)$ is a chief factor of L. It is centralised by L^n (good Frattini structure) so by Curtis's theorem it is finite-dimensional. Therefore $L/(L^n \cap M)$ is finite-dimensional, so nilpotent by hypothesis. But $M/(L^n \cap M)$ is a maximal subalgebra of $L/(L^n \cap M)$, so is an ideal; and therefore M is an ideal of L.

By corollary 4.2 it follows that L is nilpotent, which completes the induction step and the proof of the theorem. □

A very similar argument gives:

THEOREM 4.4 *Let $L \in \mathfrak{G} \cap \mathfrak{N}\mathfrak{A}$, and suppose that every finite-dimensional homomorphic image of L is nilpotent. Then L is nilpotent.*

Proof: Again we show that every maximal subalgebra is an ideal. By lemma 2.2 we have $L^{(2)} \leqq \Phi(L)$, so we may assume L^2 abelian. The proof now proceeds as above, but with L^2 in place of L^n, and using lemma 11.2.4 instead of Curtis's theorem. □

The force of these results is as follows: let $\mathfrak{X}$ be any Q-closed class of Lie algebras such that finite-dimensional soluble $\mathfrak{X}$-algebras are nilpotent. Then in characteristic $p > 0$ all finitely generated soluble $\mathfrak{X}$-algebras are nilpotent; in characteristic zero all $\mathfrak{G} \cap \mathfrak{N}\mathfrak{A} \cap \mathfrak{X}$-algebras are nilpotent. For $\mathfrak{X}$ one might take the class of algebras in which all maximal subalgebras are ideals, or algebras satisfying a general Engel condition. Note that the hypothesis that L be finitely generated cannot be dropped, since then we could take $\mathfrak{X} = \text{L}\mathfrak{N}$ and deduce that all soluble locally nilpotent algebras are nilpotent, which is false. The best we can say along these lines is the obvious:

COROLLARY 4.5 *Let L be a soluble Lie algebra over a field of characteristic $p > 0$, or nilpotent-by-abelian over a field of characteristic zero. If every finite-dimensional quotient of every subalgebra of L is nilpotent, then L is locally nilpotent.* □

In a similar spirit we may prove analogues of two theorems of Stroud [343] on nilpotency properties of $\mathfrak{G} \cap \mathfrak{A}\mathfrak{N}$-algebras.

THEOREM 4.6 *Let $L \in \mathfrak{G} \cap \mathfrak{A}\mathfrak{N}$. Then $\zeta_1(L) \cap L^n = 0$ for some integer n.*

Proof: Assume not. There exists $K \lhd L$ maximal with respect to the condition

$$\zeta_1(L/K) \cap (L/K)^n \neq K/K.$$

Replacing L by L/K we may assume the theorem true for L/J whenever $0 \neq J \lhd L$. Let $H = \nu(L)$, which is nilpotent, and $B = \zeta_1(H)$. Now $H \neq L$ or the theorem is trivially true. Further $0 \neq C = \zeta_1(L) \leqq B$. We can find $t \in L \backslash H$ such that $L + t \in \zeta_1(L/H)$. Then $L_1 = H + \langle t \rangle \lhd L$. Let $\tau : B \to L$ be defined by

$$b\tau = [b, t]$$

so that $\tau = t^*|_B$. We have $B^{\tau^n} = [B, {}_nL_1]$. Define C_n to be the kernel of τ^n. Then inductively we find that $C_n = B \cap \zeta_n(L_1)$. So the C_i form an ascending chain of ideals of L, which stops by Max-$\lhd$. Therefore $C_n = C_{n+1} = \ldots$ for some integer n. Then $B^{\tau^n} \cap C_n = 0$ by the argument of lemma 2.6. If $K = B^{\tau^n} = 0$ then $L_1 \in \mathfrak{N}$ contrary to the definition of H. Therefore $K \neq 0$, so $\zeta_1(L/K) \cap (L/K)^n = K/K$ for some n, whence $C \cap L^n \leqq K$. But $C \leqq C_n$ and $C_n \cap K = 0$, so $C \cap L^n = 0$, contrary to hypothesis. □

We now introduce a concept weaker than that of a monolith. An ideal K of a Lie algebra L is a *trunk* if $0 \neq M \lhd L$ implies $M \cap K \neq 0$. As an example, we cite the centre of a nilpotent Lie algebra.

THEOREM 4.7 *Let K be a trunk of $L \in \mathfrak{G} \cap \mathfrak{AN}$. Then $C_L(K)$ is nilpotent.*

Proof: Let $D = C_L(K)$, $F = v(L)$. We prove $D \leqq F$. If not, there exists $t \in D \backslash F$ such that $F + t \in \zeta_1(L/F)$. Let $B = \zeta_1(F)$, and put $\tau = t^*|_B$ as before. Then $B^{\tau^n} \cap C_n = 0$ for some n, defining C_n as above. Let $T = B^{\tau^n}$. If $T \neq 0$ then $T \cap K \neq 0$, but $[K, D] = 0$, so $(K \cap T)^\tau = 0$, whence $K \cap T \leqq C_1 \leqq C_n$ so $C_n \cap T = K \cap T$, a contradiction. Therefore $B^{\tau^n} = 0$, so $B = \zeta_1(F) \leqq \leqq \zeta_n(F + \langle t \rangle)$ and so $F + \langle t \rangle$ is nilpotent, a contradiction. □

5. A splitting theorem

Following Towers [232] we generalise to the class $\mathfrak{G} \cap \mathfrak{N}^2$ a splitting theorem due, in the finite-dimensional case, to Barnes and Newell [23]. Our proof is a little simpler than that of Towers.

THEOREM 5.1 *Let $L \in \mathfrak{G} \cap \mathfrak{N}^2$, and suppose that L has a minimal ideal A such that $A = C_L(A)$. Then L splits over A, and if U and V are complements of A in L there exists an automorphism $\alpha = 1 + a^*$ $(a \in A)$ such that $V = U^\alpha$. Furthermore, $\Phi(L) = F(L) = 0$.*

Proof: Let $B = v(L)$. Then $A \leqq B$ and $[A, B] = 0$, whence $B \leqq C_L(A) = A$ and $A = B$. Hence L/A is finite-dimensional nilpotent. We can find $u \in L \backslash A$ such that $[L, u] \leqq A$. Let $C = A + \langle u \rangle \lhd L$. As usual we have $[C, A] = A$. Therefore $[A, u] = A$, so that $[L, u] = A$. Thus if $x \in L$ then $[x, u] \in A = [A, u]$ so $[x, u] = [a, u]$ for some $a \in A$, from which it follows that $L = A + C_L(u)$. But $C_L(u) \cap A = 0$, since if $0 \neq a \in C_L(u) \cap A$ then $a \in \zeta_1(C) \lhd L$, so $A \leqq \leqq \zeta_1(C)$, so $[C, A] = 0$. Hence $C_L(u)$ is a complement to A in L.

Now let V be another complement. There exist elements $b \in A$, $v \in V$ such that $u = b + v$. We can find $a \in A$ such that $[u, a] = -b$, so if $\alpha = 1 + a^*$ then $u^\alpha = u + [u, a] = u - b = v$. Therefore $U^\alpha \leqq C_L(v)$. This contains v. Now V is nilpotent, and if $x \in V$ then $[x, v] = [x, a + u] \in A \cap V = 0$. So $C_L(v) \geqq V$. By dimensional considerations $V = C_L(v) = U^\alpha$.

Finally we show that $F(L) = 0$. Certainly $F(L) \leqq U$ since U is maximal. Let $0 \neq x \in U$. Then $x \notin C_L(A)$ so $[x, b] \neq 0$ for some $b \in A$. Let $\beta = 1 + b^*$, and suppose $x \in U^\beta$. Then $x = y+[y, b]$ $(y \in U)$, so $x-y \in U \cap A = 0$, and then $x = x+[x, b]$ which is a contradiction. Since U^β is maximal, we have $F(L) \leqq U \cap U^\beta = 0$. □

This theorem applies, in particular, to the Hartley algebra L. So for this we have $F(L) = 0$, giving an example of an infinite-dimensional $\mathfrak{G} \cap \mathfrak{A}\mathfrak{N}$-algebra with trivial Frattini subalgebra.

Chapter 13

Neoclassical structure theory

1. Classical structure theory

The 'classical' structure theory of finite-dimensional Lie algebras over a field of characteristic 0 provides a very powerful tool for the study of these algebras. In this chapter we generalise it to certain classes of locally finite Lie algebras. As is usual in such a progression, the finite theory is itself the main tool in obtaining the generalised theory. The material is taken from [213, 214] and is extended according to a suggestion made in [215].

We begin with a summary of the relevant parts of the classical theory, proofs of which may be found in Jacobson [98].

SUMMARY 1.1 *Let L be a finite-dimensional Lie algebra over a field $\mathfrak{k}$ of characteristic 0. Then:*

(a) *L has a unique maximal nilpotent ideal N and a unique maximal soluble ideal R (the nil radical and radical, respectively).*

(b) *L/R is semisimple, that is, has trivial radical.*

(c) *If δ is a derivation of L then $R\delta \subseteq N$. In particular $R^2 \leqq N$ so that $R \in \mathfrak{N}\mathfrak{A}$.*

(d) *Every semisimple algebra is a direct sum of simple ideals. Further, the simple algebras can be classified (at any rate if $\mathfrak{k}$ is algebraically closed).*

(e) *Every derivation of a semisimple algebra is inner.*

(f) *(Levi) If L has radical R there exists at least one Levi factor; that is, a semisimple subalgebra Λ such that $L = R \dotplus \Lambda$.*

(g) *Every semisimple subalgebra of L is contained in a Levi factor.*

(h) (*Mal'cev–Harish-Chandra*) *All Levi factors of L are conjugate under the group of automorphisms of L generated by elements* $\exp(x^*)$ *for $x \in N$.* □

In succeeding sections we shall be most interested in (d), (f), (g), (h); but some preliminary material is needed to define the relevant class of algebras and develop properties of radicals.

2. Local subideals

A subalgebra H of a Lie algebra L is a *local subideal* of L if whenever X is a finite subset of L we have

$$H \text{ si } \langle H, X \rangle.$$

We write

$$H \text{ lsi } L.$$

A class $\mathfrak{X}$ is lsi-*coalescent* if any pair of local subideals belonging to $\mathfrak{X}$ generate a local subideal in $\mathfrak{X}$. The next lemma is immediate:

LEMMA 2.1 *Let the underlying field have characteristic 0. If $\mathfrak{X}$ is coalescent and is a subclass of $\mathfrak{G}$ then $\mathfrak{X}$ is* lsi-*coalescent.*

Proof: Let $H, K \in \mathfrak{X}$ and suppose both are local subideals of L. Suppose X is a finite subset of L. Since $\mathfrak{X} \leqq \mathfrak{G}$ there exist finite sets H_0, K_0 generating H, K respectively. Then

$$H \text{ si } \langle H, K_0 \cup X \rangle = \langle \langle H, K \rangle, X \rangle$$

$$K \text{ si } \langle K, H_0 \cup X \rangle = \langle \langle H, K \rangle, X \rangle$$

so by coalescence $\langle H, K \rangle \in \mathfrak{X}$ and

$$\langle H, K \rangle \text{ si } \langle \langle H, K \rangle, X \rangle. \qquad \square$$

COROLLARY 2.2 *Over fields of characteristic 0, the classes $\mathfrak{F}$, $\mathfrak{F} \cap \mathfrak{N}$, $\mathfrak{F} \cap \mathrm{E}\mathfrak{A}$, and $\mathfrak{G}$ are* lsi-*coalescent.* $\square$

We may now define the class $\mathfrak{H}$ of *neoclassical* algebras, over a field $\mathfrak{k}$ of characteristic 0. A Lie algebra L belongs to $\mathfrak{H}$ if and only if L is generated by a set of finite-dimensional local subideals.

From corollary 2.2 we see that $\mathfrak{H}$-algebras are locally finite: indeed every finite subset of an $\mathfrak{H}$-algebra is contained in a finite-dimensional local subideal. Clearly $\mathfrak{F} \leqq \mathfrak{H}$; and by lemma 1.3.7 it follows that $\mathrm{L}\mathfrak{N} \leqq \mathfrak{H}$. If we define closure operations N, Ń as follows:

N: $\mathrm{N}\mathfrak{X}$ consists of all algebra generated by $\mathfrak{X}$-subideals,

Ń: $\acute{\mathrm{N}}\mathfrak{X}$ consists of all algebras generated by ascendant $\mathfrak{X}$-subalgebras,

then $\mathrm{N}\mathfrak{F} \leqq \acute{\mathrm{N}}\mathfrak{F} \leqq \mathfrak{H}$. In [213] the class studied was in fact $\acute{\mathrm{N}}\mathfrak{F}$. To complete the picture we introduce the class $\overline{\mathfrak{F}}$ of algebras generated by $\mathfrak{F}$-ideals. Then

$$\overline{\mathfrak{F}} \leqq \mathrm{N}\mathfrak{F} \leqq \acute{\mathrm{N}}\mathfrak{F} \leqq \mathfrak{H}.$$

Now

$$\overline{\mathfrak{F}} \cap \mathrm{L}\mathfrak{N} \leqq \mathfrak{F}\mathfrak{t}$$

$$\mathrm{N}\mathfrak{F} \cap \mathrm{L}\mathfrak{N} = \mathfrak{B}$$

$$\acute{\mathrm{N}}\mathfrak{F} \cap \mathrm{L}\mathfrak{N} = \mathfrak{G}\mathfrak{r}$$

so by the results of chapter 6 the inclusions between the four classes are strict.

From theorem 3.2.4 it is easy to see that over fields of characteristic 0 the classes $\overline{\mathfrak{F}}$, $\mathrm{N}\mathfrak{F}$, $\acute{\mathrm{N}}\mathfrak{F}$, $\mathfrak{H}$ are all QS-closed. None of them is E-closed: take a vector space V with basis $v_0, v_1, v_2, \ldots$ and let σ be the *upward shift* with $v_i\sigma = v_{i+1}$. The split extension $V \dotplus \langle \sigma \rangle$ is soluble, so in $\mathrm{E}\overline{\mathfrak{F}}$; but it is generated by two elements and has infinite dimension, so cannot be locally finite. Hence $\mathrm{E}\overline{\mathfrak{F}} \nleqq \mathrm{L}\mathfrak{F}$, which implies the non-E-closure of the classes. Note further that $\mathrm{L}\mathfrak{F}$ is not E-closed, in contrast to the situation for groups.

These closure properties will be used without further comment.

We have a simple parallel to lemma 1.3.2:

LEMMA 2.3 *If H is a local subideal of the Lie algebra L then H^{ω} is an ideal of L.*

Proof: Let $x \in L$. Then H si $\langle H, x \rangle$, so $H^{\omega} \lhd \langle H, x \rangle$, so $[H, x] \subseteq H$. □

We can give an alternative definition of the class $\mathfrak{H}$ which may be useful. Say that a subalgebra $H \leqq L$ is *serial*, written

$$H \text{ ser } L,$$

if there is a series from H to L in the sense of chapter 1 section 8. Exactly as in Hartley [291] we have:

PROPOSITION 2.4 *Let L be a locally finite Lie algebra and H a subalgebra. Then H ser L if and only if $H \cap F$ si F for every finite-dimensional subalgebra F of L. The set of serial subalgebras of L is a complete sublattice of the lattice of subalgebras of L. Further, if H ser L and θ is a homomorphism of L, then $\theta(H)$ ser $\theta(L)$.*

Proof: Argue as in [291]. The technique of Mal'cev systems used to prove 'local theorems' (cf. Robinson [332] p. 94) applies to Lie algebras by virtue of the general theory of such systems given by McLain [312]. □

3. Radicals in locally finite algebras

We begin this section with a brief discussion of Fitting classes of finite-dimensional Lie algebras. It is these which give rise to useful radicals in the finite-dimensional theory; and they turn out to be important for the locally finite case as well. A *Fitting class* is a subclass $\mathfrak{X}$ of $\mathfrak{F}$ which is $\{\mathrm{N}_0, \mathrm{I}\}$-closed. Obvious examples are $\mathfrak{F} \cap \mathfrak{N}$, $\mathfrak{F} \cap \mathrm{E}\mathfrak{A}$, $\mathfrak{F}$; or the class of semisimple $\mathfrak{F}$-algebras over fields of characteristic 0. The Fitting classes form a complete lattice. If $\mathfrak{X}$ is an $\{\mathrm{N}_0, \mathrm{I}\}$-closed class then $\mathfrak{X} \cap \mathfrak{F}$ is a Fitting class. For any Fitting class $\mathfrak{X}$ and any $\mathfrak{F}$-algebra L the sum of all the $\mathfrak{X}$-ideals of L is the unique maximal $\mathfrak{X}$-ideal of L. We call this the *$\mathfrak{X}$-radical* of L and denote it by $\rho_{\mathfrak{X}}(L)$. This usage does not conflict with our earlier definition of radicals $\rho_{\mathrm{L}\mathfrak{X}}(L)$. In this notation $\rho_{\mathfrak{N}}(L)$ is the classical nil radical and $\rho_{\mathrm{LE}\mathfrak{A}}(L)$ the (soluble) radical.

To state some of our later results in full generality we need to quote a theorem of Tuck [234] and Towers [232, 233]:

THEOREM 3.1 *If L is a finite-dimensional Lie algebra over a field of characteristic 0 and H is a subspace of L invariant under all automorphisms of L, then H is invariant under all derivations of L, so is a characteristic ideal.*

Proof: See the cited papers, or Chevalley [39]. □

COROLLARY 3.2 *If $\mathfrak{X}$ is a Fitting class over a field of characteristic 0 and $L \in \mathfrak{F}$, then $\rho_{\mathfrak{X}}(L)$ is a characteristic ideal of L.* □

(Note: we can avoid the Tuck-Towers theorem if we assume $\mathfrak{X} \leqq \mathrm{E}\mathfrak{A}$ by using part (c) of summary 1.1, as outlined in [213] p. 82.)

From this we can derive:

LEMMA 3.3 *If $\mathfrak{X}$ is a Fitting class over a field of characteristic 0 and $L \in \mathfrak{F}$, then $\rho_{\mathfrak{X}}(L)$ contains every $\mathfrak{X}$-subideal of L.*

Proof: Let H si L, $H \in \mathfrak{X}$. There is a series

$$H = H_0 \lhd H_1 \lhd \ldots \lhd H_n = L.$$

For each i, $\rho_{\mathfrak{X}}(H_i)$ ch $H_i \lhd H_{i+1}$, so that $\rho_{\mathfrak{X}}(H_i) \leqq \rho_{\mathfrak{X}}(H_{i+1})$. Therefore $H = \rho_{\mathfrak{X}}(H_0) \leqq \ldots \leqq \rho_{\mathfrak{X}}(L)$. □

COROLLARY 3.4 *Over a field of characteristic 0, if $\mathfrak{X}$ is a Fitting class and H si $L \in \mathfrak{F}$, then*

$$\rho_{\mathfrak{X}}(H) = H \cap \rho_{\mathfrak{X}}(L).$$

Proof: It is clear that $\rho_{\mathfrak{X}}(H) \supseteq H \cap \rho_{\mathfrak{X}}(L)$. The reverse inclusion follows from lemma 3.3. □

COROLLARY 3.5 *Over a field of characteristic 0, every Fitting class is coalescent, ascendantly coalescent, and* lsi-*coalescent.*

Proof: We deal with ascendant coalescence. The others are similar. Suppose H, K asc L and $H, K \in \mathfrak{X}$. Then $J = \langle H, K \rangle$ asc L, and $J \in \mathfrak{F}$ by theorem 3.2.5. By lemma 3.3 both H and K lie in the $\mathfrak{X}$-radical of J, so that $J \in \mathfrak{X}$. □

Fitting classes contained in $\mathfrak{F} \cap \mathrm{E}\mathfrak{A}$ are in a sense dual to the *formations* of Barnes and Gastineau-Hills [22]. These are defined by analogy with the group-theoretical work of Gaschütz [276]: the group-theoretic dualisation is due to Fischer [274] and is developed in Hartley [290]. Some results on finite-dimensional Fitting classes of Lie algebras are to be found in Fritz [61].

If $\mathfrak{X}$ is an s-closed Fitting class then by theorem 6.1.3 we can define the $\mathrm{L}\mathfrak{X}$-radical $\rho_{\mathrm{L}\mathfrak{X}}(L)$ of any Lie algebra L. In particular we may set

$$\sigma(L) = \rho_{\mathrm{LE}\mathfrak{A} \cap \mathrm{L}\mathfrak{F}}(L),$$

which will play the rôle of the classical soluble radical: instead of the nil radical we use the Hirsch-Plotkin radical $\rho(L)$. Although $\rho(L)$ need not contain every ascendant locally nilpotent subalgebra (chapter 6 section 3), it will provided L is locally finite. To prove this (in greater generality) we need an alternative characterisation of $\rho_{\mathrm{L}\mathfrak{X}}(L)$ for locally finite algebras L.

For any locally finite algebra L let

$$\mathfrak{F}(L)$$

denote the set of all $\mathfrak{F}$-subalgebras of L.

THEOREM 3.6 *Let L be a locally finite Lie algebra and $\mathfrak{X}$ an* s-*closed Fitting class. Then*

$$\rho_{\mathrm{L}\mathfrak{X}}(L) = \{x \in L : x \in H \in \mathfrak{F}(L) \Rightarrow x \in \rho_{\mathfrak{X}}(H)\}. \qquad (*)$$

In other words, elements of the $\mathrm{L}\mathfrak{X}$-radical of L are those which lie in the $\mathfrak{X}$-radical of every $\mathfrak{F}$-subalgebra of L which contains them.

Proof: Let R denote the right-hand side of (*). We show first that $R \lhd L$. Suppose that $x, y \in R$; $z \in L$; $\lambda \in \mathfrak{k}$. Clearly $\lambda x \in R$. Suppose that $x+y \in H \in \mathfrak{F}(L)$. Let $H' = \langle H, x, y\rangle$, which is in $\mathfrak{F}(L)$ since $L \in \mathrm{L}\mathfrak{F}$. Now $x, y \in \rho_{\mathfrak{X}}(H')$ so that $x+y \in \rho_{\mathfrak{X}}(H') \cap H \leqq \rho_{\mathfrak{X}}(H)$ since $\mathfrak{X}$ is s-closed. Thus $x+y \in R$. Now suppose that $[x, z] \in H \in \mathfrak{F}(L)$. Let $H'' = \langle H, x, z\rangle \in \mathfrak{F}(L)$. Then $[x, z] \in \rho_{\mathfrak{X}}(H'') \cap H \leqq$ $\leqq \rho_{\mathfrak{X}}(H)$. So $[x, z] \in R$, which proves that $R \lhd L$.

Next we prove that R is locally in $\mathfrak{X}$. Let $x_1, \ldots, x_n \in R$. Then $H =$ $= \langle x_1, \ldots, x_n\rangle \in \mathfrak{F}(L)$, so that $\{x_1, \ldots, x_n\} \subseteq \rho_{\mathfrak{X}}(H) \in \mathfrak{X}$. Therefore $R \in \mathrm{L}\mathfrak{X}$.

Finally let I be any $\mathrm{L}\mathfrak{X}$-ideal of L. We show that $I \leqq R$. For if $x \in I$ and $x \in H \in \mathfrak{F}(L)$ then $x \in I \cap H \lhd H$, and by s-closure $I \cap H \in \mathfrak{X}$, so that $x \in \rho_{\mathfrak{X}}(H)$. Therefore $x \in R$.

It follows that R is the unique maximal $\mathrm{L}\mathfrak{X}$-ideal of L. □

Now we have a generalisation of lemma 3.3:

THEOREM 3.7 *Over a field of characteristic 0, let $\mathfrak{X}$ be an* s-*closed Fitting class and let $L \in \mathrm{L}\mathfrak{F}$. Then $\rho_{\mathrm{L}\mathfrak{X}}(L)$ contains every serial $\mathrm{L}\mathfrak{X}$-subalgebra of L.*

Proof: Let $x \in K \operatorname{ser} L$, where $K \in \mathrm{L}\mathfrak{X}$. Suppose that $x \in H \in \mathfrak{F}(L)$. Then $K \cap H \operatorname{si} H$ and $K \cap H \in \mathfrak{X}$. By lemma 3.3 $x \in \rho_{\mathfrak{X}}(H)$, so by theorem 3.6 $x \in \rho_{\mathrm{L}\mathfrak{X}}(L)$. □

In particular, with these hypotheses $\rho(L)$ contains every ascendant $\mathrm{L}\mathfrak{N}$-subalgebra of L, and $\sigma(L)$ contains every ascendant $\mathrm{LE}\mathfrak{A}$-subalgebra.

As in corollary 3.4 we deduce:

COROLLARY 3.8 *With the above hypotheses, if $H \operatorname{ser} L$ then*

$$\rho_{\mathrm{L}\mathfrak{X}}(H) = H \cap \rho_{\mathrm{L}\mathfrak{X}}(L).$$ □

Not every Fitting class is s-closed (cf. Fritz [61], Stewart [213] p. 82). If we wish to avoid this hypothesis in theorems 3.6, 3.7, 3.8 we can rework the proofs for the class $\mathfrak{H}$ of neoclassical algebras. If $L \in \mathfrak{H}$ we define

$$\mathfrak{F}^*(L)$$

to be the set of all finite-dimensional local subideals of L. For any Fitting class $\mathfrak{X}$ we may define

$$\rho_{\mathrm{L}\mathfrak{X}}(L) = \{x \in L : x \in H \in \mathfrak{F}^*(L) \Rightarrow x \in \rho_{\mathfrak{X}}(H)\}.$$

The notation is justified by the following fact: $\rho_{L\mathfrak{X}}(L)$ is, once more, the unique maximal $L\mathfrak{X}$-ideal of L. The proof follows theorem 3.6, but with $\mathfrak{F}^*(L)$ in place of $\mathfrak{F}(L)$. Instead of $H' = \langle H, x, y \rangle$ we must use some member of $\mathfrak{F}(L)$ containing H': this is possible since $L \in \mathfrak{H}$ and $\mathfrak{F}$ is lsi-coalescent. (The characteristic is 0 since $\mathfrak{H}$ is defined only for characteristic 0.) The same goes for H''. Theorem 3.7 and corollary 3.8 carry over with the new definition: and 3.8 is now true if $H \in \mathfrak{F}^*(L)$. (This follows from proposition 2.4, but is easy to prove directly.)

Now we turn to analogues of 1.1(b). If L is a locally finite Lie algebra we say that L is *semisimple* if $\sigma(L) = 0$. We let $\mathfrak{S}$ denote the class of semisimple algebras.

THEOREM 3.9 *If L is locally finite then $L/\sigma(L)$ is semisimple.*

Proof: Certainly $L/\sigma(L)$ is locally finite. Let $R = \sigma(L)$, and suppose $S/R \lhd L/R$, where $S/R \in \mathrm{LE}\mathfrak{A}$. Then $S \lhd L$. Let $H \in \mathfrak{F}(S)$. Then $H \cap R \in \mathrm{E}\mathfrak{A}$ and $H + R/R \in \mathrm{E}\mathfrak{A}$, so $H \in \mathrm{E}\mathfrak{A}$. Therefore $S \in \mathrm{LE}\mathfrak{A}$, so that $S = R$. The theorem is proved. □

Note that our definition if semisimplicity coincides with the classical one if $L \in \mathfrak{F}$.

We can find simple analogues of 1.1(c), as follows:

LEMMA 3.10 $\mathrm{L}(\mathfrak{N}\mathfrak{A}) \leqq (\mathrm{L}\mathfrak{N})\mathfrak{A}$.

Proof: Let $L \in \mathrm{L}(\mathfrak{N}\mathfrak{A})$. We show that $L^2 \in \mathrm{L}\mathfrak{N}$. Let $x_1, \ldots, x_r \in L^2$: then for $i = 1, \ldots, r$

$$x_i = [y_{i,1}, z_{i,1}] + \ldots + [y_{i,m(i)}, z_{i,m(i)}]$$

for suitable y_{ij} and $z_{ij} \in L$. Let H be the subalgebra generated by these y_{ij}, z_{ij}. Then $H \in \mathfrak{N}\mathfrak{A}$, so that $H^2 \in \mathfrak{N}$. But H^2 contains all the x_i. Therefore $L^2 \in \mathrm{L}\mathfrak{N}$, whence $L \in (\mathrm{L}\mathfrak{N})\mathfrak{A}$. □

COROLLARY 3.11 *If L is a locally finite Lie algebra over a field of characteristic 0, then $\sigma(L)^2 \leqq \rho(L)$.*

Proof: We have $\sigma(L) \in \mathrm{L}\mathfrak{F} \cap \mathrm{LE}\mathfrak{A} \leqq \mathrm{L}(\mathfrak{F} \cap \mathrm{E}\mathfrak{A}) \leqq \mathrm{L}(\mathfrak{N}\mathfrak{A})$ by 1.1(c). By lemma 3.10 $\sigma(L)^2$ is a locally nilpotent ideal of L, so lies inside $\rho(L)$. □

Again the alternative characterisation of $\sigma(L)$ in theorem 3.6 allows us to prove a stronger result:

THEOREM 3.12 *If L is a locally finite Lie algebra over a field of characteristic 0, and if δ is a locally finite derivation of L, then $\sigma(L)\delta \subseteq \rho(L)$.*

Proof: Let $x \in \sigma(L)$, and suppose that $x\delta \in H \in \mathfrak{F}(L)$. Since δ is locally finite we can find a finite-dimensional δ-invariant subspace V of L, containing H and x. Then $\langle V \rangle$ is also δ-invariant, since δ is a derivation, and $\langle V \rangle \in \mathfrak{F}(L)$. Let $H' = \langle V \rangle$. Then $x \in \rho_{\mathrm{E}\mathfrak{A}}(H')$ so that $x\delta \in \rho_{\mathfrak{N}}(H') \cap H$ by 1.1(c); and this is contained in $\rho_{\mathfrak{N}}(H)$. By theorem 3.6, $x\delta \in \rho(L)$.

COROLLARY 3.13 *If L is locally finite over a field of characteristic 0, then* $[\sigma(L), L] \leqq \rho(L)$. □

COROLLARY 3.14 *If $\mathfrak{X} \leqq \mathrm{E}\mathfrak{A}$ is an* s*-closed Fitting class and L is a locally finite Lie algebra over a field of characteristic 0, then $\rho_{\mathrm{L}\mathfrak{X}}(L)$ is invariant under all locally finite derivations of L.*

Proof: If $\mathfrak{X} \neq (0)$ we can easily show that $\mathfrak{F} \cap \mathfrak{N} \leqq \mathfrak{X}$. Now

$$\rho(L) \leqq \rho_{\mathrm{L}\mathfrak{X}}(L) \leqq \sigma(L).$$ □

Using the Tuck-Towers theorem and arguing as in theorem 3.12 it is easy to prove corollary 3.14 without assuming $\mathfrak{X} \leqq \mathrm{E}\mathfrak{A}$. For $\mathfrak{H}$-algebras we may also drop the hypothesis of s-closure.

4. Semisimplicity

The next step is to find the structure of semisimple $\mathfrak{H}$-algebras. It is here that we really begin to exploit the local subideal structure of an $\mathfrak{H}$-algebra, obtaining far stronger results than are true for general locally finite Lie algebras. We begin with a 'folklore' result:

LEMMA 4.1 *Let L be a Lie algebra. Then the sum of the minimal ideals of L is a direct sum of a subset of them.*

Proof: Let $\mathscr{M}$ be the set of minimal ideals of L. Let $\mathscr{S}$ be the set of all subsets $\mathscr{N}$ of $\mathscr{M}$ for which

$$\sum \{M : M \in \mathscr{N}\}$$

is a direct sum. The condition for this is that

$$N \cap \sum\{M : M \neq N, M \in \mathcal{N}'\} = 0$$

for all $N \in \mathcal{N}$ and for every *finite* subset $\mathcal{N}'$ of $\mathcal{M}$. Therefore $\mathcal{S}$, partially ordered by inclusion, satisfies the hypotheses of Zorn's lemma; and we can find a maximal element $\mathcal{P}$ in $\mathcal{S}$.

If $\sum\{M : M \in \mathcal{P}\} \neq \sum\{M : M \in \mathcal{M}\}$ there must exist $N \in \mathcal{M}$ such that $N \nleq \sum\{M : M \in \mathcal{P}\}$. Since N is a minimal ideal, $N \cap \sum\{M : M \in \mathcal{P}\} = 0$, so that $\mathcal{P} \cup \{N\} \in \mathcal{S}$ contrary to the maximality of $\mathcal{P}$. Therefore $\sum\{M : M \in \mathcal{M}\}$ is equal to the direct sum $\sum\{M : M \in \mathcal{P}\}$. □

The sum of the minimal ideals of L is often called the *socle* of L.

THEOREM 4.2 *L is a semisimple $\mathfrak{H}$-algebra if and only if L is a direct sum of minimal ideals, all of which are simple, non-abelian, and finite-dimensional.*

Proof: Let L be a semisimple $\mathfrak{H}$-algebra. If $H \in \mathfrak{F}^*(L)$ then H is semisimple in the usual sense. This follows either from theorem 3.7 and proposition 2.4, or by the remark just before theorem 3.9. Therefore H is a direct sum of simple finite-dimensional ideals H_i, and each $H_i \in \mathfrak{F}^*(L)$. Since $H_i^2 = H_i$ lemma 2.3 implies that $H_i \lhd L$. Now every element of L lies in some $H \in \mathfrak{F}^*(L)$, so that L is the sum of finite-dimensional non-abelian simple ideals. These are, in particular, minimal ideals, so by lemma 4.1 L is the direct sum of a subset of them.

Conversely let L be such a direct sum. Clearly $L \in \mathfrak{H}$. But if $\sigma(L) \neq 0$ then its projection into some direct summand is non-zero, soluble, and an ideal of that direct summand, contradicting simplicity. Therefore $\sigma(L) = 0$, so $L \in \mathfrak{S}$. □

We may therefore decompose any $\mathfrak{H} \cap \mathfrak{S}$-algebra as a direct sum

$$L = \mathrm{Dr}_{\lambda \in \Lambda} L_\lambda \qquad (*)$$

where the L_λ are non-abelian simple $\mathfrak{F}$-algebras. The next result implies that this decomposition is unique. The proof is from Vasilesçu [237].

LEMMA 4.3 *Suppose that $L = \mathrm{Dr}_{\lambda \in \Lambda} L_\lambda$, where each L_λ is a non-abelian simple Lie algebra. Let I be an ideal of L. Then $I = \mathrm{Dr}_{\lambda \in M} L_\lambda$ for some subset M of Λ.*

Proof: Assume not. Then, by factoring out all L_λ which lie inside I, we may assume that $I \cap L_\lambda = 0$ for all λ. Then $[I, L_\lambda] \leqq I \cap L_\lambda = 0$, so that $I \leqq C_L(L_\lambda)$. By looking at the direct decomposition, and noting that L_λ has trivial centre, we find that

$$C_L(L_\lambda) = \sum_{\mu \neq \lambda} L_\mu.$$

Therefore

$$I \leqq \bigcup_{\lambda \in \Lambda} (\sum_{\mu \neq \lambda} L_\mu) = 0,$$

which is a contradiction. □

COROLLARY 4.4 *The decomposition* (*) *of an* $\mathfrak{H} \cap \mathfrak{S}$*-algebra* L *is unique. The summands* L_λ *are precisely the minimal ideals of* L. □

It follows from lemma 4.3 that each ideal I of an $\mathfrak{H} \cap \mathfrak{S}$-algebra has a unique *complement* $I^\perp$ such that

$$L = I \oplus I^\perp.$$

For if $I = \mathrm{Dr}_{\lambda \in M} L_\lambda$, we set $I^\perp = \mathrm{Dr}_{\lambda \in \Lambda \setminus M} L_\lambda$.

From the uniqueness clause in corollary 4.4 it is obvious that we can completely classify semisimple $\mathfrak{H}$-algebras over the field $\mathfrak{k}$, provided we can classify the simple $\mathfrak{F}$-algebras over $\mathfrak{k}$.

It is not true that all derivations of semisimple $\mathfrak{H}$-algebras are inner. Let $L \in \mathfrak{H} \cap \mathfrak{S}$ with direct decomposition (*), and let

$$\tilde{L} = \mathrm{Cr}_{\lambda \in \Lambda} L_\lambda.$$

Then $L \lhd \tilde{L}$, and every inner derivation of $\tilde{L}$ induces a derivation of L. But these are the only possibilities:

PROPOSITION 4.5 *Let* L *be a semisimple* $\mathfrak{H}$*-algebra. Then*

(a) *Every derivation of* L *is induced by an inner derivation of* $\tilde{L}$.

(b) $\tilde{L}$ *acts faithfully on* L, *so that* $\mathrm{Der}(L) \cong \tilde{L}$.

(c) *A derivation of* L *is inner if and only if it maps all but a finite number of summands in* (*) *to zero*.

(d) L *has outer derivations if and only if* $L \notin \mathfrak{F}$.

(e) *All derivations of* L *are locally finite*.

Proof: (a) Let $\delta \in \mathrm{Der}(L)$. Each L_λ is a perfect ideal of L, so is characteristic by lemmas 1.6.1 and 1.3.2. Therefore $\delta|_{L_\lambda}$ is a derivation of L_λ, so is inner by 1.1(e), so is equal to $x_\lambda{}^*$ for some $x_\lambda \in L_\lambda$. Take $x \in \tilde{L}$ such that its λth component is x_λ, and then $\delta = x^*|_L$.

(b), (c), (d) and (e) are straightforward. □

Three examples

With the notation of chapter 8 section 3 let $F = L(c, \aleph_0)$ be the set of linear transformations of finite rank of a vector space V of dimension c, for c an infinite cardinal. It is easy to check that F is locally finite. Let T be the algebra of trace zero transformations. Then the only ideals of F are $0, T, F$. Therefore F is semisimple but is not a direct sum of simple ideals. Therefore theorem 4.2 does not extend to locally finite algebras, even if we drop the condition that the summands be finite-dimensional.

One might hope to avoid this trouble by chopping off a locally soluble residual at the top as well as a radical at the bottom, since in the above example F/T is abelian. This goes wrong as well. Let B be any finite-dimensional non-abelian simple algebra, and for an infinite index set Λ take isomorphic copies B_λ of B. Let b_λ denote the image in B: of an element $b \in B$ under the isomorphism $B \to B_\lambda$. Let

$$K = \mathrm{Dr}_{\lambda \in \Lambda} B_\lambda$$

and let Δ be the *diagonal subalgebra* of $\mathrm{Cr}_{\lambda \in \Lambda} B_\lambda$, consisting of elements whose λth component is b_λ for a fixed $b \in B$. Then $\Delta \cong B$. We have a split extension

$$L = K \dotplus \Delta \leqq \mathrm{Cr}_{\lambda \in \Lambda} B_\lambda.$$

Now L is locally finite, for any finite subset of L lies inside a subalgebra of the form

$$(\textstyle\sum_{\lambda \in M} B_\lambda) + \Delta$$

for a finite index set $M \subseteq \Lambda$. Further, L has an ascending series with all factors isomorphic to B, and $L = L^2$. However, L is not a direct sum of simple ideals: for if so, by lemma 4.3 there would be an ideal I such that $L = K \oplus I$, so that $I \leqq C_L(K)$. But it is easy to see that, since Λ is infinite, $C_L(K) = 0$.

Note that L is not an $\mathfrak{H}$-algebra since Δ is not contained in a finite-dimensional local subideal.

In this example the trouble seems to be caused by the presence of infinitely many factors in a series. It may be possible to avoid bad behaviour by looking at algebras having a *finite* series with non-abelian simple factors. We have no examples to the contrary in the class $\mathrm{L}\mathfrak{F}$. But in the larger class $(\mathrm{L}\mathfrak{F})\mathfrak{F}$ suitable examples can be found. Let B be any simple $\mathfrak{F}$-algebra. Then B has a faithful representation of finite dimension n, for example the adjoint representation. Express V above as a direct sum of infinitely many n-dimensional subspaces, and copy the action of B on each of these, giving a faithful representation of B on V by transformations of infinite rank. Let C be the image of this representation, and look at $\langle T, C\rangle \leqq L(c, c^+)$. We have $T \lhd \langle T, C\rangle$, and $T \cap C = 0$, so the extension is split. But the only ideals of $\langle T, C\rangle$ are 0, T, and $\langle T, C\rangle$, since T is easily seen to be centralised only by scalar transformations. So $\langle T, C\rangle$ is a split extension of a non-abelian simple algebra by a non-abelian simple algebra, yet is not a direct sum of simple algebras.

5. Levi factors

Let L be a locally finite Lie algebra with radical $\sigma(L)$. A *Levi factor* for L is a semisimple subalgebra Λ such that

$$L = \sigma(L) + \Lambda.$$

Clearly $\sigma(L) \cap \Lambda = 0$, so the extension splits; and $\Lambda \cong L/\sigma(L)$.

An automorphism of L is *radical* if it is of the form

$$\exp(x_1^*)\ldots\exp(x_n^*)$$

where $x_1, \ldots, x_n \in \rho(L)$. Note that the x_i^* are nil derivations, so the exponentials make sense. The radical automorphisms form a group, which we denote $\mathscr{R}(L)$.

In generalising 1.1(f), (g), and (h) to $\mathfrak{H}$-algebras it is radical automorphisms which appear. We will need the following lemma:

LEMMA 5.1 *Let L be an $\mathfrak{H}$-algebra, and suppose that either H ser L or $H \in \mathfrak{F}^*(L)$. Let $\alpha \in \mathscr{R}(H)$. Then α can be extended to an element $\bar{\alpha} \in \mathscr{R}(L)$; that is, $\bar{\alpha}|_H = \alpha$.*

Proof: Let $\exp_H$ and $\exp_L$ denote exponentials of linear transformations of H and L respectively. Then we have

$$\alpha = \prod_{i=1}^{n} \exp_H(x_i^*)$$

where the $x_i \in \rho(H)$. By corollary 3.8 (if H ser L) and the remark before theorem 3.9 (if $H \in \mathfrak{F}^*(L)$) it follows that $\rho(H) \leqq \rho(L)$. Therefore the x_i^* are nil derivations of L, and we may define

$$\bar{\alpha} = \prod_{i=1}^{n} \exp_L(x_i^*).$$

This belongs to $\mathscr{R}(L)$ and extends α. □

If L is locally finite define $\mathfrak{S}(L)$ to be the set of all *finite-dimensional* semisimple subalgebras of L. We begin by proving a very specialised conjugacy lemma:

LEMMA 5.2 *Let L be an $\mathfrak{H}$-algebra, and suppose that L has a Levi factor Λ which is finite-dimensional. Let $S \in \mathfrak{S}(L)$. Then there exists $\alpha \in \mathscr{R}(L)$ such that $S^{\alpha} \leqq \Lambda$.*

Proof: Let $\langle S, \Lambda\rangle \leqq H \in \mathfrak{F}^*(L)$. We claim that Λ is a Levi factor for H. Now $L = \sigma(L) \dotplus \Lambda$, and $\sigma(H) = H \cap \sigma(L)$, so that $H/\sigma(H) = H/(\sigma(L) \cap H) \cong$ $\cong (H+\sigma(L))/\sigma(L) = L/\sigma(L) \cong \Lambda$. Hence $\dim H/\sigma(H) = \dim \Lambda$, which is finite. But $\Lambda \leqq H$, and $\Lambda \cap \sigma(H) = 0$, so that $H = \sigma(H)+\Lambda$ and Λ is a Levi factor for H.

By 1.1(g) and (h) there exists $\alpha \in \mathscr{R}(H)$ such that $S^{\alpha} \leqq \Lambda$. By lemma 5.1, α extends to an element $\bar{\alpha} \in \mathscr{R}(L)$, and clearly $S^{\bar{\alpha}} \leqq \Lambda$. □

Next we drop the dimension hypothesis on Λ, but still assume its existence.

LEMMA 5.3 *Let L be an $\mathfrak{H}$-algebra having a Levi factor Λ. Let $S \in \mathfrak{S}(L)$. Then there exists $\alpha \in \mathscr{R}(L)$ such that $S^{\alpha} \leqq \Lambda$.*

Proof: Let $s_1, \ldots, s_n$ be a basis for S, and let

$$s_i = \sigma_i + \lambda_i$$

where $\sigma_i \in \sigma(L)$, $\lambda_i \in \Lambda$. Let $\Lambda' = \langle \lambda_1, \ldots, \lambda_n\rangle \leqq \Lambda$. Since $\Lambda \in \mathfrak{H} \cap \mathfrak{S}$ it is a direct sum of simple $\mathfrak{F}$-algebras, so there exists a direct summand $\Lambda'' \lhd \Lambda$ such that $\Lambda' \leqq \Lambda''$ and $\Lambda'' \in \mathfrak{F}$. Let $L_1 = \sigma(L)+\Lambda''$. By lemma 5.2 there exists $\alpha \in \mathscr{R}(L_1)$ such that $S^{\alpha} \leqq \Lambda''$. But $L_1 \lhd L$, so by lemma 5.1 α extends to $\bar{\alpha} \in \mathscr{R}(L)$. But now $S^{\bar{\alpha}} \leqq \Lambda'' \leqq \Lambda$. □

The next result gives a very satisfactory generalisation of the theorems of Levi and Mal'cev–Harish-Chandra in the case where the radical is of finite dimension or finite codimension.

THEOREM 5.4 *Let L be an $\mathfrak{H}$-algebra. Suppose that*

$$\sigma(L)+C_L(\sigma(L))$$

is of finite codimension in L. Then

(a) *L has a Levi factor.*

(b) *If Λ is a Levi factor of L and S is any semisimple subalgebra, not necessarily finite-dimensional, then there exists $\alpha \in \mathscr{R}(L)$ such that $S^{\alpha} \leqq \Lambda$.*

(c) *Every semisimple subalgebra of L is contained in a Levi factor.*

(d) *Any two Levi factors of L are conjugate by a radical automorphism.*

Proof: (a) Let $C = C_L(\sigma(L))$. Let $H \in \mathfrak{F}^*(C)$ and let Λ_H be any Levi factor of H. Then

$$[\sigma(H), \Lambda_H] = [H \cap \sigma(L), \Lambda_H] = 0.$$

It follows that $\Lambda_H \lhd H$. Therefore Λ_H lsi L, and since $\Lambda_H^2 = \Lambda_H$ we have $\Lambda_H \lhd L$ by lemma 2.3. So

$$\Lambda = \sum \{\Lambda_H : H \in \mathfrak{F}^*(C)\}$$

is a semisimple ideal of L, and $C = \sigma(C) \oplus \Lambda$. Then $F = \sigma(L) + \Lambda = C + \sigma(L)$ is of finite codimension in L, and we may take an F-transversal $\{t_1, \ldots, t_r\}$. Then

$$\langle t_1, \ldots, t_r \rangle \leqq T \in \mathfrak{F}^*(L).$$

Let Λ' be a Levi factor for T, and put $\Lambda^* = \Lambda + \Lambda'$. Clearly Λ^* is semisimple; and $L = F + T = \sigma(L) + \Lambda + \sigma(T) + \Lambda' = \sigma(L) + \Lambda^*$. Therefore Λ^* is a Levi factor for L.

(b) Let Λ and Λ^* be as in part (a). Let S be a semisimple subalgebra of L with direct decomposition

$$S = \mathrm{Dr}_{\gamma \in \Gamma} S_\gamma$$

where each S_γ is a non-abelian simple $\mathfrak{F}$-algebra. We show first that S can be conjugated inside Λ^*.

Now $\Lambda \lhd \Lambda^*$ so there is a complement $\Lambda^\perp$ to Λ in Λ^*. By the construction of Λ^*, $\Lambda^\perp$ is finite-dimensional. Let J be any ideal of S such that J is finite-dimensional and no simple direct summand of J centralises $\sigma(L)$. Then there exists $\alpha \in \mathscr{R}(L)$ such that $J^\alpha \leqq \Lambda^*$, by lemma 5.3. The centraliser condition implies that $J \cap \Lambda = 0$. Hence

$$\dim J \leqq \dim \Lambda^\perp < \infty.$$

Thus all but a finite number of the S_γ centralise $\sigma(L)$, for otherwise we could construct such ideals J of unbounded dimension. Let $S = S_C + S_N$ where S_C is the sum of those S_γ which centralise $\sigma(L)$, and S_N is the sum of those that do not. Then S_N is finite-dimensional. We claim that $S_C \leqq \Lambda$. For let S_γ be a direct summand of S_C. By lemma 5.3 there exists $\beta \in \mathscr{R}(L)$ such that $S_\gamma^\beta \leqq \Lambda$. But β is a product of exponentials of elements x^* where $x \in \rho(L) \leqq \sigma(L)$, and S_C centralises $\sigma(L)$, so β is the identity on S_γ. Therefore $S_\gamma \leqq \Lambda$, so that $S_C \leqq \Lambda$. Now we can find $\alpha \in \mathscr{R}(L)$ such that $S_N^\alpha \leqq \Lambda^*$, and then

$$S^\alpha = S_N^\alpha + S_C^\alpha \leqq \Lambda^* + \Lambda = \Lambda^*.$$

Now let Λ^0 be any Levi factor of L. By what we have just established there exists $\beta \in \mathscr{R}(L)$ such that $\Lambda^{0\beta} \leqq \Lambda^*$. Since $\Lambda^{0\beta}$ is also a Levi factor we have $\Lambda^{0\beta} = \Lambda^*$. Then $S^{\alpha\beta^{-1}} \leqq \Lambda^0$.

(c) Let S be a semisimple subalgebra of L, and Λ a Levi factor. There exists $\alpha \in \mathscr{R}(L)$ such that $S^\alpha \leqq \Lambda$. Then $S \leqq \Lambda^{\alpha^{-1}}$ which is a Levi factor.

(d) Let Λ, Λ' be Levi factors. Then $\Lambda^\alpha \leqq \Lambda'$ for some $\alpha \in \mathscr{R}(L)$; but since Λ^α and Λ' are Levi factors we must have $\Lambda^\alpha = \Lambda'$. □

COROLLARY 5.5 *Let L be an $\mathfrak{H}$-algebra, and suppose that $\sigma(L)$ is of finite dimension or of finite codimension in L. Then the conclusions (a), (b), (c), (d) of theorem 5.4 hold.*

Proof: From corollary 1.4.3 it is immediate that the hypotheses of theorem 5.4 hold. □

However, this result is not as satisfactory as it might seem, for the proof of theorem 5.4 implies:

COROLLARY 5.6 *Let L be an $\mathfrak{H}$-algebra. Then the following are equivalent:*

(a) *$\sigma(L)$ is finite-dimensional,*

(b) *L is a direct sum of $\mathfrak{F}$-algebras, of which at most one is not non-abelian simple.* □

An easy Zorn's lemma argument shows that every $\mathfrak{H}$-algebra possesses maximal semisimple subalgebras; and it is obvious that any Levi factor must be maximal semisimple. As in [214], we can prove the existence of Levi factors in the general case by establishing the converse statement.

THEOREM 5.7 *Let L be an $\mathfrak{H}$-algebra. Then every maximal semisimple subalgebra of L is a Levi factor, in consequence of which*

(a) *Levi factors exist,*

(b) *Every semisimple subalgebra is contained in some Levi factor.*

Proof: Let Λ be a maximal semisimple subalgebra of L. If $\sigma(L)+\Lambda = L$ then Λ is a Levi factor. Otherwise there exists $H \in \mathfrak{F}^*(L)$ not contained in $\sigma(L)+\Lambda$. Now lemma 2.3 implies that $H^\omega \lhd L$. Now H/H^ω is nilpotent, so that $H/(H^\omega+\sigma(H))$ is semisimple and nilpotent, so is 0. Therefore $H = H^\omega+\sigma(H)$. It follows that H^ω is not contained in $\sigma(L)+\Lambda$, so that Λ is not a Levi factor for $H^\omega+\Lambda$. But $\sigma(H^\omega+\Lambda) \in \mathfrak{F}$ since Λ is semisimple and $H^\omega \in \mathfrak{F}$. Now we can use theorem 5.4 for $H^\omega+\Lambda$: so Λ is contained in a Levi factor Λ' of $H^\omega + \Lambda$, and the inclusion is proper. But this contradicts maximality of Λ. Parts (a) and (b) are now obvious with the aid of Zorn's lemma. □

COROLLARY 5.8 *If L is a perfect $\mathfrak{H}$-algebra then $\sigma(L)$ is locally nilpotent.*

Proof: Let Λ be a Levi factor. Then

$$L = L^2 = (\sigma(L)+\Lambda)^2 \leqq (\sigma(L))^2 + [\sigma(L), \Lambda] + \Lambda^2 \leqq \rho(L)+\Lambda$$

by corollary 3.13. □

Part (d) of theorem 5.4 cannot be extended to the general case without modification. Let K be any $\mathfrak{F}$-algebra having two distinct Levi factors Λ and Λ'. Let M be an infinite index set, let K_μ be an isomorphic copy of K for each $\mu \in M$, and let Λ_μ and Λ_μ' be the images of Λ and Λ' under such an isomorphism. Then the algebra

$$L = \mathrm{Dr}_{\mu \in M} K_\mu$$

is an $\mathfrak{H}$-algebra, and has Levi factors

$$V = \mathrm{Dr}_{\mu \in M} \Lambda_\mu$$

$$W = \mathrm{Dr}_{\mu \in M} \Lambda_\mu'.$$

Now any element of $\mathscr{R}(L)$ is the identity on all but a finite number of K_μ, so that V and W cannot be conjugate under an element of $\mathscr{R}(L)$.

All this means is that $\mathscr{R}(L)$ is too small. In this example there *does* exist an automorphism α of L such that $V^\alpha = W$. For we can find radical automorphisms α_μ of K_μ such that $\Lambda_\mu^{\alpha_\mu} = \Lambda_\mu'$, and use these coordinatewise to define α. This automorphism α is not radical, but it is *locally radical* in the following sense: given any finite subset X of L there exists a radical automorphism α_X of L such that $\alpha|_X = \alpha_X|_X$. We will show that if $\sigma(L)$ is abelian then all Levi factors are conjugate by means of locally radical automorphisms. It is an open question whether this restriction on $\sigma(L)$ is necessary. We can say a little in the general case, but first we need another definition, related to work of Rae [325]. Two subalgebras U and V of L are said to be *locally conjugate* if there is an isomorphism $\beta: U \to V$ such that for every finite subset X of U there exists a radical automorphism β_X of L with the property that $\beta|_X = \beta_X|_X$. This is obviously a weaker kind of conjugacy, since β is not required to be an automorphism of L. We will prove that Levi factors of $\mathfrak{H}$-algebras are always locally conjugate.

Let L be an $\mathfrak{H}$-algebra. Let us say that a subalgebra S of L is *tame* if S is semisimple and $\sigma(L)+S \lhd L$. Then S is tame if and only if it is a direct summand of a Levi factor of L. The little that we can prove about conjugacy will follow from the:

LEMMA 5.9 *Let L be an $\mathfrak{H}$-algebra, S a tame subalgebra of L, and Λ a Levi factor of L. Suppose that α and β are radical automorphisms of L such that S^α and S^β are subalgebras of Λ. Then $\alpha|_S = \beta|_S$.*

Proof: Let $R = \sigma(L)$. Then $R+S \lhd L$. Since radical automorphisms leave ideals of L invariant, it follows that

$$R+S^\alpha = R+S = R+S^\beta.$$

Thus for any $s \in S^\alpha$ we have $s = r+t$ where $r \in R$, $t \in S^\beta$. But then $r = s-t \in R \cap \Lambda = 0$, so that $s = t$. Hence $S^\alpha \leqq S^\beta$. Similarly $S^\beta \leqq S^\alpha$, so that $S^\alpha = S^\beta$. If we set $\gamma = \alpha\beta^{-1}$ then $S^\gamma = S$, and γ is a radical automorphism. By the definition of a radical automorphism

$$s^\gamma \equiv s \pmod{R}$$

for all $s \in S$; and since $s \in S^\gamma$ it follows that $s^\gamma = s$, whence $s^\alpha = s^\beta$. Therefore $\alpha|_S = \beta|_S$ as claimed.

THEOREM 5.10 *Let L be an $\mathfrak{H}$-algebra, having Levi factors Λ and Λ'. Then Λ and Λ' are locally conjugate. Further, if $\sigma(L)$ is abelian, there exists a locally radical automorphism α of L such that $\Lambda^\alpha = \Lambda'$.*

Proof: Let

$$\Lambda = \mathrm{Dr}_{\nu \in N} S_\nu$$

be the standard direct decomposition. For each finite subset $P \subseteq N$ let

$$S_P = \mathrm{Dr}_{\nu \in P} S_\nu.$$

Then each S_P is tame. By lemma 5.3 there is a radical automorphism α_P of L such that $S_P^{\alpha_P} \leqq \Lambda'$. By lemma 5.9 the function μ given by

$$x^\mu = x^{\alpha_P} \text{ if } x \in S_P$$

is well-defined, and gives a monomorphism $\Lambda \to \Lambda'$. We prove that μ is an epimorphism. Let M be a simple direct summand of Λ'. Then by lemma 5.3 there is a radical automorphism β of L such that $M^\beta \leqq \Lambda$. Therefore $M^\beta \leqq S_P$ for some $P \subseteq N$. But now $M^{\beta\alpha_P} \leqq S_P^{\alpha_P} \leqq \Lambda'$. But if ι is the identity automorphism of L, we have $M^\iota = M \leqq \Lambda'$. By lemma 5.9, $\beta\alpha_P|_M = \iota|_M$. Therefore $M = M^{\beta\alpha_P} \leqq S_P^{\alpha_P} \leqq \Lambda^\mu$. Hence $\Lambda' \leqq \Lambda^\mu$, so μ is an epimorphism. Therefore μ is an isomorphism, and clearly has the correct effect locally. So Λ and Λ' are locally conjugate.

Suppose now that $R = \sigma(L)$ is abelian. Each α_P is the identity on R, by definition, so we can extend μ to an automorphism of L by making it act identically on R. This gives a locally radical automorphism mapping Λ to Λ'. $\square$

Chapter 14

Varieties

Many important classes of Lie algebras may be defined by means of identities. For example, L is nilpotent of class $\leqq c$ if and only if

$$[l_1, \ldots, l_{c+1}] = 0$$

for all $l_i \in L$; and L is metabelian if and only if

$$[[l_1, l_2], [l_3, l_4]] = 0$$

for all $l_i \in L$. The study of algebraic systems satisfying identities goes back at least to Birkhoff [263] and Mal′cev [155]. A survey of the very extensive theory of varieties of groups may be found in Neumann [315], while Osborn [320] discusses the problem from the viewpoint of (usually finite-dimensional) non-associative algebras. Although much of what we say here can be interpreted in a very general context we shall confine attention to Lie algebras for ease of exposition. Many results, in any case, are peculiar to the Lie theory.

1. Verbal properties

A formal framework in which to consider identities may be set up as follows. Let $\mathscr{F}_\infty$ be the free Lie algebra (Jacobson [98], Serre [189]) over $\mathfrak{k}$ on a countably infinite set

$$X = \{x_1, x_2, x_3, \ldots\}.$$

If we let $\mathscr{F}_i$ be the free Lie algebra on $\{x_1, \ldots, x_i\}$ then

$$\mathscr{F}_1 \leqq \mathscr{F}_2 \leqq \mathscr{F}_3 \leqq \ldots \leqq \mathscr{F}_\infty.$$

An element $w \in \mathscr{F}_\infty$ is called a *word* in the *variables* $x_1, x_2, x_3, \ldots$. Each word w *involves* only finitely many variables, in the sense that $w \in \mathscr{F}_n$ for some $n \in \mathbb{N}$. In this case w is said to be a word in the variables $x_1, \ldots, x_n$.

Let n be a positive integer. If L is any Lie algebra over $\mathfrak{k}$ and $l_1, \ldots, l_n \in L$ then there is a unique homomorphism $\varphi : \mathscr{F}_n \to L$ such that $\varphi(x_i) = l_i$ for $i = 1, \ldots, n$. If w is a word in $\mathscr{F}_n$ we define

$$w(l_1, \ldots, l_n) = \varphi(w)$$

and refer to φ as *evaluation at* $l_1, \ldots, l_n$ (or *substitution of* $l_1, \ldots, l_n$ *in* w). Similar remarks hold for $\mathscr{F}_\infty$. We say that a word $w \in \mathscr{F}_n$ is an *identity* or *law* of L (or that L *satisfies the identity* $w = 0$) if

$$w(l_1, \ldots, l_n) = 0$$

for all $l_i \in L$.

Let Ω be any set of words. The *variety*

$$\mathfrak{V}_\Omega$$

corresponding to Ω is the class of all Lie algebras L such that every $w \in \Omega$ is a law of L. For instance, if Ω consists of the single word $[x_1, \ldots, x_{c+1}]$ then $\mathfrak{V}_\Omega = \mathfrak{N}_c$. It is easy to see that any variety is $\{\text{s, q, r, l}\}$-closed: and by a famous theorem of Birkhoff [263] any $\{\text{q, r}\}$-closed class is a variety.

If L is a Lie algebra and Ω is a set of words, we say that an element $y \in L$ is an *Ω-value* if

$$y = w(y_1, y_2, \ldots)$$

for some choice of $y_i \in L$, and some $w \in \Omega$. We define the:

verbal subspace $\Omega_0(L)$ spanned by the Ω-values,
verbal subalgebra $\Omega_1(L) = \langle \Omega_0(L) \rangle$,
verbal ideal $\Omega(L) = \langle \Omega_1(L)^L \rangle$.

If L belongs to $\mathfrak{V}_\Omega$ and F is any free Lie algebra (not necessarily on a countable set of generators) such that there is an epimorphism

$$\varphi : F \to L$$

then clearly $\varphi(\Omega(F)) \leqq \Omega(L) = 0$; thus there is an induced epimorphism

$$\varphi' : F/\Omega(F) \to L.$$

Conversely, every epimorphic image of $F/\Omega(F)$ must belong to $\mathfrak{V}_\Omega$. So the variety $\mathfrak{V}_\Omega$ consists precisely of the epimorphic images of the *relatively free* algebras $F/\Omega(F)$, where F is free. We also refer to these as *Ω-free* or *$\mathfrak{V}$-free* algebras, where $\mathfrak{V} = \mathfrak{V}_\Omega$.

There is an inclusion-reversing bijection between varieties and verbal ideals of $\mathscr{F}_\infty$, under which $\mathfrak{V}_\Omega$ and $\Omega(\mathscr{F}_\infty)$ correspond. It is therefore of interest to find a simple characterisation of the verbal ideals of $\mathscr{F}_\infty$. As for groups (see Neumann [315] p. 5) they turn out to be precisely the *fully invariant* ideals of $\mathscr{F}_\infty$, namely: those ideals which are invariant under all Lie endomorphisms of $\mathscr{F}_\infty$. The proof of this is quite easy: it rests on the fact that every map $X \to \mathscr{F}_\infty$ extends to a (unique) endomorphism. Hence any w-value $w(y_1, \ldots, y_n)$ in $\mathscr{F}_\infty$ is the image under an endomorphism of the word $w = w(x_1, \ldots, x_n)$. Distinct subsets of $\mathscr{F}_\infty$ may determine the same variety. Thus either of

$$\{[x_1, x_2]\} \quad \{[x_1, x_2], [x_3, x_4]\}$$

determines the variety $\mathfrak{A}$. We say that Ω' is a *consequence* of Ω, and write

$$\Omega \Rightarrow \Omega',$$

if $\mathfrak{B}_\Omega \leqq \mathfrak{B}_{\Omega'}$; or equivalently if $\Omega'(\mathscr{F}_\infty) \leqq \Omega(\mathscr{F}_\infty)$. If each of Ω and Ω' is a consequence of the other, then we say that they are *equivalent* and write

$$\Omega \Leftrightarrow \Omega'.$$

It follows that under these conditions $\Omega(\mathscr{F}_\infty) = \Omega'(\mathscr{F}_\infty)$ and $\mathfrak{B}_\Omega = \mathfrak{B}_{\Omega'}$. For any Lie algebra L and any subset S of L define the *fully invariant closure* of S to be the smallest fully invariant ideal of L containing S. Then any subset of $\mathscr{F}_\infty$ is equivalent to its fully invariant closure, and two subsets are equivalent if and only if their fully invariant closures are equal.

Among the possible words we single out the *monomials*, defined recursively by:

(a) x_i is a monomial for all $i = 1, 2, \ldots$.
(b) If m is a monomial so is λm for $\lambda \in \mathfrak{k}$.
(c) If m and n are monomials then so is $[m, n]$.

The monomials are therefore the words not involving + signs. It is clear that every word is a sum of monomials. If we fix a particular basis for $\mathscr{F}_\infty$ consisting entirely of monomials (which must be possible by lemma 1.1.1) we can assume all monomials are taken from this basis, and the representation as a sum is unique. (A particularly useful choice of such a basis arises by taking the 'basic commutators' of Hall [75]). If w is a sum of monomials v_i we say that the v_i are *w-monomials*. The *degree* $\partial_i(v)$ of a monomial v in the variable x_i is defined by

$$\partial_i(0) = 0$$

$$\partial_i(x_j) = \delta_{ij}$$

$$\partial_i([v, w]) = \partial_i(v) + \partial_i(w) \text{ unless } [v, w] = 0,$$

$$\partial_i(\lambda v) = \partial_i(v) \text{ if } 0 \neq \lambda \in \mathfrak{k}.$$

Therefore $\partial_i(v)$ is the number of times that x_i occurs in v. A word w is *homogeneous* if $w = v_1 + \ldots + v_t$, where the v_i are monomials, such that

$$\partial_i(v_k) = \partial_i(v_l)$$

for all $i = 1, 2, \ldots$ and $1 \leqq k, l \leqq t$. We may then define the degrees $\partial_i(w)$ to be the same as the $\partial_i(v_1)$. A homogeneous word w is *multilinear* if

$$\partial_i(w) = 0 \text{ or } 1$$

for all $i = 1, 2, \ldots$.

The *length* of a monomial w is

$$\textstyle\sum_{i=1}^{\infty} \partial_i(w).$$

Next we introduce a useful class of words, known as *Hall words* (or *bracket words*). If w and v are words, with

$$w = w(x_1, \ldots, x_s)$$

$$v = v(x_1, \ldots, x_t)$$

we define the *outer commutator*

$$[w, v]^0(x_1, \ldots, x_{s+t}) = [w(x_1, \ldots, x_s), v(x_{s+1}, \ldots, x_{s+t})].$$

Note that this is defined only up to equivalence of words unless we specify s and t in some canonical fashion. We may now define Hall words of weight n recursively:

0 is a Hall word of weight 0,
x_i is a Hall word of weight 1 for all i,
if w and v are Hall words of weight α and β then $[w, v]^0$ is a Hall word of weight $1 + \max(\alpha, \beta)$.

Every Hall word is a monomial and is multilinear. Among them are the words defining the lower central and derived series, for which we need the following notation:

$$\gamma_1 = x_1,$$

$$\gamma_{i+1} = [\gamma_i, x_1]^0$$

$$\delta_1 = [x_1, x_2]$$

$$\delta_{i+1} = [\delta_i, \delta_i]^0.$$

We note also the *Engel words*

$$\varepsilon_i = [x_1, {}_ix_2]$$

which are not in general Hall words.

If v and w are words with

$$v = v(x_1, \ldots, x_t)$$
$$w = w(x_1, \ldots, x_s)$$

we define

$$v \bullet w(x_1, \ldots, x_{st}) = v(w(x_1, \ldots, x_s), w(x_{s+1}, \ldots, x_{2s}), \ldots$$
$$\ldots, w(x_{(t-1)s+1}, \ldots, x_{st})).$$

It is not hard to see that

$$v \bullet w(L) = v(w(L)).$$

2. Invariance properties of verbal ideals

There are two particularly important ways of deriving consequences of a given identity w, which we shall call *homogenisation* and *linearisation*. The full theory of these processes is complicated by differences due to the cardinality and characteristic of the field, and we shall not give an exhaustive treatment. For more details consult Osborn [320].

It is clear that any word w may be written as a sum of homogeneous words

$$w = h_1 + \ldots + h_r$$

where distinct h_i have different degrees in some variable. We call the h_i the *homogeneous components* of w.

THEOREM 2.1 *Let w be a word such that $\partial_i(w) < |\mathfrak{k}|$ for all i. Then w is equivalent to the set of its homogeneous components. In particular this is true for any w if the field $\mathfrak{k}$ is infinite.*

Proof: For a fixed variable x_i express w as a sum

$$w = w_0 + \ldots + w_s$$

where $\partial_i(w_j) = j$. Pick distinct elements $\alpha_0, \ldots, \alpha_s \in \mathfrak{k}$, which is possible by hypothesis. Then

$$w(x_0, \ldots, \alpha_j x_i, \ldots, x_r) = \textstyle\sum_{k=0}^{s} \alpha_j^k w_k(x_0, \ldots, x_r).$$

Now the determinant

$$\begin{vmatrix} 1 & \alpha_0 & \alpha_0^2 & \ldots & \alpha_0^s \\ 1 & \alpha_1 & \alpha_1^2 & \ldots & \alpha_1^s \\ \vdots & \vdots & \vdots & & \vdots \\ 1 & \alpha_s & \alpha_s^2 & \ldots & \alpha_s^s \end{vmatrix}$$

is a Vandermonde determinant with value

$$\textstyle\prod_{i<j} (\alpha_i - \alpha_j)$$

so is non-zero. Consequently each w_i is a linear combination of the elements

$$w(x_0, \ldots, x_{i-1}, \alpha_j x_i, x_{i+1}, \ldots, x_r)$$

and it follows that w is equivalent to $\{w_0, \ldots, w_s\}$. If we repeat the process on each w_i using a new variable, then eventually we reduce w to its homogeneous components. □

Some restriction on the cardinality of the field is necessary. Over the field of 2 elements the algebra

$$\langle a, b \colon [a, b] = a \rangle$$

satisfies the law

$$[x_1, x_2, x_3] + [x_1, x_2, x_3, x_3]$$

but not the laws

$$[x_1, x_2, x_3],$$

$$[x_1, x_2, x_3, x_3].$$

Next we turn to the process of *linearisation*. As a motivating example, consider the law $[x, y, y]$. If we replace y by $y + \alpha z$, where $\alpha \in \mathfrak{k}$ and z is a new variable, we obtain the law

$$[x, y, y] + \alpha([x, y, z] + [x, z, y]) + \alpha^2 [x, z, z]$$

which, together with the original law, implies the law

$$[x, y, z] + [x, z, y]$$

which is multilinear. Further, if we put $y = z$ in the new law, we get

$$2[x, y, y]$$

so that in fields of characteristic $\neq 2$, the new law is equivalent to the old one. But its multilinearity often makes it more tractable.

In general we start with a law $w(x_1, \ldots, x_r)$, which for simplicity we shall assume homogeneous, and substitute $x_i + \alpha z$ in place of x_i, where $\alpha \in \mathfrak{k}$ and z is a new variable (say $z = x_{r+1}$). Collecting terms of the same degree in z we get

$$w(x_1, \ldots, x_i + \alpha z, \ldots, x_r) = \sum_{j=0}^{N} \alpha^j w_{ij}(x_1, \ldots, x_r, z)$$

where $N = \partial_i(w)$. As in theorem 2.1 it follows that for $|\mathfrak{k}| \geqq \partial_i(w)$ each w_{ij} is a consequence of w. We call the laws w_{ij} the *linearisations of* w *with respect to* x_i. Note that $w_{i0} = w$, and that w_{iN} is equivalent to w if $\alpha \neq 0$. All other w_{ij} are of smaller degree in x_i than w, though of the same length: the missing x_i's have become z's. By repeating this process we can eventually obtain laws of degree 1 in each variable; since the lengths do not change the number of variables must equal the length of w. Such a word we call a *complete linearisation* of w. We extend the use of the term 'linearisation' to any of the words occurring at intermediate stages in the process.

It is not hard to give an informal description of the words w_{ij}. Suppose w has degree N in x_i, and fix j between 0 and N. Pick a subset J of $\{1, \ldots, N\}$ with $|J| = j$. In each monomial of w the variable x_i occurs in N places. Number these from 1 to N in each monomial, in order from left to right; now replace those whose numbers come from J by z's. Do this for each possible J and add the results. Then the final expression will be w_{ij}.

For example, if $w = [x, y, y, x, y]$ (where $x = x_1$, $y = x_2$) then

$$w_{21} = [x, z, y, x, y] + [x, y, z, x, y] + [x, y, y, x, z],$$

$$w_{22} = [x, z, z, x, y] + [x, z, y, x, z] + [x, y, z, x, z].$$

There is an important consequence of this process:

THEOREM 2.2 *Over a field of characteristic zero any homogeneous word* w *is equivalent to a complete linearisation of* w. *The same is true in characteristic* $p > 0$ *provided* $p > \partial_i(w)$ *for all* i.

Proof: We show that w is equivalent to any linearisation w_{ij}. We know that $w \Rightarrow w_{ij}$ so it is enough to show that $w_{ij} \Rightarrow w$. By the informal description above, it follows that

$$w_{ij}(x_1, \ldots, x_r, x_i) = \tbinom{N}{j} w(x_1, \ldots, x_r)$$

where $N = \partial_i(w)$. The hypothesis on the field assures us that $\tbinom{N}{j} \neq 0$.

By induction on the number of steps in the linearisation process, the assertion holds for any linearisation of w, and in particular any complete linearisation. □

COROLLARY 2.3 *Over a field of characteristic zero any law is equivalent to a finite set of multilinear laws.* □

Note that in order for w to imply its linearisations it is sufficient to have $\mathfrak{k}$ of sufficiently large *cardinality:* for the linearisations, individually, to imply w it is necessary to have sufficiently large *characteristic.* It is not hard to find examples to illustrate these phenomena.

Theorem 2.2 has useful applications to changes of the coefficient field, as is shown by the next result, due to Vaughan-Lee [239]. We shall need this result later.

LEMMA 2.4 *Let $\mathfrak{k}$ be any field and let $\mathfrak{k}^*$ be an extension field of $\mathfrak{k}$. Let Ω be a set of words, and let $\mathfrak{V}$ and $\mathfrak{V}^*$ be the varieties determined by Ω over $\mathfrak{k}$ and $\mathfrak{k}^*$. Let F be $\mathfrak{V}$-free over $\mathfrak{k}$, generated by $y_1, y_2, \ldots$. If every word $w \in \Omega$ is linear in each variable it involves, then $\mathfrak{k}^* \otimes_{\mathfrak{k}} F$ is $\mathfrak{V}^*$-freely generated by $y_1, y_2, \ldots$ over $\mathfrak{k}^*$.*

Proof: It is straightforward to show that $F^* = \mathfrak{k} \otimes_{\mathfrak{k}} F$ belongs to $\mathfrak{V}^*$. To show that it is $\mathfrak{V}$-freely generated by $y_1, y_2, \ldots$ we must show that if w is a word over $\mathfrak{k}^*$ for which

$$w(y_1, \ldots, y_n) = 0$$

then w is a law in $\mathfrak{V}^*$. Now we may write

$$w = \textstyle\sum_{i=1}^{r} \alpha_i w_i$$

where the α_i belong to $\mathfrak{k}^*$ and are linearly independent over $\mathfrak{k}$, and where $w_i(x_1, \ldots, x_i)$ are words over $\mathfrak{k}$. But then

$$\textstyle\sum_{i=1}^{r} \alpha_i w_i(y_1, \ldots, y_n) = 0$$

from which it follows that

$$w_i(y_1, \ldots, y_n) = 0$$

for all $i = 1, \ldots, r$. By $\mathfrak{B}$-freeness of F each w_i is a law of the variety $\mathfrak{B}$, and hence a law of $\mathfrak{B}^*$. □

COROLLARY 2.5 *Relative freeness is preserved by field extensions (in the obvious sense) in characteristic zero.* □

In passing we note another property of relatively free Lie algebras, due to Mal'cev [155]. Free associative algebras have a natural structure as *graded* algebras (see Jacobson [98] p. 168) and contain free Lie algebras in a natural way. Now a quotient of a graded algebra by a homogeneous ideal is again graded. Words which are homogeneous in the Lie sense are also homogeneous in the associative sense. Now an associative algebra which is graded and has no elements of degree 0 is residually nilpotent, in the sense that the intersection of its powers is 0. By interpreting this fact for the Lie algebra, it is not hard to obtain:

PROPOSITION 2.6 *Let Ω be a set of homogeneous words over the field $\mathfrak{k}$. Then every Ω-free Lie algebra over $\mathfrak{k}$ is residually nilpotent. In particular every relatively free Lie algebra over an infinite field is residually nilpotent.* □

We now apply these methods to a more detailed study of $\Omega_0(L)$, $\Omega_1(L)$, and $\Omega(L)$. The main result, taken from [219], is that these are equal for infinite fields, and are all characteristic ideals. We shall give some examples to show that this result is not true in general for finite fields. The essential point to observe is:

LEMMA 2.7 *Let $\mathfrak{k}$ be any field, and let $w = w(x_1, \ldots, x_n)$ be a homogeneous word of degree m_i in x_i ($1 \leqq i \leqq n$). Let L be a Lie algebra over $\mathfrak{k}$ and δ be a derivation of L. Suppose that $|\mathfrak{k}| \geqq m_i$ for $1 \leqq i \leqq n$. Then $w_0(L)$ is δ-invariant.*

Proof: In each w-monomial the variable x_i appears in m_i different positions. Let

$$\sigma_i^k(z)w$$

be the word formed from w by replacing the kth occurrence of x_i in each monomial by the new variable z. Let

$$\sigma_i^k(a)w(b_1, \ldots, b_n)$$

be the result of substituting elements b_i, a of L for x_i, z respectively. An easy induction shows that

$$(w(y_1, \ldots, y_n))\delta = \sum_{i,k} \sigma_i^k(y_i\delta)w(y_1, \ldots, y_n)$$

for all $y_i \in L$.

But

$$\sum_k \sigma_i^k(z)w(x_1, \ldots, x_n)$$

is a linearisation of w, corresponding to w_{i1} in the notation used above. It can be written as a linear combination of elements of the form

$$w(x_1, \ldots, x_{i-1}, x_i+\alpha z, x_{i+1}, \ldots, x_n)$$

provided (as one sees by looking at determinants) that $|\mathfrak{k}| \geqq m_i$. Hence for any $t \in L$,

$$\sum_k \sigma_i^k(t)w(y_1, \ldots, y_n) \in w_0(L).$$

Summing over i finishes the proof. □

THEOREM 2.8 *If L is a Lie algebra over an infinite field $\mathfrak{k}$ and Ω is any set of words, then $\Omega_0(L)$ is invariant under all derivations of L. In consequence*

$$\Omega_0(L) = \Omega_1(L) = \Omega(L)$$

and $\Omega(L)$ is a characteristic ideal of L.

Proof: It is clear that we may assume $\Omega = \{w\}$ for a single word w, and it is not hard to see that we may further assume w to be homogeneous. Lemma 2.7 finishes the proof. □

Note that since any field has at least 2 elements, the conclusion of theorem 2.8 is valid for *any* field, finite or infinite, provided Ω consists of homogeneous words of degree $\leqq 2$ in each variable. In particular it holds for all Hall words, and for the 2-Engel word $[x, y, y]$.

If w is of degree greater than 2 in some variable then the hypothesis on $\mathfrak{k}$ cannot be removed entirely. To see this, let L be the free $\mathfrak{N}_5$-algebra on 2 generators u, v over the field $\mathfrak{k}$ of 2 elements, and let w be the 3-Engel word

$$[x, y, y, y].$$

Since L has class 5, $w_0(L)$ is spanned by all elements of the form

$$[a, b, b, b]$$

where

$$a = A[u, v]+Bu+Cv,$$

$$b = Du+Ev$$

where A, B, C, D, E are 0 or 1 in $\mathfrak{k}$. Simple but tedious calculations (using the basic commutators of Hall [75]) show that the element

$$[u, v, v, v, u]$$

of $w(L)$ does not lie in $w_0(L)$. So $w_0(L)$ is not an ideal.

Even if $w_0(L)$ is an ideal of L, it need not be characteristic. Take w as above, but now make L free $\mathfrak{N}_4$ on 3 generators u, v, w over the field $\mathfrak{k}$ of 2 elements. The derivation which is the unique extension to L of $u \mapsto u$, $v \mapsto w$, $w \mapsto v$ sends $[u, v, v, v]$ to

$$[u, v, v, v]+[u, w, v, v]+[u, v, w, v]+[u, v, v, w]$$

which can be shown not to lie in $w_0(L)$. But certainly $w_0(L)$ is central in L, so is an ideal.

The general case of non-associative algebras is studied in [219]: in particular it turns out that $w_0(A)$ is almost never an ideal unless A is a *Lie* algebra.

3. Ellipticity

In this section we continue the study of abelian-by-finite algebras of chapters 11 and 12, and prove an analogue of a theorem of Stroud [344].

Let L be a Lie algebra, and let S be a sub*set* of L. For each integer $n \geqq 0$ let $S(n)$ be the subset of L consisting of all linear combinations of $\leqq n$ elements of S. Then we have

$$0 = S(0) \subseteq S = S(1) \subseteq S(2) \subseteq S(3) \subseteq \ldots$$

and the subspace spanned by S is

$$\bigcup_{n=0}^{\infty} S(n).$$

It may happen that this subspace is equal to $S(n)$ for some n, or equivalently that $S(n) = S(n+1)$. If this is the case we say that S is an *elliptic* subset of L, of *ellipticity* $\leqq n$. If not, we call S a *parabolic* subset.

(There is a parallel concept for the subalgebra generated by S: this is the union of the chain $S[n]$ where

$$S[n] = S+S^2+\ldots+S^n$$

and one may ask whether or not the chain stops. So perhaps we should talk of *additive* and *multiplicative* ellipticity. However, for the present purposes, it is the additive structure which is relevant).

Obviously any subset which spans a finite-dimensional subspace is elliptic, with ellipticity equal to the dimension of the subspace. Again, L is an elliptic subset of itself. Our aim is to show that for suitable L the sets of w-values, for all words w, are elliptic. Such a theorem is essentially about the fine structure of $w_0(L)$ (and hence of $w(L)$ for infinite fields) and is thus somewhat alien to the usual theorems about varieties, all of which are intent on throwing $w(L)$ away.

We begin with a lemma whose statement is longer than its proof.

LEMMA 3.1 *Let $w = w(x_1, \ldots, x_n)$ be a word, and let F be a free Lie algebra on $x_1, \ldots, x_n, a_1, \ldots, a_n$. Then*

$$w(x_1+a_1, \ldots, x_n+a_n) - w(x_1, \ldots, x_n)$$
$$= w_1(x_1, \ldots, x_n, a_1) + \ldots + w_n(x_1, \ldots, x_n, a_n) +$$
$$+ v(x_1, \ldots, x_n, a_1, \ldots, a_n)$$

where each w_i is of degree 1 in a_i, and where v is a (possibly empty) linear combination of monomials, each of which involves at least two of the a_i. Further,

$$w_i(x_1, \ldots, x_n, a_i) = w(x_1, \ldots, x_{i-1}, x_i+a_i, x_{i+1}, \ldots, x_n) -$$
$$- w(x_1, \ldots, x_n) + u_i(x_1, \ldots, x_n, a_i)$$

where u_i is a (possibly empty) linear combination of monomials, each of degree at least 2 in a_i.

Proof: For the first part, expand w linearly and pick out terms of degree 0, 1, and at least 2 in the a_i. For the evaluation of w_i, put

$$a_1 = \ldots = a_{i-1} = a_{i+1} = \ldots = a_n = 0. \qquad \square$$

THEOREM 3.2 *Let L be any abelian-by-finite Lie algebra, and w any word. Then the set of w-values is elliptic in L.*

Proof: There are two cases to consider, depending on whether $\mathfrak{k}$ is infinite or finite.

First assume $\mathfrak{k}$ infinite. Let $I \lhd L$, such that I is abelian and L/I is finite-dimensional. Pick a transversal T to I in L. Then

$$|T| = \dim L/I = f$$

is finite. Consider first the case when w is homogeneous, and let $x \in w_0(L)$. Then x is a linear combination of expressions of the form

$$w(j_1+\lambda_1 t_1, \ldots, j_n+\lambda_n t_n) \tag{1}$$

where $j_i \in I$, $t_i \in T$, and $\lambda_i \in \mathfrak{k}$ ($1 \leqq i \leqq n$). By homogeneity we may assume each λ_i to be 0 or 1. By lemma 3.1 we have

$$w(j_1+\lambda_1 t_1, \ldots, j_n+\lambda_n t_n) - w(\lambda_1 t_1, \ldots, \lambda_n t_n) =$$
$$= w_1(\lambda_1 t_1, \ldots, \lambda_n t_n, j_1) + \ldots + w_n(\lambda_1 t_1, \ldots, \lambda_n t_n, j_n) \tag{2}$$

and further

$$w_i(\lambda_1 t_1, \ldots, \lambda_n t_n, j_i) = w(\lambda_1 t_1, \ldots, j_i+\lambda_i t_i, \ldots, \lambda_n t_n)$$
$$- w(\lambda_1 t_1, \ldots, \lambda_n t_n)$$

since $I \lhd L$ and $I^2 = 0$, so the expressions v and u_i of lemma 3.1 vanish. Hence $w_i(\lambda_1 t_1, \ldots, \lambda_n t_n, j_i)$ is a sum of two w-values.

From (1) and (2) we may express x as a linear combination of expressions of the form

$$w(\lambda_1 t_1, \ldots, \lambda_n t_n) \tag{3}$$
$$w_i(\lambda_1 t_1, \ldots, \lambda_n t_n, j_i^{\alpha}) \tag{4}$$

for various $j_i^{\alpha} \in I$. But the w_i are linear in a_i, so for each sequence $\lambda_1, \ldots, \lambda_n$, $t_1, \ldots, t_n$ we can collect together terms of the type (4) to obtain

$$w_i(\lambda_1 t_1, \ldots, \lambda_n t_n, k_i) \tag{5}$$

for $k_i \in I$.

Since each $\lambda_i = 0$ or 1 and each $t_i \in T$, there are at most $(2f)^n$ elements of the form (3), and $(2f)^n$ of the form (4). Hence (remembering not to count type (3) elements twice!) x is a linear combination of at most

$$(n+1)(2f)^n$$

w-values.

When w is not homogeneous we express w as a linear combination of its homogeneous components:

$$w = \sum_{j=1}^{d} w^j$$

and note that there exists e such that each w^j-value is a linear combination of at most e w-values. It then follows that if $x \in w_0(L)$ then x is a linear combination of at most

$$de(n+1)(2f)^n$$

w-values.

Now suppose that $\mathfrak{k}$ is finite, with g elements. We proceed as above, but do not reduce to the homogeneous case. We cannot assume that $\lambda_i = 0$ or 1, so we note instead that the subspace spanned by T is finite, with g^f elements. Then we collect terms according to elements of this subspace, thereby expressing x as a linear combination of at most

$$(n+1)g^{fn}$$

w-values. □

Note that the proof in [219] is not quite correct. For infinite fields, theorem 3.2 yields information about the verbal ideal $w(L)$.

It is not possible to extend theorem 3.2 from $\mathfrak{A}\mathfrak{F}$ to $\mathfrak{N}\mathfrak{F}$. In fact, the free $\mathfrak{N}_2$-algebra on countably many generators is not elliptic as regards the δ_1-values.

4. Marginal properties

Next we turn to a concept 'dual' to the verbal ideal, in the sense that the centre is 'dual' to the derived algebra. The group-theoretic version is due to Hall and is expounded in Robinson [331].

Let L be a Lie algebra and w a word. The *marginal subspace* $\hat{w}_0(L)$ is the set of all $x \in L$ such that for all $y_1, \ldots, y_n \in L$, $\alpha \in \mathfrak{k}$, $i = 1, \ldots, n$, we have

$$w(y_1, \ldots, y_{i-1}, y_i+\alpha x, y_{i+1}, \ldots, y_n) = w(y_1, \ldots, y_n). \tag{6}$$

Manifestly it *is* a subspace, and is the largest subspace V such that the w-values depend only on the cosets (mod V) to which elements of L belong. As with verbal subspaces, but dually, we define the *marginal subalgebra* $\hat{w}_1(L)$ to be the largest subalgebra of L contained in $\hat{w}_0(L)$, if such exists, and *the marginal ideal* $\hat{w}(L)$ to be the largest ideal contained in $\hat{w}_0(L)$ (which always exists).

If $w = \gamma_n$ then $\hat{w}(L)$ is easily seen to equal $\zeta_{n-1}(L)$.

PROPOSITION 4.1 *Let L be a Lie algebra over $\mathfrak{k}$, and let w be a word. If w is multilinear, or if $\mathfrak{k}$ has characteristic zero, then $\hat{w}_0(L)$ is invariant under all* *derivations of L. Thus*

$$\hat{w}_0(L) = \hat{w}_1(L) = \hat{w}(L)$$

and all three are characteristic ideals of L.

Proof: This is much the same as for lemma 2.7, except that we must linearise w completely. We then apply the given derivation to equation (6) and argue as before. □

Our next result may be regarded as a Lie-theoretic analogue of theorems of Baer [259], Schur [338], and Stroud [344].

THEOREM 4.2 *Let L be a Lie algebra over* $\mathfrak{k}$ *and let w be any word. Suppose that* $\hat{w}_0(L)$ *is of finite codimension in L. Then* $w_0(L)$ *is finite-dimensional.*

Proof: Let $T = \{t_1, \ldots, t_n\}$ be a transversal to $\hat{w}_0(L)$ in L, and let V be the subspace spanned by T. From (6) it follows that $w_0(L)$ is spanned by all elements

$$w(v_1, \ldots, v_n) \tag{7}$$

for $v_i \in V$. If $\mathfrak{k}$ is finite, there are only finitely many such sequences, and the result follows. If $\mathfrak{k}$ is infinite we may assume w homogeneous. Then we have

$$w(y_1, \ldots, y_{i-1}, y_i + \alpha t, y_{i+1}, \ldots, y_n) = \textstyle\sum_{j=0}^{m_i} \alpha^j w_{ij},$$

where the w_{ij} are linear combinations of certain elements

$$w(y_1, \ldots, y_{i-1}, y_i + \alpha_j t, y_{i+1}, \ldots, y_n)$$

for a fixed subset $\{\alpha_1, \ldots, \alpha_m\} \subseteq \mathfrak{k}$, where $m = m_i$. Inductively it follows that every element of the form (7) is a linear combination of elements $w(u_1, \ldots, u_n)$ where each u_i is a linear combination of elements of T, with coefficients taken from a certain fixed finite set of scalars. There are only finitely many such elements. □

Hall, in lectures, has introduced a convenient notation in which to express such results. For any word w and class $\mathfrak{X}$ of algebras, we let

$$\hat{\mathfrak{B}}_w \mathfrak{X}$$

denote the class of all Lie algebras L such that $L/\hat{w}(L)$ belongs to $\mathfrak{X}$. Expressions such as $\hat{\mathfrak{N}}_c \mathfrak{X}$ or $\hat{\mathfrak{A}}\mathfrak{X}$ are given the obvious meaning. In fact, if

$$\mathfrak{B} = \textstyle\bigcup_w \mathfrak{B}_w$$

for a certain set of words w, we can define

$$\hat{\mathfrak{V}}\mathfrak{X} = \bigcup_w \hat{\mathfrak{V}}_w\mathfrak{X}.$$

This is well-defined by virtue of:

LEMMA 4.3 *If a class* $\mathfrak{V}$ *is expressed as a union of varieties in two ways,*

$$\mathfrak{V} = \bigcup_{w\in W} \mathfrak{V}_w = \bigcup_{u\in U} \mathfrak{V}_u,$$

then for all $w \in W$ *there exists* $u \in U$ *such that* $\mathfrak{V}_w \leqq \mathfrak{V}_u$, *and vice versa.*

Proof: Suppose not. Then for all $u \in U$ there exists $L_u \in \mathfrak{V}_w \backslash \mathfrak{V}_u$. Then

$$L = \mathrm{Dr}_{u\in U} L_u$$

does not belong to any $\mathfrak{V}_u$, so does not belong to $\mathfrak{V}$. But it belongs to $\mathfrak{V}_w \leqq \mathfrak{V}$, a contradiction. □

COROLLARY 4.4 *If a variety* $\mathfrak{V}$ *is contained in a union of varieties*

$$\bigcup_{w\in W} \mathfrak{V}_w$$

then $\mathfrak{V} \leqq \mathfrak{V}_w$ *for some* $w \in W$. □

Theorem 4.2 may be restated in these terms: for Lie algebras over infinite fields, and for any word w, we have

$$\hat{\mathfrak{V}}_w\mathfrak{F} \leqq \mathfrak{F}\mathfrak{V}_w.$$

Over all fields,

$$\hat{\mathfrak{N}}_c\mathfrak{F} \leqq \mathfrak{F}\mathfrak{N}_c.$$

As in Hall [284] there is a partial converse to this latter inclusion, namely

$$\mathfrak{F}\mathfrak{N}_c \leqq \hat{\mathfrak{N}}_{2c}\mathfrak{F}.$$

An example given in [219] shows that, at least when $c = 1$, the subscript $2c$ cannot be reduced.

5. Hall varieties

Every *Hall variety* (determined by a single Hall word) consists purely of soluble algebras. Hall [287] noted that Hall varieties of groups provide a useful coarse classification of properties of soluble groups: not surprisingly a similar idea applies to Lie algebras. Given any class $\mathfrak{X}$ of Lie algebras we may ask the test question: which Hall varieties are contained in $\mathfrak{X}$? In fact it seems to be more informative to ask the question only for finitely generated algebras, since soluble non-finitely generated algebras do not seem to have many interesting general properties. We therefore let $\mathscr{H}$ be the set of all Hall words, and define:

$$\mathscr{H}^+ = \{\varphi \in \mathscr{H} : \mathfrak{G} \cap \mathfrak{B}_\varphi \leqq \mathfrak{X}\},$$

$$\mathscr{H}^- = \{\varphi \in \mathscr{H} : \mathfrak{G} \cap \mathfrak{B}_\varphi \nleqq \mathfrak{X}\}.$$

Clearly

$$\mathscr{H}^+ \cup \mathscr{H}^- = \mathscr{H}, \; \mathscr{H}^+ \cap \mathscr{H}^- = \varnothing. \tag{1}$$

We call the pair $(\mathscr{H}^+, \mathscr{H}^-)$ a *dichotomy*.

Of course, any pair satisfying (1) arises from some class $\mathfrak{X}$: we just put $\mathfrak{X} = \bigcup_{\varphi \in \mathscr{H}^+} \mathfrak{B}_\varphi$. What is more interesting is that for certain natural classes $\mathfrak{X}$ both $\mathscr{H}^+$ and $\mathscr{H}^-$ are particularly easy to describe explicitly. Often the set of varieties $\mathfrak{B}_\varphi$ for $\varphi \in \mathscr{H}^-$ has a single minimal member (or perhaps two), which we may call the *minimal countervariety* (*-ies*) for $\mathfrak{X}$. For a given $\mathfrak{X}$ to give rise to such a dichotomy it is necessary and sufficient that $\mathfrak{G} \cap \mathfrak{B}_\varphi \leqq \mathfrak{X}$ for all $\varphi \in \mathscr{H}^+$, whilst some $L \in \mathfrak{G} \cap \mathfrak{B}_\varphi \backslash \mathfrak{X}$ where $\mathfrak{B}_\varphi$ is the minimal countervariety. Thus the establish such a dichotomy for $\mathfrak{X}$ it suffices to prove a single theorem and exhibit a single counterexample (or, if there are n countervarieties, n counterexamples).

The dichotomies which interest us (three of which correspond to group-theoretic dichotomies of Hall [287] and Stroud [343] are as follows.

Type I(k)

$$\varphi \in \mathscr{H}^+ \Leftrightarrow \mathfrak{B}_\varphi \leqq \mathfrak{N}^k$$

$$\varphi \in \mathscr{H}^- \Leftrightarrow \mathfrak{B}_\varphi \geqq \mathfrak{A}^{k+1}$$

with minimal countervariety $\mathfrak{A}^{k+1}$. Here k is an integer $\geqq 0$, although the case $k = 0$ is not very interesting!

Type II(0)

$$\varphi \in \mathscr{H}^+ \Leftrightarrow \mathfrak{B}_\varphi \leqq \mathfrak{A}$$

$$\varphi \in \mathscr{H}^- \Leftrightarrow \mathfrak{B}_\varphi \geqq \mathfrak{N}_2.$$

Type II(1)

$$\varphi \in \mathscr{H}^+ \Leftrightarrow \mathfrak{B}_\varphi \leqq \mathfrak{A}\mathfrak{N}$$

$$\varphi \in \mathscr{H}^- \Leftrightarrow \mathfrak{B}_\varphi \geqq \hat{\mathfrak{A}}\mathfrak{A}^2$$

with minimal countervariety $\hat{\mathfrak{A}}\mathfrak{A}^2$, the centre-by-metabelian variety determined by the law $[x, [[y, z], [t, u]]]$. (These are probably part of a sequence of dichotomies:

Type II(k)

$$\varphi \in \mathscr{H}^+ \Leftrightarrow \mathfrak{B}_\varphi \leqq \mathfrak{A}\mathfrak{N}^k$$

$$\varphi \in \mathscr{H}^- \Leftrightarrow \mathfrak{B}_\varphi \geqq \hat{\mathfrak{A}}\mathfrak{A}^{k+1}$$

which certainly exists for groups. However, it is difficult to prove that $\mathscr{H}^+ \cap \mathscr{H}^- = \varnothing$ for $k \geqq 2$.)

Type III

$$\varphi \in \mathscr{H}^+ \Leftrightarrow \mathfrak{B}_\varphi \leqq \hat{\mathfrak{N}}\mathfrak{A}\mathfrak{N}$$

$$\varphi \in \mathscr{H}^- \Leftrightarrow \mathfrak{B}_\varphi \geqq \mathfrak{N}_2\mathfrak{A}.$$

Type IV

To describe this we recall the variety $\mathfrak{K}_n$ of algebras L such that $[L^2, L^n] = 0$ mentioned in chapter 11.

Then

$$\varphi \in \mathscr{H}^+ \Leftrightarrow \mathfrak{B}_\varphi \leqq \mathfrak{K}_n \text{ for some } n,$$

$$\varphi \in \mathscr{H}^- \Leftrightarrow \mathfrak{B}_\varphi \geqq \mathfrak{A}\mathfrak{N}_2 \text{ or } \mathfrak{B}_\varphi \geqq \hat{\mathfrak{A}}\mathfrak{A}^2.$$

For this there are two minimal countervarieties.

We proceed to establish the existence of these dichotomies. Given words α, β define recursively

$$[\alpha, {}_n\beta]^0 = [[\alpha, {}_{n-1}\beta]^0, \beta]^0$$

$$\alpha \bullet_n \beta \; = (\alpha \bullet_{n-1} \beta) \bullet \beta.$$

From now on let φ, χ, ψ, be arbitrary Hall words.

$I(k)$: By corollary 4.4 this corresponds to the statement

$$\varphi \Rightarrow \gamma_r \bullet_{k-1} \gamma_r \text{ or } \delta_{k+1} \Rightarrow \varphi$$

for some $r \in \mathbb{N}$. We prove this by induction on k. First let $k = 1$. Either $\varphi \Leftrightarrow \gamma_1$ or $\varphi = [\varphi_1, \varphi_2]^0$ for φ_1 and φ_2 of smaller weight. Suppose $\delta_2 \not\Rightarrow \varphi$. Inductively on the weight $\varphi_1 \Leftrightarrow \gamma_m$ and $\varphi_2 \Leftrightarrow \gamma_n$ for some m, n. If both m and $n > 1$ then $\delta_2 \Rightarrow \varphi$. Hence one of m, n is 1 and therefore $\varphi \Leftrightarrow \gamma_{\max(m,n)+1}$. This proves the case $k = 1$.

Now suppose the result true for k. Suppose $\delta_{k+1} \not\Rightarrow \varphi$. Again we may assume that $\varphi = [\varphi_1, \varphi_2]^0$. Then $\delta_{k+1} \not\Rightarrow \varphi_1$ and $\delta_k \not\Rightarrow \varphi_2$. By induction on the weight

$$\varphi_2 \Rightarrow \eta = \gamma_r \bullet_{k-1} \gamma_r$$

$$\varphi_1 \Rightarrow \gamma_s \bullet_{k-2} \gamma_s.$$

We may assume that $r = s$. Then $\varphi_1 \Rightarrow \gamma_s \bullet \eta$, $\varphi_2 \Rightarrow \eta$, so $\varphi \Rightarrow [\gamma_s \bullet \eta, \eta]^0 = = \gamma_{s+1} \bullet \eta \Rightarrow \gamma_t \bullet_{k-1} \gamma_t$ for $t = \max(s+1, r)$. This completes the induction.

We know that $\mathscr{H} = \mathscr{H}^+ \cup \mathscr{H}^-$ for type I(k). It remains to prove that $\mathscr{H}^+ \cap \mathscr{H}^- = \varnothing$. This follows from:

LEMMA 5.1 *For all k there exists a Lie algebra $L \in \mathfrak{A}^{k+1}$ which does not lie in $\mathfrak{N}^k$.*

Proof: For any Lie algebra L define $L^\natural$ as follows: let $U = U(L)$ be the universal enveloping algebra, and think of U as an abelian algebra. Let

$$L^\natural = L \dot{+} L$$

where L acts on U by right multiplication. We claim that the Fitting radical

$$\nu(L^\natural) = U.$$

Certainly $U \leqq \nu(L^\natural)$. If $\nu(L^\natural) > U$ then there exists $0 \neq x \in L$ such that $U + \langle x \rangle$ is locally nilpotent, so that for all $u \in U$

$$[u, {}_n x] = ux^n = 0$$

for some $n \in \mathbb{N}$. But U has no zero-divisors (Jacobson [98] p. 166), a contradiction.

Now let L_1 be any non-zero abelian Lie algebra, and define recursively

$$L_{i+1} = L_i^\natural.$$

Then clearly $L_{k+1} \in \mathfrak{A}^{k+1}$. But by looking at Fitting radicals, $L_{k+1} \notin \mathfrak{N}^k$. □

A construction similar to this has been used by Simonjan [193]. Notice that it shows that $L/\nu(L)$ may have any assigned isomorphism type if L is chosen suitably.

II(k): This is obvious for $k = 0$. We prove

$$[\delta_{k+1}, \gamma_1]^0 \Rightarrow \varphi \text{ or } \varphi \Rightarrow [\psi, \psi]^0$$

where $\psi = \gamma_r \bullet_{k-1} \gamma_r$ for some r. Now if $[\delta_{k+1}, \gamma_1]^0 \not\Rightarrow \varphi$ and $\varphi = [\varphi_1, \varphi_2]^0$ then $\delta_{k+1} \not\Rightarrow \varphi_1$ and $\delta_{k+1} \not\Rightarrow \varphi_2$. By type I(k), $\varphi_i \Rightarrow \psi$ $(i = 1, 2)$ for large enough r. This proves $\mathscr{H} = \mathscr{H}^+ \cup \mathscr{H}^-$.

We proved in theorem 11.1.6 that $\mathfrak{G} \cap \mathfrak{A}\mathfrak{N} \leqq$ Max-$\lhd$, and gave an example to show that there exists $L \in \mathfrak{G} \cap \hat{\mathfrak{A}}\mathfrak{A}^2$ such that $L \notin$ Max-$\lhd$. But if $\varphi \in \mathscr{H}^+ \cap \mathscr{H}^-$ then $\mathfrak{G} \cap \mathfrak{B}_\varphi \leqq$ Max-$\lhd$ and yet $L \in \mathfrak{G} \cap \mathfrak{B}_\varphi$, a contradiction.

This shows that II(1) is a genuine dichotomy. Obviously II(0) is. The case of II(k), for $k \geqq 2$, we leave open.

III: We prove

$$[\delta_2, \gamma_2]^0 = \gamma_3 \bullet \gamma_2 \Rightarrow \varphi$$

or

$$\varphi \Rightarrow [\gamma_r, \gamma_r, {}_r\gamma_1]^0$$

for some r. Suppose $[\delta_2, \gamma_2]^0 \not\Rightarrow \varphi = [\varphi_1, \varphi_2]^0$. Either

$$\varphi = [\chi_1, \chi_2, {}_r\gamma_1]^0$$

with neither $\chi_j = \gamma_1$ $(j = 1, 2)$ or else $\varphi = \gamma_r$ for some r. Then $[\delta_2, \gamma_2]^0 \not\Rightarrow$ $\not\Rightarrow [\chi_1, \chi_2]^0$ but $\gamma_2 \Rightarrow \chi_j$ and $\delta_2 \not\Rightarrow \chi_j$, whence $\chi_j = \gamma_{r_j}$ $(j = 1, 2)$.

To prove that $\mathscr{H}^+ \cap \mathscr{H}^- = \varnothing$, note that $\mathfrak{G} \cap \hat{\mathfrak{N}}\mathfrak{A}\mathfrak{N} \leqq \hat{\mathfrak{N}}$(Max-$\lhd$). It is easy to find an $\mathfrak{N}_2\mathfrak{A}$-algebra with trivial centre that does not satisfy Max-$\lhd$.

IV: We prove

$$\varphi \Rightarrow [\gamma_2, \gamma_r]^0 \text{ for some } r$$

or

$$[\gamma_3, \gamma_3]^0 \Rightarrow \varphi \text{ or } [\gamma_1, \delta_2]^0 \Rightarrow \varphi.$$

We may assume $\varphi = [\chi, \psi]^0$. If χ and ψ have length $\geqq 3$ then $[\gamma_3, \gamma_3]^0 \Rightarrow \varphi$. So without loss of generality χ has length 1 or 2. Suppose χ has length 1, so that $\chi \Leftrightarrow \gamma_1$. Then $\varphi \Leftrightarrow [\gamma_1, \psi]^0$. If $\delta_2 \Rightarrow \psi$ then $[\gamma_1, \delta_2]^0 \Rightarrow \varphi$. By I(1) we may assume $\psi \Rightarrow \gamma_r$, and then $\varphi \Rightarrow [\gamma_1, \gamma_r]^0 = \gamma_{r+1} \Rightarrow [\gamma_2, \gamma_r]^0$. If χ has length 2 then $\chi \Leftrightarrow \gamma_2$, so $\varphi \Leftrightarrow [\gamma_2, \psi]^0$. If $\psi \Rightarrow \gamma_r$ then $\varphi \Rightarrow [\gamma_2, \gamma_r]^0$. Otherwise $\delta_2 \Rightarrow \psi$ and then $[\gamma_2, \delta_2]^0 \Rightarrow \varphi$. But $[\gamma_1, \delta_2]^0 \Rightarrow [\gamma_2, \delta_2]^0$.

The Hartley algebra (example 6.3.6) lies in $\mathfrak{A}\mathfrak{N}_2$ but not in any $\mathfrak{K}_n$. The $\hat{\mathfrak{A}}\mathfrak{A}^2$-algebra of chapter 11 section 1 is obviously not in any $\mathfrak{K}_n$.

We now tabulate the type of dichotomy corresponding to various natural choices of $\mathfrak{X}$. For convenience we distinguish between characteristic 0 and $p > 0$. The proofs that these really are the relevant dichotomies involve little more than quoting previous theorems and counterexamples. Note particularly that $\mathfrak{X} = \mathrm{R}\mathfrak{F}$ is type IV for characteristic 0 but type II for characteristic $p > 0$. Frequently the test question: which dichotomy corresponds to $\mathfrak{X}$? – reveals gaps in our knowledge and leads to new conjectures. The existence of the type IV dichotomy and the fact that $\mathfrak{G} \cap \mathfrak{K}_n \leqq \mathrm{R}\mathfrak{F}$ in characteristic 0 were suggested by such a question.

class $\mathfrak{X}$	char. 0	char. p
$\mathfrak{F}$	I(1)	I(1)
$\mathfrak{D}$	I(1)	I(1)
Max-$\lhd$	II	II
$L \in \mathfrak{X} \Leftrightarrow$ all irreducible L-modules are finite-dimensional	II(0)	I(1)
$\mathfrak{F}$	IV	II
$\hat{\mathfrak{N}}$(Max-$\lhd$)	III	III

As regards dichotomies, it is wise to quote Hall [287]: '... the classification of properties adopted here is a crude one. It serves merely as a first approximation and as a convenient way of expressing a certain body of facts with brevity and precision.'

Chapter 15

The finite basis problem

A variety of Lie algebras is *finitely based* if it has a finite set of laws, such that every law is a consequence of this finite set. Equivalently, the corresponding verbal ideal of $\mathscr{F}_\infty$ must be finitely generated as a verbal (or fully invariant) ideal. We pose the question: which varieties are finitely based?

The corresponding problem has been studied for many algebraic structures. In particular, it is now known that a variety of groups need not be finitely based (Olšanskii [319], Adjan [257], Vaughan-Lee [347]).

We shall prove (following Cohen [270], Lyndon [307], and Vaughan-Lee [238]) that subvarieties of the classes of nilpotent or metabelian Lie algebras are finitely based. The proof of the latter case utilises some work of Higman [293] on partial orderings. It is also true that any subvariety of $\mathfrak{A}\mathfrak{N} \cap \mathfrak{N}\mathfrak{A}$ is finitely based, but we shall not prove this. Next we give examples, due to Vaughan-Lee [239], of a subvariety of $\mathfrak{N}_2\mathfrak{A}$ (over any field of characteristic 2) which is not finitely based, and of a finite-dimensional algebra (over any infinite field of characteristic 2) without a finite basis for its laws. Finally we show that in contrast to this, subvarieties of $\mathfrak{N}_2\mathfrak{A}$ over fields of characteristic $\neq 2$ *are* finitely based (Bryant and Vaughan-Lee [34]).

For the purposes of this chapter it is convenient to omit square brackets from Lie products and to left norm all products, so that $xy^iz^j\ldots$ will mean $[x, {}_iy, {}_jz, \ldots]$.

1. Nilpotent varieties

We shall show that any nilpotent variety is finitely based, using a proof similar to that of Lyndon [307] for groups.

LEMMA 1.1 *Let I be an ideal of $\mathscr{F}_n$, and let $r \in \mathbb{N}$. Then there exists a finite subset I_r of I which, together with $I \cap \mathscr{F}_n^{r+1}$, generates I.*

Proof: $I/(I\cap\mathscr{F}_n^{r+1}) \cong (\mathscr{F}_n^{r+1}+I)/\mathscr{F}_n^{r+1}$ which is finite-dimensional by theorem 1.2.4. □

LEMMA 1.2 *Let V be a verbal ideal of $\mathscr{F}_\infty$. If $n \geqq 1$ then V is generated, as verbal ideal, by a finite set S of words in at most n indeterminates, together with $V_{n+1} = V\cap\mathscr{F}_\infty^{n+1}$.*

Proof: We proceed by induction on n, the case $n = 1$ being trivial. Consider

$$w = \textstyle\sum_{i=1}^m \mu_i\pi_i+s \in V\cap\mathscr{F}_\infty^{n+1}$$

where the π_i have length $n+1$, and $s \in V\cap\mathscr{F}_\infty^{n+2}$. For each i_0 let X_{i_0} be the set of generators occurring in π_{i_0}, and define an endomorphism φ of $\mathscr{F}_\infty$ by

$$\varphi(x_k) = \begin{cases} 0 & x_k \in X\backslash X_{i_0} \\ x_k & x_k \in X_{i_0}. \end{cases}$$

Then

$$\varphi(w) = \textstyle\sum' \mu_i\pi_i+s'$$

where the sum $\sum'$ is taken over those i for which $x_i \in X_{i_0}$. Therefore $w' = w-\varphi(w)$ contains fewer terms than w, and w is equivalent to w' and $\varphi(w)$. Repeating this process on w' we find that w is equivalent to a set of words Ω involving at most $n+1$ generators, together with V_{n+2}. Applying another endomorphism allows us to assume that Ω lies in $\mathscr{F}_{n+1}$; and lemma 1.1 allows us to assume that Ω is finite. □

The main result follows easily:

THEOREM 1.3 *Every nilpotent variety is finitely based.*

Proof: Let $\mathfrak{V}$ be a subvariety of $\mathfrak{N}$. By corollary 14.4.4 we have $\mathfrak{V} \subseteq \mathfrak{N}_c$ for some c, so that if V is the verbal ideal corresponding to V we have

$$V \supseteq \mathscr{F}_\infty^{c+1}.$$

By lemma 1.2 it follows that V is generated by $S\cup V_{c+1}$, for a finite set of words S. But $V_{c+1} = \mathscr{F}_\infty^{c+1}$ which is generated by the single word $x_1 \ldots x_{c+1}$. Therefore V is finitely generated as a verbal ideal. □

2. Partially well ordered sets

To make further progress we must develop some techniques involving ordered sets. We begin with a theorem of Higman [293]. This is most conveniently stated and proved as a theorem about abstract algebras (in the sense of Cohn [271] p. 48) and we shall follow Higman in so presenting it.

A relation $\leqq$ on a set A is a *quasiorder* if for all $a, b \in A$

(i) $a \leqq a$
(ii) $a \leqq b \leqq c$ implies $a \leqq c$.

Let A be a quasiordered set and suppose $B \subseteq A$. The *closure* of B in A is

$$\bar{B} = \{a \in A : b \leqq a \text{ for some } b \in B\}.$$

We say that B is *closed* if $B = \bar{B}$, and is *open* if its complement is closed. We say that A is *partially well ordered* if every closed set is the closure of a finite set.

An *abstract algebra* is a set A with a set M of operations. Each $\mu \in M$ is an *r-ary* operation for some $r \in \mathbb{N}$, and maps each sequence $x = (a_1, \ldots, a_r)$ of elements $a_i \in A$ to an element

$$\mu x = \mu(a_1, \ldots, a_r) \in A.$$

We let M_r be the set of r-ary operations in M, and assume M_r is empty for all $r \geqq$ some $n \in \mathbb{N}$. We adopt the convention that $a, b, c, \ldots$ are elements of A, whereas $x, y, z, \ldots$ are sequences of elements of A. Whenever we write μx we assume that x has the correct length for μx to be defined (namely r if μ is r-ary). Sequences $x, y, z, \ldots$ may be empty. A *subalgebra* of (A, M) is an algebra (B, M) with $B \subseteq A$, such that $\mu x \in B$ for all $\mu \in M$,

$$x = (b_1, \ldots, b_r) \in B \times \ldots \times B.$$

We identify the operations μ with their restrictions to B. The algebra (A, M) is *minimal* if it has no subalgebras. Note that this does not make A empty since we allow 0-ary operations. In applications we shall assume that A is generated by a certain set of elements, and define 0-ary operations which, when applied to the empty sequence, yield the generators. This makes A minimal. However, for the proof of the theorem it is best not to use generators.

The algebra (A, M) is ordered if there is a quasiorder $\leqq$ on A such that for all $\mu \in M$; a, b, x, y we have

$$a \leqq b \Rightarrow \mu xay \leqq \mu xby.$$

The quasiorder is a *divisibility order* if for all μ, a, x, y we have

$$a \leqq \mu x a y.$$

Suppose further that each M_r is quasiordered. Then a quasiorder on A is *compatible* with that on M_r if for all $\lambda, \mu \in M_r$

$$\lambda \leqq \mu \Rightarrow \lambda x \leqq \mu x.$$

Our immediate object is to prove the following theorem of Higman [293] p. 327:

THEOREM 2.1 *Let (A, M) be a minimal algebra, with $M_r = \varnothing$ for all $r \geqq n \in \mathbb{N}$, and suppose that each M_r is partially well ordered for $r = 0, \ldots, n$. Then any divisibility order of A which is compatible with the quasiorders of the M_r is necessarily a partial well ordering of A.*

An important special case occurs if the M_r are finite for $r > 0$. We define $a \leqq b$ on M_r $(r > 0)$ to mean $a = b$. Using the connection between 0-ary operations and generators we obtain:

COROLLARY 2.2 *An abstract algebra with a finite set of r-ary operations ($r > 0$) is partially well ordered in a divisibility order if some generating set is partially well ordered.* □

We begin by listing some properties equivalent to partial well ordering.

PROPOSITION 2.3 *The following conditions on a quasiordered set A are equivalent:*

(a) *A is partially well ordered,*
(b) *Every closed subset of A is the closure of a finite set,*
(c) *The closed subsets of A satisfy the maximal condition,*
(d) *The open subsets of A satisfy the minimal condition,*
(e) *If B is any subset of A there is a finite set B_0 such that $B_0 \subseteq B \subseteq \bar{B}_0$,*
(f) *Every infinite sequence of elements of A has an infinite ascending subsequence,*
(g) *If $a_1, a_2, \ldots$ is a sequence of elements of A there exist integers $i < j$ such that $a_i \leqq a_j$,*
(h) *There exists neither an infinite strictly descending sequence of elements of A, nor an infinite set of mutually incomparable elements of A.*

Proof: By definition (a) $\Leftrightarrow$ (b). The equivalence of (b), (c), (e) is a standard phenomenon. By taking complements (c) $\Leftrightarrow$ (d). Clearly (f) $\Rightarrow$ (g) $\Rightarrow$ (h). To show that (e) $\Rightarrow$ (g) let

$$B = \{a_1, a_2, \ldots\}.$$

Since B_0 is finite there exists j such that every $b \in B_0$ is an a_i with $i < j$. Then $a_i \leqq a_j$. To show that (g) $\Rightarrow$ (b) assume we have a strictly ascending sequence

$$A_1 \subseteq A_2 \subseteq \ldots$$

of closed sets: take $a_i \in A_{i+1} \backslash A_i$ and obtain a contradiction. It is well known that (h) $\Rightarrow$ (g) (e.g. Birkhoff [264] p. 39). □

From these we derive some simple consequences.

LEMMA 2.4 *If A is partially well ordered, so is every subset and every homomorphic image.* □

Proof: For subsets use (g); for homomorphic images use (c), noting that inverse images of closed sets are closed. □

If A and B are quasiordered sets we can quasiorder $A \times B$ by

$$(a_1, b_1) \leqq (a_2, b_2) \Leftrightarrow a_1 \leqq a_2 \text{ and } b_1 \leqq b_2.$$

LEMMA 2.5 *If A and B are partially well ordered, so is $A \times B$.*

Proof: Use (f). □

The next result is a sort of 'transfinite induction' for partially well ordered sets.

LEMMA 2.6 *A proposition concerning a quasiordered set M which is true provided it is true for every proper open subset of M is true of every partially well ordered set M.*

Proof: If the proposition is false for M, it is false for a proper open subset M_1 of M, hence false for a proper open subset M_2 of M_1, and so on. This contradicts (d) of proposition 2.3. □

LEMMA 2.7 *If M is a quasiordered set such that for all $a \in M$, $M \backslash \{\bar{a}\}$ is partially well ordered, then M is partially well ordered.*

Proof: Let $E \neq \varnothing$ be a closed subset of M, and take $a \in E$. Then $E \backslash \{\bar{a}\}$ is a closed subset of $M \backslash \{\bar{a}\}$ and is therefore the closure in $M \backslash \{\bar{a}\}$ of a finite set E_0. Then E is the closure in M of the finite set $E_0 \cup \{a\}$. □

Let $(M_i)_{0 \leqq i \leqq n}$ be a sequence of partially well ordered sets. A second sequence (M'_i) is obtained from (M_i) by *descent* if for some integer r with $0 \leqq r \leqq n$ we have

(i) If $i > r$ then $M'_i = M_i$,
(ii) M'_r is a proper open subset of M_r,
(iii) For $i < r$, M'_i is any partially well ordered set.

LEMMA 2.8 *There exists no infinite chain $(M_i^{(r)})$ of sequences, $r = 1, 2, 3, \ldots$, such that for all r, $(M_i^{(r+1)})$ is obtained from $(M_i^{(r)})$ by descent.*

Proof: Use induction on n. If $n = 0$ use (d) of proposition 2.3. Assume such a chain exists. Then $M_n^{(r)}$ is constant for large r. Remove the nth terms of the sequences and apply the induction hypothesis. □

Equivalently we have another induction principle:

PROPOSITION 2.9 *Let $p(M_i)$ be a proposition concerning the sequence*

$$(M_i) = (M_1, \ldots, M_n)$$

of partially well ordered sets. If the falsity of $p(M_i)$ for any sequence of partially well ordered sets implies the falsity of $p(M'_i)$ for some (M'_i) obtained from (M_i) by descent, then $p(M_i)$ is true for any (M_i). □

We need one more lemma before proving theorem 2.1:

LEMMA 2.10 *Let (A, M) be an abstract algebra with a divisibility ordering. If X is a closed subset of A and (A_0, M) is a subalgebra, then $(A_0 \cup X, M)$ is a subalgebra.*

Proof: Let μ be an r-ary operation in M, let $a_1, \ldots, a_r$ belong to $A_0 \cup X$. Put $b = \mu a_1 \ldots a_r$. If all $a_i \in A_0$ then $b \in A_0$. On the other hand, if some $a_i \in X$ then $a_i \leqq b$ so $b \in X$. □

Proof of theorem 2.1

We use the induction principle of proposition 2.9. That is, we let (A, M) be a minimal algebra, with each M_i partially well ordered, having a compatible divisibility order which is not a partial well order. We shall construct an algebra (A', M') with the same properties, where (M_i') is obtained from (M_i) by descent.

Let

$$X = \{a \in A : A \backslash \{\bar{a}\} \text{ is not partially well ordered}\}.$$

By lemma 2.7 $X \neq A$. Since (A, M) is minimal, X is not a subalgebra. Hence there exists $\mu \in M_r$ for some r, and a sequence $z = (a_1, \ldots, a_r)$ of elements of X such that $b = \mu z$ is not an element of X. Now we define M'. If $i \neq r, r-1$ take $M_i' = M_i$. Let $M_r' = M_r \backslash \{\bar{\mu}\}$. If $r \geqq 1$ and λ is any r-ary operation, let $\lambda(a, s)$ $(a \in A,\ s = 1, \ldots, r)$ be the $(r-1)$-ary operation with

$$\lambda(a, s)xy = \lambda xay$$

where x has length $s-1$. Let $M_{r-1}^{(s)}$ be the set of operations $\lambda(a, s)$ for $\lambda \in \{\bar{\mu}\}$ and $a \in A \backslash \{\bar{a}_s\}$, with ordering induced from $\{\bar{\mu}\} \times (A \backslash \{\bar{a}_s\})$. Let M_{r-1}' be the union of M_{r-1} and the $M_{r-1}^{(s)}$ for $s = 1, \ldots, r$, with no extra order relations. Finally let (A', M') be the unique minimal subalgebra of (A, M').

Clearly M' is obtained from M by descent. It only remains to check that (A', M') has the required properties. We quasiorder A' as a subset of A: this gives a divisibility order compatible with the quasiorders of $M_0', \ldots, M_n'$. Moreover, all these are partially well ordered. This is obvious except perhaps for M_{r-1}'. But M_{r-1} is partially well ordered, and each $M_{r-1}^{(s)}$ is partially well ordered; it is then easy to see that the union of these sets, which is M_{r-1}', is partially well ordered.

Next we show that A' is not partially well ordered. To prove this we show that $A' \supseteq A \backslash \{\bar{b}\}$ which we know is not partially well ordered. Equivalently, we must show that $A = A' \cup \{\bar{b}\}$. By minimality it suffices to prove that $(A' \cup \{\bar{b}\}, M)$ is a subalgebra. By lemma 2.10 $A' \cup \{\bar{b}\}$ is closed under the operations of M', so we need prove only closure under operations of $M \backslash M'$, i.e. under λ for $\lambda \in \{\bar{\mu}\}$. Suppose that $c_1, \ldots, c_r \in A' \cup \{\bar{b}\}$, and let $d = \lambda c_1 \ldots c_r$. If $a_i \leqq c_i$ for all i, then $\mu a_1 \ldots a_r \leqq \lambda c_1 \ldots c_r$ or $d \in \{\bar{b}\}$. But if (as can happen only if $r \geqq 1$) for some s we have $c_s \in A \backslash \{\bar{c}_s\}$, then $d = \lambda(c_s, s)x$ where x is a sequence of elements of $A' \cup \{\bar{b}\}$. Since $\lambda(c_s, s) \in M_{r-1}^{(s)} \subseteq M'$ it again follows that $d \in A' \cup \{\bar{b}\}$. This proves the theorem. □

Next we introduce the set Ξ of all order-preserving injections $\mathbb{N} \backslash \{0\} \rightarrow$ $\rightarrow \mathbb{N} \backslash \{0\}$. For any quasiordered set A we let $V(A)$ be the set of all finite sequences of elements of A, ordered as follows:

$$(a_1, \ldots, a_r) \leqq (b_1, \ldots, b_s)$$

if there exists $\xi \in \Xi$ such that $\xi(r) \leqq s$ and $a_i \leqq b_{\xi(i)}$ for $i = 1, \ldots, r$.

THEOREM 2.11 *If A is partially well ordered then so is $V(A)$.*

Proof: Under the operation of juxtaposition $V(A)$ is a semigroup, and the quasiorder defined above is a divisibility order. The 1-element sequences form a set of generators, order-isomorphic with A. The result follows from corollary. 2.2. □

We need an extension of this result, which is proved in Bryant and Newman [266]; a slightly special case is proved in Vaughan-Lee [239].

Let $(A, \leqq)$ be a partially ordered set, and take $0 < k \in \mathbb{N}$. Let S be the set of all elements

$$(i_1, \dots, i_k; \alpha_1, \dots, \alpha_m)$$

where $0 < i_j \in \mathbb{N}$ for $j = 1, \dots, k$, and each $\alpha_j \in A$ for $j = 1, \dots, m$. We partially order S by setting

$$(i_1, \dots, i_k; \alpha_1, \dots, \alpha_m) \leqq (j_1, \dots, j_k; \beta_1, \dots, \beta_n)$$

if there exists $\xi \in \Xi$ such that

(i) $\xi(i_\lambda) = j_\lambda \qquad (\lambda = 1, \dots, k)$
(ii) $\xi(m) \leqq n$
(iii) $\alpha_\mu \leqq \beta_{\xi(\mu)} \qquad (\mu = 1, \dots, m)$.

THEOREM 2.12 *If A is partially well ordered than so is S.*

Proof: We show that S is order-isomorphic to a certain partially well ordered set T, which we construct.

We adjoin two new elements 0, 1 to A, and define

$$0 \leqq 1 \leqq \alpha \qquad (\alpha \in A).$$

The new set $B = \{0, 1\} \cup A$ is clearly partially well ordered. Let C be the set of $(k+1)$-tuples of elements of B. By lemma 2.5 C is partially well ordered by componentwise ordering:

$$(b_1, \dots, b_{k+1}) \leqq (c_1, \dots, c_{k+1})$$

if and only if $b_i \leqq c_i$ for $i = 1, \dots, k+1$.

If $a \in C$ we let $a(\lambda)$ denote the λth component of a. Thus if $a, b \in C$ then $a \leqq b$ if and only if $a(\lambda) \leqq b(\lambda)$ for all $\lambda = 1, \dots, k+1$.

By theorem 2.11 the set of finite sequences of elements of C is partially well ordered if we put

$$(a_1, \dots, a_m) \leqq (b_1, \dots, b_n)$$

if there exists $\xi \in \Xi$ such that $\xi(m) \leqq n$ and $a_\mu \leqq b_{\xi(\mu)}$ for $\mu = 1, \ldots, m$. In other words, we have

$$a_\mu(\lambda) \leqq b_{\xi(\mu)}(\lambda)$$

for $\lambda = 1, \ldots, k+1$, and $\mu = 1, \ldots, m$.

Let T be the set of finite sequences of elements of C with the property that $(a_1, \ldots, a_m) \in T$ if and only if exactly one of $a_1(\lambda), \ldots, a_m(\lambda)$ is 1, and the rest are 0, for each $\lambda = 1, \ldots, k$. Note that we place no restriction on the $a_i(k+1)$. Then $(T, \leqq)$ is partially well ordered.

We now show that S and T are order-isomorphic. To do this we construct functions

$$\theta : T \rightarrow S$$

$$\theta^* : S \rightarrow T$$

which are mutually inverse, and show that they preserve order. We define them as follows:

$$\theta(a_1, \ldots, a_m) = (i_1, \ldots, i_k;\ a_1(k+1), \ldots, a_m(k+1))$$

where i_λ is equal to the unique μ such that $a_\mu(\lambda) = 1$, for $\lambda = 1, \ldots, k$.

$$\theta^*(i_1, \ldots, i_k;\ \alpha_1, \ldots, \alpha_m) = (b_1, \ldots, b_m)$$

where

$$b_\mu(\lambda) = \begin{cases} \alpha_\mu & \text{if } \lambda = k+1, \\ 1 & \text{if } \lambda = 1, 2, \ldots, k \text{ and } i_\lambda = \mu, \\ 0 & \text{otherwise.} \end{cases}$$

Since $m \geqq i_1, \ldots, i_k$ exactly one of $b_1(\lambda), \ldots, b_m(\lambda)$ is 1 and the rest are 0. It follows without difficulty that θ and θ^* are mutual inverses.

Next we show that θ preserves order. Suppose that

$$(a_1, \ldots, a_m) \leqq (b_1, \ldots, b_n).$$

Then there exists $\xi \in \Xi$ such that $\xi(m) \leqq n$ and

$$a_\mu(\lambda) \leqq b_{\xi(\mu)}(\lambda) \tag{*}$$

for $\lambda = 1, \ldots, k+1$, $\mu = 1, \ldots, m$. Now for $\lambda \leqq k$ exactly one $a_i(\lambda)$ is 1 and the rest are 0, and exactly one $b_j(\lambda)$ is 1 and the rest are 0. Hence for $\lambda \leqq k$ we have $a_\mu(\lambda) = b_{\xi(\mu)}(\lambda)$ for $\mu = 1, \ldots, m$. Suppose that

$$\theta(a_1, \ldots, a_m) = (i_1, \ldots, i_k;\ a_1(k+1), \ldots, a_m(k+1))$$

$$\theta(b_1, \ldots, b_m) = (j_1, \ldots, j_k;\ b_1(k+1), \ldots, b_n(k+1)).$$

From (*) $a_\mu(k+1) \leqq b_{\xi(\mu)}(k+1)$, so to prove that

$$\theta(a_1, \ldots, a_m) \leqq \theta(b_1, \ldots, b_n)$$

it suffices to show that $j_\lambda = \xi(i_\lambda)$ for $\lambda = 1, \ldots, k$. Suppose that $i_\lambda = \mu$. Then

$$1 = a_\mu(\lambda) = b_{\xi(\mu)}(\lambda)$$

whence $j_\lambda = \xi(\mu) = \xi(i_\lambda)$. This proves that θ preserves order.

Finally we show that θ^* preserves order. Let

$$(i_1, \ldots, i_k; \alpha_1, \ldots, \alpha_m) \leqq (j_1, \ldots, j_k; \beta_1, \ldots, \beta_n).$$

Then there exists $\xi \in \Xi$ such that $\xi(i_\lambda) = j_\lambda$ for $\lambda = 1, \ldots, k$; $\xi(m) \leqq n$, and $a_\mu \leqq b_{\xi(\mu)}$ for $\mu = 1, \ldots, m$. Let the images of these two sequences under θ^* be $(a_1, \ldots, a_m)$ and $(b_1, \ldots, b_n)$ respectively. Then for $\lambda \leqq k$,

$$a_\mu(\lambda) = \delta_{\mu, i_\lambda} = \delta_{\xi(\mu), \xi(i_\lambda)} = \delta_{\xi(\mu), j_\lambda} = b_{\xi(\mu)}(\lambda).$$

Also

$$a_\mu(k+1) = \alpha_\mu \leqq \beta_{\xi(\mu)} = b_{\xi(\mu)}(k+1).$$

Hence

$$(a_1, \ldots, a_m) \leqq (b_1, \ldots, b_n)$$

and θ^* is order-preserving.

Hence S and T are order-isomorphic. But T is partially well ordered, and the theorem is proved. □

Next we develop some related ring-theoretic ideas of Cohen [270].

An *algebraic closure operation* on a set C is a function associating to each $X \subseteq C$ its *closure* $\overline{X} \subseteq C$, such that:

(i) $X \subseteq \overline{X}$,
(ii) $X \subseteq Y \Rightarrow \overline{X} \subseteq \overline{Y}$,
(iii) $\overline{\overline{X}} = \overline{X}$,
(iv) If $x \in \overline{X}$ then there exists a finite subset $X_0 \subseteq X$ such that $x \in \overline{X}_0$.

If C is a quasiordered set then the usual closure operation is an algebraic closure operation. If C is a ring then the operation 'ideal generated by' is an algebraic closure operation.

An algebraic closure operation has the *finite basis property* if every closed set is the closure of a finite subset. For a quasiordered set this property is equivalent to partial well ordering; for a ring it is the Noetherian property.

Let C be a set with an algebraic closure operation, and let P be a quasi-ordered set. We define an algebraic closure operation on $C \times P$, the *induced* operation, by setting $(c, p) \in \overline{X}$ if and only if there exist finitely many elements $(c_i, p_i) \in X$ $(1 \leqq i \leqq n)$ such that $c \in \overline{\{c_1, \ldots, c_n\}}$ and $p \in \overline{\{p_1, \ldots, p_n\}}$ (i.e. $p_i \leqq p$ for some $i = 1, \ldots, n$).

LEMMA 2.13 *If the closure operation on C has the finite basis property and if P is partially well ordered, then the induced operation on $C \times P$ has the finite basis property.*

Proof: Let X be any closed subset of $C \times P$. For any $p \in P$ let $X(p)$ be the set of first coordinates c of elements $(c, p) \in X$. Each $X(p)$ is a closed subset of C, and if $p < q$ then $X(p) \subseteq X(q)$. We order the set of pairs $(X(p), p)$ as follows: $(X(p), p) < (X(q), q)$ if and only if $X(p) = X(q)$ and $p < q$. For any p there exists q such that $(X(q), q)$ is minimal with respect to being $< (X(p), p)$, since P is partially well ordered. Further, there are only finitely many minimal elements $(X(p_i), p_i)$. For if not we could find an infinite sequence $p_1 < p_2 < \ldots$ with each $(X(p_i), p_i)$ minimal. Then if $i < j$ we have $X(p_i) \subseteq X(p_j)$, and the minimality implies that $X(p_i) \neq X(p_j)$. By the finite basis property C has the maximal condition for closed subsets, a contradiction.

Let $(X(p_1, p_1), \ldots, (X(p_r), p_r)$ be these minimal elements. By the finite basis property we can find $c_{ij} \in C$, for $i = 1, \ldots, n_j$, such that

$$X(p_j) = \overline{\bigcup_i \{c_{ij}\}}.$$

If we take any $(c, p) \in X$ then $c \in X(p)$ which is $> X(p_j)$ for some $j \leqq r$. Hence

$$X = \overline{\bigcup_{i,j} \{c_{ij}\}}$$

and $C \times P$ has the finite basis property. □

Let R be a commutative ring. Let

$$S = R[x_1, x_2, \ldots]$$

be a polynomial ring in countably many indeterminates over R. Let M be the free S-module with countably many generators $t_1, t_2, \ldots$. An ideal of S is a *Ξ-ideal* if whenever $p(x_1, x_2, \ldots) \in S$ and $\xi \in \Xi$ then $p(x_{\xi(1)}, x_{\xi(2)}, \ldots) \in S$. A submodule N of M is a *Ξ-submodule* if whenever

$$\sum p_i(x_1, x_2, \ldots) t_i \in N$$

and $\xi \in \Xi$ then

$$\sum p_i(x_{\xi(1)}, x_{\xi(2)}, \ldots) t_{\xi(i)} \in N.$$

Let $V = V(\mathbb{N}\backslash\{0\})$ as defined before theorem 2.11, with ordering $\leqq$. By theorem 2.11 V is partially well ordered. We define another ordering $\prec$ on V as follows:

$$(i_1, \ldots, i_m) \prec (j_1, \ldots, j_n)$$

if $m < n$, or if $m = n$ and for some r we have $i_r < j_r$, and $i_s = j_s$ for all $s > r$. The identity map from $(V, \leqq)$ to $(V, \preccurlyeq)$ is order-preserving; and $(V, \preccurlyeq)$ is well ordered.

THEOREM 2.14 *If R is a Noetherian domain then S has the maximal condition for Ξ-ideals.*

Proof: Define the *weight* of a monomial $ax_1^{i_1}\ldots x_n^{i_n}$ to be $(i_1, \ldots, i_n) \in V$. The *leading term* of a polynomial p is its monomial of highest weight in the ordering $\preccurlyeq$, and the weight $w(p)$ of p is the weight of its leading term. We define

$$\theta : S \to R \times V$$

by

$$\theta(p) = (b, w(p))$$

where b is the coefficient of the leading term of p. The closure operation 'ideal generated by' on R is algebraic, and R has the finite basis property since it is Noetherian. By lemma 2.13 the closure operation induced on $R \times V$ has the finite basis property. Now if I is a Ξ-ideal of S, then $\theta(I)$ is a closed subset of $R \times V$.

Therefore $\theta(I) = \overline{\{\theta(p_1), \ldots, \theta(p_r)\}}$. Now if $p \in I$ then there exists q in the Ξ-ideal generated by $p_1, \ldots, p_r$ such that $\theta(q) = \theta(p)$, so that $w(p-q) < < w(p)$. Since $(V, \preccurlyeq)$ is well-ordered it follows by induction that I is the Ξ-ideal generated by $p_1, \ldots, p_r$, which is equivalent to the statement of the theorem. □

THEOREM 2.15 *If R is a Noetherian domain then M has the maximal condition for Ξ-submodules.*

Proof: We define a map

$$\pi : M \to R \times (\mathbb{N}\backslash\{0\}) \times V$$

by

$$\pi(0) = (0, 1, 0, \ldots, 0),$$

$$\pi(p) = (a, n, w(p_1), \ldots, w(p_n))$$

if $0 \neq p = \sum_{i=1}^{n} p_i t_i$ where $p_n \neq 0$, and the leading coefficient of p_n is a. We define two orderings on $(\mathbb{N}\backslash\{0\}) \times V$. The first ($\leqq$) is the order considered immediately before lemma 2.12, and consequently is a partial well order. The second ($\preccurlyeq$) is defined by

$$(r, i_1, \ldots, i_m) \prec (s, j_1, \ldots, j_n)$$

if $r < s$ or if $r = s$ and $(i_1, \ldots, i_m) \prec (j_1, \ldots, j_n)$. This is a well order. The identity map $((\mathbb{N}\backslash\{0\}) \times V, \leqq) \rightarrow ((\mathbb{N}\backslash\{0\}) \times V, \preccurlyeq)$ is order-preserving. The dénoument now parallels that of theorem 2.14. □

3. Metabelian varieties

Our aim is to prove the following theorem of Vaughan-Lee [238], whose proof is similar to that of Cohen [270]:

THEOREM 3.1 *Any subvariety of the class of metabelian Lie algebras is finitely based.*

Recall that for this chapter we omit square brackets from Lie products and left-norm all products. Let L be the free metabelian Lie algebra generated by $x_0, x_1, \ldots$. We begin by finding a basis for L^2.

LEMMA 3.2 *L^2 has a basis consisting of all products*

$$x_{i_1} \ldots x_{i_n}$$

where $i_1 > i_2 \leqq i_3 \leqq \ldots \leqq i_n$, and $n \geqq 2$.

Proof: For brevity call such products *regular*. In any metabelian Lie algebra we have

$$0 = ab(cd) = -b(cd)a - (cd)ab$$

so that

$$(cd)ab = (cd)ba$$

for all a, b, c, d. It follows that

$$x_{i_1} x_{i_2} x_{i_3} \ldots x_{i_n} = x_{i_1} x_{i_2} x_{\sigma(i_3)} \ldots x_{\sigma(i_n)}$$

for any permutation σ of $i_3, \ldots, i_n$. By lemma 1.1.1 we know that L^2 is spanned by all $x_{i_1} \ldots x_{i_n}$ $(n \geqq 2)$: we may assume that $i_3 \leqq i_4 \leqq \ldots \leqq i_n$. If $i_2 \leqq i_3$ the element is regular, and if $i_1 \leqq i_3$ it becomes regular on multiplying by -1 and interchanging x_{i_1} and x_{i_2}. If however $i_3 < i_1, i_2$ we use the Jacobi identity to write

$$x_{i_1} x_{i_2} x_{i_3} \ldots x_{i_n} = -x_{i_2} x_{i_3} x_{i_1} \ldots x_{i_n} + x_{i_1} x_{i_3} x_{i_2} \ldots x_{i_n}$$

and then apply suitable permutations to $\{1, 4, \ldots, n\}$ and $\{2, 4, \ldots, n\}$ to make the two summands regular. This proves that L^2 is spanned by regular words.

To show that the regular words are linearly independent consider a Lie algebra A whose basis consists of sequences

$$(i_1, i_2, \ldots, i_n)$$

of integers with $i_1 > i_2 \leqq i_3 \leqq \ldots \leqq i_n$, where multiplication is defined by

$$(i_1, \ldots, i_n)(j_1, \ldots, j_m) = 0 \qquad (m, n \geqq 2),$$

$$(i_1, \ldots, i_n)(j) = (i_1, \ldots, i_k, j, i_{k+1}, \ldots, i_n)$$

where $i_k \leqq j \leqq i_{k+1}$ $(m \geqq 2, j \geqq i_2)$,

$$(i_1, \ldots, i_n)(j) = (i_1, j, i_2, \ldots, i_n) - (i_2, j, i_3, \ldots, i_k, i_1, i_{k+1}, \ldots, i_n)$$

where $i_k \leqq i_1 \leqq i_{k+1}$ $(m \geqq 2, j < i_2)$,

$$(j)(i_1, \ldots, i_n) = -(i_1, \ldots, i_n)(j),$$

$$(i)(j) = \begin{cases} (i, j) & (i > j) \\ -(j, i) & (i < j) \\ 0 & (i = j). \end{cases}$$

It is straightforward though tedious to check that A is a metabelian Lie algebra. The map which assigns to each regular product in L its sequence of subscripts extends to a homomorphism $L \to A$, from which it is evident that the regular products are linearly independent, because their images are. $\square$

With L as above, let $R = \mathfrak{k}[x_0]$ be a polynomial ring over the ground field $\mathfrak{k}$, and let

$$S = R[x_1, x_2, \ldots] = \mathfrak{k}[x_0, x_1, \ldots].$$

Our aim is to use theorem 2.15. We define an action of S on L^2 as follows: let

$$\textstyle\sum_{i=1}^m \alpha_i x_0^{i_0} \ldots x_n^{i_n} \in S \qquad (\alpha_i \in \mathfrak{k})$$

and $w \in L^2$. Put

$$ws = \sum_{i=1}^{m} \alpha_i w x_0^{i_0} \dots x_n^{i_n}$$

where by abuse of notation w is thought of as an element of S. This is well-defined since for any $u \in L^2$ we have

$$ux_i x_j = ux_j x_i$$

as in the proof of lemma 3.2. Since L^2 is spanned by regular products every element of L^2 can be written uniquely in the form

$$\sum_{i>j\geqq 0} x_i x_j r_{ij}$$

where

$$r_{ij} \in \mathfrak{k}[x_j, x_{j+1}, \dots].$$

LEMMA 3.3 *If* $r_{ij} \in \mathfrak{k}[x_j, x_{j+1}, \dots]$ *and* $w = \sum_{i>j\geqq 0} x_i x_j r_{ij}$ *then* w *is equivalent, in the variety of metabelian Lie algebras, to the set of words*

$$\{\sum_{i>j} x_i x_j r_{ij} : j = 0, 1, 2, \dots\}.$$

Proof: Let μ_j $(j = 0, 1, \dots)$ be the endomorphism of L which maps $x_0, \dots, x_j \mapsto 0$, and $x_i \mapsto x_i$ for $i > j$. Then for $j \geqq 1$

$$\mu_{j-1} w - \mu_j w = \sum_{i>j} x_i x_j r_{ij}$$

so that w implies the words on the right of this equation. Together with w, these imply $\sum_{i>0} x_i x_0 r_{i0}$. Conversely, the given set implies w. □

COROLLARY 3.4 *Every set of words of* L^2 *is equivalent to a set of words, each of the form* $\sum_i x_i x_0 r_i$, *where* $r_i \in S$.

Proof: Use lemma 3.3 and an endomorphism induced by the map which interchanges x_0 and x_j, fixing all other x_i. □

Let $\mathfrak{V}$ be a subvariety of the metabelian variety. It is determined by a fully invariant ideal I of L. If $\mathfrak{V}$ is nontrivial we must have $I \leqq L^2$. Hence by corollary 3.4 I is equivalent to a subset of the set T of words of the form $\sum_{i=1}^{m} x_i x_0 p_i$, where $p_i \in S$. Let M, as before, be the free S-module generated by $t_1, t_2, \dots$.

LEMMA 3.5 *With the obvious S-action, T is isomorphic to a* Ξ*-submodule of M.*

Proof: Define $\theta : T \to M$ by

$$\theta(\sum x_i x_0 p_i) = \sum t_i p_i.$$

Clearly θ is a module monomorphism, and its image is a Ξ-submodule. □

We may now complete the proof of theorem 3.1. By theorem 2.15 $\theta(T)$ is finitely generated as a Ξ-submodule of M. Since elements of Ξ induce endomorphisms of S, it follows that T is equivalent to a finite subset $T_0 \subseteq T$. Therefore $\mathfrak{V}$ is finitely based. □

This proof gives no information as to the structure of the subvarieties of the metabelian variety. Further information on this question is given in Vaughan-Lee [238]: we quote without proof the main results.

If $\mathfrak{k}$ has characteristic 0 then the subvarieties of $\mathfrak{A}^2$, other than 0 and $\mathfrak{A}^2$ itself, are the varieties $\mathfrak{A}^2 \cap \mathfrak{N}_c$ for $c = 1, 2, \ldots$.

If $\mathfrak{k}$ is an infinite field of characteristic $p > 0$ then any nontrivial proper subvariety of $\mathfrak{A}^2$ is locally nilpotent, and is determined by a finite set of laws of the form

$$x_{n+1} x_1^{p^{\alpha(1)}} \ldots x_n^{p^{\alpha(n)}} x_{n+1}^{p^\alpha - 1} \qquad (0 \leqq \alpha(1) \leqq \ldots \leqq \alpha(n), \alpha(1) \leqq \alpha)$$

or

$$\textstyle\sum_{i=2}^m x_i x_1^{p^\alpha} \ldots x_{i-1}^{p^\alpha} x_i^{p^\alpha - 1} x_{i+1}^{p^\alpha} \ldots x_m^{p^\alpha} x_{m+1}^{p^{\alpha(m+1)}} \ldots x_n^{p^{\alpha(n)}}$$

with $0 \leqq \alpha \leqq \alpha(m+1) \leqq \ldots \leqq \alpha(n)$, and m a multiple of p.

If $\mathfrak{k}$ is a finite field with q elements, there is a unique minimal non-locally finite subvariety of $\mathfrak{A}^2$, determined by the laws

$$x_1 x_2 (x_3 x_4)$$

$$x_2 x_3 x_1^q + x_3 x_1 x_2^q + x_1 x_2 x_3^q$$

$$x_2 x_3 x_1 x_2^{q-1} x_3^{q-1} + x_3 x_1 x_2 x_3^{q-1} x_1^{q-1} + x_1 x_2 x_3 x_1^{q-1} x_2^{q-1}.$$

This is a proper subvariety of the metabelian variety.

4. Non-finitely based varieties

Throughout this section $\mathfrak{k}$ is a field of characteristic 2. Let F be the free centre-by-metabelian Lie algebra over $\mathfrak{k}$ generated by $y_1, y_2, \ldots$. The centre-by-metabelian law

$$(x_1 x_2)(x_3 x_4) x_5$$

is equivalent, via the Jacobi identity, to the law

$$(x_1x_2x_5)(x_3x_4)+(x_1x_2)(x_3x_4x_5). \qquad (*)$$

By lemma 3.2 we know that F^2 is spanned modulo $F^{(2)}$ by regular products

$$y_{i_1}\ldots y_{i_k}$$

with $k \geqq 2$, and $i_1 > i_2 \leqq i_3 \leqq \ldots \leqq i_k$. Now $(F^2)^3 = 0$, so that $F^{(2)}$ is spanned by products

$$(y_{i_1}\ldots y_{i_k})(y_{j_1}\ldots y_{j_l})$$

where each of the two factors is regular. From (*) and the fact that $(F^2)^3 = 0$ it follows that for all $a, b, c, d, g_1, \ldots, g_n \in F$ we have

$$(abg_1\ldots g_n)(cd) = (abg_{\sigma(1)}\ldots g_{\sigma(r)})(cdg_{\sigma(r+1)}\ldots g_{\sigma(n)}) \qquad (**)$$

for any permutation σ of $\{1, \ldots, n\}$ and any integer r with $0 \leqq r \leqq n$. Hence $F^{(2)}$ is spanned by products

$$(y_{i_1}\ldots y_{i_k})(y_{i_{k+1}}y_{i_{k+2}})$$

with $k \geqq 2$, $i_1 > i_2 \leqq i_3 \leqq \ldots \leqq i_k$, $i_{k+1} > i_{k+2} \leqq i_3$, $i_2 \geqq i_{k+2}$, $i_1 \geqq i_{k+1}$ if $i_2 = i_{k+2}$.

We shall be interested in the laws

$$(y_1\ldots y_k)(y_1y_2)$$

for $k \geqq 3$. We begin by showing that these are not consequences of the centre-by-metabelian law.

LEMMA 4.1 *There exists a centre-by-metabelian Lie algebra over* $\mathfrak{k}$ *which does not satisfy the law* $(y_1\ldots y_n)(y_1y_2)$ *for any* $n \geqq 3$.

Proof: Let T be the set of all finite subsets of $\mathbb{N}$. Let A be a Lie algebra over $\mathfrak{k}$ generated by $\{a_s, b_t : s, t \in T, t \neq \varnothing\}$ subject to

$$a_sa_t = \begin{cases} 0 & \text{if } s = t \text{ or } s \cap t \neq \varnothing \\ b_{s\cup t} & \text{if } s \neq t \text{ and } s \cap t = \varnothing \end{cases}$$

$$b_sa_t = b_sb_t = 0 \text{ for all } s, t \in T.$$

Next, define derivations $g, g_1, g_2, \ldots$ of A as follows:

$$a_tg = a_t$$

$$b_tg = 0$$

$$a_t g_i = \begin{cases} 0 & \text{if } i \in T \\ a_{t \cup \{i\}} & \text{if } i \notin T \end{cases}$$

$$b_t g_i = 0.$$

Using the fact that $\mathfrak{k}$ has characteristic 2 it is easy to check (case by case, as in Vaughan-Lee [239] p. 304) that these really are derivations. Clearly they commute. We form the split extension B of A by $\langle g, g_1, g_2, \ldots \rangle$. Manifestly B is centre-by-metabelian. Furthermore,

$$(a_\varnothing g g_3 g_4 \ldots g_n)(a_\varnothing g) = a_{\{3,4,\ldots,n\}} a_\varnothing = b_{\{3,4,\ldots,n\}}$$

so that B does not satisfy the law $(y_1 \ldots y_n)(y_1 y_2)$ for any $n \geqq 3$. □

LEMMA 4.2 *The verbal closure of* $(y_1 \ldots y_n)(y_1 y_2)$ *in* F *is spanned by words of the form* $(m_1 \ldots m_n)(m_1 m_2)$ *where the* m_i *are monomials.*

Proof: Since $(y_1 \ldots y_n)(y_1 y_2)$ is in the centre of F its verbal closure is spanned by elements $(a_1 \ldots a_n)(a_1 a_2)$ with the $a_i \in F$. By linearity we may suppose that $a_3, \ldots, a_n$ are monomials. Suppose that

$$a_1 = \beta b + \gamma c$$

where $\beta, \gamma \in \mathfrak{k}$, and $b, c \in F$. Then

$$(a_1 \ldots a_n)(a_1 a_2) = \beta^2 (b a_2 a_3 \ldots a_n)(b a_2) + \gamma^2 (c a_2 a_3 \ldots a_n)(c a_2) + \\ + \beta\gamma (b a_2 a_3 \ldots a_n)(c a_2) + \beta\gamma (c a_2 a_3 \ldots a_n)(b a_2).$$

But using (**), the last two terms are equal: because of the characteristic they cancel. We may therefore suppose a_1 to be monomial. Similarly we may suppose a_2 monomial. □

Now we can prove that the laws $(y_1 \ldots y_n)(y_1 y_2)$ are independent.

THEOREM 4.3 *If* $n \geqq 3$ *the word* $(y_1 \ldots y_n)(y_1 y_2)$ *is not a consequence of the set*

$$\{(y_1 \ldots y_k)(y_1 y_2) : n \neq k \geqq 3\}.$$

Proof: We encounter difficulties if $\mathfrak{k}$ is finite, so we take an infinite field $\mathfrak{k}^* \supseteq \mathfrak{k}$. Let F^* be the free centre-by-metabelian Lie algebra over $\mathfrak{k}^*$ generated by $y_1, y_2, \ldots$. Since the law $(x_1 x_2)(x_3 x_4) x_5$ is multilinear, it follows by lemma 14.2.4 that $F^* = \mathfrak{k}^* \otimes_{\mathfrak{k}} F$.

Suppose that $(y_1 \ldots y_n)(y_1 y_2)$ is in the verbal closure in F of

$$\{(y_1 \ldots y_k)(y_1 y_2) : n \neq k \geqq 3\}.$$

By lemma 4.2,

$$(y_1 \dots y_n)(y_1 y_2) = \sum_{i=1}^{r} \alpha_i v_i$$

where the $\alpha_i \in \mathfrak{k}$, $v_i \in F$, and where each v_i is of the form

$$(m_1 \dots m_k)(m_1 m_2)$$

for $n \neq k \geqq 3$, the m_j being monomials of F. This is an identity in F, so is an identity in F^*. By theorem 14.2.1 we may assume each v_i to be of degree 2 in y_1 and y_2, and linear in $y_3, \dots, y_n$.

Consider such a product of monomials. The total degree is $n+2$, so that $k \leqq n$. Since $k \neq n$ it follows that $k < n$. But then not all m_j can be linear. Suppose that the degree of m_i is greater than 1. If $i \geqq 3$ then

$$(m_1 \dots m_k)(m_1 m_2) = 0$$

since F is centre-by-metabelian. Without loss of generality we may suppose that

$$m_1 = y_{i_1} \dots y_{i_s}$$

where $s \geqq 2$. Then using (**) we have

$$(m_1 \dots m_k)(m_1 m_2) = (y_{i_1} \dots y_{i_s} m_2 m_2 m_3 \dots m_k)(y_{i_1} \dots y_{i_s}).$$

Now we must have m_2 linear, or else the expression is zero. Therefore $m_2 = y_j$ for some j. But now the degree is 2 in each of $y_{i_1}, \dots, y_{i_s}, y_j$. Since $s \geqq 2$ this is not possible. The only remaining possibility is that

$$(y_1 \dots y_n)(y_1 y_2) = 0$$

but this too is not the case by lemma 4.1. □

COROLLARY 4.4 *Over any field of characteristic 2 there exist non-finitely based subvarieties of the centre-by-metabelian variety.*

Proof: Let $\mathfrak{V}$ be the variety determined by the centre-by-metabelian law

$$(y_1 y_2)(y_3 y_4) y_5$$

together with *all* the laws

$$(y_1 \dots y_n)(y_1 y_2)$$

for $n \geqq 3$. If this variety were determined by some finite set $\{w_1, \dots, w_r\}$ of laws then each w_i would lie in the verbal closure of a finite set of laws $(y_1 \dots y_n)(y_1 y_2)$ or $(y_1 y_2)(y_3 y_4) y_5$. Therefore any given law $(y_1 \dots y_k)(y_1 y_2)$ not occurring in any of these finite sets $(1 \leqq i \leqq r)$ would be in the verbal closure of laws $(y_1...y_n)(y_1y_2)$ $(n \neq k)$ in the free Lie algebra F. But this contradicts theorem 4.3. □

COROLLARY 4.5 *Over any field of characteristic 2 there exist uncountably many distinct centre-by-metabelian varieties.*

Proof: If S is any subset of $\{3, 4, 5, ...\}$ let $\mathfrak{V}_S$ be the variety determined by the centre-by-metabelian law, together with all laws $(y_1...y_n)(y_1y_2)$ with $n \in S$. By theorem 4.3 we have $\mathfrak{V}_S \neq \mathfrak{V}_T$ if $S \neq T$. □

Oates and Powell [318] have proved that the variety generated by a finite group is finitely based. In contrast to their result, we have another theorem of Vaughan-Lee [239]:

THEOREM 4.6 *If $\mathfrak{k}$ is any infinite field of characteristic 2, then there is a 4-dimensional centre-by-metabelian Lie algebra over $\mathfrak{k}$ which generates a non-finitely based variety.*

Proof: Let B be the free $\mathfrak{N}_2$-algebra over $\mathfrak{k}$ generated by $\{b, c\}$. There is a derivation a of B such that

$$ba = b, \; ca = c, \; (bc)a = 0.$$

Let A be the split extension $B \dotplus \langle a \rangle$. This has dimension 4, and is clearly centre-by-metabelian. Using the methods of lemma 4.2 and theorem 4.3 we see that A satisfies all of the laws $(y_1...y_n)(y_1y_2)$ $(n \geqq 3)$. We show that the variety generated by A is not finitely based.

Suppose that w is a law in A. Since A is centre-by-metabelian we may suppose that $w \in F$, where F is as above. Since $\mathfrak{k}$ is infinite we may assume (from theorem 14.2.1) that w is homogeneous, and of degree r_i in y_i. We may further assume that $r_i > 0$ for $i = 1, ..., n$. By lemma 3.2 F^2 is spanned, modulo $F^{(2)}$, by regular products, so that

$$w = \textstyle\sum_{i=2}^{n} \alpha_i y_i y_1^{r_1} ... y_{i-1}^{r_{i-1}} y_i^{r_i - 1} y_{i+1}^{r_{i+1}} ... y_n^{r_n} + u$$

where $u = u(y_1, ..., y_n) \in F^{(2)}$ and the $\alpha_i \in \mathfrak{k}$.

Suppose that $\alpha_j \neq 0$ for $2 \leqq j \leqq n$. Substitute $a+b$ for y_j and substitute a for y_i $(i \neq j)$: this yields $\alpha_j b = 0$, a contradiction. Therefore $w = u \in F^{(2)}$.

By the discussion at the beginning of this section, $F^{(2)}$ is spanned by elements

$$(y_{i_1}...y_{i_k})(y_{i_{k+1}} y_{i_{k+2}})$$

subject to certain restrictions on the ordering of the subscripts. Suppose w is not linear in each y_i. Then we may assume that $r_1 \geqq 2$. Therefore w may be written in the form

$$\sum \alpha_i (y_i y_1^{r_1-1} y_2^{r_2} \dots y_{i-1}^{r_{i+1}} y_i^{r_i-2} y_{i+1}^{r_{i+1}} \dots y_n^{r_n})(y_i y_1) +$$
$$+ \sum{}^{*} \beta_{ij} (y_i y_1^{r_1-1} y_2^{r_2} \dots y_{j-1}^{r_{j-1}} y_j^{r_j-1} y_{j+1}^{r_{j+1}} \dots y_{i-1}^{r_{i-1}} y_i^{r_i-1} y_{i+1}^{r_{i+1}} \dots y_n^{r_n})(y_j y_1)$$

where $\sum$ is over all i such that $r_i \geqq 2$, $2 \leqq i \leqq n$, and $\sum^*$ is over all (i, j) with $i > j \geqq 2$.

Suppose that $\beta_{ij} \neq 0$ for some $i > j$. Substitute $a+b$ for y_i, $a+c$ for y_j, and a for y_k $(k \neq i, j)$. This yields $\beta_{ij} bc = 0$, a contradiction. Therefore $\beta_{ij} = 0$ for all i, j, from which it follows that w is implied by the laws $(y_1 \dots y_n)(y_1 y_2)$ $(n \geqq 3)$.

We are left with the case that w is linear in each of $y_1, \dots, y_n$. Then the verbal closure of w is spanned by words $w(m_1, \dots, m_n)$ where the m_j are monomials of F. We claim that $w(m_1, \dots, m_n) = 0$ unless all m_j are of degree 1. If not we may suppose that $w(m_1, \dots, m_n) \neq 0$ where the degree of m_1 is greater than 1. Then

$$w(y_{n+1} y_1, y_2, \dots, y_n) \neq 0$$

in F, and yet $w(y_{n+1} y_1, y_2, \dots, y_n)$ is a law of A. Since w is contained in $F^{(2)}$ and is linear in the y_j, we have

$$w(y_{n+1} y_1, y_2, \dots, y_n) = \textstyle\sum_{i=3}^{n} \alpha_i (y_i y_2 \dots y_{i-1} y_{i+1} \dots y_n)(y_{n+1} y_1)$$

for $\alpha_i \in \mathfrak{k}$, not all zero. Suppose that $\alpha_j \neq 0$ and substitute b for y_j, c for y_{n+1}, a for all other variables. Then $\alpha_j bc = 0$, a contradiction.

It follows that the laws of A are determined by laws of two types:

Type I $w(y_1, \dots, y_n)$ where the verbal closure of w is spanned by words of degree n.

Type II $(y_1 \dots y_n)(y_1 y_2)$ $(n \geqq 3)$.

If the variety generated by A were finitely based then the laws of A would be determined by a finite set of words of type I and a finite set of words of type II. Let m be the maximum degree of the words in these finite sets. Then

$$(y_1 \dots y_m)(y_1 y_2) = u + v$$

where u is implied by type I words of degree $\leqq m$ and v is implied by type II words with $n \leqq m-2$. By theorem 14.2.1 we may suppose that u is a linear combination of monomials of degree $m+2$, whence $u = 0$. But then $(y_1 \dots y_m)(y_1 y_2) = v$ contrary to theorem 4.3. □

5. Class 2-by-abelian varieties

In contrast to corollary 4.4 we have the following theorem of Bryant and Vaughan-Lee [34], which generalises earlier work of Vaughan-Lee:

THEOREM 5.1 *Every subvariety of $\mathfrak{N}_2\mathfrak{A}$ over a field of characteristic not 2 is finitely based.*

We defer the proof in order to make some useful observations. Let F be the free $\mathfrak{N}_2\mathfrak{A}$-algebra on a countable generating set $\{y_1, y_2, \ldots\}$. We wish to show that F satisfies the maximal condition for fully invariant ideals. By theorem 3.1 we know that $F/F^{(2)}$ satisfies this maximal condition; so to prove the theorem it is sufficient to show that F satisfies the maximal condition for fully invariant ideals *contained in* $F^{(2)}$, or equivalently that all such ideals are finitely generated as fully invariant ideals.

We prove this using the order-theoretic techniques of section 2. Let A denote the set of all infinite sequences of non-negative integers, all but a finite number of which are 0. If $\alpha \in A$ we write $\alpha(r)$ for the rth term of α. We order A by putting $\alpha < \alpha'$ if for some n, $\alpha(n) < \alpha'(n)$, but for all $r > n$, $\alpha(r) = \alpha'(r)$.

Let S be the set of all *hexads*

$$(i, j, k, l, \alpha, \beta)$$

where i, j, k, l are positive integers and $\alpha, \beta \in A$. We Let $\ll$ denote lexicographic ordering on S, induced from the orderings on $\mathbb{N}$ and A. Then $(S, \ll)$ is well-ordered.

Let P be the set of ordered pairs (a, b) of non-negative integers, and write

$$(a, b) \leqq^* (a', b')$$

if $a \leqq a'$ and $b \leqq b'$. By lemma 2.5 $(P, \leqq^*)$ is partially well ordered.

If

$$s = (i, j, k, l, \alpha, \beta)$$

$$s' = (i', j', k', l', \alpha', \beta')$$

are hexads we put

$$s \leqq_1 s'$$

if there exists $\xi \in \Xi$ (as defined before theorem 2.11) such that

$$\xi(i) = i', \ \xi(j) = j', \ \xi(k) = k', \ \xi(l) = l',$$

$$\alpha(r) \leqq \alpha'(\xi(r)), \ \beta(r) \leqq \beta'(\xi(r))$$

for all r. If $\alpha(r) = \beta(r) = 0$ for all $r > m$ then (α, β) may be identified with the finite sequence

$$(\alpha(1), \beta(1), \ldots, \alpha(m), \beta(m))$$

of elements of P. It follows from theorem 2.12 that $(S, \leqq_1)$ is partially well ordered.

Finally we put

$$s \preccurlyeq s'$$

with s, s' as above, if

$$s \leqq_1 s'$$

and

$$\textstyle\sum_{r=1}^{\infty} \alpha(r) \equiv \sum_{r=1}^{\infty} \alpha'(r) \pmod 2.$$

Since the second condition partitions S into two subsets, it is easy to check that $(S, \preccurlyeq)$ is partially well ordered.

This completes the order-theoretic preliminaries, and we return to the algebras. From section 4 it follows that $F^{(2)}$ is spanned by elements

$$y_i y_j y_1^{\alpha(1)} \ldots y_m^{\alpha(m)} (y_k y_l y_1^{\beta(1)} \ldots y_m^{\beta(m)})$$

where the $\alpha(r)$ and $\beta(r)$ are non-negative integers. From the Jacobi identity it is spanned by elements

$$y_i y_j y_1^{\alpha(1)} \ldots y_m^{\alpha(m)} (y_k y_l) y_1^{\beta(1)} \ldots y_m^{\beta(m)}.$$

We identify such an element with the hexad

$$(i, j, k, l, \alpha, \beta)$$

where $\alpha(r) = \beta(r) = 0$ if $r > m$. Thus every non-zero element of $F^{(2)}$ has an expression

$$\lambda_1 s_1 + \ldots + \lambda_n s_n$$

where the λ_i are non-zero elements of the underlying field $\mathfrak{k}$ and $s_1 \gg s_2 \gg \ldots \gg s_n$. The element s_1 is called the *weight* of the expression.

The theorem will follow easily once we have proved:

LEMMA 5.2 *Let $h' \in F^{(2)}$ have an expression of weight s' and suppose that s is a hexad with $s' \preccurlyeq s$. Then there is an element h in the fully invariant closure of h' having an expression of weight s.*

Proof: Let n be any positive integer. Define the endomorphism θ of F by

$$\theta(y_r) = y_r + y_r y_n$$

for $r \in \mathbb{N}$. By induction we may show that for any monomial a of F

$$\theta(a) \equiv a + a y_n \pmod{F^{(2)}}.$$

Let p and q be positive integers: define endomorphisms η_p and η_{pq} by

$$\eta_p(y_n) = y_p$$

$$\eta_p(y_r) = y_r \qquad (r \neq n)$$

$$\eta_{pq}(y_n) = y_p + y_q$$

$$\eta_{pq}(y_r) = y_r \qquad (r \neq n).$$

Let w be the element of $F^{(2)}$ identified with the hexad $(i, j, k, l, \alpha, \beta)$ and suppose we choose n greater than i, j, k, l and such that $\alpha(r) = \beta(r) = 0$ for all $r > n$. Since $(F^2)^3 = 0$ we find that

$$\theta(w) = (y_i y_j y_1^{\alpha(1)} \ldots y_m^{\alpha(m)} + y_i y_j y_1^{\alpha(1)} \ldots y_m^{\alpha(m)} y_n)$$
$$(y_k y_l + y_k y_l y_n) y_1^{\beta(1)} \ldots y_m^{\beta(m)}.$$

It follows that

$$w_1 = \theta(w) - w y_n - w$$
$$= y_i y_j y_1^{\alpha(1)} \ldots y_m^{\alpha(m)} y_n (y_k y_l y_n) y_1^{\beta(1)} \ldots y_m^{\beta(m)}.$$

Let $w_2 = \eta_{pq}(w_1) - \eta_p(w_1) - \eta_q(w_1)$. From the Jacobi identity we find that

$$w_2 = -2 y_i y_j y_1^{\alpha(1)} \ldots y_m^{\alpha(m)} y_p y_q (y_k y_l) y_1^{\beta(1)} \ldots y_m^{\beta(m)} +$$
$$+ y_i y_j y_1^{\alpha(1)} \ldots y_m^{\alpha(m)} y_p (y_k y_l) y_1^{\beta(1)} \ldots y_m^{\beta(m)} y_q +$$
$$+ y_i y_j y_1^{\alpha(1)} \ldots y_m^{\alpha(m)} y_q (y_k y_l) y_1^{\beta(1)} \ldots y_m^{\beta(m)} y_p.$$

Since $\mathrm{char}(\mathfrak{k}) \neq 2$, the expression for w_2 has weight

$$w(p, q) = (i, j, k, l, \alpha', \beta)$$

where

$$\alpha'(r) = \alpha(r) \qquad (r \neq p \text{ or } q)$$

$$\alpha'(p) - \alpha(p) = 2 \qquad (p = q)$$

$$\alpha'(p) - \alpha(p) = \alpha'(q) - \alpha(q) = 1 \qquad (p \neq q).$$

Also, if v is a hexad with $w \gg v$ then $w(p, q) \gg v(p, q)$.

Let $g \in F^{(2)}$ have an expression of weight w and take n sufficiently large that y_n does not occur in this expression. Let g_1, g_2 be defined as for w above. Then g_2 lies in the fully invariant closure of g and has an expression of weight $w(p, q)$.

Now suppose $h' \in F^{(2)}$ has an expression of weight

$$s' = (i', j', k', l', \alpha', \beta')$$

and suppose that

$$s = (i, j, k, l, \alpha, \beta)$$

satisfies $s' \preccurlyeq s$. Let $\xi \in \Xi$ be the map that shows $s' \preccurlyeq s$, and let ζ be the endomorphism of F defined by

$$\zeta(y_r) = y_{\xi(r)}$$

for all $r \in \mathbb{N}$. Then we see that $h'' = \zeta(h')$ has an expression of weight

$$(i, j, k, l, \alpha'', \beta'')$$

where for all r, $\alpha''(r) \leqq \alpha(r)$ and $\beta''(r) \leqq \beta(r)$, and

$$\sum (\alpha(r) - \alpha''(r)) \equiv 0 \pmod 2.$$

For some m we have $\beta(r) = \beta''(r) = 0$ for all $r > m$. Take

$$g = h'' y_1^{\beta(1) - \beta''(1)} \ldots y_m^{\beta(m) - \beta''(m)}.$$

Then g belongs to the fully invariant closure of h' and has an expression of weight

$$u = (i, j, k, l, \alpha'', \beta).$$

By induction on $\sum (\alpha(r) - \alpha''(r))$ the lemma will follow if we can prove the case where this sum is 2. But then, by a suitable choice of p, q (which may be equal) we have $s = u(p, q)$. As already shown, there is an element h in the fully invariant closure of g having an expression of weight s. □

Proof of theorem 5.1

Let I be a fully invariant ideal of F, contained in $F^{(2)}$. We must show that I is finitely generated. Since $(S, \preccurlyeq)$ is partially well ordered there is a finite subset G of I such that for any $h^* \in I$ having an expression of weight s there exists $h' \in G$ having an expression of weight $s' \preccurlyeq s$. By lemma 5.2 there is an element h of the fully invariant closure of h' with an expression of weight s. Therefore there is a non-zero $\lambda \in \mathfrak{k}$ for which $h^* + \lambda h$ is either 0, or has an expression of weight strictly smaller than s in the well-ordering $\geqslant$. By induction on the well-ordering it follows that I is the fully invariant closure of G. $\square$

Chapter 16

Engel conditions

In chapter 7 we defined the class $\mathfrak{E}$ of Engel Lie algebras by: $L \in \mathfrak{E}$ if to each (ordered) pair $x, y \in L$ there corresponds a positive integer $n = n(x, y)$ such that $[x, {}_n y] = 0$.

The observation that $\mathrm{L}\mathfrak{N} \leqq \mathfrak{E}$ leads to the natural question: for what classes $\mathfrak{X}$ of Lie algebras do we have

$$\mathfrak{X} \cap \mathfrak{E} \leqq \mathfrak{N} \text{ or } \mathfrak{X} \cap \mathfrak{E} \leqq \mathrm{L}\mathfrak{N}?$$

This problem and its group-theoretic counterpart has been the subject of investigation in numerous papers, among them those of Zorn [253], Heineken [79], Higgins [89], Gruenberg [72], and Kostrikin [109–115].

In this chapter and the next two chapters we present an account of the main classes which have been shown to have the required property.

The class $\mathfrak{E}$ has three major families of subclasses, these being:

- $\mathfrak{E}^*$: defined by $L \in \mathfrak{E}^*$ if to each $x \in L$ there corresponds a positive integer $n = n(x)$ such that $[L, {}_n x] = 0$.
- $^*\mathfrak{E}$: defined by $L \in {}^*\mathfrak{E}$ if $L = \bigcup_{n=1}^{\infty} L(g_{n,1})$ or equivalently to each $x \in L$ there corresponds a positive integer $n = n(x)$ such that $[x, {}_n y] = 0$ for all $y \in L$.
- $\mathfrak{E}_n$: defined by $L \in \mathfrak{E}_n$ if $[x, {}_n y] = 0$ for all $x, y \in L$ or, equivalently $L = L(g_{n,1})$. This is the class of *n-Engel* algebras.

Evidently $\mathfrak{E}_n \leqq \mathfrak{E}_{n+1}$ for each n and

$$\bigcup_{n=1}^{\infty} \mathfrak{E}_n \leqq {}^*\mathfrak{E} \cap \mathfrak{E}^*.$$

Of these classes the classes $\mathfrak{E}_n$ have been the most fruitful for investigation. Several results have been obtained about $\mathfrak{E}_n$ for small n. In section 1 we give a brief account of these results and investigate the classes $\mathfrak{E}_2$ and $\mathfrak{E}_3$ fully.

Perhaps the most remarkable result about the classes $\mathfrak{E}_n$ is that of Kostrikin (see chapter 17) that over fields of characteristic zero or $p > n-1$,

$$\mathfrak{E}_n \leqq \mathrm{L}\mathfrak{N}.$$

In section 2 we show that this result cannot be extended to $\mathfrak{E}$ by proving that over any field $\mathfrak{E}^* \not\leq \text{L}\mathfrak{N}$ and so $\mathfrak{E} \not\leq \text{L}\mathfrak{N}$.

By theorem 10.1.3 we have

$$\mathfrak{N} \cap \text{Fin-}\mathfrak{A} \leqq \mathfrak{F} \cap \mathfrak{N}.$$

A result of Zorn [253] p. 404 states that

$$\mathfrak{E}^* \cap \text{Max} \leqq \mathfrak{N},$$

which of course extends the well known result that a finite-dimensional Engel algebra is nilpotent. In the first half of section 3 we prove a generalisation of these results, that

$$\mathfrak{E} \cap \text{Fin-}\mathfrak{A} \leqq \mathfrak{F} \cap \mathfrak{N}.$$

We know that in general the class $\text{L}\mathfrak{N}$ is not closed under extensions. In the second half of section 3 we prove that

$$\{\text{L}, \acute{\text{E}}\}\mathfrak{A} \cap \mathfrak{E} = \text{L}\mathfrak{N}.$$

This shows that the class $\text{L}\mathfrak{N}$ is '$\{\text{L}, \acute{\text{E}}\}$-closed with respect to $\mathfrak{E}$'. It also extends the result of Gruenberg [72] theorem 1′ that a soluble Engel algebra is locally nilpotent. Another conclusion is that if $L \in \mathfrak{E}$ then $\rho(L/\rho(L)) = 0$, which is the basis of the proof of Kostrikin's theorem mentioned above.

In section 4 we consider left and right Engel elements in a Lie algebra in the spirit of Gruenberg [279, 280]. We find that in contrast with groups the set of right Engel elements of a Lie algebra always forms a subalgebra, which is an Engel algebra.

1. The second and third Engel conditions

By theorem 2 of Higgins [89] or corollary 7.3.3 we have $\mathfrak{E}_2 \leq \mathfrak{N}_3$. However if $L \in \mathfrak{E}_2$ and $a, x, y \in L$ then by equation 7.13 it follows that

$$[a, x, y] = -[a, y, x].$$

From this and the Jacobi identity, $[a, x, y]+[x, y, a]+[y, a, x] = 0$, we deduce that $3[a, x, y] = 0$. Thus if L is defined over a field of characteristic $\neq 3$ then $[a, x, y] = 0$ and so $L \in \mathfrak{N}_2$. This result will not hold for fields of characteristic 3. To show this it is enough to construct a 3-generator $\mathfrak{E}_2$-algebra defined over a field of characteristic 3 and which is not in $\mathfrak{N}_2$. Now

$\mathfrak{E}_2$ is a variety so we may construct the free $\mathfrak{E}_2$-algebra on any given set of generators and in particular for a set of 3 generators.

Let $\mathfrak{k}$ be a field and let F be the free Lie algebra over $\mathfrak{k}$ on three generators u, v, w. By the theory of *basic commutators* (see Hall [75]) F has a basis consisting of $u, v, w, [u, v], [u, w], [v, w], [u, v, u], [u, v, v], [u, w, u], [u, w, v], [u, w, w], [v, w, u], [v, w, v], [v, w, w]$, together with basic commutators of weight at least 4; the basic commutators of weight at least 4 form a basis for F^4.

Let N be the ideal of F generated all elements of the form $[a, b, b]$ for $a, b \in F$. Then $E = F/N$ is the free 3-generator $\mathfrak{E}_2$-algebra over $\mathfrak{k}$. Since $\mathfrak{E}_2 \leqq \mathfrak{N}_3$ we have $L \in \mathfrak{N}_3$ and so

$$F^4 < N.$$

Now let I be the subspace of F spanned by the basic commutators of weight at least 4 together with the elements $[u, v, u], [u, v, v], [u, w, u], [u, w, w], [v, w, v], [v, w, w]$ and

$$x = [u, v, w] + [u, w, v]. \tag{*}$$

Evidently

$$x = [u, v+w, v+w] - [u, v, v] - [u, w, w] \in N,$$

and so

$$F^4 < I \leqq N \text{ and } I \lhd F.$$

By the Jacobi identity we have

$$[u, v, w] + [v, w, u] + [w, u, v] = 0$$

and so $x = [u, v, w] + [u, w, v] = -[v, w, u] + 2[u, w, v]$. Thus $[u, w, v] \notin I$ and so $F/I \notin \mathfrak{N}_2$.

Suppose now that $\mathfrak{k}$ has characteristic 3. We will show that $L/I \in \mathfrak{E}_2$, so that $N \leqq I$ and hence $L \in \mathfrak{E}_2$ and $L \notin \mathfrak{N}_2$. Employing the Jacobi identity we have

$$\begin{aligned} y = [v, w, u] + [v, u, w] &= -[w, u, v] - [u, v, w] + [v, u, w] \\ &= [u, w, v] - 2[u, v, w] \\ &= [u, w, v] + [u, v, w] = x. \end{aligned}$$

From above we also have

$$\begin{aligned} x = -[v, w, u] + 2[u, w, v] &= [w, v, u] - 2[w, u, v] \\ &= [w, v, u] + [w, u, v] = z. \end{aligned}$$

So $x = y = z = [u, v, w]+[u, w, v] = [v, w, u]+[v, u, w] = [w, u, v]+[w, v, u]$, when $\mathfrak{k}$ has characteristic 3. Let $a, b \in F$. Then $a = a_0+a_1$ and $b = b_0+b_1$, where a_0, b_0 are linear combinations of u, v, w and $a_1, b_1 \in F^2$. Hence

$$[a, b, b] = [a_0, b_0, b_0]+c,$$

where $c \in F_4 \leqq I$. Let $b_0 = \lambda u+\mu v+\nu w$. Now

$$[u, b_0, b_0] = [u, \lambda u+\mu v+\nu w, \lambda u+\mu v+\nu w]$$
$$= \lambda\mu[u, v, u]+\lambda\nu[u, w, u]+\mu\nu x \in I.$$

Similarly $[v, b_0, b_0] \in I$ and $[w, b_0, b_0] \in I$ and so $[a_0, b_0, b_0] \in I$, whence $[a, b, b] \in I$ for all $a, b \in F$. Thus as $I \lhd F$ we have $N \leqq I \leqq N$. Therefore $N = I$ and $L = F/I$.

Evidently $F^3 = \langle [u, v, w]\rangle+I$, L has dimension 7 and $L \notin \mathfrak{N}_2$. So we have proved

PROPOSITION 1.1 (a) *Over any field of characteristic* $\neq 3$,

$$\mathfrak{N}_2 = \mathfrak{E}_2.$$

(b) *Over any field of characteristic* 3

$$\mathfrak{N}_2 < \mathfrak{E}_2 < \mathfrak{N}_3.$$ □

From chapter 7 we know that $\mathfrak{D}_n \leqq \mathfrak{D}_{n,1} \leqq \mathfrak{E}_n$. Trivially $\mathfrak{D}_1 = \mathfrak{D}_{1,1} = \mathfrak{N}_1$. By proposition 1.1 (a) it follows that over fields of characteristic $\neq 3$,

$$\mathfrak{D}_2 = \mathfrak{D}_{2,1} = \mathfrak{N}_2.$$

Consider the Lie algebra L constructed above. Then for any $a \in L$ we have $[a, {}_nL, a] = 0$ for all $n = 0, 1, 2, \ldots$. Thus $\langle a\rangle \lhd \langle a^L\rangle \lhd L$ and $L \in \mathfrak{D}_{2,1}$. On the other hand if $H = \langle u+I, v+I\rangle$ then

$$H = \langle u+I\rangle+\langle v+I\rangle+\langle [u, v]+I\rangle$$

and $[v, w, u]+I \in [L, {}_2H]$ but $[v, w, u]+I \notin H$. Thus $L \notin \mathfrak{D}_{2,2}$ and in particular $L \notin \mathfrak{D}_2$.

If K is any $\mathfrak{D}_{2,2}$-algebra then $K \in \mathfrak{E}_2$. Hence if K is defined over $\mathfrak{k}$, then any 3-generator subalgebra of K is isomorphic to L/M for some ideal M of L. If $M = 0$ then $L \cong M \in \mathfrak{D}_{2,2}$, which is impossible. Hence $M \neq 0$ which implies that M intersects the centre of L non-trivially. Therefore since the centre of L is 1-dimensional ($\zeta_1(L) = L^3 = \langle u, v, w]+I\rangle$), $L/M \in \mathfrak{N}_2$ and so $K \in \mathfrak{N}_2$. So we have:

COROLLARY 1.2 (a) *Over any field* $\mathfrak{D}_{2,1} = \mathfrak{E}_2$.
(b) *Over fields of characteristic not 3,*

$$\mathfrak{N}_2 = \mathfrak{E}_2 = \mathfrak{D}_{2,1} = \mathfrak{D}_{2,2} = \mathfrak{D}_{2,3} = \ldots = \mathfrak{D}_2.$$

(c) *Over fields of characteristic 3,*

$$\mathfrak{E}_2 = \mathfrak{D}_{2,1} > \mathfrak{D}_{2,2} = \mathfrak{D}_{2,3} = \ldots = \mathfrak{D}_2 = \mathfrak{N}_2. \qquad \square$$

LEMMA 1.3 (*Higgins*) *Let* L *be a Lie algebra and let* $x, y, z \in L$. *Then for any* $n \in \mathbb{N}$

$$\sum_{i=0}^{n} [y, {}_ix, z, {}_{n-i}x] = \sum_{j=0}^{n} (-1)^{j+1} \binom{n+1}{j+1} [z, {}_jx, y, {}_{n-j}x].$$

Proof: Evidently $[y, {}_ix, z] = -[z, [y, {}_ix]]$

$$= -\sum_{j=0}^{i} (-1)^j \binom{i}{j} [z, {}_jx, y, {}_{i-j}x].$$

Operate on both sides by x^{*n-i} and sum over i. Then we have

$$\sum_{i=0}^{n} [y, {}_ix, z, {}_{n-i}x] = \sum_{i=0}^{n} \sum_{j=0}^{i} (-1)^{j+1} \binom{i}{j} [z, {}_jx, y, {}_{n-j}x]$$

$$= \sum_{j=0}^{n} (-1)^{j+1} \left\{ \sum_{i=j}^{n} \binom{i}{j} \right\} [z, {}_jx, y, {}_{n-j}x]$$

$$= \sum_{j=0}^{n} (-1)^{j+1} \binom{n+1}{j+1} [z, {}_jx, y, {}_{n-j}x]. \qquad \square$$

COROLLARY 1.4 (*Higgins*) *Let* L *be a Lie algebra over a field of characteristic 0 or* $p > n-1$. *If* $y \in L(g_{n,1})$ *then for any* $x, z \in L$,

(a) $\sum_{i=0}^{n-1} [y, {}_ix, z, {}_{n-1-i}x] = 0$

(b) $\sum_{j=0}^{n-1} (-1)^{j+1} \binom{n}{j+1} x^{*j} y^{*} x^{*n-1-j} = 0$

(c) $y^{*n} = 0$.

Proof: (a) is obtained by picking out the terms with coefficient λ from $y(x+\lambda z)^{*n} = [y, {}_n x+\lambda z] = 0$ (where λ is a scalar). (c) is obtained from (a) by putting $x = y$ to get $[y, z, {}_{n-1}y] = 0$ so that $[z, {}_ny] = 0$, whence $y^{*n} = 0$. Let u be the expression on the left hand side of (b). Then by lemma 1.3 and

(a) we have

$$zu = \sum_{i=0}^{n-1} [y, {}_i x, z, {}_{n-1-i} x] = 0, \text{ so } u = 0. \qquad \square$$

Remark 1. Let L be a Lie algebra over a field of characteristic 0 or $p > n-1$ and let A, B be subspaces of L and θ a linear homomorphism of L. Suppose that $[a, {}_n b]\theta = 0$ for all $a \in A$, $b \in B$. Then by lemma 1.3 and the proof of corollary 1.4 it is not hard to show that for any $a, a_1 \in A$ and $b, b_1 \in B$,

$$a \sum_{i=0}^{n-1} b^{*i} b_1{}^* b^{*n-1-i} \theta = 0 \tag{1}$$

$$b_1 \sum_{j=0}^{n-1} (-1)^{j+1} \binom{n}{j+1} b^{*j} a^* b^{*n-1-j} \theta = 0. \tag{2}$$

If also $A \subseteq B$ then

$$[b, {}_n a]\theta = 0 \tag{3}$$

and equations (1) and (2) hold if we replace a with b and a_1 with b_1.

In particular if $n = 2$ then

$$[a, b, b_1]\theta = -[a, b_1, b]\theta \tag{*}$$

and if $A \subseteq B$, $[b, a, a_1]\theta = -[b, a_1, a]\theta$ and in particular putting $b_1 = a_1$ in (*) we have $[a, b, a_1]\theta = -[a, a_1, b]\theta$. But by the Jacobi identity we have (for θ is linear)

$$\begin{aligned} 0 &= [a, b, a_1]\theta + [b, a_1, a]\theta + [a_1, a, b]\theta \\ &= [a, b, a_1]\theta - [b, a, a_1]\theta - [a, b, a_1]\theta \\ &= 3[a, b, a_1]\theta \end{aligned}$$

So if the characteristic is not 3 then

$$[a, b, a_1]\theta = [a, a_1, b]\theta = 0. \tag{4}$$

PROPOSITION 1.5 *Over any field of characteristic prime to 2 and 5,*

$$\mathfrak{E}_3 \leqq \mathfrak{N}_5$$

Proof: Let $L \in \mathfrak{E}_3$ and let $x, y \in L$. Then by corollary 1.4

$$y^* x^{*2} + x^* y^* x^* + x^{*2} y^* = 0$$

and

$$3y^* x^{*2} - 3x^* y^* x^* + x^{*2} y^* = 0.$$

If the characteristic is 3 then $x^{*2}y^* = 0$ so that if $H = \{[z, {}_2x] : z \in L$ and $x \in L\}$ then $H \subseteq Z = \zeta_1(L)$. Hence $L/Z \in \mathfrak{E}_2 \leqq \mathfrak{N}_3$ and so $L \in \mathfrak{N}_4$.

Suppose that the characteristic is not 3. Then from the given equations we obtain (since division by 2 is possible)

$$x^{*2}y^* = -3x^*y^*x^* \text{ and } y^*x^{*2} = 2x^*y^*x^*.$$

Hence for any $z \in L$,

$$\begin{aligned} 3[z, {}_2x, {}_2y] &= (3[z, x])x^*y^{*2} = (3[z, x])2y^*x^*y^* \\ &= z(3x^*y^*x^*)2y^* \\ &= z(-x^{*2}y^*)2y^*. \end{aligned}$$

Therefore $5x^{*2}y^{*2} = 0$ and so $x^{*2}y^{*2} = 0$ (characteristic $\neq 5$). By equation (4) (taking $\theta = y^{*2}$ and $A = B = L$) we see that for any $x_0, x_1 \in L$ we have

$$[x_0, x_1, x]y^{*2} = 0.$$

Therefore $L^3 \leqq L(g_{2,1}) \leqq \zeta_3(L)$ by theorem 7.3.2. Hence $L \in \mathfrak{N}_5$. □

As $\mathfrak{D}_{3,1} \leqq \mathfrak{E}_3$ the restriction on characteristic 2 in proposition 1.5 is necessary. A result of chapter 18 will show that over fields of characteristic 5, $\mathfrak{E}_3 \nleqq \mathfrak{N}$.

Suppose that $L \in \mathfrak{E}_4$ over a field of characteristic prime to 2, 3, 5, 7. Then from corollary 1.4 we deduce that (see Higgins [89] p. 12ff.) for any $x, y, z \in L$,

$$x^{*3}y^{*3}z^{*3} = 0.$$

Using this Heineken shows in [79] that $L \in \mathfrak{N}_{29} \cap \mathfrak{A}^5$. In view of theorem 7.4.1 the restriction on characteristic 3 is necessary and so are 2 and 5 since $\mathfrak{E}_3 \leqq \mathfrak{E}_4$. However the restriction on characteristic 7 is shown not to be necessary by Walkup [241]. Specifically he proves:

THEOREM 1.6 *Over any field of characteristic prime to 2, 3, 5,*

$$\mathfrak{E}_4 \leqq \mathfrak{N}_9 \cap \mathfrak{A}^4.$$ □

2. A non-locally nilpotent Engel algebra

Techniques developed in the fundamental paper of Golod and Šafarevič [70] have been used by Golod [69] to construct associative nil algebras (over any field) which are not locally nilpotent (in the associative sense). In a footnote

Golod remarks that the same methods can be used to construct Lie algebras with the Engel condition which are not locally nilpotent. In fact a slightly stronger remark can be made as we now show.

We recall that if A is an associative algebra then $[A]$ is the Lie algebra obtained from A by defining

$$[a, b] = ab-ba$$

for $a, b \in A$. A simple induction on j shows that for all $a, b \in A$ we have

$$[a, {}_jb] = \sum_{i=0}^{j} (-1)^i b^i a b^{j-i}. \qquad (*)$$

Thus if A is nil then to each $b \in A$ there corresponds $n = n(b)$ such that $b^n = 0$. Hence by (*) we have for any $a \in A$,

$$[a, {}_{2n}b] = 0.$$

So $[A] \in \mathfrak{E}^*$ whenever A is a nil algebra (the converse is false as can be seen by considering the polynomial algebra $\mathfrak{k}[t]$).

Suppose that B is a nil algebra generated (as an associative algebra) by a finite number of elements $b_1, \ldots, b_m$. Let H be the (Lie) subalgebra of $[B]$ generated by $b_1, \ldots, b_m$. We claim that if $H \in \mathfrak{N}_c$ then B is nilpotent (as an associative algebra). To prove this we use induction on c. If $c = 1$ then B is a commutative algebra. Since B is nil we can find r such that $b_i^r = 0$ for $i = 1, \ldots, m$. Hence

$$b_{i_1} \ldots b_{i_{m(r-1)+1}} = 0 \qquad (b_{i_j} \in \{b_1, \ldots, b_m\})$$

since in any such product some b_i occurs at least r times. Thus $x_1 \ldots x_{m(r-1)+1} = 0$ for all $x_i \in B$ so B is nilpotent. Let $c > 1$ and assume that the result holds for $c-1$. Let $H_1 = \zeta_1(H)$. Then H_1 is finite dimensional with basis $a_1, \ldots, a_k$ say. Furthermore we can find n such that $a_i^n = 0$ for all i. Since the a_i commute with each b_j and so with all elements of B we have $a_{i_1} \ldots a_{i_{k(n-1)+1}} = 0$ (as above) and

$$\begin{aligned} &a_{i_1} y_1 a_{i_2} y_2 \ldots a_{i_{k(n-1)+1}} y_{k(n-1)+1} \\ &\quad = a_{i_1} \ldots a_{i_{k(n-1)+1}} y_1 \ldots y_{k(n-1)+1} = 0, \end{aligned}$$

for all $y_i \in B$. Hence any product of $k(n-1)+1$ elements of $H_1 + H_1 B$ is zero so $C = H_1 + H_1 B$ is an (associative) nilpotent ideal of B. Now $(H+C)/C \in \mathfrak{N}_{c-1}$ and so by induction B/C is nilpotent, whence B is nilpotent. This completes our induction on c and proves our claim.

Evidently if B is nilpotent then $H \in \mathfrak{N}_c$ for some c and $[B] \in \mathfrak{N}$. Hence if A is a nil algebra then A is locally nilpotent if and only if $[A]$ is locally nilpotent.

So we have proved

THEOREM 2.1 *Let A be an associative nil algebra. Then*

(1) $[A] \in \mathfrak{E}^*$
(2) $[A] \in \mathrm{L}\mathfrak{N}$ *if and only if A is locally nilpotent (as an associative algebra).* □

From the work of Golod [69] (see also Herstein [86]) and theorem 2.1 we have

COROLLARY 2.2 *Over any field there exists a Lie algebra in $\mathfrak{E}^*$ which is not locally nilpotent. Hence*

$$\mathfrak{E}^* \nleqq \mathrm{L}\mathfrak{N}$$

and in particular

$$\mathfrak{E} \nleqq \mathrm{L}\mathfrak{N}.$$ □

By corollary 1.4(iii) we have over fields of characteristic 0,

$$*\mathfrak{E} \leqq \mathfrak{E}^*.$$

The inclusion is strict since over any field

$$\mathfrak{E}^* \cap \mathrm{L}\mathfrak{N} \nleqq *\mathfrak{E}.$$

For consider the Lie algebra M with basis

$$\{t_{ij} : i < j,\ i, j \in \mathbb{N}\}$$

and multiplication

$$\begin{aligned}[t_{ij}, t_{mn}] &= t_{in} && \text{if } m = j,\\ &= -t_{mj} && \text{if } n = i\\ &= 0 && \text{otherwise.}\end{aligned}$$

Then $t_{ij}^M \in \mathfrak{A}$ for all i, j and so $M \in \mathfrak{B}$, whence $M \in \mathfrak{E}^* \cap \mathrm{L}\mathfrak{N}$. However for each n we have

$$[t_{01,n}(t_{12} + \ldots + t_{n,n+1})] = t_{0,n+1} \neq 0$$

and so $M \notin *\mathfrak{E}$.

3. Finiteness conditions on Engel algebras

By theorem 10.1.3 we know that a locally nilpotent algebra satisfying Fin-$\mathfrak{A}$ is finite dimensional. Thus

$$\text{L}\mathfrak{N} \cap \text{Fin-}\mathfrak{A} = \mathfrak{F} \cap \mathfrak{N}.$$

In [253] Zorn proved that

$$\mathfrak{E}^* \cap \text{Max} \leqq \mathfrak{F} \cap \mathfrak{N}.$$

Our object in this section is to generalise both results by proving that

$$\mathfrak{E} \cap \text{Fin-}\mathfrak{A} = \mathfrak{F} \cap \mathfrak{N}.$$

Our approach is based on Amayo [10], where it is shown that this result applies to more general types of Lie rings.

First we establish another characterisation of Lie algebras in the classes $\mathfrak{E}$ and $\mathfrak{E}_n$.

Let A, B, C be subsets of a Lie algebra L and suppose that $C \subseteq I_L(B)$. Then clearly by the Jacobi identity

$$[A, B, C] \subseteq [A, B] + [A, C, B]$$

and

$$[A, C, B] \subseteq [A, B] + [A, B, C].$$

LEMMA 3.1 *Let A, B, C be subsets of a Lie algebra L such that $C \subseteq I_L(B)$. If $m_1, \ldots, m_k, n_1, \ldots, n_k \in \mathbb{N}$ then*

(a) $[A, {}_{m_1}B, {}_{n_1}C] \subseteq \sum_{i=0}^{n_1} [A, {}_iC, {}_{m_1}B]$

(b) $[A, {}_{n_1}C, {}_{m_1}B] \subseteq \sum_{i=0}^{n_1} [A, {}_{m_1}B, {}_iC]$

(c) $[A, {}_{m_1}B, {}_{n_1}C, \ldots, {}_{m_k}B, {}_{n_k}C] \subseteq \sum_{i=0}^{n_1+\ldots+n_k} [A, {}_iC, {}_{m_1+\ldots+m_k}B]$.

Proof: (a) and (b) follow by induction on m_1+n_1, using the inequalities established prior to the lemma.

(c) follows from (a) by induction on k. □

Under the hypothesis of lemma 3.1 it follows that if $[A, {}_mB] = 0$ then $[A^C, {}_mB] = 0$.

Let $r \in \mathbb{N}$ and let $A, B, C \subseteq L$. Clearly

$$[A, {}_rB+C] = \sum [A, {}_{m_1}B, {}_{n_1}C, \ldots, {}_{m_k}B, {}_{n_k}C], \qquad (*)$$

where the summation is taken over all sets of non-negative integers $m_1, \ldots, m_k$, $n_1, \ldots, n_k$ for which $\sum m_i + \sum n_i = r$, $k > 0$ and $m_2, \ldots, m_k, n_1, \ldots, n_{k-1} \neq 0$ (if $k > 1$).

PROPOSITION 3.2 *Let L be a Lie algebra and $A, B, C \subseteq L$. If $C \subseteq I_L(B)$ and $[A^B, {}_mB] = [A^B, {}_nC] = 0$ then*

$$[A^B, {}_{mn}B+C] = 0.$$

Proof: If $m = 0$ or $n = 0$ then $A^B = 0$ and the result is trivial. Assume that $m, n > 0$. Let $A_1 = A^B$. By (*)

$$[A_1, {}_{mn}B+C] = \sum [A_1, {}_{m_1}B, {}_{n_1}C, \ldots, {}_{m_k}B, {}_{n_k}C],$$

with $\sum m_i + \sum n_i = mn$ and if $k > 1$ then $m_2, \ldots, m_k, \ldots, n_{k-1} > 0$. Consider a typical term

$$X = [A_1, {}_{m_1}B, {}_{n_1}C, \ldots, {}_{m_k}B, {}_{n_k}C].$$

Now by lemma 3.1 we have

$$[A_1^C, {}_mB] = 0 \text{ and } X \subseteq \textstyle\sum_{i=0}^{\Sigma n_j} [A_1, {}_iC, {}_{\Sigma m_j}B].$$

Hence if $\sum m_i \geqq m$ then $X = 0$. Suppose that $\sum m_i \leqq m-1$. Since $m_2, \ldots, m_k$ are non-zero we have $k-1 \leqq \sum m_i \leqq m-1$ and so $k \leqq m$ (true even if $k = 1$ since $m > 0$). Now $\sum n_i = mn - \sum m_i \geqq mn-(m-1) = m(n-1)+1 \geqq$ $\geqq k(n-1)+1$ and so some $n_i \geqq n$. If $i = 1$ then $[A_1, {}_{m_1}B, {}_{n_1}C] = 0$ so $X = 0$. If $i > 1$ then $Y = [A_1, {}_{m_1}B, {}_{n_1}C, \ldots, {}_{m_i}B] \subseteq A_1^C$ (by induction and lemma 3.1(b)) and so $[Y, {}_{n_i}C] = 0$, whence $X = 0$. Thus in all cases $X = 0$ and so $[A^B, {}_{mn}B+C] = 0$. □

Let $A, B, C \subseteq L$ and suppose that $[B, C] \subseteq \mathfrak{k}(B \cap C)$ (where L is defined over the field $\mathfrak{k}$). Then it follows by lemma 3.1(a) and (b) that

$$[A, {}_{m_1}B, {}_{n_1}C, \ldots, {}_{m_k}B, {}_{n_k}C] \subseteq (\textstyle\sum_{i=0}^{\Sigma n_j} [A, {}_{\Sigma m_j}B, {}_iC])$$

and

$$[A, {}_{m_1}B, {}_{n_1}C, \ldots, {}_{m_k}B, {}_{n_k}C] \subseteq (\textstyle\sum_{i=0}^{\Sigma m_j} [A, {}_{\Sigma n_j}C, {}_iB]).$$

Hence if $[A, {}_mB] = [A, {}_nC] = 0$ then $[A, {}_{m+n}B+C] = 0$. An induction on r now gives us

COROLLARY 3.3 *Let $A, B_1, \dots, B_r$ be subsets of a Lie algebra L such that $[A, {}_{m_i}B_i] = 0$ for $i = 1, \dots, r$ and $[B_1+\dots+B_i, B_{i+1}] \subseteq (B_1+\dots+B_i)\cap B_{i+1}$ for $i = 1, \dots, r-1$. Then*

$$[A,{}_{m_1+\dots+m_r}B_1+\dots+B_r] = 0. \qquad \square$$

We can now prove:

THEOREM 3.4 (a) *$L \in \mathfrak{E}$ if and only if to each finite subset A and each $\mathfrak{G}\cap\mathfrak{N}$-subalgebra H of L there corresponds a positive integer $n = n(A, H)$ such that $[A, {}_nH] = 0$.*

(b) *$L \in \mathfrak{E}^*$ if and only if to each $\mathfrak{G}\cap\mathfrak{N}$-subalgebra H of L there corresponds $n = n(H)$ such that $[L, {}_nH] = 0$.*

(c) *$L \in \mathfrak{E}_n$ if and only if to each pair r, c of positive integers there corresponds a positive integer $h = h(n, r, c)$ such that for any $\mathfrak{G}_r\cap\mathfrak{N}_c$-subalgebra H of L, $[L, {}_hH] = 0$; and such that $h(n, 1, c) = n$.*

Proof: (a) If the condition is satisfied then clearly $L \in \mathfrak{E}$. Conversely let $L \in \mathfrak{E}$ and let $H = \langle x_1, \dots, x_r\rangle \in \mathfrak{N}_c$. Let A be a finite subset of L. We use induction on c to show that for some $n = n(A, H)$ we have $[A, {}_nH] = 0$.

If $c = 1$, then since A is finite we can find n_i such that $[A, {}_{n_i}x_i] = 0$, whence $[A, {}_{n_i}\langle x_i\rangle] = 0$ for $i = 1, \dots, r$. Now $H \in \mathfrak{A}$ so that $H = \langle x_1\rangle+\dots+\langle x_r\rangle$ and so by corollary 3.3 $[A, {}_nH] = 0$ where $n = n_1+\dots+n_r$.

Let $c > 1$ and assume inductively that the result is true for $c-1$ in place of c. Put $B_i = \langle x_i^H\rangle$ for $i = 1, \dots, r$. Then $B_i \lhd H$, $B_i \in \mathfrak{F}$ (since $H \in \mathfrak{F}$) and $B_i \in \mathfrak{N}_{c-1}$, for $B_i \leqq \langle x_i\rangle + H^2 \in \mathfrak{N}_{c-1}$. Thus $B_i \in \mathfrak{G}\cap\mathfrak{N}_{c-1}$ and so by induction there exists $m_i = m_i(A, B_i)$ such that $[A, {}_{m_i}B_i] = 0$. By corollary 3.3 we have $[A, {}_nH] = 0$, where $n = m_1+\dots+m_r$. This completes our induction on c and proves (a).

(b) That the given condition implies $L \in \mathfrak{E}^*$ is trivial. Conversely suppose that $L \in \mathfrak{E}^*$. Consider $H \leqq L$ with $H \in \mathfrak{G}\cap\mathfrak{N}_c$ and induct on c as in (a). For $c = 1$, by the definition of $\mathfrak{E}^*$ we can find $n_i = n_i(x_i)$ with $[L, {}_{n_i}x_i] = 0$, whence, as $H = \langle x_1, \dots, x_r\rangle$, by corollary 3.3 $[L, {}_nH] = 0$ where $n = n_1+\dots+n_r$. The inductive step follows as in (a).

(c) The condition obviously implies that $L \in \mathfrak{E}_n$ since $h(n, 1, c) = n$. Conversely suppose that $L \in \mathfrak{E}_n$. Now $\mathfrak{G}_r\cap\mathfrak{N}_c \leqq \mathfrak{F}_m$ where $m = r+\dots+r^c$ and so if $H \in \mathfrak{G}_r\cap\mathfrak{N}_c$ then $H \in \mathfrak{F}_m$. Since L is an H-module then by lemma 11.4.7 we have $L = L^*(H) = L_h(H)$, where $h = n^m$ (for $L = L^*(x) = L_n(x)$ for all $x \in L$ and so for all $x \in H$). Therefore $[L, {}_hH] = 0$ as required. $\square$

COROLLARY 3.5 *Let $L \in \mathfrak{E}$ and suppose that $H < L$. If $H \in \mathfrak{G} \cap \mathfrak{N}$ then there exists $K \in \mathfrak{G} \cap \mathfrak{N}$ such that $H < K \leqq I_L(H)$.*

Proof: Let A be a finite subset of L with $A \nsubseteq H$. By theorem 3.4(a) we can find n such that $[A, {}_nH] = 0$. Let k be minimal with respect to $[A, {}_kH] \subseteq H$. Then $0 < k \leqq n$ since $A \nsubseteq H$. So we can find $x \in [A, {}_{k-1}H] \subseteq I_L(H)$ such that $x \notin H$ (since by definition of k we have $[A, {}_{k-1}H] \subseteq I_L(H)$ and $[A, {}_{k-1}H] \nsubseteq H$). Let $K = \langle x \rangle + H$ and let $X = \langle x \rangle$. Now $K \in \mathfrak{F}$ and so for some m, $[K, {}_mH] = [K, {}_mX] = 0$ (by theorem (3.4), whence $[K, {}_{m.m}K] = 0$ by proposition 3.2 (for $K^H = K$ and $X \subseteq I_L(H)$). Hence $K \in \mathfrak{F} \cap \mathfrak{N} = \mathfrak{G} \cap \mathfrak{N}$. □

Now $L \in$ Fin-$\mathfrak{A}$ if every abelian subalgebra is finite-dimensional. Thus if $H \leqq L$ then $H \in$ Fin-$\mathfrak{A}$. Suppose that $L \in \mathfrak{E} \cap$ Fin-$\mathfrak{A}$. By Zorn's lemma L has a maximal locally nilpotent subalgebra H. So $H \in \mathrm{L}\mathfrak{N} \cap$ Fin-$\mathfrak{A} \leqq \mathfrak{F} \cap \mathfrak{N}$, by theorem 10.1.3. If $H \neq L$ then we can find $K \in \mathfrak{G} \cap \mathfrak{N}$ with $H < K$, by corollary 3.5. This contradicts the definition of H as a maximal locally nilpotent subalgebra of L. Hence $H = L$ and so $L \in \mathfrak{F} \cap \mathfrak{N}$. Conversely if $L \in \mathfrak{F} \cap \mathfrak{N}$ then $L \in$ Fin-$\mathfrak{A}$ and $L \in \mathfrak{E}$. So we have proved:

THEOREM 3.6 $\mathfrak{E} \cap$ Fin-$\mathfrak{A} = \mathfrak{F} \cap \mathfrak{N}$. □

Evidently if $L \in \mathrm{E}\mathfrak{A} \cap$ Fin-$\mathfrak{A}$ then $L \in \mathfrak{F} \cap \mathrm{E}\mathfrak{A}$ by theorem 10.1.1(b) (or theorem 9.3.4). Thus *if $H \lhd L$ and $H \in \mathrm{E}\mathfrak{A}$ then $L \in$ Fin-$\mathfrak{A}$ if and only if $H \in \mathfrak{F} \cap \mathrm{E}\mathfrak{A}$ and $L/H \in$ Fin-$\mathfrak{A}$*, as can easily be checked. (In general Fin-$\mathfrak{A}$ is not Q-closed.)

Clearly Max $\leqq$ Fin-$\mathfrak{A}$, Min $\leqq$ Fin-$\mathfrak{A}$ and $\mathfrak{F} \leqq$ Fin-$\mathfrak{A}$.

COROLLARY 3.7 (a) $\mathfrak{E} \cap$ Max $= \mathfrak{E} \cap$ Min $= \mathfrak{E} \cap$ Fin-$\mathfrak{A} = \mathfrak{E} \cap \mathfrak{F} = \mathfrak{F} \cap \mathfrak{N}$.
(b) $\acute{\mathrm{E}}\mathfrak{E} \cap$ Fin-$\mathfrak{A} = \mathfrak{F} \cap \mathrm{E}\mathfrak{A}$.

Proof: (a) follows from our remarks above and theorem 3.6.
(b) Let $L \in \acute{\mathrm{E}}\mathfrak{E} \cap$ Fin-$\mathfrak{A}$. Then L has an ascending $\mathfrak{E}$-series

$$0 = L_0 \lhd L_1 \lhd \ldots L_\lambda = L$$

for some ordinal λ. Suppose that $L \notin \mathfrak{F}$. Let $\alpha \leqq \lambda$ be an ordinal minimal with respect to $L_\alpha \notin \mathfrak{F}$. If α is not a limit ordinal then $\alpha - 1$ exists. But then as $L_\alpha \in$ Fin-$\mathfrak{A}$ and $L_{\alpha-1} \in \mathfrak{F} \cap \mathrm{E}\mathfrak{A}$ we have $L_\alpha/L_{\alpha-1} \in$ Fin-$\mathfrak{A}$; by definition of L we have $L_\alpha/L_{\alpha-1} \in \mathfrak{E}$ and so by theorem 3.6 $L_\alpha/L_{\alpha-1} \in \mathfrak{F} \cap \mathfrak{N}$, whence $L_\alpha \in \mathfrak{F}$, a contradiction. If α is a limit ordinal then $L_\beta \in \mathfrak{F}$ and so $L_\beta \in \mathfrak{F} \cap \mathrm{E}\mathfrak{A}$ for all $\beta \leqq \alpha$. Now $L_\alpha = \bigcup_{\beta < \alpha} L_\beta$ and so $L_\alpha \in \acute{\mathrm{E}}(\mathfrak{F} \cap \mathrm{E}\mathfrak{A}) \cap$ Fin-$\mathfrak{A} \leqq \mathfrak{F} \cap \mathrm{E}\mathfrak{A}$ (by theorem 9.3.4), another contradiction. Hence $L \in \mathfrak{F}$ and so $L \in \mathfrak{F} \cap \mathrm{E}\mathfrak{A}$. The converse is trivial. □

In the proof of corollary 3.7(b) we have used the fact that $\acute{\mathrm{E}}\mathfrak{E}$ is s-closed and that $\mathfrak{F} \cap \acute{\mathrm{E}}\mathfrak{E} \leqq \mathfrak{F} \cap \mathrm{E}\mathfrak{A}$ (which follows by part (a)).

It follows from corollary 3.7(b) that

$$\mathrm{L}\acute{\mathrm{E}}\mathfrak{A} \cap \text{Fin-}\mathfrak{A} \leqq \mathrm{L}(\mathfrak{F} \cap \mathrm{E}\mathfrak{A}) \cap \text{Fin-}\mathfrak{A} \tag{5}$$

and by the remarks preceeding corollary 3.7 we have

$$\mathrm{L}(\mathfrak{N}\mathfrak{A}) \cap \text{Fin-}\mathfrak{A} = (\mathrm{L}\mathfrak{N})\mathfrak{A} \cap \text{Fin-}\mathfrak{A} = (\mathfrak{F} \cap \mathfrak{N})(\mathfrak{F} \cap \mathfrak{A}). \tag{6}$$

For fields of characteristic zero we have $\mathrm{L}(\mathfrak{F} \cap \mathrm{E}\mathfrak{A}) \leqq \mathrm{L}(\mathfrak{N}\mathfrak{A})$ so we can deduce theorem 10.1.4 from equations (5) and (6). We have not determined whether for fields of non-zero characteristic it is true that $\mathrm{L}(\mathfrak{F} \cap \mathrm{E}\mathfrak{A}) \cap \text{Fin-}\mathfrak{A} \leqq \mathfrak{F} \cap \mathrm{E}\mathfrak{A}$.

It is clear that $\mathfrak{E}$ is not an E-closed class. Indeed the non-abelian two dimensional soluble algebra shows that the sum of an abelian ideal and an abelian subalgebra need not be an Engel algebra. Under extra hypotheses we can show that the sum of an Engel ideal and a $\mathrm{L}\mathfrak{N}$-ideal is an Engel algebra.

Let $H \leqq L$ and $K \leqq I_L(H)$. We say that K *acts on* H *by nilpotent derivations* if for each $y \in K$ there is a positive integer $n = n(y)$ such that $[H, {}_n y] = 0$. We say that K acts on H by nil derivations if to each pair of elements $x \in H$ and $y \in K$ there is a positive integer $n = n(x, y)$ such that $[x, {}_n y] = 0$. Evidently if $H + K \in \mathfrak{E}^*$ (resp. $\mathfrak{E}$) then $H, K \in \mathfrak{E}^*$ (resp. $\mathfrak{E}$) and K acts on H by nilpotent (resp. nil) derivations. We obtain a partial converse if we make the restriction $H \in \mathrm{L}\mathfrak{N} \cap \mathfrak{E}^*$ (resp. $\mathrm{L}\mathfrak{N} \cap \mathfrak{E} = \mathrm{L}\mathfrak{N}$), as we now prove.

THEOREM 3.8 *Let L be a Lie algebra, $H \leqq L$ and $K \leqq I_L(H)$. If $H \in \mathrm{L}\mathfrak{N}$ then*

(a) *$H + K \in \mathfrak{E}^*$ if and only if $H, K \in \mathfrak{E}^*$ and K acts on H by nilpotent derivations. In this case $H + P \in \mathrm{L}\mathfrak{N}$ for every $\mathrm{L}\mathfrak{N}$-subalgebra P of K.*

(b) *$H + K \in \mathfrak{E}$ if and only if $K \in \mathfrak{E}$ and K acts on H by nil derivations. In this case $H + P \in \mathrm{L}\mathfrak{N}$ for every $\mathrm{L}\mathfrak{N}$-subalgebra P of K.*

Proof: (a) Suppose that $H, K \in \mathfrak{E}^*$ and K acts on H by nilpotent derivations. Let P be a locally nilpotent subalgebra of K and $Y \leqq P$ with $Y \in \mathfrak{G}$. Then $Y \in \mathfrak{G} \cap \mathfrak{N}$. Evidently for each $y \in K$ we can find $n = n(y)$ such that $[H, {}_n y] = 0 = [K, {}_n y]$ and so $[H + K, {}_n y] = 0$. Using this and induction on the nilpotency class of Y it follows (see the proof of theorem 3.4(b)) by corollary 3.3 that $[H + K, {}_{n'} Y] = 0$ for some $n' = n'(Y)$. Now let A be a finite subset of H. Then as $[A, {}_{n'} Y] = 0$,

$$A^Y = \sum_{i \leqq n'-1} [A, {}_i Y]$$

is a finite-dimensional subspace of H and so $X = \langle A^Y \rangle \in \mathfrak{G} \cap \mathfrak{N}$ since $H \in \mathrm{L}\mathfrak{N}$. Since $H \in \mathfrak{E}^*$ it follows by theorem 3.4(b) that $[H, {}_m X] = 0$ for some

$m = m(X)$. As $[K, X] \subseteq [K, H] \subseteq H$ then $[H+K, {}_{m+1}X] = 0$. By definition Y idealises X and so by proposition 2.2 we have

$$[H+K, {}_{(m+1)n'}X+Y] = 0$$

and in particular $X+Y \in \mathfrak{N}_{(m+1)n'}$. Thus $H+K \in \mathfrak{E}^*$ and $H+P \in \mathrm{L}\mathfrak{N}$, for every element of $H+K$ is contained an a subalgebra of the form $X+Y$ and every finitely generated subalgebra of $H+P$ is contained in one like $X+Y$.

(b) Let $K \in \mathfrak{E}$ and K act on H by nil derivations and let P be a $\mathrm{L}\mathfrak{N}$-subalgebra of K. Let $Y \in \mathfrak{G}$, $Y \leqq P$, with $Y \in \mathfrak{G} \cap \mathfrak{N}$. An induction on the nilpotency class of Y and corollary 3.3 (see the proof of theorem 3.4(a)) shows that for each finite subset (or finite-dimensional subspace) B of $H+K$ there exists $n = n(B, Y)$ such that $[B, {}_nY] = 0$. Now let A be a finite subset of H. Then $[A, {}_{n'}Y] = 0$ for some n' and so as in part (a) above we have $X = \langle A^Y \rangle \in \mathfrak{G} \cap \mathfrak{N}$ and $X \leqq H$. Suppose that B is a finite subset (or finite-dimensional subspace) of $H+K$. Then $[B, X]+X \subseteq H$ and $[B, X]+X$ is finite-dimensional so $\langle [B, X], X \rangle \in \mathfrak{G} \cap \mathfrak{N}_c$ for some c since $H \in \mathrm{L}\mathfrak{N}$. Let $m = c+1$. Then $[B, {}_mX] = [[B, X], {}_cX] = 0$ so $[B^X, {}_mX] = 0$. Furthermore $B^X = B+[B, X]+ + \ldots + [B, {}_cX]$, is a finite-dimensional subspace of $H+K$ and so we can find $n = n(B^X, Y)$ such that $[B^X, {}_nY] = 0$. Since Y idealises X we have by proposition 2.2 that

$$[B^X, {}_{mn}X+Y] = 0,$$

where mn depends on B, X and Y. This shows that $H+K \in \mathfrak{E}$. Furthermore if we take $X+Y$ for B above we see that $X+Y \in \mathfrak{N}_{mn}$ (since $X+Y$ is a finite-dimensional subspace of $H+K$). Hence $H+P \in \mathrm{L}\mathfrak{N}$; for every finite subset of $H+P$ is contained in a subalgebra of the form $X+Y$.

The 'only if' halves of (a) and (b) follow by our preceding remarks. □

Clearly if K si L and $K \in \mathfrak{E}^*$ (resp. K asc L and $K \in \mathfrak{E}$) then K acts on L by nilpotent (resp. nil) derivations. So it follows from theorem 3.8 that:

COROLLARY 3.9 *If H is a locally nilpotent ideal of L and K* si *L resp. K* asc *L then $H+K \in \mathfrak{E}^*$ (resp. $\mathfrak{E}$) if and only if $H, K \in \mathfrak{E}^*$ (resp. $\mathfrak{E}$). In either case if $P \leqq H+K$ then $P \in \mathrm{L}\mathfrak{N} \Leftrightarrow H+P \in \mathrm{L}\mathfrak{N} \Leftrightarrow (H+P) \cap K \in \mathrm{L}\mathfrak{N}$, and $\rho(H+K) = H+\rho(K)$.* □

Suppose that H is a locally nilpotent subalgebra of L and $K \leqq I_L(H)$ with $K \in \mathfrak{E}$ and K acting on H by nil derivations. Then $H+K \in \mathfrak{E}$ by theorem 3.8. Now let $X \leqq K$ with $X^2 \leqq H$, and let $x_1, \ldots, x_r \in X$. Since $X^2 \in \mathrm{L}\mathfrak{N}$ and $\langle x_1 \rangle$ acts on X^2 by nil derivations then $X_1 = X^2 + \langle x_1 \rangle \in \mathrm{L}\mathfrak{N}$ (theorem 3.8(b)). Inductively suppose that $X_i = \langle X^2, x_1, \ldots, x_i \rangle \in \mathrm{L}\mathfrak{N}$. As $\langle x_{i+1} \rangle$ idealises X_i and acts on X_i by nil derivations then $X_{i+1} = X_i + \langle x_{i+1} \rangle \in \mathrm{L}\mathfrak{N}$.

Hence $X_r = \langle X^2, x_1, \ldots, x_r\rangle \in \mathrm{L}\mathfrak{N}$ and so $\langle x_1, \ldots, x_r\rangle \in \mathfrak{N}$. So $X \in \mathrm{L}\mathfrak{N}$. A simple induction on d now shows that

$$\text{if } X^{(d)} \leqq H \text{ then } X \in \mathrm{L}\mathfrak{N}, \text{ whence } H+X \in \mathrm{L}\mathfrak{N} \tag{*}$$

by theorem 3.8(b). In particular suppose that $P \leqq K$ and $P/P\cap H \in \mathrm{LE}\mathfrak{A}$. Then by (*) for every $\mathfrak{G}$-subalgebra X of P we have $H+X \in \mathrm{L}\mathfrak{N}$ and therefore $H+P \in \mathrm{L}\mathfrak{N}$; and $P \in \mathrm{L}\mathfrak{N}$.

Corollary 3.10 *Let $H \leqq L$, $K \leqq I_L(H)$ and suppose that $H \in \mathrm{L}\mathfrak{N}$, $K \in \mathfrak{E}$, and K acts on H by nil derivations. If $P \leqq K$ then*

$$H+P \in \mathrm{L}\mathfrak{N} \Leftrightarrow P/P\cap H \in \mathrm{L\acute{E}}\mathfrak{A} \Leftrightarrow P/P\cap H \in \{\mathrm{L}, \mathrm{\acute{E}}\}\mathfrak{A}.$$

In particular

$$\{\mathrm{L}, \mathrm{\acute{E}}\}\mathfrak{A}\cap\mathfrak{E} = \mathrm{L}\mathfrak{N}.$$

Proof: Let $X \leqq K$ with $X/X\cap H \in \mathrm{\acute{E}}\mathfrak{A}$. Then X has an ascending series $X\cap H = X_0 \lhd X_1 \lhd \ldots X_\lambda = X$ with $X_{\alpha+1}/X_\alpha \in \mathfrak{A}$ for all α. Now $X_0 \in \mathrm{L}\mathfrak{N}$. If $X_\alpha \in \mathrm{L}\mathfrak{N}$ then, as $X_{\alpha+1}/X_\alpha \in \mathfrak{A}$ and $X_{\alpha+1}$ acts on X_α by nil derivations, by the remarks above we have $X_{\alpha+1} \in \mathrm{L}\mathfrak{N}$. If $\mu \leqq \lambda$ is a limit ordinal and $X_\alpha \in \mathrm{L}\mathfrak{N}$ for all $\alpha < \mu$, then clearly $X_\mu = \bigcup_{\alpha<\mu} X_\alpha \in \mathrm{L}\mathfrak{N}$. Hence $X_\alpha \in \mathfrak{N}$ for all α and so $X = X_\lambda \in \mathrm{L}\mathfrak{N}$. Similarly if $X/X\cap H \in \mathrm{\acute{E}LE}\mathfrak{A}$ then $X \in \mathrm{L}\mathfrak{N}$.

Now suppose that $P/P\cap H \in \mathrm{L\acute{E}}\mathfrak{A}$. If X is a $\mathfrak{G}$-subalgebra of P then $X/X\cap H \in \mathrm{\acute{E}}\mathfrak{A}$ and so from above $X \in \mathrm{L}\mathfrak{N}$, so $X \in \mathfrak{N}$. Thus $P \in \mathrm{L}\mathfrak{N}$ and so by our preceeding remarks $H+P \in \mathrm{L}\mathfrak{N}$. This proves the first equivalence.

For the second we note that $\{\mathrm{L}, \mathrm{\acute{E}}\}\mathfrak{A} = \bigcup_\alpha (\mathrm{L\acute{E}})^\alpha\mathfrak{A}$. We have shown that $\mathrm{L\acute{E}}\mathfrak{A}\cap\mathfrak{E} \leqq \mathrm{L}\mathfrak{N}$ and $\mathrm{\acute{E}LE}\mathfrak{A}\cap\mathfrak{E} \leqq \mathrm{L}\mathfrak{N}$. Assume inductively that $(\mathrm{L\acute{E}})^\alpha\mathfrak{A}\cap\mathfrak{E} \leqq \mathrm{L}\mathfrak{N}$ and let $X \in (\mathrm{L\acute{E}})^{\alpha+1}\mathfrak{A}\cap\mathfrak{E}$. Then if Y is a $\mathfrak{G}$-subalgebra of X we can find $Y_1 \leqq X$ such that $Y \leqq Y_1 \in \mathrm{\acute{E}}((\mathrm{L\acute{E}})^\alpha\mathfrak{A})\cap\mathfrak{E}$. Hence $Y_1 \in \mathrm{\acute{E}}((\mathrm{L\acute{E}})^\alpha\mathfrak{A}\cap\mathfrak{E})$, since $\mathfrak{E}$ is $\{\mathrm{S}, \mathrm{Q}\}$-closed. Thus $Y_1 \in \mathrm{\acute{E}L}\mathfrak{N}$, whence $Y_1 \in \mathrm{L}\mathfrak{N}$ and so $X \in \mathrm{L}\mathfrak{N}$ by the induction on α. At limit ordinals μ we have $(\mathrm{L\acute{E}})^\mu\mathfrak{A}\cap\mathfrak{E} = \bigcup_{\alpha<\mu}((\mathrm{L\acute{E}})^\alpha\mathfrak{A}\cap\mathfrak{E}) \leqq \mathrm{L}\mathfrak{N}$ if $(\mathrm{L\acute{E}})^\alpha\mathfrak{A}\cap\mathfrak{E} \leqq \mathrm{L}\mathfrak{N}$ for all $\alpha < \mu$. Hence $\{\mathrm{L}, \mathrm{\acute{E}}\}\mathfrak{A}\cap\mathfrak{E} \leqq \mathrm{L}\mathfrak{N}$. Thus if $P/P\cap H \in \{\mathrm{L}, \mathrm{\acute{E}}\}\mathfrak{A}$ then as $K \in \mathfrak{E}$ we have $P/P\cap H \in \mathrm{L}\mathfrak{N}$, so $P \in \mathrm{L}\mathfrak{N}$ and $H+P \in \mathrm{L}\mathfrak{N}$. □

By corollary 3.10 we have

$$\mathfrak{E}\cap\{\mathrm{L}, \mathrm{\acute{E}}\}\mathfrak{A} = \mathrm{L}\mathfrak{N}. \tag{7}$$

This implies of course that with respect to $\mathfrak{E}$ the class of locally nilpotent algebras is E-closed. So we deduce:

Corollary 3.11 *If $L \in \mathfrak{E}$ then $\rho(L/\rho(L)) = 0$.* □

Equation (7) also implies the result of Gruenberg [72] that a locally soluble Engel algebra is locally nilpotent; but is more general, since it also implies that an Engel algebra with an ascending $\mathrm{L}\mathrm{E}\mathfrak{A}$-series is locally nilpotent.

In the language of section 6.1 we can also say that:

If $L \in \mathfrak{E}$ *then* $\rho(L) = \rho_{\mathfrak{X}}(L)$ *for* $\mathfrak{X} = \mathrm{E}\mathfrak{A}, \mathfrak{F}$ *or* $\mathrm{E}\mathfrak{A} \cap \mathfrak{F}$.

A property of Gruenberg algebras

Let L be a Lie algebra and G a group of automorphisms of L. If $H \leqq L$ then $H^G = \sum_{g \in G} H^g$. Evidently if $H^G \subseteq I_L(H)$ then $H^G \subseteq I_L(H^g)$ for all $g \in G$ and so $H^G \leqq L$. Furthermore

$$\textit{if } H^G \subseteq I_L(H) \text{ then } H^G \in \acute{\mathrm{E}}\mathrm{Q}(H), \tag{8}$$

where (H) is the class of algebras generated by H. The proof of (8) is trivial. In particular if $H \in \acute{\mathrm{E}}\mathfrak{X}$ and $\mathfrak{X} = \mathrm{Q}\mathfrak{X}$ then $H^G \in \acute{\mathrm{E}}\mathrm{Q}\acute{\mathrm{E}}\mathfrak{X} = \acute{\mathrm{E}}\mathfrak{X}$ (for $\mathrm{Q}\acute{\mathrm{E}} \leqq \acute{\mathrm{E}}\mathrm{Q}$).

We recall that if $L \in \mathfrak{E}$ over a field of characteristic zero and $x \in L$ then $\exp(x^*)$ is a well defined automorphism of L; and that if A and B are subsets of L then

$$A^B = \sum_{e \in \exp(\mathrm{ad}_B(L))} A^e. \tag{9}$$

LEMMA 3.12 *Suppose that* $\mathfrak{X}$ *is a* Q*-closed class of Lie algebras and* L *is an Engel algebra over a field of characteristic zero. Then*

(a) *If* H asc L *and* $H \in \acute{\mathrm{E}}\mathfrak{X}$ *then* $H^L \in \acute{\mathrm{E}}\mathfrak{X}$.

(b) L *is generated by ascendant* $\acute{\mathrm{E}}\mathfrak{X}$*-subalgebras if and only if* $L \in \acute{\mathrm{E}}\mathfrak{X}$.

(c) $L \in \acute{\mathrm{E}}\mathfrak{X}$ *if and only if every non-trivial homomorphic image of* L *has a non-trivial ascendant* $\mathfrak{X}$*-subalgebra.*

Proof: (a) Let $H \lhd^{\lambda} L$ and use transfinite induction on λ. For $\lambda = 0$ or 1 the result is trivial.

Let $\lambda > 0$ and assume that the result is true for all $\alpha < \lambda$. If λ is not a limit ordinal then $\lambda - 1$ exists and we have

$$H \lhd^{\lambda-1} H_{\lambda-1} \lhd H_{\lambda} = L.$$

Put $K = H_{\lambda-1}$. Then $K \in \mathfrak{E}$ and $H \lhd^{\lambda-1} K$ and so by induction $K_0 = H^K \in \acute{\mathrm{E}}\mathfrak{X}$. But $H^L = (H^K)^L = K_0^L = \sum_{e \in \exp(\mathrm{ad}(L))} K_0^e$, by (9). Evidently $H^L \leqq K$ and $K_0 \lhd K$ and so by (8) we have $H^L \in \acute{\mathrm{E}}\mathfrak{X}$.

If λ is a limit ordinal then we have an ascending series

$$H = H_0 \lhd H_1 \lhd \dots H_{\lambda} = \bigcup_{\alpha < \lambda} H_{\alpha} = L.$$

For each $\alpha > 0$ define $K_\alpha = H^{H_\alpha}$. It is readily checked that we have an ascending series

$$H = K_1 \lhd K_2 \lhd \ldots K_\lambda = \bigcup_{\alpha<\lambda} K_\alpha = H^L.$$

Furthermore for each $\alpha < \lambda$ we have $H_\alpha \in \mathfrak{E}$ and $H \lhd^\alpha H_\alpha$ and so by induction $K_\alpha \in \acute{\mathrm{E}}\mathfrak{X}$. Hence $L \in \acute{\mathrm{E}}(\mathrm{Q}\acute{\mathrm{E}}\mathfrak{X}) = \acute{\mathrm{E}}\mathfrak{X}$. This completes the induction on λ and proves (a).

Evidently if J is a sum of $\mathfrak{X}$-ideals of L then $J \in \acute{\mathrm{E}}(\mathrm{Q}\acute{\mathrm{E}}\mathfrak{X}) = \acute{\mathrm{E}}\mathfrak{X}$, since $\mathfrak{X}$ is Q-closed. Thus if $K = \langle H : H \text{ asc } L \text{ and } H \in \acute{\mathrm{E}}\mathfrak{X}\rangle$ then by (a), K^L is a sum of $\acute{\mathrm{E}}\mathfrak{X}$-ideals of L so $K^L \in \acute{\mathrm{E}}\mathfrak{X}$. But by equation (9) $K^L = \langle K^e : e \in \exp(\mathrm{ad}(L))\rangle =$ $= \langle H^e : H \text{ asc } L,\ H \in \acute{\mathrm{E}}\mathfrak{X} \text{ and } e \in \exp(\mathrm{ad}(L))\rangle$ and so $K^L \leqq K \in \acute{\mathrm{E}}\mathfrak{X}$.

(b) If L is generated by ascendant $\acute{\mathrm{E}}\mathfrak{X}$-subalgebras then from the above we have $L \in \acute{\mathrm{E}}\mathfrak{X}$ and conversely.

(c) One implication is trivial since $\mathfrak{X}$ is Q-closed. Suppose that every non-trivial homomorphic image of L contains a non-trivial ascendant $\mathfrak{X}$-subalgebra. Let K be the join of all the ascendant $\mathfrak{X}$-subalgebras of L. From above we have $K \lhd L$ and $K \in \acute{\mathrm{E}}\mathfrak{X}$. Clearly if H/K asc L/K and $H/K \in \mathfrak{X}$ then H asc L and $H \in \acute{\mathrm{E}}\mathfrak{X}$, so $H \leqq K$. Therefore we must have $K = L \in \acute{\mathrm{E}}\mathfrak{X}$. □

By theorem 6.5.4 we have that over fields of characteristic zero, $L = \gamma(L)$ if and only if $L \in \acute{\mathrm{E}}\mathfrak{A} \cap \mathrm{L}\mathfrak{N}$. By (7) we have $\acute{\mathrm{E}}\mathfrak{A} \cap \mathfrak{E} = \acute{\mathrm{E}}\mathfrak{A} \cap \mathrm{L}\mathfrak{N}$. From this and lemma 3.12 it is not hard to deduce:

THEOREM 3.13 *If $L \in \mathfrak{E}$ over a field of characteristic zero then*

(a) *$\rho(L)$ contains every ascendant $\mathrm{L}\mathfrak{N}$-subalgebra of L.*

(b) *$\gamma(L) \in \acute{\mathrm{E}}\mathfrak{A}$ and contains every ascendant $\acute{\mathrm{E}}\mathfrak{A}$-subalgebra of L. In particular $\gamma(L/\gamma(L)) = 0$.* □

Remark 2. Of course, if $L \in \mathfrak{E}$ over a field of characteristic zero then $\gamma(L) \lhd L$.

It is not very hard to show that if $L \in \mathfrak{E}$ and A is an abelian subalgebra of L then X asc $(A+X)$ for every $\mathfrak{G} \cap \mathfrak{N}$-subalgebra X of $I_L(A)$. From this it follows that (over any field) *if H asc $L \in \mathfrak{E}$ and $H \in \acute{\mathrm{E}}(\lhd)\mathfrak{A}$ then every finitely generated subalgebra of H is ascendant in H and so in L.* However we have not been able to determine whether over fields of characteristic $p > 0$, $L \in \acute{\mathrm{E}}\mathfrak{A} \cap \mathfrak{E}$ $= \acute{\mathrm{E}}\mathfrak{A} \cap \mathrm{L}\mathfrak{N}$ implies that every element generates an ascendant subalgebra of L.

4. Left and right Engel elements

Let L be a Lie algebra and $x \in L$. We say that x is a *right Engel element* if for each $y \in L$ we can find $n = n(x, y) \geqq 0$ such that

$$[x, {}_n y] = 0.$$

If n can be chosen independently of y (i.e. $x \in L(g_{n,1})$) we say that x is a *right n-Engel element*; if x is a right n-Engel element for some n we say that x is a *bounded right Engel element*. The sets of right Engel, right n-Engel and bounded right Engel elements are denoted by

$$\mathfrak{r}(L),\ \mathfrak{r}_n(L) = L(g_{n,1}),\ \text{and}\ \bar{\mathfrak{r}}(L) = \bigcup_n \mathfrak{r}_n(L)$$

respectively.

We say that x is a *left Engel element* if for each $y \in L$ there exists $n = n(x, y)$ such that

$$[y, {}_n x] = 0.$$

We observe that here the variable y appears on the *left*. If n can be chosen to be independent of y (i.e. $[L, {}_n x] = 0$) we say that x is a *left n-Engel element*; if x is a left n-Engel element for some n we say that x is a *bounded left Engel element*. The sets of left Engel, left n-Engel and bounded left Engel elements are denoted by

$$\mathfrak{e}(L),\ \mathfrak{e}_n(L)\ \text{and}\ \bar{\mathfrak{e}}(L) = \bigcup_n \mathfrak{e}_n(L)$$

respectively.

PROPOSITION 4.1 *Let L be a Lie algebra.*

(a) $\bar{\mathfrak{r}}(L) \leqq \mathfrak{r}(L) \leqq L$; $\bar{\mathfrak{r}}(L) \in {}^*\mathfrak{E}$ *and* $\mathfrak{r}(L) \in \mathfrak{E}$.

(b) $\mathfrak{r}_n(L) \subseteq \mathfrak{r}_{f(n)}(L) \leqq L$ *where* $f(n) = n$ *if* L *has characteristic* 0 *or* $p \geqq n$ *and* $f(n) = p^{n'}$ *if* L *has characteristic* $p < n$, *where* $p^{n'-1} < n \leqq p^{n'}$.

(c) $\mathfrak{r}_n(L) \lhd L$ *if* L *has characteristic* 0 *or* $p > n-1$ *and in this case* $\mathfrak{r}_n(L) \subseteq \mathfrak{e}_n(L)$.

(d) $\zeta_\omega(L) \leqq \bar{\mathfrak{r}}(L) \cap \bar{\mathfrak{e}}(L)$ *and* $\zeta_*(L) \leqq \mathfrak{r}(L) \cap \mathfrak{e}(L)$.

(e) $\rho(L) \subseteq \mathfrak{e}(L)$, $\{x : \langle x \rangle \text{ asc } L\} \subseteq \mathfrak{e}(L)$ *and* $\{x : \langle x \rangle \text{ si } L\} \subseteq \bar{\mathfrak{e}}(L)$.

Proof: (a) Trivially $\bar{\mathfrak{r}}(L)$ and $\mathfrak{r}(L)$ are subspaces of L. If $[x, {}_n y] = [z, {}_m y] = 0$ then $[[x, z], {}_{n+m} y] = 0$. Hence $\mathfrak{r}(L) \leqq L$ and $[\mathfrak{r}_n(L), \mathfrak{r}_m(L)] \subseteq \mathfrak{r}_{n+m}(L)$ so that $\bar{\mathfrak{r}}(L) \leqq L$. By definition $\bar{\mathfrak{r}}(L) \in {}^*\mathfrak{E}$, and $\mathfrak{r}(L) \in \mathfrak{E}$.

(b) follows from corollary 1.4 and the fact that if d is a derivation of L and L has characteristic $p > 0$ then d^p is also a derivation.

(c) follows from lemma 7.3.1 and corollary 1.4.

(d) If $x \in \zeta_n(L)$ then $[L, {}_n x] = 0$ and $[x, {}_n y] = 0$ for all $y \in L$. Hence $\zeta_n(L) \leqq \mathfrak{r}_n(L) \cap \mathfrak{e}_n(L)$. A transfinite induction on α shows that if $x \in \zeta_\alpha(L)$ then for each $y \in L$ there is an n depending on x and y such that $[x, {}_n y] = [y, {}_n x] = 0$.

(e) is straightforward. □

We observe that if $\mathfrak{r}_n(L) \leqq L$ then $\mathfrak{r}_n(L) \in \mathfrak{E}_n$. In any case for any n we can find $f(n)$ such that

$$\langle \mathfrak{r}_n(L) \rangle \leqq \mathfrak{r}_{f(n)}(L) \in \mathfrak{E}_{f(n)},$$

by proposition 4.1(b).

We note that if $\mathfrak{r}(L) \lhd L$ then $\mathfrak{r}(L/\mathfrak{r}(L)) = 0$. We also have the equivalences

$$L \in \mathfrak{E} \Leftrightarrow L = \mathfrak{r}(L) \Leftrightarrow L = \mathfrak{e}(L).$$

Corresponding to the various 'Engel subsets' we define the classes:

$$*\mathfrak{E}_n: \ L \in *\mathfrak{E}_n \text{ if } L = \langle \mathfrak{r}_n(L) \rangle;$$

$$\mathfrak{E}_n^*: \ L \in \mathfrak{E}_n^* \text{ if } L = \langle \mathfrak{e}_n(L) \rangle;$$

$$\mathfrak{E}_\omega^*: \ L \in \mathfrak{E}_\omega^* \text{ if } L = \langle \bar{\mathfrak{e}}(L) \rangle;$$

$$\mathfrak{E}_*: \ L \in \mathfrak{E}_* \text{ if } L = \langle \mathfrak{e}(L) \rangle.$$

By our remarks above we have

$$*\mathfrak{E}_n \leqq \mathfrak{E}_{f(n)}.$$

(i) Over any field of characteristic $p > 0$,

$$\mathfrak{A}\mathfrak{N}_2 \cap \mathfrak{E}_p^* \nleqq \mathfrak{N}.$$

For consider the split extension $A + \langle x, y \rangle$ where $A = \mathfrak{k}[t]$ with $t^p = 0$ and

$$x: a \mapsto ta \text{ and } y: a \mapsto \mathrm{d}a/\mathrm{d}t \text{ for all } a \in A.$$

Then $1, x, y$ are left p-Engel elements. The result follows from lemma 3.1.1. We note that $[x, y] = z: a \mapsto a$ for all $a \in A$ and so z is not a left Engel element.

(ii) Over any field of characteristic zero,

$$\mathfrak{F}_3 \cap \mathfrak{E}_3^* \nleqq \mathfrak{N}.$$

For the 3-dimensional simple Lie algebra (see Jacobson [98] p. 13) has a basis consisting of left 3-Engel elements. From (i) and (ii) we conclude that

$$\mathfrak{E}^* < \mathfrak{E}^*_\omega \text{ and } \mathfrak{E} < \mathfrak{E}_*$$

and that $\mathfrak{e}(L)$, $\mathfrak{e}_n(L)$ and $\bar{\mathfrak{e}}(L)$ need not be subspaces nor be closed under Lie products in general.

In spite of these disadvantages $\mathfrak{e}(L)$ is a useful object. For if $x \in \mathfrak{e}(L)$ then x^* is a nil derivation and so $\exp(x^*)$ can be defined. Furthermore if $x \in \mathfrak{e}(L)$ then $\mu x \in \mathfrak{e}(L)$ for all scalars μ.

Evidently if H asc L then $\mathfrak{e}(H) \subseteq \mathfrak{e}(L)$ and if H si L then $\bar{\mathfrak{e}}(H) \subseteq \bar{\mathfrak{e}}(L)$. Thus $\mathfrak{e}(L)$ contains every ascendant $\mathrm{L}\mathfrak{N}$-subalgebra of L and $\bar{\mathfrak{e}}(L)$ contains every nilpotent subideal of L. For fields of characteristic zero (ii) above shows that these inclusions are proper in general.

Suppose that L is a Lie algebra over a field of characteristic zero and H asc L. Let $\{H_\alpha : 0 \leqq \alpha \leqq \lambda\}$ be an ascending series from H to L. Suppose that $x \in I_L(H) \cap \mathfrak{e}(L)$. Define for each α,

$$K_\alpha = \bigcap_{n=0}^{\infty} H_\alpha^{\exp(nx^*)}.$$

Then $K_0 = H$ and $K_\lambda = L$. If β is a limit ordinal and $y \in K_\beta$ then as $\{y\}^{\langle x \rangle}$ is finite dimensional we have $\{y\}^{\langle x \rangle} \subseteq H_\alpha$ for some $\alpha < \beta$, and so $\{y\}^{\langle x \rangle} \subseteq K_\alpha$, whence $y \in K_\alpha$ and $K_\beta = \bigcup_{\alpha<\beta} K_\alpha$. So following the proof of theorem 6.5.4 we see that $\{K_\alpha : 0 \leqq \alpha \leqq \lambda\}$ is an ascending series from H to L whose terms are idealised by x.

THEOREM 4.2 *Let L be a Lie algebra over a field of characteristic zero.*

(a) *If $L \in \acute{\mathrm{E}}\mathfrak{A}$ then $\rho(L) \subseteq \mathfrak{e}(L) = \gamma(L)$.*
(b) *If $L \in \mathrm{E}\mathfrak{A}$ then $\bar{\mathfrak{e}}(L) = \beta(L)$ and $\bar{\mathfrak{r}}(L) = \zeta_\omega(L)$.*
(c) *If $L \in \acute{\mathrm{E}}\mathrm{L}\mathfrak{N}$ then $\mathfrak{e}(L) \leqq L$ and $\mathfrak{e}(L) = \rho(\mathfrak{e}(L))$.*

Proof: Let $\mathfrak{X}$ be one of the classes $\mathfrak{A}$, $\mathrm{L}\mathfrak{N}$, $\mathrm{LE}\mathfrak{A}$ and let $L \in \acute{\mathrm{E}}\mathfrak{X}$. Then L has as ascending $\mathfrak{X}$-series

$$0 = H_0 \lhd H_1 \lhd \ldots H_\lambda = L.$$

Let $x \in \mathfrak{e}(L)$. Since x idealises 0 then this series can be replaced by

$$0 = K_0 \lhd K_1 \lhd \ldots K_\lambda = L \qquad (*)$$

where each term is idealised by x. If $\mathfrak{X} = \mathfrak{A}$ we have $K_{\alpha+1}^2 \leqq H_{\alpha+1}^2 \leqq H_\alpha$ and $K_{\alpha+1}^2$ is invariant under x so by definition $K_{\alpha+1}^2 \leqq K_\alpha$. Thus (*) is an $\mathfrak{A}$-series for L. If $\mathfrak{X} = \mathrm{L}\mathfrak{N}$ or $\mathrm{LE}\mathfrak{A}$, let B be a finite subset of $K_{\alpha+1}$ and $C =$

$= \langle B^{\langle x\rangle}\rangle$. Then $B^{\langle x\rangle}$ is a finite dimensional subspace of $K_{\alpha+1}$ and so for some c we have $C^{c+1} \leqq H_\alpha$ (resp. $C^{(c)} \leqq H_\alpha$). But C and so C^{c+1} and $C^{(c)}$ are invariant under x so $C^{c+1} \leqq K_\alpha$ (resp. $C^{(c)} \leqq K_\alpha$), whence $K_{\alpha+1}/K_\alpha \in \mathfrak{X}$. Thus in all cases (*) is an $\mathfrak{X}$-series for L whose terms are invariant under x.

(a) If $\mathfrak{X} = \mathfrak{A}$ then following the proof of theorem 6.5.4 we see that $\langle K_\alpha, x\rangle$ asc $\langle K_{\alpha+1}, x\rangle$ for all α, whence $\langle x\rangle$ asc L and so $x \in \gamma(L)$. So we have

$$\rho(L) \subseteq \mathfrak{e}(L) \subseteq \gamma(L) \subseteq \mathfrak{e}(L).$$

(b) If $L \in \mathfrak{A}^d$ then it is readily checked that for any $x \in \mathfrak{e}_n(L)$ we have $(L^{(i)}+\langle x\rangle) \lhd^n (L^{(i-1)}+\langle x\rangle)$, whence $\langle x\rangle \lhd^{nd} L$. Thus $\bar{\mathfrak{e}}(L) \subseteq \beta(L) \subseteq \bar{\mathfrak{e}}(L)$. By theorem 7.3.5 we have $\mathfrak{r}_n(L) \leqq \zeta_{(n^d-1)/(n-1)}(L)$, so $\bar{\mathfrak{r}}(L) \leqq \zeta_\omega(L) \leqq \bar{\mathfrak{r}}(L)$.

(c) Assume temporarily that L has countable dimension. Now L has the series in (*) with terms idealised by x and factors in $\mathrm{L}\mathfrak{N}$. For each α we have that $\langle x\rangle$ acts by nil derivations on $K_{\alpha+1}/K_\alpha$ and so by theorem 3.8, $\langle K_{\alpha+1}, x\rangle/K_\alpha \in \mathrm{L}\mathfrak{N}$. Now in a $\mathrm{L}\mathfrak{N}$-algebra of countable dimension every element generates an ascendant subalgebra. Thus

$$\langle K_\alpha, x\rangle \text{ asc } \langle K_{\alpha+1}, x\rangle$$

and so $\langle x\rangle$ asc L, for all $x \in \mathfrak{e}(L)$.

Now we abandon the assumption that L has countable dimension and let $x, y \in \mathfrak{e}(L)$, $z \in L$ and X be any countable-dimensional subalgebra of L which contains x, y, z. Clearly $X \in \acute{\mathrm{E}}\mathrm{L}\mathfrak{N}$ and $x, y \in \mathfrak{e}(X)$. From above we have that $\mathfrak{e}(X) = \gamma(X)$. Thus $x+y$, $[x, y] \in \mathfrak{e}(X)$ so we can find n and m such that $[z, {}_n x+y] = [z, {}_m[x, y]] = 0$. This proves that $\mathfrak{e}(L) \leqq L$. Suppose that X above is chosen to be a finitely generated subalgebra of $\mathfrak{e}(L)$. Then $\langle x\rangle$ asc X for all $x \in X$, so $X \in \mathfrak{G} \cap \mathfrak{G}\mathfrak{r} \leqq \mathfrak{N}$. Hence $\mathfrak{e}(L) \in \mathrm{L}\mathfrak{N}$ and so $\mathfrak{e}(L) = \rho(\mathfrak{e}(L))$. □

Let $L \in \mathrm{L}\acute{\mathrm{E}}\mathrm{L}\mathfrak{N}$ over a field of characteristic zero and let $x \in \mathfrak{e}(L)$. From the proof of theorem 4.2 it follows that $\langle x\rangle$ asc X for every finitely generated subalgebra X of L which contains x. Hence $\mathfrak{e}(L) \leqq L$ and $\mathfrak{e}(L) \in \mathrm{L}\mathfrak{N}$.

So from this and theorem 4.2 we have:

THEOREM 4.3 *Over fields of characteristic zero*

(a) *If* $L \in \mathrm{L}\acute{\mathrm{E}}\mathrm{L}\mathfrak{N}$ *then* $\mathfrak{e}(L) \leqq L$ *and* $\mathfrak{e}(L) \in \mathrm{L}\mathfrak{N}$.
(b) $\mathfrak{E}_* \cap \acute{\mathrm{E}}\mathfrak{A} = \mathfrak{G}\mathfrak{r} \cap \acute{\mathrm{E}}\mathfrak{A} = \mathrm{L}\mathfrak{N} \cap \acute{\mathrm{E}}\mathfrak{A}$.
(c) $\mathfrak{E}_* \cap \mathrm{L}\acute{\mathrm{E}}\mathrm{L}\mathfrak{N} = \mathrm{L}\mathfrak{N}$. □

The set $\mathfrak{r}(L)$ of right Engel elements is not easy to work with, even over fields of characteristic zero. We have not determined whether it is true in general that $\mathfrak{r}(L) \subseteq \mathfrak{e}(L)$. Thus the methods employed in proving theorems

4.2 and 4.3 may not work when dealing with $\mathfrak{r}(L)$. We give below one instance when it is true that $\mathfrak{r}(L) \subseteq \mathfrak{e}(L)$.

Let L be a Lie algebra (over any field), $H \leqq L$ and let

$$x \in I_L(H) \cap \mathfrak{r}(L).$$

We claim that if $H \in \mathfrak{A}^d$ for some d then $x \in \mathfrak{e}(H+\langle x \rangle)$. For this we use induction on d.

If $d = 1$ let $a \in H$. Then for some m we have

$$0 = [x, {}_m a+x] = [x, a, {}_{m-1} a+x] = [x, a, {}_{m-1} x]$$

and so $[a, {}_m x] = 0$, since $H^2 = 0$ and $[x, a] \in H$. Let $d > 1$ and assume inductively that the result is true for $d-1$. Evidently x idealises H^2 and so $x \in \mathfrak{e}(H^2+\langle x \rangle)$. Let $K = (H+\langle x \rangle)/H^2$. Then $H^2+x \in I_K(H/H^2) \cap \mathfrak{r}(K)$ and so by the first part given $a \in H$ we can find m such that $[a, {}_m x] \in H^2$. Thus as $x \in \mathfrak{e}(H^2+\langle x \rangle)$ we can find n such that $[[a, {}_m x], {}_n x] = 0$, whence $x \in \mathfrak{e}(H+\langle x \rangle)$ and our induction is complete. So we have that

$$\text{if } x \in I_L(H) \cap \mathfrak{r}(L) \text{ and } H \in \text{E}\mathfrak{A} \text{ then } x \in \mathfrak{e}(\langle H, x \rangle). \tag{10}$$

PROPOSITION 4.4 *Let L be a Lie algebra.*

(a) *If $L \in \text{LE}\mathfrak{A}$ then $\mathfrak{r}(L) \subseteq \mathfrak{e}(L)$ and $\mathfrak{r}(L) \in \text{L}\mathfrak{N}$.*
(b) *If $L \in \text{LE}\mathfrak{A}$ over a field of characteristic zero then $\mathfrak{r}(L) \lhd \mathfrak{e}(L) \in \text{L}\mathfrak{N}$.*

Proof: (a) follows from (10), (7) and the fact that $\mathfrak{r}(L) \in \mathfrak{E}$.
(b) follows from (a) and theorem 4.3. □

Chapter 17

Kostrikin's theorem

1. The Burnside problem

In 1902 Burnside [267, 268] posed a problem which nowadays is formulated in at least two different ways:

(a) *Let n be a positive integer. Is every group of exponent n locally finite?*

(b) *Let m and n be positive integers. Is the order of a finite m-generator group of exponent n bounded by a function of m and n?*

Problem (b) is usually called the *restricted Burnside problem.* Problem (a) is known to have an affirmative answer for $n = 2, 3, 4$, or 6 (see Hall [282] or Stewart [341] for references) and a negative answer for odd $n \geqq 697$ (Novikov and Adjan [316], Adjan [255], Britton [265]). Nonetheless, Kostrikin [114] showed that the restricted Burnside problem is answered affirmatively for all *prime* n and for all m. The proof is based on a connection between groups of prime exponent and Lie rings satisfying Engel conditions, and the major part of the proof is devoted to a remarkable theorem about Lie algebras. (In fact, Lie rings: but the two interpretations are essentially equivalent.) By a judicious combination of ideas, suggested by his earlier work [109, 111] on the cases $n = 5, 7$, and a brilliant display of computational virtuosity, Kostrikin proved (essentially) the following theorem:

THEOREM 1.1 *Let L be a Lie algebra over the field $\mathfrak{k}$ satisfying the nth Engel condition $[x, {}_ny] = 0$ $(x, y \in L)$. If either*

(a) *$\mathfrak{k}$ has characteristic 0 or*

(b) *$\mathfrak{k}$ has characteristic $p \geqq n$*

then L is locally nilpotent. □

The proof of this theorem is given in [114], but with many computational details omitted. To give a complete proof would take us too far afield in the present work, and we have therefore compromised. Our aim in this chapter will be to prove a weaker of Kostrikin's theorem, in which condition (b) is replaced by $p \geqq 2n$. We shall then sketch the proof of the full theorem. The advantage of the weaker form is that the proof becomes considerably shorter, while still following closely the lines of the full theorem: moreover it leads to the same result in characteristic 0 and to a significant result about Lie algebras in general. The disadvantage is that the weaker form has no implications as regards the restricted Burnside problem, which requires case (b) when $n = p - 1$.

To see why this should be, we must make brief reference to the connection between groups of prime exponent and Lie algebras with Engel condition. Let G be any group, with lower central series $G_n = \gamma_n(G)$, so that

$$G = G_1 \rhd G_2 \rhd G_3 \rhd \dots.$$

Letting round brackets denote group commutators, it may be shown that

$$(G_i, G_j) \subseteq G_{i+j}$$

for all $i, j \geqq 1$. Therefore each quotient group

$$L_i = G_i/G_{i+1}$$

is abelian. Write L_i additively, and form the direct sum

$$L = \mathrm{Dr}_{i=1}^{\infty} L_i$$

which is also an abelian group. We make this the additive group of a Lie ring by defining multiplication as follows: given 'homogeneous' elements

$$x' = xG_{i+1} \in L_i$$

$$y' = yG_{j+1} \in L_j$$

we set

$$[x', y'] = (x, y)G_{i+j+1} \in L_{i+j}.$$

This operation is extended to the whole of L using the distributive law.

It is a consequence of commutator calculations in G that L satisfies the identities

$$[x, x] = 0$$

$$[[x, y], z] + [[y, z], x] + [[z, x], y] = 0,$$

the latter following from the *Witt identity*

$$(x, y^{-1}, z)^y (y, z^{-1}, x)^z (z, x^{-1}, y)^x = 1$$

in G. Furthermore, the distributive laws hold. Thus L is a *Lie ring*.

Now suppose that G is a group of exponent p, so that $x^p = 1$ for every $x \in G$. Then each L_i is an elementary abelian p-group, and L is a Lie algebra over $\mathbb{Z}_p$. The crucial point to observe is the result of Higman [91] that L *satisfies the* $(p-1)$*-th Engel condition*

$$[x, {}_{p-1}y] = 0.$$

(For proofs of these facts, see Hall [282] p. 329, Higman [91], Wiegold [243].)

Suppose now that we know Kostrikin's theorem over the field $\mathbb{Z}_p$ when $n = p-1$. Let m be any integer, and let $B_{m,p} = G$ be the m-generator relatively free group for the identity $x^p = 1$. Define L as above. Then it may be shown that L is finitely generated. By Kostrikin's theorem, L is nilpotent, so $L^t = 0$ for some integer t. It then follows that $\gamma_t(G) = \gamma_{t+1}(G)$. The order g of $G/\gamma_t(G)$ is finite, and depends only m on and p. But every finite m-generator group of exponent p is a homomorphic image of $G/\gamma_s(G)$ for some s, and hence of $G/\gamma_t(G)$: therefore the order of such a finite group is bounded by a function of m, p.

At the present time this is the only known method by which this group-theoretic result may be proved. Kostrikin's theorem therefore represents an important application of Lie algebras to the study of finite groups.

2. Basic computational results

Throughout most of this chapter we shall be working with algebras having trivial centre, which are therefore isomorphic to their algebras of inner derivations. The inner derivations generate an associative algebra, and the Engel condition

$$[x, {}_n y] = 0$$

implies that

$$y^{*n} = 0$$

where y^* is the inner derivation corresponding to y. To begin with, we study the effect of linearising this relation.

Consider the free associative algebra over a field $\mathfrak{k}$ on two generators u, v. Fix n, and suppose that $\mathfrak{k}$ has characteristic 0 or $p > n$.

Define $\begin{Bmatrix} u & v \\ i & n-i \end{Bmatrix}$ by:

$$(u+v)^n = \sum_{i=0}^{n} \begin{Bmatrix} u & v \\ i & n-i \end{Bmatrix} \tag{1}$$

where $\begin{Bmatrix} u & v \\ i & n-i \end{Bmatrix}$ has degree i in u. Thus

$$\begin{Bmatrix} u & v \\ 0 & n \end{Bmatrix} = v^n$$

$$\begin{Bmatrix} u & v \\ 1 & n-1 \end{Bmatrix} = \sum_{i=1}^{n} v^i uv^{n-1-i}.$$

Since

$$(u+v)^{n+1} = (u+v)^n(u+v)$$

we have

$$\begin{Bmatrix} u & v \\ i & n-i+1 \end{Bmatrix} = \begin{Bmatrix} u & v \\ i-1 & n-i+1 \end{Bmatrix} u + \begin{Bmatrix} u & v \\ i & n-i \end{Bmatrix} v \tag{2}$$

whence induction on n yields

$$\begin{Bmatrix} u & v \\ 1 & n-1 \end{Bmatrix} = \sum_{j=1}^{n-1} jv^{j-1}(uv-vu)v^{n-j-1} + nv^{n-1}u. \tag{3}$$

The usual linearisation process (as in chapter 14) yields:

LEMMA 2.1 *For each k with $0 \leqq k \leqq n$, the element $\begin{Bmatrix} u & v \\ k & n-k \end{Bmatrix}$ is a linear combination of nth powers of elements $\alpha u+\beta v$, where $\alpha, \beta \in \mathfrak{k}$.* □

From this we may prove a lemma allowing us to express products of elements of an associative algebra as sums of powers of Lie elements.

LEMMA 2.2 *Let $u_1, \ldots, u_s$ be elements of an associative $\mathfrak{k}$-algebra. Then $u_1 \ldots u_s$ may be expressed as a linear combination of rth powers ($r = 1, 2, \ldots, s$) of elements in the Lie algebra generated by $u_1, \ldots, u_s$.*

Proof: We use induction on s, the case $s = 1$ being trivial. So we may assume that

$$u_1 \ldots u_{s-1} = \sum v^{s-1} + \sum b^{s-2} + \ldots + \sum c, \tag{4}$$

where $v, b, \ldots, c$ are in the Lie algebra generated by $u_1, \ldots, u_{s-1}$, and $\sum$ represents $\mathfrak{k}$-linear combinations. If we let (*) denote a linear combination of powers of the desired kind, then multiplying (4) by $u = u_s$ and using induction gives

$$u_1 \ldots u_s = \sum v^{n-1} + (*). \tag{5}$$

Now (3) implies that

$$v^{n-1}u = \frac{1}{n}\begin{Bmatrix} u & v \\ 1 & n-1 \end{Bmatrix} - \frac{1}{n}\sum_{j=1}^{n-1} jv^{j-1}(vu-uv)v^{n-j-1}.$$

Each term in the summation is a product of $(j-1)+1+(n-j-1) = n-1$ Lie elements, so by induction

$$v^{n-1}u = \frac{1}{n}\begin{Bmatrix} u & v \\ 1 & n-1 \end{Bmatrix} + (*).$$

But by lemma 2.1, $\begin{Bmatrix} u & v \\ 1 & n-1 \end{Bmatrix}$ is also of type (*). By (4) the lemma follows. □

Next, a simple result of which we make heavy use.

LEMMA 2.3 *If L is a Lie algebra, $x, y \in L$, and $r \geqq 1$ is an integer, then*

$$(xy^{*r})^* = \sum_{i=0}^{r} (-1)^i \binom{r}{i} y^{*i} x^* y^{*r-i}.$$

Proof: Use induction on r. □

Finally we have a theorem of Kostrikin [113] relating associative and Lie nilpotence in an associative algebra.

THEOREM 2.4 *Let A be an associative algebra over a field $\mathfrak{k}$. Suppose that there exists $n \in \mathbb{N}$ such that $x^n = 0$ for all $x \in A$, with $n <$ char($\mathfrak{k}$). If $[A]$ is nilpotent as a Lie algebra then A is nilpotent as an associative algebra.*

Proof: We use induction on the nilpotency class c of $[A]$.

If $c = 1$ then A is commutative. Linearising the identity

$$x^n = 0$$

as in chapter 14 section 2 gives as complete linearisation

$$(n!)\, x_1 x_2 \ldots x_n = 0$$

since A is commutative. Now $n <$ char($\mathfrak{k}$), so A satisfies this complete linearisation, and $A^n = 0$.

Now we assume the theorem true for $c-1$ and prove it for c. Let $Z = \zeta_1([A])$; let B be the associative subalgebra of A generated by Z; let I be the associative ideal of A generated by Z. Then B is commutative, so by the case $c = 1$ above, $B^n = 0$. Now B is generated by elements which commute with A, so B is central in A. Therefore I is spanned by elements bx where $b \in B$, $x \in A$. But

$$(\sum b_1 x_1)(\sum b_2 x_2)\dots(\sum b_n x_n) = \sum (b_1\dots b_n)(x_1\dots x_n)$$
$$= 0$$

by centrality of B and the fact that $B^n = 0$. Therefore $I^n = 0$. Now $[A/I] \cong [A]/[I]$ which is a quotient of $[A]/\zeta_1([A]) \in \mathfrak{N}_{c-1}$. By induction,

$$A^t \subseteq I$$

for some $t \in \mathbb{N}$, whence

$$A^{nt} = 0. \qquad \square$$

Clearly the proof shows that $A^{n^c} = 0$. This is best possible for $c = 1$, but probably can be improved for larger c.

3. The existence of an element of order 2

An element x of a Lie algebra L will be said to have *order* t if $x^{*t} = 0$, or equivalently, $[L, {}_t x] = 0$. If L satisfies the nth Engel condition then every element of L has order n. Kostrikin's theorem 2 asserts that much more is true:

THEOREM 3.1 *Let $L \neq 0$ be a Lie algebra over $\mathfrak{k}$ satisfying the nth Engel condition, where either*

(a) *$\mathfrak{k}$ has characteristic 0,*

(b) *$\mathfrak{k}$ has characteristic $p > n$.*

Then L has an element $c \neq 0$ of order 2.

The proof of this will occupy the rest of this section, and (with the author's permission) is taken largely without alteration from Wiegold's notes [243].

If L has non-zero centre we may take c to be central. So we may assume $\zeta_1(L) = 0$. Then we may identify each $x \in L$ with the inner derivation x^*, and embed L in the associative algebra generated by all the x^* ($x \in L$). In this algebra we have

$$[x, y] = xy - yx$$

for all $x, y \in L$. If $u_1, \ldots, u_s$ are elements of L, then the equation

$$u_1 \ldots u_s = 0$$

means the same as

$$[L, u_1, \ldots, u_s] = 0.$$

Hence if A and B are strings of elements of L, it follows that

$$A u_1 \ldots u_s B = 0.$$

Likewise, if $A_1, \ldots, A_k$ are strings of elements of L, all equal to 0, then

$$A_1 + \ldots + A_k = 0.$$

We write

$$[xy^r] = [x, {}_r y]$$

and omit commas whenever no confusion can arise. Thus, for example,

$$[u[bu]^m b^2] = [u, {}_m[b, u], b, b].$$

The plan of the proof is as follows. First we find a non-zero element b of order 3. For arbitrary $u \in L$ we put

$$g_m = g_m(u) = [u[bu]^m b^2]$$

and show that

$$g_0^2 = [ub^2]^2 = b^2 u^2 b^2, \; g_m^2 = b^2(u^2 b^2)^{m+1}.$$

Now

$$g_{n-1}(u) = 0$$

so that

$$b^2(u^2 b^2)^t = 0$$

for some fixed t, $0 \leqq t \leqq n$. If $t = 0$ then $b^2 = 0$. If $t = 1$, then $[ub^2]^2 = 0$ for all u. If $b^2 \neq 0$ we can choose u to make $[ub^2] \neq 0$. Finally if $t > 1$ we show that there is an element b_0 (actually equal to a certain $g_m(f)$) of order at most 3, with $b_0 \neq 0$, such that

$$b_0^2(u^2 b_0^2)^s = 0 \qquad s = [t/2]$$

and hence by an inductive process we find b_1 such that

$$b_1 \neq 0,\ b_1^3 = 0,\ b_1^2 u b_1^2 = 0$$

for all u. Then either $b_1^2 = 0$ or we can find u such that $0 \neq [ub_1^2]$ has order 2.

We now embark upon the proof.

STEP I: *Finding an element of order 3*

Suppose $v^m = 0$, $4 \leqq m \leqq n$. We show that

$$[uv^{m-1}]^{m-1} = 0$$

for all $u \in L$. An inductive process (which we leave till the end) yields an element of order 3.

By lemma 2.3,

$$[uv^{m-1}] = \sum_{j=0}^{m-1} \alpha_j v^j uv^{m-1-j}$$

for suitable $\alpha_j \in \mathfrak{k}$. Therefore

$$[uv^{m-1}]^{m-1} = \sum \alpha_{j_1} \ldots \alpha_{j_{m-1}} (v^{j_1} uv^{m-1-j_1}) \ldots (v^{j_{m-1}} uv^{m-1-j_{m-1}})$$

where the summation is over all j_r with $0 \leqq j_r \leqq m-1$.

I(a): Any term such that $m-1-j_r+j_{r+1} = 0$ *for some* r *is* 0.

In such a case, $j_r = m-1$ and $j_{r+1} = 0$ so the term is of the form

$$\ldots(v^{m-1}u)(uv^{m-1})\ldots$$

and since $m-1 \geqq 3$ at least one '...' must be non-empty: hence the term must be premultiplied by $v^{m-1}u$ or postmultiplied by uv^{m-1}; for otherwise it is zero since $v^m = 0$. Therefore the term is

$$\ldots v^{m-1} uv^{m-1} \ldots .$$

But

$$0 = v^{m-2}[uv^m] = v^{m-2} \sum_{i=0}^{m} (-1)^i \binom{m}{i} v^i uv^{m-i}$$

$$= -mv^{m-1}uv^{m-1}.$$

But $m \leqq n < p$, so $v^{m-1}uv^{m-1} = 0$ and I(a) is proved.

I(b): $[uv^{m-1}]^{m-1}$ *is a scalar multiple of* $v^{m-1}(uv^{m-2})^{m-1}$.

From I(a), $[uv^{m-1}]^{m-1}$ is a linear combination of terms like

$$v^{i_1}uv^{i_2}u \ldots uv^{i_{m-1}}uv^{i_m} \tag{6}$$

where

$$i_1+\ldots+i_m = (m-1)^2$$

and

$$i_2 \neq 0,\ i_3 \neq 0,\ \ldots,\ i_{m-1} \neq 0,$$
$$i_r \leqq m-1 \text{ for all } r.$$

Now since $v^m = 0$ we have

$$0 = [uv^m] = \textstyle\sum_{i=1}^{m-1} (-1)^i \binom{m}{i} v^i uv^{m-i}$$

so that vuv^{m-1} is a linear combination of terms

$$v^2uv^{m-2}, \ldots, v^{m-1}uv.$$

Hence we can express (6) as a linear combination of similar terms, but where $i_m \leqq m-2$. For if $i_m = m-1$ the term becomes

$$v^{i_1}uv^{i_2}u\ldots uv^{i_{m-1}-1}$$

plus a linear combination of $v^2uv^{m-2}, \ldots, v^{m-1}uv$.

Therefore $[uv^{m-1}]^{m-1}$ is a linear combination of terms like (6) except that $i_m \leqq m-2$, where the total degree in v remains unchanged. Apply this process to the next power of v to the left, and delete anything involving v^m: this expresses (6) as a linear combination of similar terms, but with

$$i_m \leqq m-2,\ i_{m-1} \leqq m-2.$$

Continuing this process we arrive at a linear combination of terms (6) with

$$i_m \leqq m-2,\ i_{m-1} \leqq m-2,\ \ldots,\ i_2 \leqq m-2,\ i_1 \leqq m-1.$$

None of these inequalities is strict, or else

$$i_1+\ldots+i_m < (m-1)+(m-1)(m-2) = (m-1)^2,$$

a contradiction. Therefore $[uv^{m-1}]^{m-1}$ is a scalar multiple of $v^{m-1}(uv^{m-2})^{m-1}$ as claimed.

I(c): $v^{m-1}(uv^{m-2})^{m-1} = 0$.

We show that some segment of $v^{m-1}(uv^{m-2})^{m-1}$ is zero. Certainly we have

$$0 = v^{m-3}[uv^m] = v^{m-3} \textstyle\sum_{i=1}^{m-1} (-1)^i \binom{m}{i} v^i uv^{m-i}$$
$$= \binom{m}{2} v^{m-1}uv^{m-2} - mv^{m-2}uv^{m-1}.$$

Also,

$$0 = (-1)^{m-1}[uv^m]v^{m-3} = (-1)^{m-1}\sum_{i=1}^{m-1}(-1)^i\binom{m}{i}v^iuv^{m-i}v^{m-3}$$

$$= mv^{m-1}uv^{m-2} - \binom{m}{2}v^{m-2}uv^{m-1}.$$

Now the determinant

$$\begin{vmatrix} \binom{m}{2} & -m \\ m & -\binom{m}{2} \end{vmatrix} = \frac{m^2}{4}(m-3)(m+1)$$

so if $m < n$ or if $m = n < p-1$ we have

$$v^{m-1}uv^{m-2} = 0$$

and the result follows. This leaves the case $m = n = p-1$. Now in I(a) we showed that $v^{m-1}uv^{m-1} = 0$ for all $u \in L$, so that

$$0 = v^{m-1}[uv^{m-3}u]v^{m-1}$$

$$= v^{m-1}[uv^{m-3}]uv^{m-1} - v^{m-1}u[uv^{m-3}]v^{m-1}$$

$$= v^{m-1}\sum_{i=0}^{m-3}(-1)^i\binom{m-3}{i}v^iuv^{m-3-i}uv^{m-1} -$$

$$- v^{m-1}u\sum_{i=0}^{m-3}(-1)^i\binom{m-3}{i}v^iuv^{m-3-i}v^{m-1}.$$

In the first sum, the only terms which might not be 0 is that with $i = 0$: in the second, with $i = m-3$. Now $m = p-1$ and the theorem is trivial for $p = 2$, so p may be assumed odd: therefore $(-1)^{m-3} = -1$ and we have

$$0 = 2v^{m-1}uv^{m-3}uv^{m-1}.$$

Since p is odd,

$$0 = v^{m-1}uv^{m-4}.vuv^{m-1}.$$

But $[uv^m] = 0$ so that

$$mvuv^{m-1} = \sum_{i=2}^{m-1}(-1)^i\binom{m}{i}v^iuv^{m-i}$$

and it follows that

$$0 = v^{m-1}uv^{m-4}\sum_{i=2}^{m-1}(-1)^i\binom{m}{i}v^iuv^{m-i}.$$

Since $v^m = 0 = v^{m-1}uv^{m-1}$, this gives

$$0 = \binom{m}{2} v^{m-1}uv^{m-2}uv^{m-2}$$

and then $v^{m-1}(uv^{m-2})^2 = 0$, and I(c) is proved.

Now we finish step I. We know that $[uv^{m-1}]^{m-1} = 0$. If all elements of L have order $\leqq 3$ we are finished. If not, take v of order m, with $4 \leqq m \leqq n$: certainly some such v exists since $v^n = 0$ for all $v \in L$. Either v has order $m-1$ or $[uv^{m-1}] \neq 0$ has order $m-1$ for suitable u. Inductively we find an element of order 3.

STEP II: *There is an element of order 2.*

If $p = 2$ or 3 then L is nilpotent by propositions 16.1.1 and 16.1.5, and we are finished. So *from now on we may assume* $p \geqq 5$.

We take an element $b \in L$ such that $b \neq 0$, $b^3 = 0$, which exists by step I. For the moment we develop properties of this b.

For any $u \in L$,

$$0 = [ub^3] = 3b^2ub - 3bub^2$$

so that as $p \geqq 5$,

$$bub^2 = b^2ub \tag{7}$$

so that

$$b^2ub^2 = 0 \tag{8}$$

on multiplying by b. Putting $[bu^3]$ for u and expanding these leads to

$$b^2ubu^2b^2 = b^2u^2bub^2 = bub^2u^2b^2 = b^2u^2b^2ub. \tag{9}$$

Next, using $b^3 = 0$ several times,

$$[ub^2]^2 = (ub^2 - 2bub + b^2u)(ub^2 - 2bub + b^2u)$$

and using (7) and (8) we get

$$[ub^2]^2 = b^2u^2b^2. \tag{10}$$

Again,

$$[u[bu]b^2]^2 = [[bu^2]b^2]^2 = b^2[bu^2]^2b^2$$

from (10), so expanding and using (7) and (9) we get

$$[u[bu]b^2]^2 = b^2u^2b^2u^2b^2. \tag{11}$$

The identities (10) and (11) are the first two steps of a general process. We define

$$g_m = g_m(u) = [u[bu]^m b^2]$$

for $m \geqq 0$.

II(a): If $m \geqq 2$ then $g_m = [g_{m-2}u^2b^2]$.

Firstly, $[bu]b^2 = bub^2 = b^2ub = -b^2[bu]$ by (7) so moving $m-2$ of the $[bu]$ to the right gives

$$g_m = (-1)^m[u[bu]^2b^2[bu]^{m-2}].$$

Now

$$\begin{aligned} g_2 = [u[bu]^2b^2] &= -[bu^2[bu]b^2] \\ &= -[bu^2bub^2] + [bu^3b^3] \\ &= -[bu^2bub^2] \\ &= [ububub^2]. \end{aligned}$$

Therefore

$$g_m = (-1)^m[ububub^2[bu]^{m-2}].$$

But $[ubub] = -[b[ub]u] = [ub^2u]$ so that

$$g_m = (-1)^m[ub^2u^2b^2[bu]^{m-2}]. \tag{12}$$

But now

$$b^2u^2b^2[bu] = -b^2u^2b^2ub$$

$$[bu]b^2u^2b^2 = bub^2u^2b^2$$

so that by (9)

$$b^2u^2b^2[bu] = -[bu]b^2u^2b^2. \tag{13}$$

So $b^2u^2b^2$ and $[bu]$ anticommute, and we get

$$\begin{aligned} g_m &= (-1)^m(-1)^{m-2}[u[bu]^{m-2}b^2u^2b^2] \\ &= [g_{m-2}u^2b^2]. \end{aligned}$$

II(b): $(g_m(u))^2 = b^2(u^2b^2)^{m+1}$.

This we already know for $m = 0, 1$ by (10) and (11), so we set up an inductive proof. Firstly,

$$\begin{aligned}[ub^2]^3 &= b^2u^2b^2[ub^2] \\ &= b^2u^2b^2(ub^2-2bub+b^2u) \\ &= 0\end{aligned}$$

since $b^3 = 0$ and (8) holds. So in (10) we may put $[ub^2]$ in place of b, and $[u[bu]^{m-2}]$ in place of u, which yields

$$[u[bu]^{m-2}[ub^2]^2]^2 = [ub^2]^2[u[bu]^{m-2}]^2[ub^2]^2.$$

By (13), (10), and II(a) the left-hand side of this is equal to g_m^2. Therefore

$$\begin{aligned}g_m^2 &= [ub^2]^2[u[bu]^{m-2}]^2[ub^2]^2 \\ &= b^2u^2b^2[u[bu]^{m-2}]^2b^2u^2b^2 \\ &= b^2u^2[u[bu]^{m-2}b^2]^2u^2b^2 \\ &= b^2u^2g_{m-2}^2u^2b^2\end{aligned}$$

and II(b) follows by induction.

II(c): Next observe that

$$g_{n-1}(u) = (-1)^{n-1}[ub^2[bu]^{n-1}]$$

since $[bu]$ and b^2 anticommute. Now

$$\begin{aligned}0 = [b[bu]^n] &= [b[bu][bu]^{n-1}] \\ &= [ub^2[bu]^{n-1}]\end{aligned}$$

so that

$$g_{n-1}(u) = 0.$$

By II(b), there is a fixed t with $0 \leqq t \leqq n$ such that

$$b^2(u^2b^2)^t = 0 \tag{14}$$

for all $u \in L$.

If $t = 0$ or 1 we have $b^2 = 0$ or $[ub^2]^2 = 0$ and the theorem follows as remarked just before step I. So we assume $t > 1$ and find $b_0 \neq 0$ such that $b_0^3 = 0$ and $b_0^2(u^2b_0^2)^s = 0$, where $s = [t/2]$. As remarked earlier, an inductive process yields an element of order 2.

It remains to construct b_0. We assume $b^2 \neq 0$ (or else we are finished). Then there must exist $f \in L$ and $m \geqq 0$ such that $g_{m+1}(f) = 0$, $g_m(f) \neq 0$. Define

$$a = [f[bf]^m].$$

Clearly $[ab^2]^3 = 0$ and $g_m(f) = [ab^2] \neq 0$. In fact $g_m(f)$ will be the desired b_0. However, the proof is as yet some distance away, and we need several other facts.

II(d): $[ab^2ab] = 0$.

First we show that $g_{m+2}(f) = 0$. By (12)

$$\begin{aligned} g_{m+2}(f) &= (-1)^{m+2}[fb^2f^2b^2[bf]^m] \\ &= -[g_{m+1}(f)[bf]] \\ &= 0. \end{aligned}$$

Next,

$$[ab^2ab] = [f[bf]^m b^2(\textstyle\sum_{i=0}^m [bf]^i f[bf]^{m-i})b].$$

Terms with $i > 0$ give

$$\begin{aligned} [f[bf]^m b^2[bf]\ldots] &= -[f[bf]^{m+1}b^2\ldots] \\ &= -[g_{m+1}(f)\ldots] \\ &= 0 \end{aligned}$$

so that

$$[ab^2ab] = [f[bf]^m b^2 f[bf]^m b].$$

Therefore

$$[ab^2ab] = [f[bf]^m b^2 f(bf-fb)[bf]^{m-1}b].$$

Now

$$f(bf-fb) = fbf - f^2b = (fb-bf)f + bf^2 - f^2b.$$

Since $b^3 = 0$,

$$[ab^2ab] = [f[bf]^m b^2[fb]f\ldots] - [f[bf]^m b^2 f^2 b[bf]^{m-1}b].$$

The first term starts with $[g_{m+1}(f)\ldots]$ so is 0. Hence

$$\begin{aligned} [ab^2ab] &= -[f[bf]^m b^2 f^2 b[bf]^{m-1}b] \\ &= -[f[bf]^m b^2 f^2 b^2 f[bf]^{m-2}b] + [f[bf]^m b^2 f^2 bfb[bf]^{m-2}b]. \end{aligned}$$

But

$$[f[bf]^m b^2 f^2 b^2] = [g_m(f) f^2 b^2] = g_{m+2}(f) = 0.$$

Breaking up the second term yields

$$[ab^2ab] = [f[bf]^m b^2 f^2 bfb^2 f[bf]^{m-3} b] - [f[bf]^m b^2 f^2 bfbfb[bf]^{m-3} b].$$

Now $bfb^2 = b^2fb$ by (7) so the first terms starts with $g_{m+2}(f)$ and is 0. Continuing the process of breaking $[bf]$ into $bf - fb$ and deleting zero terms leaves just a term

$$\pm [f[bf]^m b^2 f^2 bfbfb \ldots bfb^2].$$

Now $b^2fb = bfb^2$, so we slide the b^2 to the front: the term now starts with $g_{m+2}(f)$ so is 0. Thus $[ab^2ab] = 0$.

II(e): We continue to accumulate identities. Let

$$w = [ab^2ab] = 0.$$

Expanding we get

$$0 = w = b^2a^2b + ba^2b^2 + 2abab^2 + 2b^2aba - 4babab.$$

Multiplying on the right by b gives

$$b^2a^2b^2 = 2b^2abab. \tag{15}$$

Also,

$$0 = wab^2 = b^2a^2bab^2 + 2b^2aba^2b^2 - 4bababab^2.$$

But by (7)

$$2bababab^2 = 2b^2ababab = b^2a^2b^2ab$$

and by (9)

$$b^2a^2b^2ab = 0.$$

Now let h be any element of L. Since $b^2hb^2 = 0$ by (8), and using (7), (15)

$$0 = whb^2 = 2bab^2ahb^2 - b^2a^2b^2hb.$$

Finally,

$$0 = bawhb^2 = b^2a^2b^2ahb^2 - bab^2a^2b^2hb.$$

But $bab^2a^2b^2 = 0$ by (8) and we get

$$b^2a^2b^2ahb^2 = 0.$$

But

$$b^2ahb^2 = b^2hab^2 + b^2[ah]b^2 = bhab^2$$

so that

$$b^2a^2b^2ahb^2 = b^2a^2b^2hab^2 = 0 \tag{16}$$

for all $h \in L$.

A similar argument gives

$$b^2hab^2a^2b^2 = b^2ahb^2a^2b^2 = 0 \tag{17}$$

for all $h \in L$.

II(f): Now define

$$b_0 = g_m(f) = [ab^2] \neq 0.$$

We know that $b_0^3 = 0$. We must consider

$$H = b_0^2(h^2b_0^2)^s$$

where $s = [t/2]$, for arbitrary $h \in L$: we want $H = 0$ and then the theorem will follow inductively.

From (10)

$$b_0^2 = [ab^2]^2 = b^2a^2b^2$$

so that

$$H = b^2a^2b^2(h^2b^2a^2b^2)^s.$$

Define

$$\delta = 2s+1-t = \begin{cases} 0 & (t \text{ odd}) \\ 1 & (t \text{ even}) \end{cases}.$$

Take any scalar λ. By (14), with $u = h+\lambda a$,

$$(b^2a^2)^\delta b^2\{h^2b^2 + \lambda(ah+ha)b^2 + \lambda^2a^2b^2\}^{2s+1-\delta} = 0. \tag{18}$$

When the left hand side is expanded, the coefficient of λ^k has degree $2\delta+k$ in a and $2(2s+1-\delta)$ in h: therefore

$$H_{2\delta} + \lambda H_{2\delta+1} + \ldots + \lambda^{2(s+1-\delta)}H_{2s+2} + \ldots = 0,$$

where H_k is homogeneous of degree k in a and $4s+2-k$ in h. Note that $H_{2\delta} = 0$ by (14). We show that $H_{2s+2} = H$ and that $H_k = 0$ if $k > 2s+2$.

II(g): First we do the case $\delta = 0$. Consider

$$b^2\{h^2b^2 + \lambda(ah+ha)b^2 + \lambda^2a^2b^2\}^{2s+1}.$$

We consider the terms of degree $2s$ in h. Clearly in any such term, S say, there are $2k$ occurrences of $(ah+ha)b^2$, $s-k$ of h^2b^2, and $2s+1-(s-k) = s-k+1$ of a^2b^2. Now $S = 0$ if any a^2b^2 is preceded by a^2b^2 or $(ah+ha)b^2$, by (17). So any non-zero contribution to H_{2s+2} has the form

$$S = b^2a^2b^2\ldots h^2b^2a^2b^2\ldots h^2b^2a^2b^2\ldots h^2b^2a^2b^2\ldots$$

where '...' represents a string of terms $(ah+ha)b^2$. But in fact no such strings occur, since $b^2a^2b^2(ah+ha)b^2 = 0$ by (16). The only possibility is $k = 0$, and then

$$H_{2s+2} = b^2a^2b^2(h^2b^2a^2b^2)^s = H.$$

The case $\delta = 1$ goes much the same way, and it is not hard to see that $H_k = 0$ if $k > 2s+2$. So we have the system of equations

$$H_{2\delta+1} + \lambda H_{2\delta+2} + \ldots + \lambda^{2(s-\delta)}H_{2s+1} + \lambda^{2(s-\delta)+1}H = 0$$

for any λ. The determinant of this system is Vandermonde, and since

$$2s+2-2\delta-1 = 2s+1-2\delta = t-2\delta \leqq t \leqq n < p$$

it may be made non-zero by suitable choice of the λ's. Thus

$$H = b_0^2(h^2b_0^2)^s = 0.$$

The theorem now follows as in the introduction. □

4. Elements which generate abelian ideals

Suppose that L satisfies the nth Engel condition, with the usual restrictions on the characteristic. We now develop a method for constructing abelian ideals in L, which is needed both in our proof of the weak theorem and in the proof of the full theorem of Kostrikin.

LEMMA 4.1 *Let $c \in L$. Then the ideal C generated by c is abelian if and only if*

$$c^*x^{*r}c^* = 0 \tag{19}$$

for all $x \in L$ and all $r = 0, 1, \ldots, n-1$.

Proof: If C is abelian then $c \in \zeta_1(C)$. Conversely, if $c \in \zeta_1(C)$ then $c \in \zeta_1(C) \text{ ch } C \lhd L$ so that $C = \zeta_1(C)$ and is abelian. To prove C abelian it is therefore sufficient to prove that

$$c^* y_1^* \ldots y_s^* c^* = 0$$

for all $y_1, \ldots, y_s \in L$. Now by lemma 2.2 $y_1^* \ldots y_s^*$ may be expressed as a sum of rth powers of elements z^*, for $z \in L$. Hence

$$c^* y_1^* \ldots y_s^* c^* = \sum_z c^* z^{*r} c^*.$$

For $r = 0, \ldots, n-1$ the terms $c^* z^{*r} c^*$ vanish by hypothesis: for $r \geqq n$ they vanish by the Engel condition.

Conversely, if C is abelian, then $c^* x^{*r} c^* = 0$ for all $r = 0, \ldots, n-1$. □

We seek to weaken the hypothesis of this lemma. For any $m = 1, 2, \ldots, \left[\frac{n-1}{2}\right]$ we write

$$c_{(m)}$$

for any element with the property

$$c_{(m)}^* x^{*r} c_{(m)}^* = 0$$

for all $x \in L$, and for $r = 0, 1, \ldots, 2m-1$. The reason for considering such elements comes out of Kostrikin's inductive method of proof. The point is:

LEMMA 4.2 *If $c = c_{(m)}$ where $m = \left[\frac{n-1}{2}\right]$ then the ideal generated by c is abelian.*

Proof: By definition, $c^* x^{*r} c^* = 0$ for $0 \leqq r \leqq n-3$, since $2\left[\frac{n-1}{2}\right] - 1 \geqq$ $\geqq n-3$. We have to prove the same thing for $r = n-2, n-1$. The condition for $r \leqq n-3$ implies that

$$\begin{Bmatrix} c^* & x^* \\ 2 & n-2 \end{Bmatrix} = c^* x^{*n-2} c^*$$

and the left hand side is zero: which deals with $r = n-2$. Finally,

$$c^* x^{*n-1} = \begin{Bmatrix} c^* & x^* \\ 1 & n-1 \end{Bmatrix} - \sum_{i=1}^{n-1} x^{*i} c^* x^{*n-i-1}$$

so that

$$c^*x^{*n-1}c^* = -\sum_{i=1}^{n-1} x^{*i}(c^*x^{*n-i-1}c^*) = 0.$$

The lemma now follows by lemma 4.1. □

In fact we can get away with even less: all we need prove is that

$$c^*_{(m)}x^{*2s}c^*_{(m)} = 0 \tag{20}$$

for all $x \in L$ and for $s = 0, 1, \ldots, m-1$. To see this, suppose (20) holds. We prove (19) by induction on r. For r even this is obvious, so we take $r = 2k+1$. Now

$$\begin{aligned}
0 &= (xc^*_{(m)}x^{*2k}c^*_{(m)})^* \\
&= (xc^*_{(m)}x^{*2k})^*c^*_{(m)} - c^*_{(m)}(xc^*_{(m)}x^{*2k})^* \\
&= \sum_{i=0}^{2k}(-1)^i\binom{2k}{i}x^{*i}(xc^*_{(m)})^*x^{*2k-i}c^*_{(m)} - \\
&\quad - \sum_{i=0}^{2k}(-1)^i\binom{2k}{i}c^*_{(m)}x^{*i}(xc^*_{(m)})^*x^{*2k-i} \\
&= \sum_{i=0}^{2k}(-1)^i\binom{2k}{i}x^{*i}(x^*c^*_{(m)} - c^*_{(m)}x^*)x^{*2k-i}c^*_{(m)} - \\
&\quad - \sum_{i=0}^{2k}(-1)^i\binom{2k}{i}c^*_{(m)}x^{*i}(x^*c^*_{(m)} - c^*_{(m)}x^*)x^{*2k-i} \\
&= -2c^*_{(m)}x^{*2k+1}c^*_{(m)}
\end{aligned}$$

since by hypothesis the other terms go out. Since $p \geqq 5$ the induction step holds.

Theorem 3.1 says that there exists a non-zero $c_{(1)}$. So we already have a proof of:

THEOREM 4.3 *The full Kostrikin theorem is true for $p = 5$, and the restricted Burnside problem has an affirmative solution for $p = 5$.*

Proof: There exists $c_{(1)}$. But if $p = 5$ then $n \leqq 4$, and $\left[\frac{n-1}{2}\right] = \left[\frac{3}{2}\right] = 1$. So $c_{(1)}$ generates an abelian ideal of L. Now we argue as follows: if L satisfies the 4th Engel condition then $L/\rho(L)$ has trivial Hirsch-Plotkin radical, by corollary 16.3.11. But $L/\rho(L)$ has trivial centre by the same token: hence theorem 3.1 implies the existence of $c_{(1)} \neq 0$. But $c_{(1)}$ generates an abelian ideal, which is a contradiction unless $L = \rho(L)$, which implies L locally nilpotent. □

This device of throwing away $\rho(L)$ is of major importance in the proof of Kostrikin's theorem: it allows us to assume $\rho(L) = 0$ if we wish. *We may also now assume that* $p \geqq 7$. The next step is to prove:

THEOREM 4.4 *Let L satisfy the nth Engel condition with the usual restrictions on characteristic. If* $7 \leqq 2m+3 < p$, *and there exists in L an element* $c_{(m)} \neq 0$, *then there exists an element* $c_{(m+1)} \neq 0$.

The proof of this will take the rest of this section. It enables us to find $c_{(m)} \neq 0$ for $m = \left[\frac{n-1}{2}\right]$ provided we can find $c_{(2)} \neq 0$. This yields an abelian ideal, which combines with the argument involving the Hirsch-Plotkin radical to give a contradiction. However, the proof of the existence of $c_{(2)}$ is exceptionally complicated, and we shall make do with a special case which gives the weaker theorem. But we still need theorem 4.4.

Obviously we may assume $\zeta_1(L) = 0$, since otherwise we take $c_{(m+1)} \neq 0$ in this centre. So we may once more omit *'s and identify $x \in L$ with $x^* \in \mathrm{Der}(L)$. The idea of the proof is straightforward: if $c_{(m)}$ is not already a $c_{(m+1)}$ then there is an element $a \in L$ such that

$$c_0 = [c_{(m)}a^{2m+1}c_{(m)}]$$

is another $c_{(m)}$. If c_0 is not a $c_{(m+1)}$ there is an element $b \in L$ such that

$$[c_0 b^{2m+1} c_0]$$

is a $c_{(m+1)} \neq 0$.

For the proof, set $c = c_{(m)}$, so that

$$c \neq 0, \ cu^\alpha c = 0 \qquad (\alpha = 0, \ldots, 2m-1).$$

From lemma 4.1 we see that

$$cu_1 \ldots u_\alpha c = 0 \tag{21}$$

for $u_1, \ldots, u_\alpha \in L$ and $\alpha = 0, \ldots, 2m-1$. Now

$$\begin{aligned} cu_1 \ldots u_i u_j \ldots u_{2m} c &= cu_1 \ldots u_j u_i \ldots u_{2m} c + \\ &\quad + cu_1 \ldots [u_i u_j] \ldots u_{2m} c \\ &= cu_1 \ldots u_j u_i \ldots u_{2m} c \end{aligned}$$

by (21), so that

$$cu_1 \ldots u_{2m}c = cu_{\pi(1)} \ldots u_{\pi(2m)}c \tag{22}$$

for any permutation π of $\{1, \ldots, 2m\}$. Arguments of a similar type yield

$$cvu^{2m-1+\nu}c = \textstyle\sum_{i=0}^{\nu} \alpha_i cu^{2m-1+\nu-i}vu^i c \tag{23}$$

$$cu^{2m-1+\nu}vc = \textstyle\sum_{i=0}^{\nu} \beta_i cu^i vu^{2m-1+\nu-i}c \tag{24}$$

for $\nu \geqq 0$, where $u, v \in L$ and the coefficients $\alpha_i, \beta_i \in \mathfrak{k}$. It is irrelevant to our purposes what the exact values are.

Next we note the following identities, which depend on the fact that $m \geqq 2$.

$$\begin{aligned} A &= ca^{2m}ca^{2m-1}uc = 0 \\ B_1 &= ca^{2m}ca^{2m}uc = 0 \\ B_2 &= ca^{2m+1}ca^{2m-1}uc = 0 \\ \bar{A} &= cua^{2m-1}ca^{2m}c = 0 \\ \bar{B}_1 &= cua^{2m}ca^{2m}c = 0 \\ \bar{B}_2 &= cua^{2m-1}ca^{2m+1}c = 0. \end{aligned} \tag{25}$$

The proofs of these are as follows:

$$0 = c[ca^{2m+2}]a^{2m-3}uc = \binom{2m+2}{2}A,$$

$$0 = (-1)^{\delta}ca^{\delta}[ca^{2m+3}]a^{2m-3-\delta}uc = \binom{2m+3}{3+\delta}B_1 - \binom{2m+3}{2+\delta}B_2$$

(for $\delta = 0, 1$). The other three are similar.

LEMMA 4.5 *For all $a, u \in L$ we have*

$$[ca^{2m+1}c]u^{2m}ca^{2m}c = ca^{2m}cu^{2m}[ca^{2m+1}c] = 0.$$

Proof: Let

$$K_1 = [ca^{2m+1}]cu^{2m}ca^{2m}c,$$

$$K_2 = c[ca^{2m+1}]u^{2m}ca^{2m}c.$$

Then

$$[ca^{2m+1}c]u^{2m}ca^{2m}c = K_1 - K_2.$$

Using (23) we have

$$\begin{aligned} K_1 &= [ca^{2m+1}]c[cu^{2m}]a^{2m}c \\ &= \alpha_0[ca^{2m+1}]ca^{2m}[cu^{2m}]c + \alpha_1[ca^{2m+1}]ca^{2m-1}[cu^{2m}]ac \\ &= \alpha_0[ca^{2m+1}]ca^{2m}cu^{2m}c - 2m\alpha_1[ca^{2m+1}]ca^{2m-1}ucu^{2m-1}ac. \end{aligned}$$

But the initial segments

$$[ca^{2m+1}]ca^{2m}c, \quad [ca^{2m+1}]ca^{2m-1}uc$$

can be written as linear combinations of A and B_2, so are zero, whence

$$K_1 = 0.$$

Similarly,

$$K_2 = \alpha_0 cu^{2m}[ca^{2m+1}]ca^{2m}c + \alpha_1 cu^{2m-1}[ca^{2m+1}]uca^{2m}c$$

which is also zero.

The other identity is proved symmetrically. □

LEMMA 4.6 *Let* $c_0 = [ca^{2m+1}c]$, *and let* $u, v \in L$. *Then*

$$T = c_0 u^{2m} c_0 v^{2m} c_0 = 0.$$

Proof: Clearly

$$T = 2c_0 u^{2m} ca^{2m+1} cv^{2m} c_0$$

since

$$c_0 = 2ca^{2m+1}c - (2m+1)ca^{2m}ca - (2m+1)aca^{2m}c$$

and each of

$$(c_0 u^{2m} ca^{2m} c)av^{2m} c_0, \quad c_0 u^{2m} a(ca^{2m} cv^{2m} c_0)$$

is zero.

Now

$$S = c_0 u^{2m} ca^{2m+1} c = -(2m+1)aS_1 + S_2 - S_3$$

where

$$\begin{aligned} S_1 &= ca^{2m} cu^{2m} ca^{2m+1} c \\ S_2 &= ca^{2m+1} cu^{2m} ca^{2m+1} c \\ S_3 &= c[ca^{2m+1}]u^{2m} ca^{2m+1} c. \end{aligned}$$

From (24) and (25)

$$\begin{aligned}
S_1 &= ca^{2m}[cu^{2m}]ca^{2m+1}c \\
&= \beta_0 c[cu^{2m}]a^{2m}ca^{2m+1}c + \beta_1 ca[cu^{2m}]a^{2m-1}ca^{2m+1}c \\
&= \beta_0 cu^{2m}(ca^{2m}ca^{2m+1}c) - 2m\beta_1 cu^{2m-1}a(cua^{2m-1}ca^{2m+1}c) \\
&= 0.
\end{aligned}$$

Unfortunately S_2 and S_3 may not be zero. We introduce monomials

$$\begin{aligned}
R_1 &= cau^{2m}ca^{2m}ca^{2m+1}c \\
R_2 &= ca^2u^{2m-1}cua^{2m-1}ca^{2m+1}c \\
R_3 &= cu^{2m}ca^{2m+1}ca^{2m+1}c \\
R_4 &= cu^{2m-1}acua^{2m}ca^{2m+1}c \\
R_5 &= cu^{2m-2}a^2cu^2a^{2m-1}ca^{2m+1}c,
\end{aligned}$$

and

$$R = cu^{2m-2}a^3cu^2a^{2m-2}ca^{2m+1}c.$$

Then

$$\begin{aligned}
S_2 = {} & \beta_0 c[cu^{2m}]a^{2m+1}ca^{2m+1}c + \beta_1 ca[cu^{2m}]a^{2m}ca^{2m+1}c + \\
& + \beta_2 ca^2[cu^{2m}]a^{2m-1}ca^{2m+1}c,
\end{aligned}$$

which equals

$$\beta_0 R_3 + \beta_1(\alpha R_4 + R_1) + \beta_2(\beta R_5 + \gamma R_2)$$

for suitable $\alpha, \beta, \gamma \in \mathfrak{k}$.

Now we remark that $R_1 = R_2 = 0$ by (25), so that S_2 is a linear combination of R_3, R_4, R_5.

Further,

$$\begin{aligned}
0 &= cu^{2m-2}[cu^2a^{2m+1}]ca^{2m+1}c \\
&= cu^{2m-2}[cu^2 - 2ucu + u^2c,\ a^{2m+1}]ca^{2m+1} \\
&= X - 2Y + Z,
\end{aligned} \tag{26}$$

say. We have

$$\begin{aligned}
X &= cu^{2m-2}[cu^2,\ a^{2m+1}]ca^{2m+1}c \\
&= cu^{2m-2}\textstyle\sum_{i=0}^{2m+1}(-1)^i\dbinom{2m+1}{i}a^icu^2a^{2m+1-i}ca^{2m+1}c.
\end{aligned}$$

The only summands that are not known to be zero are where $i \geqq 2$ and $i \leqq 3$, and we get

$$X = \binom{2m+1}{2} R_5 - \binom{2m+1}{3} R.$$

Similarly, using (25),

$$Y = -(2m+1)R_4$$

$$Z = R_3.$$

Hence by (26) R_5 is a linear combination of R_3, R_4, R. Next,

$$0 = cu^{2m-2}a^3cu^2a^{2m-4}[ca^{2m+3}]c$$

$$= cu^{2m-2}a^3cu^2a^{2m-4}.\sum_{i=0}^{2m+3}\binom{2m+3}{i}(-1)^i a^i ca^{2m+3-i}c.$$

Only the terms with $i = 2, 3$ count: they yield

$$R = \alpha cu^{2m-2}a^3cu^2a^{2m-1}ca^{2m}c,$$

and the important thing for us here is that R ends with the string $\ldots ca^{2m}c$. We show that the same is true of R_3 and R_4 provided that $p > 2m+5$ (and leave the case $p = 2m+5$ until later).

For any $x \in L$,

$$0 = c\ldots cxa^{2m-3}[ca^{2m+4}]c$$

$$= c\ldots cxa^{2m-3}.\sum_{i=0}^{2m+4}(-1)^i a^i ca^{2m+4-i}c$$

$$= \binom{2m+4}{2} c\ldots cxa^{2m-1}ca^{2m+2}c - \binom{2m+4}{3} c\ldots cxa^{2m}ca^{2m+1}c +$$

$$+ \binom{2m+4}{4} c\ldots cxa^{2m+1}ca^{2m}c. \tag{27}$$

Further,

$$0 = c\ldots cxa^{2m-4}[ca^{2m+4}]ac$$

$$= c\ldots cxa^{2m-4}.\sum_{i=0}^{2m+4}(-1)^i\binom{2m+4}{i} a^i ca^{2m+4-i}ac$$

$$= -\binom{2m+4}{3} c\ldots cxa^{2m-1}ca^{2m+2}c + \binom{2m+4}{4} c\ldots cxa^{2m}ca^{2m+1}c -$$

$$- \binom{2m+4}{5} c\ldots cxa^{2m+1}ca^{2m}c. \tag{28}$$

Multiply (27) by $\binom{2m+4}{3}$ and (28) by $\binom{2m+4}{2}$ and add, which gives

$$\alpha c\ldots cxa^{2m}ca^{2m+1}c = \beta c\ldots cxa^{2m+1}ca^{2m}c \tag{29}$$

where

$$\alpha = \binom{2m+4}{3}^2 - \binom{2m+4}{4}\binom{2m+4}{2}$$

$$= \frac{(2m+4)^2(2m+3)^2(2m+2)(2m+5)}{144}$$

and $\beta \in \mathfrak{k}$. Since $p > 2m+5$, $\alpha \neq 0$. Putting $x = u$ and $x = a$ in (29) allows us to express R_4 and R_3 as monomials ending in $\ldots ca^{2m}c$.

Similarly, starting with the expression

$$S_3 = \alpha_0 cu^{2m}[ca^{2m+1}]ca^{2m+1}c + \alpha_1 cu^{2m-1}[ca^{2m+1}]uca^{2m+1}c$$

we express S_3 as a linear combination of R_3 and R_4. Hence for $p > 2m+5$ the monomial S is a sum of terms $c\ldots c\ldots ca^{2m}c$, and so T is a sum of terms

$$c\ldots c\ldots ca^{2m}cv^{2m}c_0$$

which are 0 by lemma 4.5. Hence $T = 0$ in this case.

To finish the proof we must deal with the case $p = 2m+5$. The above work shows that S is a linear combination of R, R_3, R_4. Therefore T is a linear combination of $Rv^{2m}c_0$ (which is 0 and may be deleted), $R_3v^{2m}c_0$, and $R_4v^{2m}c_0$. Both of these end in $\ldots U$ where

$$U = U(v) = ca^{2m+1}cv^{2m}[ca^{2m+1}c].$$

We show that $U = 0$ by a series of transformations similar to those above. We have

$$U = S_3' - (2m+1)S_1'a + S_2'$$

where

$$S_1' = ca^{2m+1}cv^{2m}ca^{2m}c$$

$$S_2' = ca^{2m+1}cv^{2m}ca^{2m+1}c$$

$$S_3' = ca^{2m+1}cv^{2m}[ca^{2m+1}]c.$$

Using (25) we find that

$$S_1' = 0.$$

Now

$$S_2' = \alpha_0 R_3' + \alpha_1 \beta R_4' + \alpha_2 \delta R_5'$$

where $\beta, \delta \in \mathfrak{k}$ and

$$R_3' = ca^{2m+1}ca^{2m+1}cv^{2m}c$$

$$R_4' = ca^{2m+1}ca^{2m}vca^{2m-1}c$$

$$R_5' = ca^{2m+1}ca^{2m-1}v^2cv^{2m-2}a^2c.$$

If we put

$$R' = ca^{2m+1}ca^{2m-2}v^2ca^3v^{2m-2}c$$

then, as for R_5, we can get R_5' as a linear combination of R', R_3', and R_4'. We also, as for R, get an alternative expression

$$R' = \alpha ca^{2m}ca^{2m-1}v^2ca^3v^{2m-2}c.$$

Then S_2' is a linear combination of R_3', R_4', R'. Similarly S_3' is a linear combination of R_3' and R_4'.

Using these, we express T as a linear combination of terms like

$$cu^{2m}(ca^{2m+1}ca^{2m+1}ca^{2m+1}c)v^{2m}c$$

$$cu^{2m}(ca^{2m+1}ca^{2m+1}ca^{2m}vc)v^{2m-1}ac$$

$$cu^{2m-1}a(cua^{2m}ca^{2m+1}ca^{2m+1}c)v^{2m}c$$

$$cu^{2m-1}a(cua^{2m}ca^{2m+1}ca^{2m}vc)v^{2m-1}ac$$

where the bracketing is to emphasise that each term has the shape

$$W = c\ldots cxa^{2m}ca^{2m+1}ca^{2m}yc\ldots c$$

where $x = a$ or u, and $y = a$ or v.

Now if $p = 2m+5$ then $n \leqq 2m+4$. By linearisation we know that

$$0 = c\ldots cxa^{2m-3}\begin{Bmatrix} c & a \\ 1 & n-1 \end{Bmatrix} a^{2m+4-n}yc\ldots c$$

and using (25) we find that the right hand side is equal to $2W$. Hence $W = 0$, and therefore $T = 0$. $\square$

Now we begin to construct $c_{(m+1)}$. We have $0 \neq c = c_{(m)}$ and $7 \leqq \leqq 2m+3 < p$. If $cu^{2m}c = 0$ for all $u \in L$ then we may take $c_{(m+1)} = c$. We show that there exists $a \in L$ such that

$$c_0 = [ca^{2m+1}c] \neq 0.$$

For suppose not, so that $[ca^{2m+1}c] = 0$ for all $u \in L$. Now for any $v \in L$,

$$[vcu^{2m}c] = \sum [cu_i^{2m+1}c] + \sum_{r \leqq 2m} [cu_j^r c]$$

for elements $u_i, u_j \in L$, by lemma 2.2. The first sum is zero. But for $r \leqq 2m$,

$$[cu^r c] = \sum_{i=0}^{r} (-1)^i \binom{r}{i} u^i cu^{r-i}c - \sum_{i=0}^{r} (-1)^i \binom{r}{i} cu^i cu^{r-i}$$

which is obviously 0 for $r < 2m$. For $r = 2m$ it becomes $cu^{2m}c - cu^{2m}c = 0$ also. Hence $[vcu^{2m}c] = 0$ for any $v \in L$, so that $cu^{2m}c = 0$ for all $u \in L$, a contradiction. Hence we can find a.

LEMMA 4.7 *The element c_0 has the properties of a $c_{(m)}$.*

Proof: We have, for all $u \in L$,

$$c_0 u^\alpha c_0 = \{2ca^{2m+1}c - (2m+1)aca^{2m}c - (2m+1)ca^{2m}ca\}u^\alpha \times$$
$$\times \{2ca^{2m+1}c - (2m+1)aca^{2m}c - (2m+1)ca^{2m}ca\}$$

and this is zero if $\alpha \leqq 2m-3$. But also

$$0 = c[ca^{2m+2}]u^{2m-2}c$$
$$= c \,.\, \sum_{i=0}^{2m+2} (-1)^i \binom{2m+2}{i} a^i ca^{2m+2-i} u^{2m-2} c$$
$$= \binom{2m+2}{2} ca^{2m}ca^2 u^{2m-2}c,$$

while

$$c_0 u^{2m} c_0 = (2m+1)^2 ca^{2m}cau^{2m-2}aca^{2m}c$$

which is therefore zero. □

If this c_0 is not a $c_{(m+1)}$ then the previous argument shows that there exists $b \in L$ such that

$$c' = [c_0 b^{2m+1} c_0] \neq 0.$$

We show that c' is a $c_{(m+1)}$. By the above lemma, it is a $c_{(m)}$, so it remains to show that

$$H = c'u^{2m}c'$$

is zero. Once more the proof involves extensive computations. First note that

$$c' = 2c_0 b^{2m+1} c_0 - (2m+1)c_0 b^{2m} c_0 b - (2m+1)bc_0 b^{2m} c_0. \tag{30}$$

From lemma 4.5, expanding the second c' in H, we find that

$$H = 2c'u^{2m}c_0b^{2m+1}c_0 - (2m+1)c'u^{2m}bc_0b^{2m}c_0$$
$$= 2H_1 - (2m+1)H_2,$$

say. We work on the two summands separately. From lemma 4.6,

$$0 = -[bc_0b^{2m}c_0u^{2m}c_0]$$
$$= [c_0b^{2m+1}c_0u^{2m}c_0]$$
$$= [c_0b^{2m+1}c_0u^{2m}]c_0 - c_0[c_0b^{2m+1}c_0u^{2m}]$$
$$= \sum_{i=0}^{2m}(-1)^i\binom{2m}{i}u^ic'u^{2m-i}c_0 - \sum_{i=0}^{2m}(-1)^i\binom{2m}{i}c_0u^ic'u^{2m-i}.$$

Using (30) and the fact that c_0 is a $c_{(m)}$ we see that

$$0 = c'u^{2m}c_0 + \alpha uc_0b^{2m}c_0(bu^{2m-1})c_0 - c_0u^{2m}c' +$$
$$+ \beta c_0(u^{2m-1}b)c_0b^{2m}c_0u.$$

However, the bracketed terms bu^{2m-1} and $u^{2m-1}b$ are sums of powers of elements of L, up to the $2m$th power. These terms therefore make no contribution if the power is $< 2m$, since c_0 is a $c_{(m)}$. But they also make no contribution for a $2m$th power, by lemma 4.6. Hence

$$c'u^{2m}c_0 = c_0u^{2m}c',$$

and therefore

$$H_1 = c_0u^{2m}c'b^{2m+1}c_0$$
$$= c_0u^{2m}\{2c_0b^{2m+1}c_0 - (2m+1)bc_0b^{2m}c_0 - (2m+1)c_0b^{2m}c_0b\} \times$$
$$\times b^{2m+1}c_0.$$

This is a sum of three terms. The second contains $\bar{B}_2$ so is 0 by (25), the third is 0 by lemma 4.6. Thus

$$H_1 = 2c_0u^{2m}c_0b^{2m+1}c_0b^{2m+1}c_0.$$

But now,

$$0 = c_0u^{2m}c_0[c_0b^{2m+3}]b^{2m-1}c_0$$
$$= c_0u^{2m}c_0 . \sum_{i=0}^{2m+3}(-1)^i\binom{2m+3}{i}b^ic_0b^{2m+3-i}b^{2m-1}c_0 =$$

$$= \binom{2m+3}{1} c_0 u^{2m} c_0 b^{2m+2} c_0 b^{2m} c_0 -$$

$$- \binom{2m+3}{2} c_0 u^{2m} c_0 b^{2m+1} c_0 b^{2m+1} c_0. \tag{31}$$

Similarly,

$$0 = c_0 u^{2m} c_0 b [c_0 b^{2m+3}] b^{2m-2} c_0$$

$$= \sum_{i=0}^{2m+3} (-1)^i \binom{2m+3}{i} c_0 u^{2m} c_0 b^{i+1} c_0 b^{2m+3-i} b^{2m-2} c_0$$

$$= -\binom{2m+3}{2} c_0 u^{2m} c_0 b^{2m+2} c_0 b^{2m} c_0 +$$

$$+ \binom{2m+3}{3} c_0 u^{2m} c_0 b^{2m+1} c_0 b^{2m+1} c_0. \tag{32}$$

The system of equations (31) and (32) has determinant

$$\tfrac{1}{12}(2m+3)^2(2m+2)(2m+4) \neq 0$$

so that

$$H_1 = 0.$$

Further,

$$c_0 u^{2m} c_0 b^{2m+2} c_0 b^{2m} c_0 = 0. \tag{33}$$

Now we go back to formula (3) of section 2. If in this we set $n = 2m+1$, $v = u$, $u = [c_0 b^{2m+2}]$ we get

$$-c_0 [c_0 b^{2m+2}] u^{2m} c_0 b^{2m} c_0 = -c_0 u^{2m} [c_0 b^{2m+2}] c_0 b^{2m} c_0.$$

The left hand side of this is

$$(m+1)Y = -c_0 \sum_{i=0}^{2m+2} (-1)^i \binom{2m+2}{i} b^i c_0 b^{2m+2-i} u^{2m} c_0 b^{2m} c_0$$

$$= (m+1)\{2c_0 b^{2m+1} c_0 b u^{2m} c_0 b^{2m} c_0 -$$

$$- (2m+1) c_0 b^{2m} c_0 b^2 u^{2m} c_0 b^{2m} c_0\}$$

while the right hand side is

$$-c_0 u^{2m} \sum_{i=0}^{2m+2} (-1)^i \binom{2m+2}{i} b^i c_0 b^{2m+2-i} c_0 b^{2m} c_0$$

$$= -c_0 u^{2m} c_0 b^{2m+2} c_0 b^{2m} c_0$$

$$= 0$$

by (33). Therefore $Y = 0$, or equivalently

$$2c_0b^{2m+1}c_0bu^{2m}c_0b^{2m}c_0 = (2m+1)c_0b^{2m}c_0b^2u^{2m}c_0b^{2m}c_0. \tag{34}$$

Now, using (3), we see that

$$H_2 = [c_0b^{2m+1}c_0]u^{2m}bc_0b^{2m}c_0$$

differs from

$$X = [c_0b^{2m+1}c_0]bu^{2m}c_0b^{2m}c_0$$

only by terms of the form

$$[c_0b^{2m+1}c_0]x^ic_0b^{2m}c_0$$

for $i \leqq 2m$. The terms with $i < 2m-1$ are zero since c_0 is a $c_{(m)}$, that with $i = 2m$ is zero by lemma 4.5, and that with $i = 2m-1$ is equal to

$$-(2m+1)c_0b^{2m}c_0bx^{2m-1}c_0b^{2m}c_0$$

which is 0 on applying the sums-of-powers formula. So $H_2 = X$. But expanding c' by (30) and then using (34) gives us

$$X = -(2m+1)bc_0b^{2m}c_0bu^{2m}c_0b^{2m}c_0.$$

Now

$$[c_0bu^{2m}]c_0 = \textstyle\sum_{i=0}^{2m}(-1)^i\binom{2m}{i}u^i[c_0, b]u^{2m-i}c_0$$

$$= \textstyle\sum_{i=0}^{2m}(-1)^i\binom{2m}{i}u^ic_0bu^{2m-i}c_0 - \sum_{i=0}^{2m}(-1)^i\binom{2m}{i}u^ibc_0u^{2m-i}c_0$$

$$= c_0bu^{2m}c_0 - 2muc_0bu^{2m-1}c_0 - bc_0u^{2m}c_0.$$

Thus

$$\begin{aligned} c_0b^{2m}c_0bu^{2m}c_0b^{2m}c_0 &= c_0b^{2m}[c_0bu^{2m}]c_0b^{2m}c_0 + \\ &\quad + 2mc_0b^{2m}uc_0bu^{2m-1}c_0b^{2m}c_0 + \\ &\quad + c_0b^{2m+1}c_0u^{2m}c_0b^{2m}c_0 \\ &= c_0b^{2m}xc_0b^{2m}c_0 \end{aligned}$$

for suitable $x \in L$,

$$= c_0xb^{2m}c_0b^{2m}c_0$$

by (3)

$$= 0$$

since $\bar{B}_1 = 0$ by (25). Multiplying by $-(2m+1)b$ on the left gives $X = 0$, hence $H_2 = 0$. Therefore $H = 0$, so c' is a $c_{(m+1)}$. □

5. Algebras generated by elements of order 2

The inductive nature of the proof of Kostrikin's theorem becomes clearer if we introduce two statements $\mathscr{E}_{n,p}$ and $\mathscr{F}_{n,p}$ (where p is 0 or a prime, and in the latter case, $n \leqq p-1$) as follows:

$\mathscr{E}_{n,p}$: Any Lie algebra over a field of characteristic p satisfying the nth Engel condition is locally nilpotent.

$\mathscr{F}_{n,p}$: Any Lie algebra over a field of characteristic p satisfying the nth Engel condition and generated by elements of order 2 is locally nilpotent.

The inductive step takes the form

$$\mathscr{E}_{n-1,p} \Rightarrow \mathscr{F}_{n,p} \Rightarrow \mathscr{E}_{n,p}.$$

The subject of this section is the first of these implications:

THEOREM 5.1 *If $p = 0$ and n is arbitrary, or if p is prime and $n \leqq p-1$, then*

$$\mathscr{E}_{n-1,p} \Rightarrow \mathscr{F}_{n,p}.$$

Proof: We may assume $\mathscr{E}_{n-1,p}$. So suppose L satisfies the nth Engel condition and is generated by elements of order 2. We want to prove that L is locally nilpotent. It is sufficient to consider the case where L is finitely generated. Since $L/\rho(L)$ has trivial Hirsch-Plotkin radical we may assume further that $\rho(L) = 0$. To reach a contradiction we use theorem 4.4 and lemma 4.2: if we can find in L an element $c_{(2)} \neq 0$ then L has a non-zero abelian ideal. Therefore $L = \rho(L)$ is locally nilpotent.

So now we may assume that L is finitely generated, hence generated by a finite set

$$\{x_0, \ldots, x_d\}$$

where each x_i has order 2. (This follows since L is generated by elements of order 2, so each of a finite set of generators is expressible in terms of finitely many elements of order 2.) Going by induction on d, we may assume that

$$M = \langle x_1, \ldots, x_d \rangle$$

is locally nilpotent, hence nilpotent. Our aim, in this situation, is to construct $c_{(2)} \neq 0$. In fact (as suggested by a later lemma of Kostrikin [114]) we find an element $c \neq 0$ such that $cu^2cv^2c = 0$ for all $u, v \in L$, and derive a contradiction if this is not a $c_{(2)}$.

STEP I: *If a and b are of order 2 then $[ab]$ is of order 2.*

We have $a^2 = b^2 = 0$. Further,

$$\begin{aligned} 0 &= [ab^2] \\ &= ab^2 - 2bab + b^2a \\ &= 2bab \end{aligned}$$

so $bab = 0$. But then

$$\begin{aligned} [ab]^2 &= (ab - ba)^2 \\ &= abab - ab^2a - ba^2b + baba \\ &= 0. \end{aligned}$$

Suppose now that we let $f_1, \ldots, f_q$ be the left-normed commutators in $x_1, \ldots, x_d$. There are finitely many of these since M is nilpotent. By the above, each f_i has order 2, and furthermore elements like $[f_if_j]$, $[[f_if_j][f_kf_l]]$, etc. also have order 2.

STEP II: *Let $a_0, a_1, a_2, a_3, a_4 \in L$ be such that $a_i^2 = 0$ $(i > 0)$,*

$$c = [a_0a_1a_2a_3a_4] \neq 0$$

and that for all permutations π of $\{1, 2, 3, 4\}$

$$c = [a_0a_{\pi(1)}a_{\pi(2)}a_{\pi(3)}a_{\pi(4)}].$$

Then

$$c^2 = 0 = cu^2cv^2c$$

for all $u, v \in L$.

As in step I, $a_iua_i = 0$ for all $i > 0$. So

$$[cua_i] = [a_0a_{\pi(1)}a_{\pi(2)}a_{\pi(3)}a_iua_i] = 0$$

for suitable π. Similarly

$$ca_i = 0 = a_ic \tag{35}$$

for $i > 0$, so that

$$0 = [cua_i] = [[cu]a_i]$$

and we get

$$cua_i + a_i uc = 0. \tag{36}$$

Now for $i, j, k > 0$,

$$\begin{aligned} c[a_i u^2]a_j a_k &= c(a_i u^2 - 2ua_i u + u^2 a_i)a_j a_k \\ &= -2cua_i ua_j a_k + cu^2 a_i a_j a_k \end{aligned}$$

by (35). Using (36) this reduces to

$$cu^2 a_i a_j a_k = 0 \tag{37}$$

and similarly we get

$$a_i a_j a_k u^2 c = 0. \tag{38}$$

The same kind of argument on $c[a_i u^2]a_j v a_k a_l$ for $i, j, k, l > 0$ leads to

$$cu^2 a_i a_j v a_k a_l = 0 = a_k a_l v a_i a_j u^2 c. \tag{39}$$

Expanding the middle c in cu^2cv^2c gives

$$\sum \pm\, cu^2 a_{\pi(0)} a_{\pi(1)} a_{\pi(2)} a_{\pi(3)} a_{\pi(4)} v^2 c$$

where π is a permutation of $\{0, 1, 2, 3, 4\}$. Using (37), (38), and (39) we see that each term of the sum is 0, so that

$$cu^2cv^2c = 0.$$

Now we show $c^2 = 0$. Now

$$c = [a_0 a_1 a_2 a_3]a_4 - a_4[a_0 a_1 a_2 a_3]$$

so that

$$\begin{aligned} c^2 &= c[a_0 a_1 a_2 a_3]a_4 && \text{(by (35))} \\ &= c[a_0 a_1 a_2]a_3 a_4 && \text{(similarly)} \\ &= -a_3[a_0 a_1 a_2]ca_4 \\ &= 0. \end{aligned}$$

STEP III: *c is a $c_{(2)}$, or $[cb^3c]$ is a $c_{(2)}$ for some $b \in L$.*

If c is not, then (as in section 4) for some $b \in L$ we have

$$[cb^3c] \neq 0.$$

By step I every element of L is a sum of elements of order 2, so it is enough to assume

$$b = b_1 + b_2 + b_3$$

where the b_i have order 2. We may then show that

$$[cb^3c]u^2[cb^3c] = 0$$

and so $[cb^3c]$ is a $c_{(2)}$.

STEP IV: *There exist elements a_i as required in step II.*

Note that this finishes the proof of the main theorem of this section, by steps II and III.

We take $c_0 \neq 0$ in the centre of M (which is nilpotent), and then

$$[c_0 f_i] = 0$$

for all the left-normed commutators f_i in M. If we had $[c_0 x_0] = 0$ then c_0 would be central in L, contrary to $\zeta_1(L) = 0$. Therefore $[c_0 x_0] \neq 0$. By theorem 2.4 M generates a nilpotent subalgebra of the associative algebra in which we have embedded L. So the set of elements of the form

$$[c_0 x_0 f_{i_1} \ldots f_{i_\alpha}] \neq 0 \tag{40}$$

is finite, and there exists

$$h = [c_0 x_0 f_{i_1} \ldots f_{i_s}] \neq 0$$

for which s is maximal. (There need not be a unique such h). By definition,

$$[hx_i] = 0$$

for $i > 0$. If h is not invariant under permutations of $i_1, \ldots, i_s$ then some product

$$[c_0 x_0 f_{i_1} \ldots [f_{i_k} f_{i_{k+1}}] \ldots f_{i_s}]$$

is non-zero. This element is still a linear combination of terms (40) with $\alpha = s$, and $[f_{i_k} f_{i_{k+1}}]$ has order 2. Treat $[f_{i_k} f_{i_{k+1}}]$ as a single element: if the new product is not permutation-invariant we find another involving an extra commutation. Eventually we get to

$$c_1 = [c_0 x_0 a_1 a_2 \ldots a_m] \neq 0$$

invariant under all permutations of $\{1, \ldots, m\}$. This c_1 is a linear combination of terms (40) with $\alpha = s$, so

$$[c_1 x_i] = 0$$

for $i > 0$. Each a_i is a Lie monomial in the f_j, so has order 2. And $[c_1 x_0] \neq 0$ or else c_1 is central in L.

Now $m \neq 0$, since if $m = 0$ then

$$[c_1 x_0] = [c_0 x_0 x_0] = 0.$$

And $m \neq 1$, for if $m = 1$ then

$$[c_1 x_0] = [c_0 x_0 a_1 x_0] = 0.$$

On the other hand if $m \geqq 3$ we have elements as in step II. For the a_i may be permuted without changing c_1, and c_0 commutes with all the a_i. Putting $-x_0$ for a_0, c_0 for a_1, a_1 for a_2, a_2 for a_3, a_3 for a_4 we are at step II if $m = 3$. If $m > 3$ we put $-[x_0 c_0 \ldots a_{m-4}]$ for a_0, a_{m-4+i} for a_i $(i > 0)$ and again arrive at step II.

Hence $m = 2$. So

$$c_1 = [c_0 x_0 a_1 a_2] \neq 0, \quad [c_1 x_i] = 0 \quad (i > 0)$$

and the a_i have order 2 and permute in c_1. Now we can start again with c_1 in place of c_0, getting

$$c_2 = [c_0 x_0 a_1 a_2 x_0 b_1 b_2] \neq 0, \quad [c_2 x_i] = 0 \quad (i > 0).$$

Now repeat with c_2. The process never stops, or else we get to step II. So for arbitrarily large τ we get a product

$$c_\tau = [c_0 x_0 a_1 a_2 x_0 b_1 b_2 \ldots x_0 e_1 e_2] \neq 0$$

with τ occurrences of x_0, such that $[c_\tau, x_i] = 0$ $(i > 0)$. Therefore

$$[c_\tau x_0] \neq 0$$

or else c is central in L. Now we write all the $a_i, b_i, \ldots, e_i$ as linear combinations of f_j and expand: some monomial in this is non-zero, since the whole sum is, and we get

$$u_\tau = [c_0 x_0 f_{\mu_1} f_{\nu_1} x_0 f_{\mu_2} f_{\nu_2} x_0 \ldots x_0 f_{\mu_\tau} f_{\nu_\tau} x_0] \neq 0 \tag{41}$$

for arbitrarily large τ.

Next,

$$[a[x_0 b_1 b_2] x_0] = [a x_0 b_1 b_2 x_0]$$

on expanding the middle term and using $x_0^2 = 0$. So we have

$$u_\tau = [c_0 y_1 \ldots y_\tau x_0]$$

where

$$y_i = [x_0 f_{\mu_i} f_{\nu_i}].$$

Consider the subalgebra $Y = \langle y_i \rangle$. In fact the y_i can take at most $q(q-1)$ values, since the f_j take at most q values; therefore Y is finitely generated. We let I be the verbal ideal of Y corresponding to the $(n-1)$th Engel word. Then by $\mathscr{E}_{n-1,p}$ it follows that Y/I is locally nilpotent, hence nilpotent since it is finitely generated. Therefore for sufficiently large τ, we have

$$y_1 \ldots y_\tau \in I.$$

Now we note that for all u, v

$$\begin{aligned} u^{n-1}v &= \textstyle\sum_{i=0}^{n-2} u^i[uv]u^{n-2-i} + vu^{n-1} \\ &= \begin{Bmatrix} u & [uv] \\ n-2 & 1 \end{Bmatrix} + vu^{n-1} \\ &= vu^{n-1} + \textstyle\sum_k u_k^{n-1} \end{aligned}$$

by linearisation. It follows that

$$y_1 \ldots y_\tau = \textstyle\sum_j z_j^{k_j} w_j^{n-1},$$

where $z_j, w_j \in Y$. The y_i commute with x_0, so $[w_j, x_0] = 0$. Then

$$n.y_1 \ldots y_\tau x_0 = \textstyle\sum_j z_j^{k_j} . \sum_{i=0}^{n-1} w_j^i x_0 w_j^{n-1-i}$$

since x_0 commutes with the w_j,

$$\begin{aligned} &= \textstyle\sum_j z_j^{k_j} . \begin{Bmatrix} w_j & x_0 \\ n-1 & 1 \end{Bmatrix} \\ &= 0. \end{aligned}$$

Therefore

$$y_1 \ldots y_\tau x_0 = 0$$

for all sufficiently large τ, so that $u_\tau = 0$. But this contradicts (41). The theorem is proved. □

6. A weakened form of Kostrikin's theorem

At this point we make a detour from the line of argument followed by Kostrikin [114], since we are in a position to prove the weaker form of the theorem mentioned in the introductory section:

THEOREM 6.1 *Let L be a Lie algebra over a field $\mathfrak{k}$ satisfying the nth Engel condition, where either*

(a) *$\mathfrak{k}$ has characteristic 0*

or (b) *$\mathfrak{k}$ has characteristic $p \geqq 2n$.*

Then L is locally nilpotent.

Proof: We use induction on n. The result is already known for $n \leqq 4$. We may assume the truth of proposition $\mathscr{E}_{n-1,p}$. As usual we may use theorem 16.3.11 to reduce to the case $\rho(L) = 0$, $L \neq 0$, and aim at a contradiction. By theorem 3.1 there exists a non-zero element of order 2 in L. Hence

$$K = \langle x \in L : x \text{ has order } 2 \rangle \neq 0.$$

Now theorem 5.1 implies the truth of $\mathscr{F}_{n,p}$, which means that K is locally nilpotent.

Let $y \in L$. Condition (b) implies that y^* is $[p/2]$-nilpotent as a result of the nth Engel condition. By theorem 1.4.7 $\exp(\lambda y^*)$ is an automorphism of L for all $\lambda \in \mathfrak{k}$. Obviously this implies that K is invariant under all $\exp(\lambda y^*)$ and hence by lemma 1.4.8 K is invariant under y^*. Therefore K is an ideal of L. But $0 \neq K$ is locally nilpotent, which contradicts triviality of the Hirsch-Plotkin radical of L. This completes the inductive step. □

The example of chapter 1 section 4, and others like it, show that we cannot in general assert that $\exp(\lambda y^*)$ is an automorphism if all we know is that y^* is $(p-1)$-nilpotent. So it seems unlikely that the above method can be adapted to solve the restricted Burnside problem. But very possibly the condition $p \geqq 2n$ might be improved a little, by making a more careful analysis.

7. Sketch proof of Kostrikin's theorem

Since exponentials are no help for the full theorem 1.1 we return to Kostrikin's proof. This also proceeds by induction on n, and reduces to the case $\rho(L) = 0$,

$L \neq 0$. The contradiction comes by finding an element $c_{(2)} \neq 0$ and using the results of section 4 to obtain a non-zero abelian ideal. It therefore remains only to construct $c_{(2)}$. Unfortunately the computations required for this are even more complex than those needed so far: further, we have to use everything so far proved in this chapter (with the exception of section 6). To give full details would require some 40 pages (cf. Wiegold [244]). The following outline may give some idea of how the arguments go. For the moment we take $n \leqq p-1$.

STEP I: Let $c \neq 0$ be of order 2. Suppose also that we have c_0 of order 2, where $cc_0 = c_0c = 0$. Let $c_1 = [c_0a^2c]$ for $a \in L$. Then $c_1^2 = 0$. If in addition $[c_0u^2cv^2c] = 0$ then

$$c_1u^2c_1v^2c_1 = 0.$$

STEP II: Suppose there exists $m \geqq 2$ such that

$$cu_1^2cu_2^2c\ldots cu_m^2c = 0$$

for all $u_i \in L$. Then there exists $c_{(2)} \neq 0$.

STEP III: Let $c_0 = [hc_1] \neq 0$ where $c_1^2 = 0$ and h satisfies one of the following conditions:

(a) $h^2 = 0$,
(b) $h = [c_1a^3]$,
(c) $h = [ac_2]$ where $c_2^2 = [c_1c_2] = 0$ and $a \in L$.

If further

$$[c_0u^2c_1uvc_1] = 0$$

for all $u, v \in L$, then there exists $c_{(2)} \neq 0$.

From these we get

STEP IV: If $c = c_{(1)}$ and for all $u, v \in L$

$$[cu^3c][cv^3c] = 0 \tag{42}$$

then there exists $c_{(2)} \neq 0$.

Now we proceed by contradiction, assuming no $c_{(2)}$ exists. Then there cannot exist c satisfying (42), from which it follows that there exist c_1, c_2 satisfying

$$c_1c_2 \neq 0,\ c_1^2 = c_2^2 = [c_1c_2] = 0 \tag{43}$$

where we take

$$c_1 = [cu^3c], \; c_2 = [cv^3c].$$

We try to improve c_1 and c_2 to get extra equations

$$c_i u^\alpha c_1 c_2 = 0 \quad (i = 1, 2; \alpha = 2, 3, \ldots, n-1).$$

The most important step in this direction is:

STEP V: If there is no $c_{(2)}$ then we can find c_1, c_2 such that (43) holds, and further

$$c_1 u^2 c_1 c_2 = 0 = c_2 u^2 c_1 c_2$$

for all $u \in L$.

This is the heart of the proof. We need to use theorem 5.1: the computations become extremely heavy.

STEP VI: Look at $N = \langle c^L \rangle$ where $c = [ae^2]$, $e = c_1 + c_2$, and $a \in L$. If N has non-zero centre we can find an element $c_{(2)}$ in L, since $\zeta_1(N)$ is an abelian ideal. Hence we may assume $\zeta_1(N) = 0$, and everything we have proved for L therefore holds good in N. We then construct in L an element $e_0 \neq 0$ of order 2, such that

$$[e_0 u^3 e_0][e_0 v^3 e_0] = 0$$

for all $u, v \in L$. By step IV this is a contradiction.

STEP VII: All that remains is the case $n = p$. Let L be finitely generated, satisfying the pth Engel condition over a field of characteristic p. Let I be the ideal of L generated by all elements $[xy^{p-1}]$, for $x, y \in L$. Then L/I satisfies the $(p-1)$th Engel condition, so $L^r \leqq I$ for some r. But for any $z \in L$,

$$\begin{aligned}[xy^{p-1}z] &= -[z[xy^{p-1}]] \\ &= -\textstyle\sum_{i=0}^{p-1} zy^i xy^{p-1-i} \\ &= \textstyle\sum [zu_k^p] \\ &= 0.\end{aligned}$$

Hence the generators of I are central in L, so $I \leqq \zeta_1(L)$, so $L^{r+1} = 0$. □

For more of the computational details consult Kostrikin [114] or Wiegold [244].

It is an open question whether we can improve the conclusion that L is locally nilpotent to read 'L is nilpotent' in characteristic 0. The theorems of Razmyslov [177] and Bachmuth, Mochizuki, and Walkup [18] to be proved in the next chapter shows that for characteristic $p \geqq 5$ no such improvement is possible if $n \geqq p-2$.

Chapter 18

Razmyslov's theorem

For some time after Kostrikin's work it was not known to what extent local nilpotence is essential in the conclusion of the theorem; and correspondingly it was not known whether locally nilpotent groups of exponent p need be nilpotent. The first results in this direction, due to Bachmuth, Mochizuki, and Walkup [18] dealt with the case $p = 5$. They constructed a Lie algebra of characteristic 5 satisfying the 3rd Engel condition which is not nilpotent, and a non-nilpotent locally nilpotent group of exponent 5. Independently Razmyslov [177] proved similar results for every prime $p \geqq 5$. In this case the relevant Engel condition is the $(p-2)$th. The object of this chapter is to prove Razmyslov's theorem.

1. The construction

Our aim is to prove the following theorem of Razmyslov:

THEOREM 1.1 *For every prime $p \geqq 5$ there exists a Lie algebra L satisfying the $(p-2)$th Engel condition which is not nilpotent (or soluble).*

Note that if L were soluble then by a theorem of Higgins [89] it would be nilpotent, so it suffices to find a non-nilpotent algebra L. We construct L as follows:

Let $\mathfrak{k}$ be a field of characteristic $p \geqq 5$. Let A be the associative algebra over $\mathfrak{k}$ generated by elements $x_1, x_2, \ldots$ subject to the following relations:

$$(v_i \circ v_j)v_k = v_k(v_i \circ v_j) \tag{1}$$

where '$\circ$' is defined by

$$v_i \circ v_j = v_i v_j + v_j v_i$$

and either

$$v_i = x_i,\ v_j = x_j,\ v_k = x_k$$

or

$$v_i = [x_i, x_t],\ v_j = x_j,\ v_k = x_k$$

for $i, j, k, t \geqq 1$.

$$a^n = 0 \tag{2}$$

for all a belonging to the Lie subalgebra of $[A]$ generated by the x_i, for a fixed $n < p$.

$$(x_i \circ x_j)x_i = -\gamma x_i^2 x_j \tag{3}$$

for all $i, j \geqq 1$, where γ is a fixed element of $\mathfrak{k}$.

$$w = 0 \tag{4}$$

where w is any monomial of degree $\geqq 3$ in some x_i.

LEMMA 1.2 *Let A be generated by $x_1, x_2, \ldots$ subject to relations (1), (2), (3), (4); let L be the Lie algebra generated by $x_1, x_2, \ldots$. Then*

(a) *Relation (1) holds for all $v_i, v_j, v_k \in L$.*
(b) *L satisfies the nth Engel condition.*
(c) *If L is nilpotent then A is nilpotent.*

Proof: We start with (a). From (1) we have

$$(x_i \circ x_j)x_k = x_k(x_i \circ x_j) \tag{5}$$

so the elements

$$x_i \circ x_j = x_i x_j + x_j x_i$$

and in particular

$$\tfrac{1}{2} x_i \circ x_i = x_i^2$$

lie in the centre C of A. Since

$$\begin{aligned}[x_i, x_j, x_k] &= x_i x_j x_k + x_k x_j x_i - x_j x_i x_k - x_k x_i x_j \\ &= -2(x_i \circ x_k)x_j + 2x_j(x_i \circ x_k) + x_k(x_i \circ x_j) - (x_i \circ x_j)x_k \\ &= 2x_i(x_j \circ x_k) - 2x_j(x_i \circ x_k)\end{aligned} \tag{6}$$

it follows that every element of L is a linear combination of elements of the form

$$vc$$

where $v = x_i$ or $[x_i, x_j]$ and $c \in C$. It therefore suffices to prove that (1) holds when v_i and v_j are either x_i's or $[x_i, x_j]$'s, and when v_k is an x_k.

If we commutate (1) with x_s we get

$$([v_i, x_s] \circ v_j)v_k + (v_i \circ [v_j, x_s])v_k + (v_i \circ v_j)[v_k, x_s] =$$
$$= v_k([v_i, x_s] \circ v_j) + v_k(v_i \circ [v_j, x_s]) + [v_k, x_s](v_i \circ v_j).$$

If in this equation we set $v_j = x_j$, $v_k = x_k$, and replace v_i by a commutator of length 2 in the x_i, then use of (1) and (6) yields

$$(v_i \circ [x_j, x_s])x_k = x_k(v_i \circ [x_j, x_s])$$

which proves part (a).

Now we note that part (a) implies that for all $v \in L$ we have $v^2 \in C$. Linearising (2) gives

$$\textstyle\sum_{j=0}^{n-1} v^j u v^{n-1-j} = 0$$

for $u, v \in L$. Since $v^2 \in C$ this can be written as

$$vuv^{n-2} = \alpha uv^{n-1}$$

for a certain $\alpha \in \mathfrak{k}$. But now

$$\begin{aligned}[u, {}_n v] &= \textstyle\sum_{j=0}^{n} (-1)^j \binom{n}{j} v^j u v^{n-j} \\ &= \alpha_1 uv^n + \alpha_2 vuv^{n-1} \\ &= (\alpha_1 + \alpha\alpha_2)uv^n \\ &= 0\end{aligned}$$

where again α_1 and α_2 are certain elements of $\mathfrak{k}$. This proves part (b). Part (c) follows as in theorem 17.2.4. □

In view of part (c) of the above lemma, theorem 1.1 will follow if we can prove that with $n = p-2$, and with a suitable choice of γ, the algebra A is non-nilpotent as an associative algebra.

2. Proof of non-nilpotence

From now on we take $n = p-2$. The choice of γ is left open at the moment. Our object is to prove A non-nilpotent: our proof is of course combinatorial. We begin by finding a canonical form for elements of A, as a computational aid.

LEMMA 2.1 *Every monomial $u = x_{i_1}\ldots x_{i_m}$ of A can be expressed as a linear combination of monomials of the same degree as u in each x_i, and having the form*

$$x_{j_1}^2\ldots x_{j_s}^2 . x_{k_1}\ldots x_{k_t} \tag{7}$$

where $s, t \geqq 0$ and the j's and k's are pairwise distinct.

Proof: By (4) no x_i has degree $\geqq 3$, or else $u = 0$ and the lemma is obvious. Suppose x_i has degree 2 in u. If u has x_i^2 as a subword we can move it to the front, since squares are central. Suppose u has a subword $x_i v x_i$, where v is non-empty and contains no x_i. Then $v = x_j v'$ (where v' may be empty). Then

$$\begin{aligned} x_i v x_i &= x_i x_j v' x_i \\ &= -x_j x_i v' x_i + v'(x_i \circ x_j) x_i \\ &= -x_j x_i v' x_i - \gamma x_i^2 v' x_j. \end{aligned} \tag{8}$$

Working inductively on the segment $x_i v' x_i$ we eventually collect together the two occurrences of x_i, and move the x_i^2 which results to the front. Once all x_i of degree 2 have been moved, only the x_{k_i} remain. □

This expression for u is not, in general, unique. However, in an important special case, and with the right choice of γ, it *is* unique. Say that $u = x_{i_1}\ldots x_{i_m}$ is a *2-word* if every x_i has degree 0 or 2 in u. Linear combinations of 2-words we call *2-elements*. The next lemma tells us that if we take $\gamma = -\frac{2}{3}$, the expression of 2-words is unique.

LEMMA 2.2 *Suppose $\gamma = -\frac{2}{3}$. Then every 2-word in A can be expressed in the form*

$$\alpha x_{j_1}^2\ldots x_{j_m}^2 \quad (\alpha \in \mathfrak{k}) \tag{9}$$

and this expression is unique up to permutation of the factors $x_{j_i}^2$.

This, of course, is the crucial result; and in particular the proof of theorem 1.1 is immediate. For by uniqueness the words

$$x_1^2x_2^2\ldots x_k^2$$

are all non-zero. Since k may be arbitrarily large, it follows that A is non-nilpotent when $\gamma = -\frac{2}{3}$. By lemma 1.2(c) it follows that L is non-nilpotent; by lemma 1.2(b) it satisfies the $(p-2)$-th Engel condition. □

It remains to prove lemma 2.2. The proof is moderately long, and breaks up into three stages; at each stage more of the defining relations are brought into play.

STEP I: Let A_0 be the associative algebra over $\mathfrak{k}$ generated by $x_1, x_2, \ldots$ subject to the relations

$$\begin{aligned} x_i^2x_j &= x_jx_i^2 \\ x_i^3 &= 0 \end{aligned} \tag{10}$$

and

$$x_ivx_i = x_i^2\{\gamma \textstyle\sum_{t=1}^{s}(-1)^t x_{j_1}\ldots x_{j_{t-1}}x_{j_{t+1}}\ldots x_{j_s}x_{j_t} + (-1)^s v\} \tag{11}$$

where

$$v = x_{j_1}\ldots x_{j_s}$$

which last is an immediate consequence of (8) and so holds in A. These relations (10) and (11) are sufficient for us to prove lemma 2.1 with A_0 in place of A. If $u = x_{i_1}\ldots x_{i_m} \in A_0$ is expressed as a linear combination of words of the form (7) and of the same degrees as u, we call this expression a *canonical form* for u in A_0. The name is justified by:

LEMMA 2.3 *The canonical form of a monomial $u \in A_0$ is unique, up to permutations of the $x_{j_t}^2$ terms.*

Proof: We denote by π_i the operation which replaces a subword x_ivx_i of u by the right hand side of (11), and let $\pi(v)$ be the expression in curly brackets in (11). We let τ be the operation which sends a subword $x_ix_j^2$ to $x_j^2x_i$, and let δ be the operation which deletes a word having x_i^3 as a subword. The lemma will follow if we can show that these operations commute, regarded as functions defined on the free associative algebra generated by $x_1, x_2, \ldots$. For then any two sequences of operations which lead to a canonical form must lead to the *same* canonical form. As far as δ is concerned, this is obvious:

it is not much more obscure for τ, given the form of the required relation:

$$\pi_\alpha(x_\alpha y_1 \dots y_s x_\beta^2 z_1 \dots z_t x_\alpha) = x_\beta^2 \pi_\alpha(x_\alpha y_1 \dots y_s z_1 \dots z_t x_\alpha)$$

(where the y_i, z_j are certain of the generators $x_1, x_2, \dots$). All that remains is to prove that π_α and π_β commute. The proof is straightforward but cumbersome, proceeds by induction, and splits into two cases:

(A) $u = x_\alpha x_1 \dots x_s x_\beta y_1 \dots y_t x_\alpha z_1 \dots z_k x_\beta$,

(B) $u = x_\alpha x_1 \dots x_s x_\beta y_1 \dots y_t x_\beta z_1 \dots z_k x_\alpha$.

First consider case (A). Proceed by induction on s, and note the recurrence relation

$$\pi(x_1 v) = -\gamma v x_1 - x_1 \pi(v).$$

If $s \geqq 1$ then

$$\pi_\alpha(u) = x_\alpha^2\{-\gamma x_2 \dots x_s x_\beta y_1 \dots y_t x_1 z_1 \dots z_k x_\beta\} - $$
$$- x_1 \pi_\alpha(x_\alpha x_2 \dots x_s x_\beta y_1 \dots y_t x_\alpha z_1 \dots z_k x_\beta).$$

Further,

$$\begin{aligned}\pi_\beta \pi_\alpha(u) &= -\gamma x_\alpha^2 x_\beta^2 x_2 \dots x_s \pi(y_1 \dots y_t x_\alpha z_1 \dots z_k) - \\ &\quad - x_1 \pi_\beta \pi_\alpha(x_\alpha x_2 \dots x_s x_\beta y_1 \dots y_t x_\alpha z_1 \dots z_k x_\beta) \\ &= -\gamma x_\alpha^2 x_\beta^2 x_2 \dots x_s \pi(y_1 \dots y_t x_1 z_1 \dots z_k) - \\ &\quad - x_1 x_\beta^2 \pi(x_\alpha x_2 \dots x_s \pi(y_1 \dots y_t x_\alpha z_1 \dots z_k)).\end{aligned}$$

In the same way,

$$\pi_\alpha \pi_\beta(u) = x_\beta^2 \pi_\alpha(x_\alpha x_1 \dots x_s \pi(y_1 \dots y_t x_\alpha z_1 \dots z_k)).$$

Now we set

$$\pi(y_1 \dots y_t x_\alpha z_1 \dots z_k) = \textstyle\sum_i u_i x_\alpha v_i$$

and deduce that

$$\begin{aligned}\pi_\alpha(\textstyle\sum_i x_\alpha x_1 \dots x_s u_i x_\alpha v_i) &= -\gamma x_\alpha^2 . \textstyle\sum_i x_2 \dots x_s u_i x_1 v_i - \\ &\quad - x_1 \pi_\alpha(\textstyle\sum_i x_\alpha x_2 \dots x_s u_i x_\alpha v_i) \\ &= -\gamma x_\alpha^2 x_2 \dots x_s \pi(y_1 \dots y_t x_1 z_1 \dots z_k) - \\ &\quad - x_1 \pi_\alpha(x_\alpha x_2 \dots x_s \pi(y_1 \dots y_t x_\alpha z_1 \dots z_k)).\end{aligned}$$

Therefore

$$\pi_\beta\pi_\alpha(u) = \pi_\alpha\pi_\beta(u).$$

Next, suppose $s = 0$. We use induction on t. If $t \geqq 1$ then

$$\pi_\alpha(x_\alpha x_\beta y_1 \ldots y_t x_\alpha z_1 \ldots z_k x_\beta) = -\gamma x_\alpha^2 y_1 \ldots y_t x_\beta z_1 \ldots z_k x_\beta +$$
$$+ \gamma x_\alpha^2 x_\beta y_2 \ldots y_t y_1 z_1 \ldots z_k x_\beta + x_\beta y_1 \pi_\alpha(x_\alpha y_2 \ldots y_t x_\alpha z_1 \ldots z_k x_\beta),$$

$$\pi_\beta\pi_\alpha(u) = -\gamma x_\alpha^2 x_\beta^2 y_1 \ldots y_t \pi(z_1 \ldots z_k) + \gamma x_\alpha^2 x_\beta^2 \pi(y_2 \ldots y_t y_1 z_1 \ldots z_k) -$$
$$- \gamma x_\alpha^2 x_\beta \pi(y_2 \ldots y_t) z_1 \ldots z_k y_1 - y_1 x_\alpha^2 x_\beta^2 \pi(\pi(y_2 \ldots y_t) z_1 \ldots z_k),$$

$$\pi_\beta(u) = -\gamma x_\beta^2 x_\alpha y_2 \ldots y_t x_\alpha z_1 \ldots z_k y_1 - x_\alpha y_1 \pi_\beta(x_\beta y_2 \ldots y_t x_\alpha z_1 \ldots z_k x_\beta),$$

$$\pi_\alpha\pi_\beta(u) = -\gamma x_\alpha^2 x_\beta^2 \pi(y_2 \ldots y_t) z_1 \ldots z_k y_1 - \pi_\alpha(\textstyle\sum_i x_\alpha y_1 u_i x_\alpha v_i)$$
$$= -\gamma x_\alpha^2 x_\beta^2 \pi(y_2 \ldots y_t) z_1 \ldots z_k y_1 + y_1 \pi_\alpha(\textstyle\sum_i x_\alpha u_i x_\alpha v_i) +$$
$$+ \gamma \textstyle\sum_i x_\alpha^2 u_i y_1 v_i$$
$$= -\gamma x_\alpha^2 x_\beta^2 \pi(y_2 \ldots y_t) z_1 \ldots z_k y_1 +$$
$$+ y_1 \pi_\alpha \pi_\beta(x_\alpha x_\beta y_2 \ldots y_t x_\alpha z_1 \ldots z_k x_\beta) +$$
$$+ \gamma x_\alpha^2 \pi_\beta(x_\beta y_2 \ldots y_t y_1 z_1 \ldots z_k x_\beta).$$

But now the inductive assumption gives

$$y_1 \pi_\alpha \pi_\beta(x_\alpha y_\beta y_2 \ldots y_t x_\alpha z_1 \ldots z_k x_\beta) =$$
$$= -y_1 x_\alpha^2 x_\beta^2 \pi(\pi(y_2 \ldots y_t) z_1 \ldots z_k) - \gamma x_\alpha^2 x_\beta^2 y_1 \ldots y_t \pi(z_1 \ldots z_k)$$
$$= y_1 \pi_\beta \pi_\alpha(x_\alpha x_\beta y_2 \ldots y_t x_\alpha z_1 \ldots z_k x_\beta)$$

which completes the induction step in this case.

Lastly, if $s = t = 0$ then

$$\pi_\alpha\pi_\beta(u) = \pi_\alpha\{x_\beta^2(-\gamma x_\alpha z_1 \ldots z_k x_\alpha - x_\alpha^2 \pi(z_1 \ldots z_k))\}$$
$$= (-1-\gamma) x_\alpha^2 x_\beta^2 \pi(z_1 \ldots z_k)$$
$$= \pi_\beta\pi_\alpha(u).$$

Case (B) is similar. If $s \geqq 1$, then

$$\pi_\alpha(u) = x_\alpha^2(-\gamma x_2 \ldots x_s x_\beta y_1 \ldots y_t x_\beta z_1 \ldots z_k x_1) -$$
$$- x_1 \pi_\alpha(x_\alpha x_2 \ldots x_s x_\beta y_1 \ldots y_t x_\beta z_1 \ldots z_k x_\alpha),$$

$$\begin{aligned}\pi_\beta\pi_\alpha(u) &= -\gamma x_\alpha^2 x_\beta^2 x_2\ldots x_s\pi(y_1\ldots y_t)z_1\ldots z_k x_1 - \\ &\quad - x_1\pi_\beta\pi_\alpha(x_\alpha x_2\ldots x_s x_\beta y_1\ldots y_t x_\beta z_1\ldots z_k x_\alpha) \\ &= -\gamma x_\alpha^2 x_\beta^2 x_2\ldots x_s\pi(y_1\ldots y_t)z_1\ldots z_k x_1 - \\ &\quad - x_1 x_\beta^2\pi_\alpha\{x_\alpha x_2\ldots x_s\pi(y_1\ldots y_t)z_1\ldots z_k x_\alpha\}.\end{aligned}$$

Further,

$$\begin{aligned}\pi_\alpha\pi_\beta(u) &= x_\beta^2(x_\alpha x_1\ldots x_s\pi(y_1\ldots y_t)z_1\ldots z_k x_\alpha) \\ &= -\gamma x_\alpha^2 x_\beta^2 x_2\ldots x_s\pi(y_1\ldots y_t)z_1\ldots z_k x_1 - \\ &\quad - x_1 x_\beta^2\pi_\alpha\{x_\alpha x_2\ldots x_s\pi(y_1\ldots y_t)z_1\ldots z_k x_\alpha\}\end{aligned}$$

which is the same.

Next suppose $s = 0$, $t \geqq 1$. We use case (A).

$$\begin{aligned}\pi_\beta\pi_\alpha(u) &= -\gamma x_\alpha^2\pi_\beta(y_1\ldots y_t x_\beta z_1\ldots z_k x_\beta) - \\ &\quad - \pi_\beta\{x_\beta\pi_\alpha(x_\alpha y_1\ldots y_t x_\beta z_1\ldots z_k x_\alpha)\} \\ &= -\gamma x_\alpha^2 x_\beta^2 y_1\ldots y_t\pi(z_1\ldots z_k) - \\ &\quad - \pi_\beta\pi_\alpha(x_\beta x_\alpha y_1\ldots y_t x_\beta z_1\ldots z_k x_\alpha) \\ &= -\gamma x_\alpha^2 x_\beta^2 y_1\ldots y_t\pi(z_1\ldots z_k) - \\ &\quad - \pi_\alpha\pi_\beta(x_\beta x_\alpha y_1\ldots y_t x_\beta z_1\ldots z_k x_\alpha) \\ &= -\gamma x_\alpha^2 x_\beta^2 y_1\ldots y_t\pi(z_1\ldots z_k) + \\ &\quad + \gamma x_\alpha^2 x_\beta^2 y_1\ldots y_t\pi(z_1\ldots z_k) + \\ &\quad + \pi_\alpha\{x_\alpha\pi_\beta(x_\beta y_1\ldots y_t x_\beta z_1\ldots z_k x_\alpha)\} \\ &= \pi_\alpha\pi_\beta(u).\end{aligned}$$

Finally, if $s = t = 0$, then $u = x_\alpha x_\beta^2 z_1\ldots z_k x_\alpha$ and we use the fact that

$$\pi_\alpha(x_\alpha x_\beta^2 z_1\ldots z_k x_\alpha) = x_\beta^2\pi_\alpha(x_\alpha z_1\ldots z_k x_\alpha)$$

together with centrality of x_α^2, x_β^2 to get $\pi_\alpha\pi_\beta(u) = \pi_\beta\pi_\alpha(u)$. This proves the lemma. □

STEP II: Let I_0 be the ideal of A_0 generated by the elements

$$[x_i \circ x_j, x_k] \tag{12}$$

$$[[x_i, x_t] \circ x_j, x_k] \tag{13}$$

which correspond to identities in A. If any of the subscripts in (12) or (13) are equal, then using canonical forms we see that the corresponding elements are 0 in A_0. So I_0 is generated by elements (12), (13) with pairwise distinct subscripts. Let B_0 be the subalgebra of A_0 generated by 2-words in $x_1, x_2, \ldots,$ and let B_1 be the image of B_0 under the natural map

$$A_0 \to A_1 = A_0/I_0.$$

LEMMA 2.4 *The algebra B_1 is isomorphic to B_0 under the natural map.*

Proof: It is sufficient to prove that $B_0 \cap I_0 = 0$. Using the canonical form in A_0 we see that 2-elements in I_0 are obtained by multiplying elements (12) and (13) on left or right by x_i's which occur in these elements, and that this process yields a generating set for $I_0 \cap B_0$. Using the fact that $\gamma = -\frac{2}{3}$ we easily find these relations:

$$x_k[x_i, x_j]x_k = (2\gamma+1)x_k^2[x_1, x_j] = -(1+\gamma)[x_i, x_j]x_k^2$$

$$[x_i, x_t]\circ x_j = x_i \circ [x_t, x_j] + (x_i \circ x_j)x_t - x_t(x_i \circ x_j)$$

$$x_j[[x_i, x_t]\circ x_j, x_k] = -\gamma x_j^2[x_i, x_t]x_k + \gamma x_j^2([x_i, x_t]\circ x_k) -$$

$$- \gamma x_j^2 x_k[x_i, x_j] = 0$$

$$x_i[[x_i, x_t]\circ x_j, x_k] = x_i[[x_i \circ x_j, x_t], x_k] = \gamma[x_k \circ x_j, x_t]x_i^2$$

$$x_k[[x_i, x_t]\circ x_j, x_k] = \gamma x_k^2\{[x_i \circ x_j, x_t] - [x_t \circ x_j, x_i]\}$$

together with similar expressions when the elements (13) are multiplied on the right by x_i, x_j, or x_k. Hence the 2-elements of I_0 are generated by elements (12), multiplied on the left or right by x_i, x_j, or x_k. But all such products are zero. Therefore $I_0 \cap B_0 = 0$. □

STEP III: Consider the ideal I_1 of A_1 generated by elements

$$\sum_\sigma v_{\sigma(1)} \cdots v_{\sigma(n)} \tag{14}$$

where $n = p-2$, σ runs over all permutations of $\{1, \ldots, n\}$, and the v_i are commutators of arbitrary length in $x_1, x_2, \ldots$ (modulo I_0). Then A_1/I_1 is isomorphic with the original algebra A if we identify the generators. Now relation (1) holds in A_1. Let L_1 be the Lie algebra generated by the x_i (modulo I_0). Then every element of L_1 can be expressed as a linear combination of elements vc, where v is a commutator of length 1 or 2 in the x_i, and c is in the centre of A_1. Thus L_1 is generated by elements (14) where the v_i are commutators of length $\leqq 2$.

Let B be the image of B_1 under the natural map

$$A_1 \to A_1/I_1 \cong A.$$

LEMMA 2.5 *The algebra B is isomorphic to B_1 under the natural map.*

Proof: For $v_1, \ldots, v_n$ as in (14) write

$$S(v_1, \ldots, v_n) = \sum_\sigma v_{\sigma(1)} \ldots v_{\sigma(n)}.$$

Commutating (14) with x_t gives

$$[S(v_1, \ldots, v_n), x_t] = \sum_{j=1}^n S(v_1, \ldots, v_{j-1}, [v_j, x_t], v_{j+1}, \ldots, v_n) \tag{15}$$

so I_1 is generated by elements (14) with the v's of length 1. Moreover, if $v_n = v_{n-1} = x_k$ in (15), then

$$2S(v_1, \ldots, v_{n-2}, [x_k, x_t], x_k) = [S(v_1, \ldots, v_{n-2}, x_k, x_k), x_t] - \\ - \sum_{j=1}^{n-2} S(v_1, \ldots, v_{j-1}, [v_j, x_t], v_{j+1}, \ldots, v_{n-2}, x_k, x_k).$$

Thus elements (14) where the commutators v_j are of length $\leqq 2$ and $v_n = x_k$, $v_{n-1} = [x_k, x_t]$ all lie in the ideal generated by elements (14) for which $v_n = v_{n-1} = x_k$. But in A_1 we have:

$$\begin{aligned} S(v_1, \ldots, v_{n-1}, x_k)x_k &= \\ &= \frac{n-1}{2} \sum_{j=1}^{n-1} S(v_1, \ldots, v_{j-1}, v_{j+1}, \ldots, v_{n-1})(v_j \circ x_k)x_k + \\ &\quad + S(v_1, \ldots, v_{n-1})x_k^2 \\ &= \left(\frac{n-1}{2}(-\gamma) + 1\right) S(v_1, \ldots, v_{n-1})x_k^2 \\ &= \left(\frac{p-3}{2}\left(\tfrac{2}{3}\right) + 1\right) S(v_1, \ldots, v_{n-1})x_k^2 \\ &= \frac{p}{3} S(v_1, \ldots, v_{n-1})x_k^2 = 0. \end{aligned} \tag{16}$$

Also,

$$\begin{aligned} -\gamma x_t^2 S(v_1, \ldots, v_{n-2}, x_k, x_k) &= S(v_1, \ldots, v_{n-2}, x_t, x_k)(x_t \circ x_k) \\ &= S(v_1, \ldots, v_{n-2}, x_t, x_k)x_k x_t + S(v_1, \ldots, v_{n-2}, x_k, x_t)x_t x_k \\ &= 0. \end{aligned} \tag{17}$$

It follows that $I_1 \cap B_1$ is generated by 2-elements obtained from (14), in which the v_j are commutators which have no x_i in common (modulo I_0). Put $v_n = x_k$ in (15), multiply by x_k, and use (17): this gives

$$S(v_1, \ldots, v_{n-1}, x_k)(x_t \circ x_k) = -\gamma x_k^2 S(v_1, \ldots, v_{n-1}, x_t)$$
$$= S(v_1, \ldots, v_{n-1}, [x_k, x_t])x_k.$$

Hence 2-elements in I_1 are generated by elements (14) for which the v_i have length 1. Now in A_0 we have the relation

$$x_j u = u x_j$$

where $x_j u$ is a 2-word, from which it follows that $I_1 \cap B_1$ is generated by 2-elements obtained by multiplying (14) on the right by generators occurring in these expressions. But from (16), using the symmetry of $S(v_1, \ldots, v_n)$, it follows that these generators are 0. Hence $I_1 \cap B_1 = 0$. □

The proof of lemma 2.2 (and with it that of theorem 1.1 as noted above) follows immediately: the canonical form for 2-words is unique in A_0, but 2-words map isomorphically into B_1 and hence into $B \subseteq A$. Hence the canonical form in A is unique. □

Razmyslov [177] goes on to show that A satisfies the identity

$$x^p = 0.$$

It follows that the group $G \subseteq A^0$ (adjoint group) generated by $1+x_1, 1+x_2, \ldots$ is of exponent p, but not nilpotent.

Some open questions

The questions that follow have not, to the best of our knowledge and at the time of writing, been answered. They represent possible lines for further research. Some are probably very easy, others extremely difficult.

1. Do there exist Lie algebras satisfying Max that are not finite-dimensional?
2. Do there exist Lie algebras satisfying Min that are not finite-dimensional?
3. If the answers to (1) and (2) are both yes, what happens for Max $\cap$ Min?
4. Every Lie algebra L has a unique ideal I maximal subject to the condition

   $$L \text{ si } K \Rightarrow I \lhd K.$$

   Certainly $I \geqq L^{\omega}$. Do we always have equality? If not, characterise I.
5. If H asc L but H is not finite-dimensional, is it necessarily the case that $H^{\omega} \lhd L$?
6. Is Max-$\lhd$ coalescent (persistence is sufficient) in characteristic zero?
7. If $\mathfrak{X}$ is $\{\mathrm{N}_0, \mathrm{I}\}$-closed, is $\mathfrak{G} \cap \mathfrak{X}$ coalescent in characteristic zero?
8. Give necessary and sufficient conditions on a pair (H, K) of Lie algebras, over the same field of characteristic zero, such that whenever L is a Lie algebra with H si L, K si L, then $\langle H, K\rangle$ si L. A reasonable guess is that H and K should both lie in the class $\mathfrak{C}$. This seems to be true for H and K abelian.
9. Is $\mathfrak{G}$ ascendantly coalescent in characteristic zero?
10. Is Max-si ascendantly coalescent in characteristic zero?
11. Is the sum of two locally soluble ideals of a Lie algebra necessarily locally soluble?
12. Can a non-trivial Baer algebra have trivial Fitting radical?
13. What is the best possible bound on the nilpotency class of $\langle (N+N\delta)^L \rangle$, where N is an $\mathfrak{N}_c$-ideal of L over a field of characteristic zero and δ is a derivation of L?
14. If $x \in L$ implies $\langle x \rangle$ asc L, is L necessarily locally nilpotent? (This is true for characteristic zero).

15. Find, or prove non-existence of, a *useful* 'wreath product' for Lie algebras. (See Robinson [331] for the group-theoretic wreath product.) It is not clear what 'useful' should mean in this context.
16. In characteristic zero, is $\mathfrak{D} = \mathfrak{N}$?
17. Is $\mathfrak{D}_{n,1} \leqq \mathfrak{N}$ for all n? Is it $\leqq \mathfrak{N}_{f(n)}$ for some function f?
18. Exactly which classes $\mathfrak{D}_{n,r}$ are nilpotent?
19. Suppose $x \in L$ implies $\langle x^L \rangle \in \mathfrak{N}_n$. Is $L \in \mathfrak{N}$?
20. We know that $\mathfrak{D}_1 = \mathfrak{N}_1$, $\mathfrak{D}_2 = \mathfrak{N}_2$. Is $\mathfrak{D}_3 = \mathfrak{N}_3$? If not, what is the best upper bound for the nilpotency class of algebras in $\mathfrak{D}_3$? If it is, do we have $\mathfrak{D}_n = \mathfrak{N}_n$ for all n?
21. In characteristic zero, if $L \in \mathrm{E}\mathfrak{A} \cap \text{Max-}\lhd \cap \text{Min-}\lhd$, need L be nilpotent-by-finite? If not, can we say anything interesting about the structure of L?
22. If $L \in$ Min-si, over a field of characteristic zero, is $w(L)$ an ideal of L?
23. Is $\lhd \mathfrak{N}_2$-Fin equal to $\lhd \mathfrak{A}$-Fin?
24. Are there any inclusions between Max-$\lhd \mathfrak{A}$, Max-$\lhd \mathfrak{N}$, Max-$\lhd \mathrm{E}\mathfrak{A}$; Min-$\lhd \mathfrak{A}$, Min-$\lhd \mathfrak{N}$, Min-$\lhd \mathrm{E}\mathfrak{A}$?
25. If L is locally soluble over a field of characteristic $p > 0$ and L has infinite dimension, must L have an infinite-dimensional abelian subalgebra?
26. Same question as (25), but with L locally finite.
27. Do there exist infinite-dimensional Lie algebra satisfying Max-u? Max-cu? Max-pu?
28. Over a field of characteristic $p > 0$, suppose that L has derived length exactly d, whereas every proper subalgebra has derived length $< d$. Is L finite-dimensional?
29. Do algebras in $\mathfrak{G} \cap \mathfrak{A}\mathfrak{N}$ possess Cartan subalgebras? Is there a reasonable formation theory?
30. In characteristic zero, if $L \in \mathfrak{G} \cap \mathfrak{N}^2$, is $\psi(L)$ nilpotent? In particular, if $L \in \mathfrak{G} \cap \mathfrak{N}$, does $U(L)$ have the chief annihilator property?
31. Do subalgebras of $\mathfrak{G} \cap \mathfrak{A}^2$-algebras have good Frattini structure?
32. Are Levi factors of $\mathfrak{H}$-algebras conjugate by locally radical automorphisms?
33. Does there exist a formation theory for locally soluble $\mathfrak{H}$-algebras? For $\overline{\mathfrak{F}}$-algebras?
34. If L is locally finite over a field of characteristic zero, and semisimple, is L necessarily locally finite-dimensional-and-semisimple?
35. Classify simple locally finite Lie algebras in characteristic zero, perhaps under extra hypotheses such as countability of dimension.
36. Which Hall words give rise to equal varieties?
37. Which Hall varieties are finitely based?
38. Which dichotomy corresponds to stuntedness?

39. Is the variety generated by a finite-dimensional Lie algebra finitely based, in characteristic $\neq 2$ (especially zero)?
40. In characteristic $p > 0$, suppose that $L \in \acute{\mathrm{E}}\mathfrak{A} \cap \mathrm{L}\mathfrak{N}$. Is it true that $x \in L$ implies $\langle x \rangle$ asc L?
41. Is the best bound for the nilpotency class of an $\mathfrak{E}_4$-algebra (in characteristic $\neq 2$, 3, or 5) 9 or 8?
42. In characteristic zero, do we have $\mathfrak{E}_n \leqq \mathfrak{N}_{f(n)}$ for a suitable function f? In particular, what happens for $\mathfrak{E}_5$?
43. Can Kostrikin's proof be simplified?
44. In general, is $\mathfrak{r}(L) \subseteq \mathfrak{e}(L)$?

References

The references are divided into two sections. The first comprises a fairly comprehensive bibliography of material related to the algebraic theory of infinite-dimensional Lie algebras, but including finite-dimensional material (such as the theory of the universal enveloping algebra) which seems to be of importance for the infinite-dimensional theory. The second lists all other material referred to in the text, in particular the group-theoretic papers motivating much of the work on Lie algebras.

Papers in languages other than English, French, German, and Italian are marked (*).

Section 1: Lie algebras

R. K. AMAYO

[1] Infinite-dimensional Lie algebras, M.Sc. thesis, Univ. of Warwick (1970).

[2] Infinite-dimensional Lie algebras, Ph.D. thesis, Univ. of Warwick (1972).

[3] Soluble subideals of Lie algebras, Compositio Math. 25 (1972) 221–232.

[4] A note on finite-dimensional subideals of Lie algebras, Bull. London Math. Soc. 5 (1973) 49–53.

[5] The derived join theorem and coalescence in Lie algebras, Compositio Math. 27 (1973) 119–123.

[6] Locally coalescent classes of Lie algebras, Compositio Math. 27 (1973) 107–117.

[7] Lie algebras in which every finitely generated subalgebra is a subideal, Tôhoku J. Math. (to appear).

[8] Lie algebras in which every n-generator subalgebra is an n-step subideal, J. Algebra (to appear).

[9] Finiteness conditions on soluble Lie algebras, J. London Math. Soc. (to appear).

[10] Engel Lie rings with chain conditions, Pacific J. Math. (to appear).

R. K. AMAYO and I. N. STEWART

[11] Finitely generated Lie algebras, J. London Math. Soc. (2) 5 (1972) 697–703.

[12] Descending chain conditions for subideals of Lie algebras of prime characteristic (to appear).

S. A. AMITSUR

[13] Derivations in simple rings, Proc. London Math. Soc. (3) 7 (1957) 87–112.

T. ANDERSON

[14] Hereditary radicals and derivations of algebras, Canad. J. Math. 21 (1969) 372–377.

S. ANDREADAKIS

[15] On a Lie algebra of polynomials, Bull. Soc. Math. Grèce (NS) 7 (1966) 148–153.

K. K. ANDREEV

[16] Nilpotent groups and Lie algebras, Algebra i Logika 7 (1968) 4–14. (*)

[17] Nilpotent groups and Lie algebras II, Algebra i Logika 8 (1969) 625–635. (*)

S. BACHMUTH, H. Y. MOCHIZUKI, and D. W. WALKUP

[18] A nonsolvable group of exponent 5, Bull. Amer. Math. Soc. 76 (1970) 638–640.

JU. A. BAHTURIN

[19] Two remarks on varieties of Lie algebras, Mat. Zametki 4 (1968) 387–398 (*), translated in Math. Notes 4 (1968) 725–730.

[20] On the approximation of Lie algebras, Mat. Zametki 12 (1972) 713–716 (*), translated in Math. Notes 12 (1972).

D. W. BARNES

[21] Lie algebras, lecture notes, Univ. of Tübingen (1968).

D. W. BARNES and H. GASTINEAU-HILLS

[22] On the theory of soluble Lie algebras, Math. Z. 106 (1968) 343–354.

D. W. BARNES and M. L. NEWELL

[23] Some theorems on saturated homomorphs of soluble Lie algebras, Math. Z. 115 (1970) 179–187.

B. BAUMSLAG

[24] Free Lie algebras and free groups, J. London Math. Soc. (2) 4 (1972) 523–532.

W. E. BAXTER

[25] Lie simplicity of a special class of associative rings, Proc. Amer. Math. Soc. 7 (1956) 855–863.

[26] Lie simplicity of a special class of associative rings II, Trans. Amer. Math. Soc. 87 (1958) 63–75.

G. BIRKHOFF

[27] Representability of Lie algebras and Lie groups by matrices, Ann. Math. (2) 38 (1937) 526–532.

L. A. BOKUT'

[28] Imbedding of algebras into algebraically closed algebras, Dokl. Akad. Nauk SSSR 145 (1962) 963–964. (*)

[29] Embedding Lie algebras into algebraically closed Lie algebras, Algebra i Logika 1 (1962) 47–53. (*)

[30] A basis for free polynilpotent Lie algebras, Algebra i Logika 2 (1963) 13–19. (*)

[31] Unsolvability of the identity problem for Lie algebras, Dokl. Akad. Nauk SSSR 206 (1972) 1277–1279 (*), translated in Soviet Math. Dokl. 13 (1972) 1388–1391.

N. BOURBAKI

[32] Groupes et algèbres de Lie, Hermann, Paris 1968.

S. G. BRAZIER

[33] Stability and parasolubility of Lie rings, thesis, Univ. of Warwick (1973).

R. M. BRYANT and M. R. VAUGHAN-LEE

[34] Soluble varieties of Lie algebras, Quart. J. Math. Oxford (2) 23 (1972) 107–112.

É. CARTAN

[35] Oeuvres complètes, Gauthier-Villars, Paris 1955.

P. CARTIER

[36] Démonstration algébrique de la formule de Hausdorff, Bull. Soc. Math. France 84 (1956) 241–249.

[37] Remarques sur la theorème de Birkhoff-Witt, Ann. Scuola Norm. Sup. Pisa (3) 12 (1958) 1–4.

B. CHANG

[38] On Engel rings of exponent $p-1$ over GF(p), Proc. London Math. Soc. (3) 11 (1968) 203–212.

C. CHEVALLEY

[39] Seminaire Chevalley Vols. I, II: Classifications des groupes de Lie algébriques, Paris 1956–8.

C-Y. CHAO

[40] Uncountably many non-isomorphic nilpotent Lie algebras, Proc. Amer. Math. Soc. 13 (1962) 903–906.

[41] Some characterisations of nilpotent Lie algebras, Math. Z. 103 (1968) 40–42.

K-T. Chen

[42] Linear independence of exponentials of Lie elements, An. Acad. Brasil Ci. 31 (1959) 507–509.

P. M. Cohn

[43] A non-nilpotent Lie ring satisfying the Engel condition and a non-nilpotent Engel group, Proc. Cambridge Philos. Soc. 51 (1955) 401–405.

[44] A remark on the Birkhoff-Witt theorem, J. London Math. Soc. 38 (1963) 197–203.

[45] Subalgebras of free associative algebras, Proc. London Math. Soc. (3) 14 (1964) 618–632.

N. Conze

[46] Idéaux primitifs de l'algèbre enveloppante d'une algèbre de Lie nilpotente et orbites dans l'espace dual, C.R.Acad. Sci. Paris Ser. A 267 (1968) 325–327.

I. Cuculescu

[47] On the formula of the differential of the exponential mapping of a Lie algebra, Stud. Cerc. Mat. 22 (1970) 223–227.

C. W. Curtis

[48] Noncommutative extensions of Hilbert rings, Proc. Amer. Math. Soc. 4 (1953) 945–955.

[49] On Lie algebras of algebraic linear transformations, Pacific J. Math. 6 (1956) 453–466.

R. L. Davis

[50] Torsion in Engel modules, Proc. Amer. Math. Soc. 10 (1959) 679–687.

C. E. Dididze

[51] Non-associative free sums of algebras with an amalgamated subalgebra, Soobšč. Akad. Nauk Gruzin SSR 18 (1957) 11–17. (*)

[52] Non-associative free sums of algebras with an amalgamated subalgebra, Mat. Sb. (NS) 43 (85) (1957) 379–396. (*)

[53] Subalgebras of non-associative free sums of algebras with arbitrary amalgamated subalgebra, Mat. Sb. (NS) 54 (96) (1961) 381–384 (*), translated in Amer. Math. Soc. Translations (2nd ser.) 50 (1966) 183–187.

[54] Free sums of Ω-algebras with an amalgamated Ω-subalgebra, Sakharth. SSR Mecn. Akad. Moambe 50 (1968) 531–534. (*)

J. Dixmier

[55] Représentations irréductibles des algèbres de Lie nilpotentes, Ann. Acad. Brasil. Cienc. (1963).

[56] Sur l'algèbre enveloppante d'une algèbre de Lie nilpotente, Arch. Math. 10 (1959) 321–326.

M. P. Drazin and K. W. Gruenberg

[57] Commutators in associative rings, Proc. Cambridge Philos. Soc. 49 (1953) 590–594.

E. B. Dynkin

[58] Normed Lie algebras and analytic groups, Uspehi Mat. Nauk (NS) 5 (35) (1950) 135–186. (*)

M. Eichler

[59] A new proof of the Baker-Campbell-Hausdorff formula, J. Math. Soc. Japan 20 (1968) 23–25.

I. Fischer and R. R. Struik

[60] Nil algebras and periodic groups, Amer. Math. Monthly 75 (1968) 611–623.

F-J. Fritz

[61] Schunk and Fitting classes of groups and Lie algebras, M.Sc. dissertation, Univ. of Warwick (1972).

References: Section 1

A. T. GAĬNOV

[62] Free commutative and free anticommutative products of algebras, Dokl. Akad. Nauk SSSR 133 (1960) 1275–1278 (*), translated in Soviet Math. Dokl. 1 (1961) 956-959.

[63] Free commutative and free anticommutative products of algebras, Sibirsk. Mat. Ž. 3 (1962) 805–833. (*)

M. B. GAVRILOV

[64] Varieties of associative algebras, C. R. Acad. Bulgare Sci. 21 (1968) 989–992. (*)

[65] T-ideals with the element $[[x_1, x_2], x_3[x_4, x_5]]$, C.R. Acad. Bulgare Sci. 21 (1968) 1153–1156. (*)

[66] On T-ideals containing the element $[x_1, x_2, x_3, x_4, [x_5, x_6]]$, Bülgar Akad. Nauk Otd. Mat. Fiz. Nauk Izv. Mat. Inst. 11 (1970) 269–271. (*)

V. M. GLUŠKOV

[67] Locally nilpotent groups without torsion, complete over simple topological fields, Mat. Sb. (NS) 37 (79) (1955) 477–506. (*)

[68] Exact triangular representations of Lie Z-algebras, Dokl. Akad. Nauk SSSR (NS) 100 (1955) 617–620. (*)

E. S. GOLOD

[69] On nil-algebras and finitely approximable p-groups, Izv. Akad. Nauk SSSR Ser. Mat. 28 (1964) 273–276 (*), translated in Amer. Math. Soc. Translations (2nd ser.) 48 (1965) 103–106.

E. S. GOLOD and I. R. ŠAFAREVIČ

[70] On class field towers, Izv. Akad. Nauk SSSR Ser. Mat. 28 (1964) 261–272 (*), translated in Amer. Math. Soc. Translations (2nd ser.) 48 (1965) 91–102.

V. E. GOVOROV

[71] Algebras freely generated by finite amalgams, Mat. Sb. (NS) 50 (92) (1960) 241–246 (*), translated in Amer. Math. Soc. Translations (2nd ser.) 50 (1966) 288–294.

K. W. GRUENBERG

[72] Two theorems on Engel groups, Proc. Cambridge Philos. Soc. 49 (1953) 377–380.

V. GUILLEMIN

[73] A Jordan-Hölder decomposition for a certain class of infinite dimensional Lie algebras, J. Diff. Geom. 2 (1968) 313–345.

[74] Infinite dimensional primitive Lie algebras, J. Diff. Geom. 4 (1970) 257–282.

M. HALL

[75] A basis for free Lie rings and higher commutators in free groups, Proc. Amer. Math. Soc. 1 (1950) 575–581.

B. HARTLEY

[76] Locally nilpotent ideals of a Lie algebra, Proc. Cambridge Philos. Soc. 63 (1967) 257–272.

N. HEEREMA

[77] A group of a Lie algebra, J. Reine Angew. Math. 244 (1970) 112–118.

H. HEINEKEN

[78] Endomorphismenringe und Engelsche Elemente, Arch. Math. 13 (1962) 29–37.

[79] Liesche Ringe mit Engelbedingung, Math. Ann. 149 (1962–3) 232–236.

I. N. HERSTEIN

[80] On the Lie ring of a simple ring, Proc. Nat. Acad. Sci. USA 40 (1954) 305–306.

[81] On the Lie ring of a division ring, Ann. of Math. (2) 60 (1954) 571–575.

[82] On the Lie and Jordan rings of a simple associative ring, Amer. J. Math. 77 (1955) 279–285.

[83] The Lie ring of a simple associative ring, Duke Math. J. 22 (1955) 471–476.
[84] Lie and Jordan systems in simple rings with involution, Amer. J. Math. 78 (1956) 629–649.
[85] Lie and Jordan structures in simple associative rings, Bull. Amer. Math. Soc. 67 (1961) 517–531.
[86] Topics in ring theory, Chicago 1969.

J-C. Herz
[87] Pseudo-algèbres de Lie I, C.R. Acad. Sci. Paris 236 (1953) 1935–1937.
[88] Pseudo-algèbres de Lie II, C.R. Acad. Sci. Paris 236 (1953) 2289–2291.

P. J. Higgins
[89] Lie rings satisfying the Engel condition, Proc. Cambridge Philos. Soc. 50 (1954) 8–15.
[90] Baer invariants and the Birkhoff-Witt theorem, J. Algebra 11 (1969) 469–482.

G. Higman
[91] On finite groups of exponent 5, Proc. Cambridge Philos. Soc. 52 (1956) 381–390.
[92] Groups and rings having automorphisms without nontrivial fixed elements, J. London Math. Soc. 32 (1957) 321–334.

G. Hochschild and J-P. Serre
[93] Cohomology of Lie algebras, Ann. of Math. (2) 57 (1953) 591–603.

K. H. Hoffman
[94] Lie algebras with subalgebras of co-dimension one, Illinois J. Math. 9 (1965) 636–643.

J. F. Hurley
[95] Ideals in Chevalley algebras, Trans. Amer. Math. Soc. 137 (1969) 245–258.
[96] Composition series in Chevalley algebras, Pacific J. Math. 32 (1970) 429–434.
[97] Extensions of Chevalley algebras, Duke Math. J. 38 (1971) 349–356.

N. Jacobson
[98] Lie algebras, Interscience, New York 1962.

S. A. Jennings
[99] The group ring of a class of infinite nilpotent groups, Canad. J. Math. 7 (1955) 169–187.

V. G. Kac
[100] Simple irreducible graded Lie algebras of finite growth, Izv. Akad. Nauk SSSR Ser. Mat. 32 (1968) 1323–1367. (*)
[101] The classification of the simple Lie algebras over a field with nonzero characteristic, Izv. Akad. Nauk SSSR Ser. Mat. 34 (1970) 385–408. (*)

I. L. Kantor
[102] Infinite dimensional simple graded Lie algebras, Dokl. Akad. Nauk SSSR 179 (1968) 534–537 (*), translated in Soviet Math. Dokl. 9 (1968) 409–412.

I. Kaplansky
[103] Lie algebras, Lectures in modern mathematics, vol. I pp. 115–132, Wiley, New York 1963.
[104] Lie algebras and locally compact groups, Chicago 1971.
[105] Infinite-dimensional Lie algebras, (preprint) (1973).

J. Knopfmacher
[106] Universal envelopes for non-associative algebras, Quart. J. Math. Oxford (2) 13 (1962) 264–282.
[107] Extensions in varieties of groups and algebras, Acta Math. 115 (1966) 17–50.
[108] On the isomorphism problem for Lie algebras, Proc. Amer. Math. Soc. 18 (1967) 898–901.

References: Section 1

A. I. Kostrikin

[109] Solution of a weakened problem of Burnside for exponent 5, Izv. Akad. Nauk SSSR Ser. Mat. 19 (1955) 233–244. (*)

[110] On Lie rings satisfying the Engel condition, Dokl. Akad. Nauk SSSR (NS) 108 (1956) 580–582. (*)

[111] Lie rings satisfying the Engel condition, Izv. Akad. Nauk SSSR Ser. Mat. 21 (1957) 515–540 (*), translated in Amer. Math. Soc. Translations (2nd ser.) 45 (1965) 191–220.

[112] On the connection between periodic groups and Lie rings, Izv. Akad. Nauk SSSR Ser. Mat. 21 (1957) 289–310 (*), translated in Amer. Math. Soc. Translations (2nd ser.) 45 (1965) 165–189.

[113] On local nilpotency of Lie rings that satisfy Engel's condition, Dokl. Akad. Nauk SSSR (NS) 118 (1958) 1074–1077. (*)

[114] The Burnside problem, Izv. Akad. Nauk SSSR Ser. Mat. 23 (1959) 3–34 (*), translated in Amer. Math. Soc. Translations (2nd ser.) 36 (1964) 63–100.

[115] Lie algebras and finite groups, Proc. Int. Congr. Math. Stockholm 1962, 264–269 (*), translated in Amer. Math. Soc. Translations (2nd ser.) 31 (196) 40–46.

[116] Theorem on semi-simple Lie p-algebras, Mat. Zametki 2 (1967) 465–474 (*), translated in Math. Notes 2 (1967) 773–778.

[117] Squares of adjoint endomorphisms in simple Lie p-algebras, Izv. Akad. Nauk SSSR Ser. Mat. 31 (1967) 445–487 (*), translated in Math. USSR-Izv. 1 (1967) 435–473.

A. I. Kostrikin and I. R. Šafarevič

[118] Graded Lie algebras of finite characteristic, Izv. Akad. Nauk SSSR Ser. Mat. 33 (1969) 251–322 (*), translated in Math. USSR-Izv. 3 (1969) 237–304.

E. F. Krause and K. W. Weston

[119] An algorithm related to the restricted Burnside group of prime exponent, Computational problems in abstract algebra pp. 185–187, Pergamon, Oxford 1970.

[120] On the Lie algebra of a Burnside group of exponent 5, Proc. Amer. Math. Soc. 27 (1971) 463–470.

G. P. Kukin

[121] The Cartesian subalgebra of the free Lie sum of Lie algebras, Algebra i Logika 9 (1970) 701–713 (*), translated in Algebra and Logic 9 (1970) 422–430.

[122] Primitive elements of free Lie algebras, Algebra i Logika 9 (1970) 458–472 (*), translated in Algebra and Logic 9 (1970).

[123] Subalgebras of a free Lie sum of Lie algebras with an amalgamated subalgebra, Algebra i Logika 11 (1972) 59–86 (*), translated in Algebra and Logic 11 (1972).

A. G. Kuroš

[124] The present status of the theory of rings and algebras, Uspehi Mat. Nauk (NS) 6 (42) (1951) 3–15. (*)

E. N. Kuz'min

[125] Engel's theorem for binary Lie algebras, Dokl. Akad. Nauk SSSR 176 (1967) 771–773 (*), translated in Soviet Math. Dokl. 8 (1967) 1191–1193.

[126] Anticommutative algebras satisfying Engel's condition, Sibirsk. Mat. Ž. 8 (1967) 1026–1034. (*)

[127] A locally nilpotent radical of Mal'cev algebras satisfying the n-th Engel condition, Dokl. Akad. Nauk SSSR 177 (1967) 508–510 (*), translated in Soviet Math. Dokl. 8 (1967) 1434–1436.

[128] Locally finite Mal'cev algebras, Algebra i Logika 6 (1967) 27–30. (*)

[129] Algebraic sets in Mal'cev algebras, Algebra i Logika 7 (1968) 42–47. (*)

V. N. LATYŠEV

[130] On algebras with identity relations, Dokl. Akad. Nauk SSSR 146 (1962) 1003–1006 (*), translated in Soviet Math. Dokl. 3 (1962) 1423–1427.

[131] On the choice of basis in a T-ideal, Sibirsk. Mat. Ž. 4 (1963) 1122–1127. (*)

[132] On Lie algebras with identical relations, Sibirsk. Mat. Ž. 4 (1963) 821–829. (*)

[133] On zero-divisors and nil-elements in a Lie algebra, Sibirsk. Mat. Ž. 4 (1963) 830–836. (*)

[134] On the finiteness of the number of generators of a T-ideal with an element $[x_1, x_2, x_3, x_4]$, Sibirsk. Mat. Ž. 6 (1965) 1432–1434. (*)

[135] The Specht property of the T-ideal $T = \{[x_1, \ldots, x_{n-2}, [x_{n-1}, x_n]]\}$, Dokl. Akad. Nauk SSSR 207 (1972) 777–780 (*), translated in Soviet Math. Dokl. 13 (1972) 1604–1608.

M. LAZARD

[136] Sur les algèbres enveloppantes universelles de certaines algèbres de Lie, C.R. Acad, Sci. Paris 234 (1952) 788–791.

[137] Sur certaines suites d'éléments dans les groupes libres et leurs extensions, C.R. Acad. Sci. Paris 236 (1953) 36–38.

[138] Problemes d'extension concernant les N-groupes; inversion de la formule de Hausdorff, C.R. Acad. Sci. Paris 237 (1953) 1377–1379.

[139] Sur les groupes nilpotents et les anneaux de Lie, Ann. Sci. Ecole Norm. Sup (3) 71 (1954) 101–190.

[140] Sur les algèbres enveloppantes universelles de certaines algèbres de Lie, Publ. Sci. Univ. Algér. Sér. A 1 (1954) 281–294.

G. LEGER

[141] Characteristically nilpotent Lie algebras, Duke Math. J. 26 (1959) 623–628.

[142] A note on free Lie algebras, Proc. Amer. Math. Soc. 15 (1964) 517–518.

[143] A particular class of Lie algebras, Proc. Amer. Math. Soc. 16 (1965) 293–296.

E. M. LEVIČ

[144] On simple and strictly simple rings, Latvijas PSR Zinātņu Akad. Vēstis Fiz. Tehn. Zinātņu Sēr. 6 (1965) 53–58. (*)

E. M. LEVIČ and A. I. TOKARENKO

[145] A note on locally nilpotent torsion-free groups, Sibirsk. Mat. Ž. 11 (1970) 1406–1408 (*), translated in Siberian Math. J. 11 (1970) 1033–1034.

J. LEWIN

[146] On Schreier varieties of linear algebras, Trans. Amer. Math. Soc. 132 (1968) 553–562.

N. LIMIĆ

[147] Nilpotent locally convex Lie algebras and Lie field structures, Commn. Math. Phys. 14 (1969) 89–107.

R. C. LYNDON

[148] Identities in finite algebras, Proc. Amer. Math. Soc. 5 (1954) 8–9.

[149] Burnside groups and Engel rings, Proc. Sympos. Pure Math. 1 pp. 4–14, Amer. Math. Soc., Providence R.I. 1959.

Š-S LYU

[150] On the splitting of infinite algebras, Mat. Sb. (NS) 42 (84) (1957) 327–352. (*)

A. I. MAL'CEV

[151] On normed Lie algebras over the field of rational numbers, Dokl. Akad. Nauk SSSR (NS) 62 (1948) 745–748. (*)

[152] Nilpotent torsion-free groups, Izv. Akad. Nauk SSSR Ser. Mat. 13 (1949) 201–202. (*)

[153] On a class of homogeneous spaces, Izv. Akad. Nauk SSSR Ser. Mat. 13 (1949) 9–32. (*)

[154] Generalised nilpotent algebras and their adjoint groups, Mat. Sb. (NS) 25 (67) (1949) 347–366. (*)
[155] On algebras defined by identities, Mat. Sb. (NS) 26 (68) (1950) 19–33. (*)
[156] On a representation of nonassociative rings, Uspehi Mat. Nauk (NS) 7 (47) (1952) 181–185. (*)

E. I. MARSHALL
[157] The Frattini subalgebra of a Lie algebra, J. London Math. Soc. 42 (1967) 416–422.

O. MARUO
[158] Pseudo-coalescent classes of Lie algebras, Hiroshima Math. J. 2 (1972) 205–214.

J. C. MCCONNELL
[159] The intersection theorem for a class of noncommutative rings, Proc. London Math. Soc. (3) 17 (1967) 487–498.
[160] A note on the Weyl algebra A_n, (preprint) Univ. of Leeds 1972.
[161] Representations of solvable Lie algebras and the Gel'fand-Kirillov conjecture, (preprint) Univ. of Leeds 1973.

J. C. MCCONNELL and J. C. ROBSON
[162] Homomorphisms and extensions of modules over certain differential polynomial rings, J. Algebra 26 (1973) 319–342.

P. MCINERNEY
[163] Soluble Lie rings, Ph.D. thesis, Univ. of Warwick 1973.

R. V. MOODY
[164] A new class of Lie algebras, J. Algebra 10 (1968) 211–230.
[165] Euclidean Lie algebras, Canad. J. Math. 21 (1969) 1432–1454.

S. MORAN
[166] The Lie ring associated with certain groups, Rozprawy Mat. 44 (1965).

T. MORIMOTO and N. TANAKA
[167] The classification of the real primitive infinite Lie algebras, J. Math. Kyoto Univ. 10 (1970) 207–243.

Y. NOUAZÉ
[168] Remarques sur 'Idéaux premiers de l'algèbre enveloppante d'une algèbre de Lie nilpotente', Bull. Sci. Math. (2) 91 (1967) 117–124.

Y. NOUAZÉ and P. GABRIEL
[169] Idéaux premiers de l'algèbre enveloppante d'une algèbre de Lie nilpotente, J. Algebra 6 (1967) 77–99.

T. NAGANO
[170] On transitive infinite Lie algebras, Sûgaku 18 (1966) 65–74. (*)

V. A. PARFENOV
[171] Varieties of Lie algebras, Algebra i Logika 6 (1967) 61–73. (*)
[172] A certain property of ideals of a free Lie algebra, Sibirsk. Mat. Ž. 10 (1969) 940–944 (*), translated in Siberian Math. J. 10 (1969) 690–693.
[173] The weakly soluble radical of Lie algebras, Sibirsk. Mat. Ž. 12 (1971) 171–176 (*), translated in Siberian Math. J. 12 (1971) 123–127.

B. I. PLOTKIN
[174] Algebraic sets of elements in groups and Lie algebras, Uspehi Mat. Nauk 13 (1958) 133–138. (*)

M. S. PUTCHA
[175] On Lie rings satisfying the fourth Engel condition, Proc. Amer. Math. Soc. 28 (1971) 355–357.

D. QUILLEN

[176] On the endomorphism ring of a simple module over an enveloping algebra, Proc. Amer. Math. Soc. 21 (1969) 171–172.

YU. P. RAZMYSLOV

[177] On Engel Lie algebras, Algebra i Logika 10 (1971) 33–44 (*), translated in Algebra and Logic 10 (1971).

R. RENTSCHLER

[178] Sur la centre du quotient de l'algèbre enveloppante d'une algèbre de Lie nilpotente par un idéal premier, C.R. Acad. Sci. Paris Sér. A-B 268 (1969) A689–A692.

JU. M. RJABUHIN

[179] On the theory of radicals of non-associative rings, Mat. Issled. 3 (1968) 86–99. (*)

A. N. RUDAKOV

[180. Automorphism groups of infinite dimensional simple Lie algebras, Izv. Akad. Nauk SSSR Ser. Mat. 33 (1969) 748–764 (*), translated in Math. USSR-Izv. 3 (1969) 707–722.

J. DE RUITER

[181] Algebraic connection theory of *L*-modules, thesis, Univ. of Groningen (1972).

[182] An improvement of a result of I. N. Stewart, Compositio Math. 25 (1972) 329–333.

H. SAMELSON

[183] Notes on Lie algebras, Van Nostrand Reinhold, New York 1969.

R. D. SCHAFER

[184] An introduction to non-associative algebras, Academic Press, New York 1966.

E. SCHENKMAN

[185] A theory of subinvariant Lie algebras, Amer. J. Math. 73 (1951) 433–474.

[186] Infinite Lie algebras, Duke Math. J. 19 (1952) 529–535.

[187] On the derivation algebra and the holomorph of a nilpotent algebra, Mem. Amer. Math. Soc. 14 (1955) 15–22.

G. B. SELIGMAN

[188] On Lie algebras of prime characteristic, Amer. Math. Soc., Providence R.I. 1956.

[189] Modular Lie algebras, Springer, Berlin 1967.

J-P. SERRE

[190] Lie algebras and Lie groups, Benjamin, New York 1965.

[191] Algèbres de Lie sémisimples complexes, Benjamin, New York 1966.

L. A. SIMONJAN

[192] On two radicals of Lie algebras, Dokl. Akad. Nauk SSSR 157 (1964) 281–283 (*), translated in Soviet Math. Dokl. 5 (1964) 941–944.

[193] Certain radicals of Lie algebras, Sibirsk. Mat. Ž. 6 (1965) 1101–1107. (*)

[194] Certain examples of Lie groups and algebras, Sibirsk. Mat. Ž. 12 (1971) 837–843 (*), translated in Siberian Math. J. 12 (1971) 602–606.

A. I. ŠIRŠOV

[195] On the representation of Lie rings as associative rings, Uspehi Mat. Nauk (NS) 8 (57) (1953) 173–175. (*)

[196] Subalgebras of free Lie algebras, Mat. Sb. (NS) 33 (75) (1953) 441–452. (*)

[197] Subalgebras of free commutative and free anticommutative algebras, Mat. Sb. (NS) 34 (76) (1954) 81–88. (*)

[198] Some questions in the theory of rings close to associative, Uspehi Mat. Nauk 13 (1958) 3–26. (*)

[199] On free Lie rings. Mat. Sb. (NS) 45 (87) (1958) 113–122. (*)

[200] On the bases of a free Lie algebra, Algebra i Logika 1 (1962) 14–19. (*)

[201] Some algorithm problems for Lie algebras, Sibirsk. Mat. Ž. 3 (1962) 292–296. (*)
[202] On a hypothesis of the theory of Lie algebras, Sibirsk. Mat. Ž. 3 (1962) 297–301. (*)

A. L. Šmelkin

[203] Free polynilpotent groups, Dokl. Akad. Nauk SSSR 151 (1963) 73–75. (*)

W. Specht

[204] Gesetze in Ringen, Math. Z. 52 (1950) 557–589.

I. N. Stewart

[205] Subideals of Lie algebras, Ph.D. thesis, Univ. of Warwick (1969).
[206] The minimal condition for subideals of Lie algebras, Math. Z. 111 (1969) 301–310.
[207] Lie algebras, Lecture notes in mathematics 127, Springer, Berlin 1970.
[208] An algebraic treatment of Mal'cev's theorems concerning nilpotent Lie groups and their Lie algebras, Compositio Math. 22 (1970) 289–312.
[209] Infinite-dimensional Lie algebras in the spirit of infinite group theory, Compositio Math. 22 (1970) 313–331.
[210] Bounds for the dimensions of certain Lie algebras, J. London Math. Soc. (2) 3 (1971) 731–732.
[211] A property of locally finite Lie algebras, J. London Math. Soc. (2) 3 (1971) 334–340.
[212] The Lie algebra of endomorphisms of an infinite-dimensional vector space, Compositio Math. 25 (1972) 79–86.
[213] Structure theorems for a class of locally finite Lie algebras, Proc. London Math. Soc. (3) 24 (1972) 79–100.
[214] Levi factors of infinite-dimensional Lie algebras, J. London Math. Soc. (2) 5 (1972) 488.
[215] Author-abstract of [213], Zbl. 225 (1972) 109.
[216] Baer and Fitting radicals in groups and Lie algebras, Arch. der Math. 23 (1972) 385–386.
[217] Tensorial extensions of central simple algebras, J. Algebra 25 (1973) 1–14.
[218] Central simplicity and Chevalley algebras, Compositio Math. 26 (1973) 111–118.
[219] Verbal and marginal properties of non-associative algebras, Proc. London Math. Soc. (to appear).
[220] Finiteness conditions in soluble groups and Lie algebras, Bull. Austral. Math. Soc. 9 (1973) 43–48.
[221] A note on 2-step subideals of Lie algebras, Compositio Math. (to appear).
[222] Adjoint groups and the Mal'cev correspondence, (preprint) Univ. of Warwick 1973.

E. L. Stitzinger

[223] The Frattini subalgebra of a Lie algebra, J. London Math. Soc. (2) 2 (1970) 429–438.
[224] Frattini subalgebras of a class of solvable Lie algebras, Pacific J. Math. 34 (1970) 177–182.

H. Sunouchi

[225] Infinite Lie rings, Tôhoku Math. J. (2) 8 (1956) 291–307.

D. Tamari

[226] On the embedding of Birkhoff-Witt rings in quotient fields, Proc. Amer. Math. Soc. 4 (1953) 197–202.

S. Tôgô

[277] Derivations of Lie algebras, J. Sci. Hiroshima Univ. Ser. A-I Math. 28 (1964) 133–158.
[228] On a class of Lie algebras, J. Sci. Hiroshima Univ. Ser. A-I Math. 32 (1968) 55–83.
[229] A theorem on characteristically nilpotent algebras, J. Sci. Hiroshima Univ. Ser. A-I Math. 33 (1969) 209–212.
[230] Radicals of infinite-dimensional Lie algebras, Hiroshima Math. J. 2 (1972) 179–203.

S. Tôgô and N. Kawamoto
[231] Ascendantly coalescent classes and radicals of Lie algebras, Hiroshima Math. J. 2 (1972) 253–261.
D. A. Towers
[232] A Frattini theory for algebras, Ph.D. thesis, University of Leeds (1972).
[233] A Frattini theory for algebras, Proc. London Math. Soc. (3) 27 (1973) 440–462.
W. Tuck
[234] Frattini theory for Lie algebras, thesis, Univ. of Sydney (1969).
W. Unsin
[235] Lie-Algebren mit Idealisatorbedingung, thesis, Univ. of Erlangen-Nürnberg (1972).
F-H. Vasilesçu
[236] Radical d'une algèbre de Lie, C.R. Acad. Sci. Paris 274 (1972) 536–538.
[237] On Lie's theorem in operator algebras, Trans. Amer. Math. Soc. 172 (1972) 365–372.
M. R. Vaughan-Lee
[238] Some varieties of Lie algebras, D.Phil. thesis, Oxford (1968).
[239] Varieties of Lie algebras, Quart. J. Math. Oxford (2) 21 (1970) 297–308.
W. von Waldenfels
[240] Zur Charakterisierung Liescher Elemente in freien Algebren, Arch. der Math. 17 (1966) 44–48.
D. W. Walkup
[241] Lie rings satisfying Engel conditions, thesis, Univ. of Wisconsin (1963).
F. Wever
[242] Über reduzierte freie Liesche Ringe, Math. Z. 56 (1952) 312–325.
J. Wiegold
[243] Kostrikin's proof of the restricted Burnside conjecture for prime exponent, (duplicated notes) Australian National Univ., Canberra 1965.
[244] Personal notes on Kostrikin's theorem.
R. L. Wilson
[245] Nonclassical simple Lie algebras, Bull. Amer. Math. Soc. 75 (1969) 987–991.
[246] Irreducible Lie algebras of infinite type, Proc. Amer. Math. Soc. 29 (1971) 243–249.
E. Witt
[247] Über freie Ringe und ihre Unterringe, Math. Z. 58 (1953) 113–114.
[248] Treue Darstellung beliebiger Liescher Ringe, Collectanea Math. 6 (1953) 107–114.
[249] Die Unterringe der freien Lieschen Ringe, Math. Z. 64 (1956) 195–216.
L. S. Wollenberg
[250] Derivations of the Lie algebra of polynomials under Poisson bracket, Proc. Amer. Math. Soc. 20 (1969) 315–320.
Y-H. Xu
[251] Generalized Lie algebras, Acta Math. Sinica 15 (1965) 188–205 (*), translated in Chinese Math. Acta 6 (1965) 495–512.
G. Zappa
[252] Anelli di Lie e gruppi nilpotenti, Atti VII Congr. Un. Mat. Italiana, Genova 1963, 278–287.
M. Zorn
[253] On a theorem of Engel, Bull. Amer. Math. Soc. (2) 43 (1937) 401–404.
A. I. Žukov
[254] Nonassociative free decompositions of algebras with a finite number of generators, Mat. Sb. (NS) 26 (68) (1950) 471–478. (*)

Section 2: Other material

S. I. Adjan

[255] Identités dans les groupes, Actes Congr. Internat. Math. (1970) 1, 263–267.

[256] Infinite irreducible systems of group identities, Dokl. Akad. Nauk SSSR 190 (1970) 499–501 (*), translated in Soviet Math. Dokl. 11 (1970) 113–115.

[257] Infinite irreducible systems of group identities, Izv. Akad. Nauk SSSR Ser. Mat. 34 (1970) 715–734. (*)

M. F. Atiyah and I. G. MacDonald

[258] Introduction to commutative algebra, Addison-Wesley, Reading Massachusetts 1969.

R. Baer

[259] Endlichkeitskriterien für Kommutatorgruppen, Math. Ann. 124 (1952) 161–177.

[260] Nil-gruppen, Math. Z. 62 (1955) 402–437.

[261] Meta-ideals, Report of a conference on linear algebra (1956) 33–52, Nat. Acad. Sci. USA, Washington D.C. 1957.

[262] Lokal Noethersche Gruppen, Math. Z. 66 (1957) 341–363.

G. Birkhoff

[263] On the structure of abstract algebras, Proc. Cambridge Philos. Soc. 31 (1935) 433–454.

[264] Lattice theory, Amer. Math. Soc. Colloquium Publications 25, New York 1948.

J. L. Britton

[265] (to appear)

R. M. Bryant and M. F. Newman

[266] Some finitely based varieties of groups, (to appear).

W. Burnside

[267] On an unsettled question in the theory of discontinuous groups, Quart. J. Math. 33 (1902) 230–238.

[268] Theory of groups of finite order, Dover, New York 1955.

C. Chevalley

[269] Fundamental concepts of algebra, Academic Press, New York 1956.

D. E. Cohen

[270] On the laws of a metabelian variety, J. Algebra 5 (1967) 267–273.

P. M. Cohn

[271] Universal algebra, Harper, New York 1965.

C. W. Curtis and I. Reiner

[272] Representation theory of finite groups and associative algebras, Interscience, New York 1962.

M. DRUKKER, D. J. S. ROBINSON and I. N. STEWART

[273] The subnormal coalescence of some classes of groups of finite rank, J. Austral. Math. Soc. (to appear).

B. FISCHER

[274] Habilitationsschrift, Univ. of Frankfurt.

L. FUCHS

[275] Abelian groups, Pergamon, Oxford 1967.

W. GASCHÜTZ

[276] Zur Theorie der endlichen auflösbaren Gruppen, Math. Z. 80 (1963) 300–305.

V. M. GLUŠKOV

[277] On some questions in the theory of nilpotent and locally nilpotent torsion-free groups, Mat. Sb. 30 (1952) 79–104. (*)

A. W. GOLDIE

[278] Some aspects of ring theory, Bull. London Math. Soc. 1 (1969) 129–154.

K. W. GRUENBERG

[279] The Engel elements of a soluble group, Illinois J. Math. 3 (1959) 151–168.

[280] The upper central series in soluble groups, Illinois J. Math. 5 (1961) 436–466.

M. HALL

[281] Solution of the Burnside problem for exponent 6, Illinois J. Math. 2 (1958) 764–786.

[282] The theory of groups, Macmillan, New York 1959.

P. HALL

[283] Finiteness conditions for soluble groups, Proc. London Math. Soc. (3) 4 (1954) 419–436.

[284] Finite-by-nilpotent groups, Proc. Cambridge Philos. Soc. 52 (1956) 611–616.

[285] Nilpotent groups, Canad. Math. Congr. Summer. Sem. Univ. Alberta 1957, republished as 'The Edmonton notes on nilpotent groups,' Q.M.C. Mathematics Notes, London 1969.

[286] On the finiteness of certain soluble groups, Proc. London Math. Soc. (3) 9 (1959) 595–622.

[287] The Frattini subgroups of finitely generated groups, Proc. London Math. Soc. (3) 11 (1961) 327–352.

[288] On non-strictly simple groups, Proc. Cambridge Philos. Soc. 59 (1963) 531–553.

P. HALL and C. R. KULATILAKA

[289] A property of locally finite groups, J. London Math. Soc. 39 (1964) 235–239.

B. HARTLEY

[290] On Fischer's dualization of formation theory, Proc. London Math. Soc. (3) 19 (1969) 193–207.

[291] Serial subgroups of locally finite groups, Proc. Cambridge Philos. Soc. 71 (1972) 199–201.

H. HEINEKEN and I. J. MOHAMED

[292] A group with trivial centre satisfying the normalizer condition, J. Algebra 10 (1968) 368–376.

G. HIGMAN

[293] Ordering by divisibility in abstract algebras, Proc. London Math. Soc. (3) 2 (1952) 326–336.

K. A. HIRSCH

[294] Über lokal-nilpotente Gruppen, Math. Z. 63 (1955) 290–294.

References: Section 2

N. Jacobson

[295] Structure of rings, Amer. Math. Soc. Colloquium Publications 37, Providence R.I. 1964.

M. I. Kargapolov

[296] Some problems in the theory of nilpotent and soluble groups, Dokl. Akad. Nauk SSSR 127 (1959) 1164–1166. (*)

[297] On the completion of locally nilpotent groups, Sibirsk Mat. Ž. 3 (1962) 695–700. (*)

[298] On the π-completion of locally nilpotent groups, Algebra i Logika 1 (1962) 5–13. (*)

[299] On a problem of O. Yu. Schmidt, Sibirsk. Mat. Ž. 4 (1963) 232–235. (*)

O. H. Kegel and B. A. F. Wehrfritz

[300] Strong finiteness conditions in locally finite groups, Math. Z. 117 (1970) 309–324.

C. R. Kulatilaka

[301] Infinite abelian subgroups of some infinite groups, J. London Math. Soc. 39 (1964) 240–244.

A. G. Kuroš

[302] The theory of groups vol. I, Chelsea, New York 1960.

[303] The theory of groups vol. II, Chelsea, New York 1960.

A. G. Kuroš and S. N. Černikov

[304] Soluble and nilpotent groups, Uspehi Mat. Nauk (NS) 2 (1947) 18–59 (*), translated in Amer. Math. Soc. Translations (1st ser.) 1 (1962) 283–338.

S. Lang

[305] Algebra, Addison-Wesley, Reading Massachusetts 1965.

J. C. Lennox and J. E. Roseblade

[306] Centrality in finitely generated soluble groups, J. Algebra 16 (1970) 399–435.

R. C. Lyndon

[307] Two notes on nilpotent groups, Proc. Amer. Math. Soc. 3 (1952) 579–583.

D. H. McLain

[308] A characteristically simple group, Proc. Cambridge Philos. Soc. 50 (1954) 641–642.

[309] A class of locally nilpotent groups, Ph.D. thesis, Univ. of Cambridge (1956).

[310] Remarks on the upper central series of a group, Proc. Glasgow Math. Soc. 3 (1956) 38–44.

[311] On locally nilpotent groups, Proc. Cambridge Philos. Soc. 52 (1956) 5–11.

[312] Local theorems in universal algebras, J. London Math. Soc. 34 (1959) 177–184.

[313] Finiteness conditions in locally soluble groups, J. London Math. Soc. 34 (1959) 101–107.

B. H. Neumann

[314] On ordered groups, Amer. J. Math. 71 (1949) 1–18.

H. Neumann

[315] Varieties of groups, Springer, Berlin 1967.

P. S. Novikov and S. I. Adjan

[316] Infinite periodic groups, Izv. Akad. Nauk SSSR Ser. Mat. 32 (1968) 212–244, 251–524, 709–731 (*), translated in Math. USSR-Izv. 2 (1968) 209–236, 241–479, 665–685.

[317] Commutative subgroups and the conjugacy problem in free periodic groups of odd order, Izv. Akad. Nauk SSSR Ser. Mat. 32 (1968) 1176–1190 (*), translated in Math. USSR-Izv. 2 (1968) 1131–1144.

S. Oates and M. B. Powell

[318] Identical relations in finite groups, J. Algebra 1 (1964) 11–39.

A. Ju.Ol'šanskiĭ

[319] The finite basis problem for identities in groups, Izv. Akad. Nauk SSSR Ser. Mat. 34 (1970) 376–384. (*)

J. M. Osborn

[320] Varieties of algebras, Advances in Math. 8 (1972) 163–369.

R. E. Phillips and C. R. Combrink

[321] A note on subsolvable groups, Math. Z. 92 (1966) 349–352.

B. I. Plotkin

[322] On some criteria of locally nilpotent groups, Uspehi Mat. Nauk 9 (1954) 181–186 (*), translated in Amer. Math. Soc. Translations (2nd ser.) 17 (1961) 1–7.

[323] Generalised soluble and nilpotent groups, Uspehi Mat. Nauk 13 (1958) 89–172 (*), translated in Amer. Math. Soc. Translations (2nd ser.) 17 (1961) 29–115.

D. Quillen

[324] Complete Hopf algebras, (duplicated notes).

A. Rae

[325] A class of locally finite groups, Proc. London Math. Soc. (3) 23 (1971) 459–476.

D. J. S. Robinson

[326] Joins of subnormal subgroups, Illinois J. Math. 9 (1965) 144–168.

[327] On the theory of subnormal subgroups, Math. Z. 89 (1965) 30–51.

[328] Finiteness conditions for subnormal and ascendant abelian subgroups, J. Algebra 10 (1968) 333–359.

[329] Infinite soluble and and nilpotent groups, Q.M.C. Mathematics notes, London 1968

[330] A theorem on finitely generated hyperabelian groups, Invent. Math. 10 (1970) 38–43.

[331] Finiteness conditions and generalized soluble groups Part I, Springer, Berlin 1972.

[332] Finiteness conditions and generalized soluble groups Part II, Springer, Berlin 1972.

J. E. Roseblade

[333] On certain subnormal coalition classes, J. Algebra 1 (1964) 132–138.

[334] A note on subnormal coalition classes, Math. Z. 90 (1965) 373–375.

[335] On groups in which every subgroup is subnormal, J. Algebra 2 (1965) 402–412.

[336] The derived series of a join of subnormal subgroups, Math. Z. 117 (1970) 57–69.

J. E. Roseblade and S. E. Stonehewer

[337] Subjunctive and locally coalescent classes, J. Algebra 8 (1968) 423–435.

I. Schur

[338] Über die Darstellung der endlichen Gruppen durch gebrochene lineare Substitutionen, J. Reine Angew. Math. 127 (1904) 20–50.

O. J. Schmidt

[339] Abstract theory of groups, Kiev 1916.

A. G. R. Stewart

[340] On the class of certain nilpotent groups, Proc. Roy. Soc. Ser. (A) 292 (1966) 374–379.

I. N. Stewart

[341] The Burnside problem, Eureka (1973) 44–48.

S. E. Stonehewer

[342] The join of finitely many subnormal subgroups, Bull. London Math. Soc. 2 (1970) 77–82.

P. W. Stroud

[343] Topics in the theory of verbal subgroups, Ph.D. thesis, Univ. of Cambridge (1966).

[344] A property of verbal and marginal subgroups, Proc. Cambridge Philos. Soc. 61 (1965) 41–48.

S. P. STRUNKOV
[345] Normalizers and abelian subgroups of certain classes of groups, Izv. Akad. Nauk SSSR Ser. Mat. 31 (1967) 657–670 (*), translated in Math. USSR-Izv. 1 (1967) 639–650.
R. G. SWAN
[346] Representations of polycyclic groups, Proc. Amer. Math. Soc. 18 (1967) 573–574.
M. R. VAUGHAN-LEE
[347] Uncountably many varieties of groups, Bull. London Math. Soc. 2 (1970) 280–286.
H. WIELANDT
[348] Eine Verallgemeinerung der invarianten Untergruppen, Math. Z. 45 (1939) 209–244.
D. I. ZAIČEV
[349] Stably soluble groups, Izv. Akad. Nauk SSSR Ser. Mat. 33 (1969) 765–780 (*), translated in Math. USSR-Izv. 3 (1969) 723–736.
O. ZARISKI and P. SAMUEL
[350] Commutative algebra vol. I, Van Nostrand, Princeton N.J. 1960.
[351] Commutative algebra vol. II, Van Nostrand, Princeton N.J. 1960.
H. ZASSENHAUS
[352] The theory of groups, Chelsea, New York 1958.

Notation index

Classes

Notation index

Closure operations

Relations

Special substructures

Others

Subject index

A

B

C